Geology of the c...
Llanilar and Rhayader

This memoir describes the geology of a corridor across central Wales extending from Cardigan Bay in the west, across the Cambrian Mountains, to Llandrindod Wells in the east. As such, it provides a detailed, east–west transect across the central part of the Lower Palaeozoic Welsh Basin onto the western edge of the Midland Platform.

The exposed basin fill is dominated by a variety of mud- and sand-dominated turbidite systems of latest Ordovician to early Silurian age. Using data from the mapping, and palaeontological and sedimentological studies, the memoir presents the first detailed sedimentary architecture for this part of the basin. It details the characteristics of each turbidite system and discusses the depositional processes involved. Factors governing the evolution of the various systems include the direction and rate of sediment supply, eustasy, and intrabasinal growth faulting resulting from the reactivation of basement structures. Similar, but structually dislocated, turbidite systems of mid- to late-Ordovician age are exposed near the basin margin. Mid-Ordovician to early Silurian shelf sequences, including the richly fossiliferous mudstones and volcanic tuffs of the Ordovician Builth Inlier, are also described. Computer modelling of the regional gravity and aeromagnetic data-bases has provided east–west profiles across the district, and located the position and approximate depths of several of the basement faults.

The structural history of the region, dominated by the low-grade metamorphism and deformation of the basin fill during the Acadian Orogeny (early to mid-Devonian), receives comprehensive treatment. Several of the late stage faults contain lead, zinc and silver ore; the disused mines form part of the Central Wales Mining Field. The geological setting, mineralogy and genesis of these miner-alised veins is described. Glacial deposits are widespread across the district, and include tills derived from the local Welsh ice and the Irish Sea ice sheet.

Cover photograph

View looking north up the Elan valley and the Craig Goch reservoir. (MN 27928) (Photographer: T P Cullen)

Caban Coch viewed from the south-eastern shore of Caban-coch reservoir.

The hillside exposes a section through the Caban Conglomerate Formation, an ancient submarine nested channel system of early Llandovery age. Two sequences of turbidite conglomerates make up the lower two-thirds of the hill above the dam. (A14706).

BRITISH GEOLOGICAL SURVEY

J R DAVIES
C J N FLETCHER
R A WATERS
D WILSON
D G WOODHALL and
J A ZALASIEWICZ

CONTRIBUTORS

Palaeontology
H F Barron
S G Molyneux
A W A Rushton
S P Tunnicliff

Metamorphism
R J Merriman
B Roberts
S R Hirons

Geophysics
R M Carruthers

Economic geology
T B Colman

Isotope geochemistry
J A Evans
B Spiro
I G Swainbank

Fluid inclusion studies
T J Shepherd

Hydrogeology
W M Edmunds

Satellite imagery
E A O'Connor

Geology of the country around Llanilar and Rhayader

Memoir for 1:50 000 Geological Sheets 178 and 179 (England and Wales)

London: The Stationery Office   1997

© *NERC copyright 1997*

First published 1997

ISBN 0 11 884510 1

Bibliographical reference

DAVIES, J R, FLETCHER, C J N, WATERS, R A, WILSON, D, WOODHALL, D G, and ZALASIEWICZ, J A. 1997. Geology of the country around Llanilar and Rhayader. *Memoir of the British Geological Survey*, Sheets 178 and 179 (England and Wales).

Authors

J R Davies
British Geological Survey, Aberystwyth

C J N Fletcher
R A Waters
D Wilson
British Geological Survey, Keyworth

D G Woodhall
British Geological Survey, Edinburgh

J A Zalasiewicz
University of Leicester

Contributors

S G Molyneux
A W A Rushton
S P Tunnicliff
R J Merriman
R M Carruthers
T B Colman
T J Shepherd
E A O'Connor
British Geological Survey, Keyworth

H F Barron
British Geological Survey, Edinburgh

W M Edmunds
British Geological Survey, Wallingford

B Roberts
S R Hirons
Birkbeck College, University of London

J A Evans
B Spiro
IG Swainbank
NERC Isotope Geosciences Laboratory, Keyworth

Other publications of the Survey dealing with this district and adjoining districts

BOOKS

British Regional Geology
South Wales, 3rd edition, 1970

Memoirs
Geology of the country around Aberystwyth and Machynlleth, Sheet 163, 1986

Mineral Reconnaissance Programme Reports
No. 5 Preliminary mineral reconnaissance of Central Wales, 1976
No. 92 A mineral reconnaissance survey of the Llandrindod Wells/Builth Wells Ordovician inlier, Powys, 1987

British Geological Survey Technical Report
WD/91/60C The mineral waters at Llandrindod Wells and Central Wales, 1991

MAPS

1:625 000
Great Britain South, Solid geology, 1979
Great Britain South, Quaternary geology 1977
Aeromagnetic map (south sheet), 1965

1:250 000
Mid-Wales and the Marches, 1990
Mid-Wales and the Marches, Bouguer gravity anomaly, 1986
Mid-Wales and the Marches, Aeromagnetic anomaly, 1980
Cardigan Bay, 1982
Cardigan Bay including part 52N 08W Waterford, Seabed sediments, 1988
Cardigan Bay including part 52N 08W Waterford, Quaternary geology, 1990
Cardigan Bay, Bouguer gravity anomaly, 1980
Cardigan Bay, Aeromagnetic anomaly, 1980

1:100 000
Central Wales Mining Field, 1974

1:50 000
Aberaeron (Sheet 177) Solid Geology and Solid and Drift Geology, 1994
Aberystwyth (Sheet 163) Solid, 1984
Aberystwyth (Sheet 163) Solid with Drift, 1989
Montgomery (Sheet 165) Solid with Drift Geology, 1994
Montgomery (Sheet 165) Solid and Drift Geology, 1994

1:25 000
Parts SO 05, 06, 15, 16 Llandrindod Wells Ordovician Inlier, Solid, 1977

Printed in the UK for the Stationery Office
Dd 296603 C6 03/97

CONTENTS

FIGURES

PLATES

TABLES

PREFACE

This memoir describes an east–west transect across the southern part of the Lower Palaeozoic Welsh Basin. It is the first complete published account of the geology of the district, and should be read in conjunction with the two published 1:50 000 geological sheets. The survey was undertaken as part of a 'rapid mapping' programme which was initiated in 1986 with the objective of completing the geological map coverage of central Wales in the shortest possible time, and to gain a fuller understanding of the complex geology. The geological maps were compiled at 1:25 000 scale and are available from the British Geological Survey. The memoir also summarises the results of several multidisciplinary studies undertaken by specialist units during the project. This research included sedimentology, palaeontology, low-grade metamorphism, geophysics, remote sensing, and isotopic geochemistry.

The district is mainly underlain by Ordovician and Silurian mudstones and sandstones. Most were deposited in deep water, but those in the south-east corner of the district accumulated in a shallower shelf setting located on the edge of the Midland Platform. The memoir describes the stratigraphy of the sedimentary sequence and interprets the various mechanisms by which the deep-water sediments were deposited. The mapping, together with a large number of new biostratigraphical dates, has made it possible to reconstruct the sedimentary architecture and event stratigraphy of much of the basin fill. The survey has also highlighted the role played by contemporaneous fault movements in defining the basin margin and controlling the distribution of basinal facies; these movements appear largely to reflect the reactivation of underlying basement fractures.

Geophysical modelling of the regional gravity and aeromagnetic data has suggested the profile of the basin, and identified the position and depth of the basement fractures. Lead, silver and zinc ores were extensively mined in the district, and the memoir described the structural setting, isotopic geochemistry, fluid compositions and genesis of this mineralisation.

The district is a crucial water catchment area of the British Isles, and has supplied Birmingham with drinking water since the beginning of this century. Formerly, the natural springs at Llandrindod Wells were used for medicinal purposes.

This memoir and the accompanying geological maps will provide a comprehensive basis for future research into the formation, development and deformation of the Welsh and other ancient sedimentary basins. They also contain the essential information required for planning and development proposals, engineering projects, pollution studies, and a conservation surveys of the district.

Peter J Cook, CBE, DSc, CGeol, FGS
Director

British Geological Survey
Kingsley Dunham Centre
Keyworth
Nottingham
NG12 5GG

LIST OF 1:25 000 MAPS

The following list shows the 1:25 000 maps included, partly or wholly, within the area of Sheets 178 (Llanilar) and 179 (Rhayader) of the Geological Map of England and Wales. All the maps are on National Grid lines, lying within the 100 kilometre squares of SN and SO. The surveyors are R Cave, J R Davies, C J N Fletcher, B A Hains, A J Reedman, R A Waters, D Wilson, D G Woodhall and J A Zalasiewicz. Uncoloured dyeline copies of all the maps can be purchased from the British Geological Survey, Keyworth, Nottingham NG12 5GG.

SN 55	DGW	1990
SN 56	DGW, AJR	1989–90
SN 57	RAW, DW, JRD	1990
SN 65	DGW, DW	1989–90
SN 66	DW, DGW, JAZ	1987-90
SN 67	RAW, JRD	1987-90
SN 75	CJNF, DGW	1987–88
SN 76	CJNF, DGW, RC, JRD, RAW	1985–90
SN 77	CJNF, RAW, RC, BAH	1972–90
SN 85	CJNF, JAZ, DW	1986–89
SN 86	DW, CJNF, RAW, JAZ	1986–89
SN 87	JRD, CJNF, AJR	1986–87
SN 95	CJNF, RAW	1988
SN 96	RAW, CJNF, JAZ, DW	1986–89
SN 97	JAZ, JRD, RAW	1986–87
SO 05	RAW, DGW	1988–89
SO 06	DW, RAW, JRD, DGW, RC, JAZ	1987–89
SO 07	RC, JAZ, DW, RAW, JRD	1986–89

NOTES

Throughout the memoir the word 'district' refers to the area covered by the 1:50 000 geological sheets 178 (Llanilar) and 179 (Rhayader).

National Grid references are given in square brackets and are prefixed by 100 km grid square identification letters SN or SO.

Numbers preceded by the letter A refer to the BGS collection of photographs; other photographs were taken by members of the field staff.

Numbers preceded by the letter E refer to the BGS sliced rock collection.

The authorship of fossil names is given in the fossil index.

Enquiries concerning geological data for the district should be addressed to the Manager, National Geosciences Data Centre, Keyworth.

Welsh words

The following list gives a translation of the most common Welsh words used in this memoir:

afon	river
bryn	hill
coed	wood
craig/graig	rock
cwm	valley or cirque
esgair	ridge
llyn	lake
nant	stream
mynydd	large hill or mountain
pant	small valley or hollow
waun	bog

ACKNOWLEDGEMENTS

The geological survey of the Llanilar and Rhayader sheets was made principally by Drs J R Davies, C J N Fletcher, R A Waters, D Wilson, D G Woodall and J A Zalasiewicz, with smaller areas by Drs R Cave and A J Reedman, between 1986 and 1990 under the direction of Dr R A B Bazley, Regional Geologist. The memoir has been written by members of the field staff, and has incorporated contributions from many specialists within the British Geological Survey. The introduction was written by Dr Fletcher, the Ordovician chapter by Drs Fletcher, Wilson and Woodhall, the Silurian chapters by Drs Davies, Waters and Zalasiewicz, structure and metamorphism by Dr Fletcher, geophysics by Mr R M Carruthers, economic geology by Dr Fletcher and Quaternary deposits by Drs Wilson and Woodhall. The memoir was compiled by Drs Fletcher and Waters (Project Leaders), and edited by Dr R W Gallois and Mr J I Chisholm.

The palaeontological identifications were undertaken by Drs S G Molyneux, A W A Rushton, D E White and J A Zalasiewicz and Mr H F Barron and Mr S P Tunnicliff. Mudrocks were analysed for illite crystallinity by Dr B Roberts and Mr S R Hirons of Birkbeck College, University of London; the results were interpreted by Dr Roberts and Mr R J Merriman. The modelling of the regional gravity and aeromagnetic anomalies and the measurement of geophysical parameters along ground traverses were undertaken by Mr R M Carruthers, Mr R B Evans and Dr A J W McDonald. The mineral deposits of the district have been examined in collaboration with Dr T B Colman. Isotopic analyses were carried out by the NERC Isotope Geosciences Laboratory; Dr J A Evans was responsible for the Rb/Sr studies on low-grade mudrocks, Dr I G Swainbank for the lead isotopes of galena, and Dr B Spiro for the stable isotopes of sulphides. Fluid inclusions in quartz were studied by Dr T J Shepherd. The enhanced satellite image of the district was provided and interpreted by Dr E A O'Connor. The hydrogeology of the district was described from a report by Dr W M Edmunds.

We gratefully acknowledge the co-operation of all the landowners in the district during the geological survey. In particular, we would like to thank Welsh Water, Severn Trent Water, the Forestry Commission, the National Trust, Tarmac Roadstone Ltd and Hendre Quarries.

ONE

Introduction

This memoir describes the geology of the district covered by the 1:50 000 Llanilar (178) and Rhayader (179) geological sheets, which provide an east–west transect across the southern part of the Lower Palaeozoic Welsh Basin (Figure 1). The district lies within the counties of Dyfed and Powys and extends inland from the cliffs fringing Cardigan Bay to the spa town of Llandrindod Wells (Figure 2a). The sparse population is scattered in many small villages, and in the towns of Rhayader, Tregaron and Llandrindod Wells.

The topography ranges from low undulating hills along the narrow coastal belt to a dissected plateau, part of the Cambrian Mountains, which lie across the central part of the district. The upland area has an average altitude of approximately 500 m, with a maximum of 610 m at Pen y Garn [SN 799 771] to the north of Cwmyst-

wyth. Farther east the elevations gradually decrease and the relief becomes more subdued. To the west of the Cambrian Mountains the Ystwyth, Rheidol and Wyre rivers flow westwards to the sea; to the east the River Wye and its tributaries the Elan, Claerwen and Ithon drain mainly south-eastwards.

Farming, forestry, water resources and tourism are the principal industries of the district. The Elan valley reservoirs, to the west and south-west of Rhayader, provide the main water supply for Birmingham. The Caban-coch, Carreg-ddu, Penygarreg and Craig Goch reservoirs were commissioned in 1906, and the Claerwen reservoir in 1952. Quarrying of hard-rock aggregate is also undertaken on a small scale. The district forms a substantial part of the Central Wales Mining Field from which lead, zinc and silver ores were extracted.

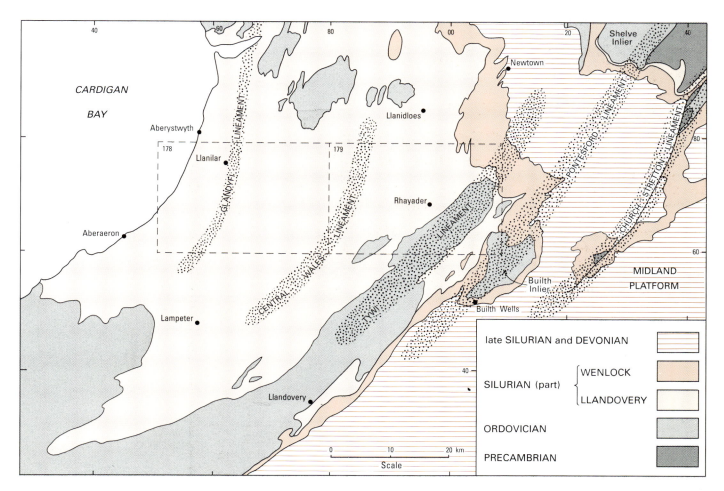

Figure 1   Sketch map illustrating the regional geological setting of the district.

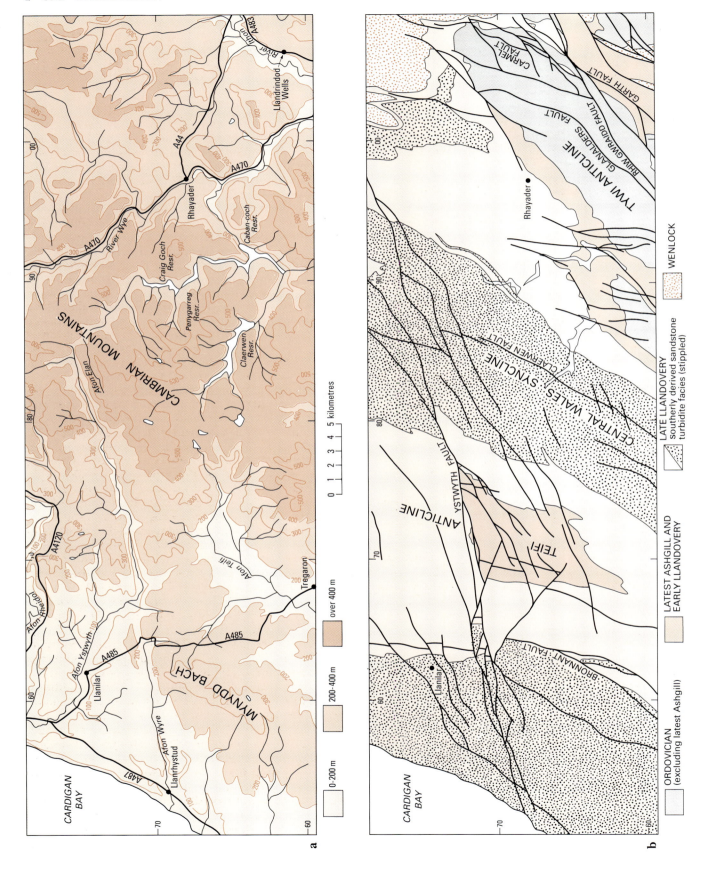

a

CARDIGAN
BAY

Llanrhystud

Llanilar

Afon Rheidol

Afon Ystwyth

Afon Wyre

A487

A485

A485

A4120

MYNYDD BACH

Tregaron

Afon Teifi

CAMBRIAN MOUNTAINS

Aber-Elan

Afon Elan

Claerwen
Resr.

Penygarreg
Resr.

Craig Goch
Resr.

Caban-coch
Resr.

Rhayader

River Wye

A470

A470

A44

Llandrindod
Wells

River Ithon

A483

60

70

80

90

00

90

80

70

60

| 0-200 m | 200-400 m | over 400 m |

0 1 2 3 4 5 kilometres

b

CARDIGAN
BAY

Llanilar

BRONNANT FAULT

YSTWYTH FAULT

TEIFI
ANTICLINE

CLAERWEN FAULT

CENTRAL WALES SYNCLINE

Rhayader

RHIW GWRAIDD FAULT

GLANALDERS
FAULT

CARMEL
FAULT

GARTH FAULT

TYWI ANTICLINE

60

70

80

90

00

ORDOVICIAN
(excluding latest Ashgill)

LATEST ASHGILL AND
EARLY LLANDOVERY

LATE LLANDOVERY
southerly derived sandstone
turbidite facies (stippled)

WENLOCK

REGIONAL TECTONIC SETTING

The Lower Palaeozoic Welsh Basin was founded on 'ensialic' basement developed by the accretion of volcanic arcs during late Precambrian to earliest Cambrian times (Thorpe, 1979). The sedimentary, volcanic, metamorphic and granitic rocks forming this basement are exposed around the margins of the Welsh Basin, in Anglesey, Pembrokeshire and the Welsh Borderland (Figure 1). The basin was the site of enhanced subsidence from Cambrian to late Silurian times, and was bordered to the east by the Midland Platform, a relatively stable area characterised by shelf sedimentation and periodic emergence, and to the north-west by a hypothetical landmass in the Irish Sea.

Until mid-Ordovician times, the Welsh Basin formed part of the Eastern Avalonia microcontinent (Soper and Hutton, 1984), which was incorporated in the continent of Gondwana (Cocks and Fortey, 1982). At that time Gondwana extended across the South Pole and was separated from Laurentia (the North American continental plate) by the Iapetus Ocean. The Iapetus oceanic plate was subducted southwards beneath Eastern Avalonia during Tremadoc to Caradoc times (Kokelaar et al., 1984), and the Welsh Basin was formed in a back-arc setting, with the generation of basaltic magmas (Dewey, 1969; Howells et al., 1991). Eastern Avalonia split off from Gondwana in the mid-Ordovician, and drifted northwards towards Laurentia during the remainder of the Ordovician and early Silurian. Following the cessation of volcanic activity, the late Ordovician and Silurian sequences were deposited in a series of sub-basins. The plate-tectonic setting of the Welsh Basin during the Silurian is not clear; fore-arc, passive-margin, intracontinental-rift and strike-slip models have been proposed (Okada and Smith, 1980; Davies and Cave, 1976; Woodcock, 1984a).

Within Eastern Avalonia, basement faults and fractures strongly influenced the sedimentary depositional patterns (James and James, 1969; Woodcock, 1984a, 1990; Smith, 1987b; Wilson et al., 1992) and the distribution of the volcanic centres (Rast, 1969; Dunkley, 1979; Campbell et al., 1988; Kokelaar, 1988; Howells et al., 1991). Tectonic lineaments identified across the Welsh Basin are an expression of the movements on these basement faults and their propagation into the cover sequence. Four such lineaments are present in the district: the Glandyfi, Central Wales, Tywi and Pontesford lineaments (Figure 1). It has been suggested that the closure of Iapetus Ocean and the 'docking' between the continental plates occurred during mid- to late Silurian times (Leggett et al., 1983, McKerrow and Soper, 1989; Soper and Woodcock, 1990), although alternative models suggest closure during late Ordovician times (Murphy and Hutton, 1986; Pickering, 1987). Con-

tinued convergence of Eastern Avalonia and Laurentia was marked in Wales by basin inversion. The final oblique collision resulted in crustal shortening across the basin during the climax of the Acadian Orogeny in late early to mid-Devonian times.

OUTLINE OF GEOLOGICAL HISTORY

The oldest rocks of the district are exposed in the Builth Ordovician Inlier, which is situated along the eastern margin of the Welsh Basin (Figure 1). There, black mudstones of Llanvirn and Llandeilo (mid-Ordovician) age are interbedded with reworked tuffs, and subordinate subaqueous ash-flow tuffs and lavas. The mudstones contain abundant graptolites and trilobites, and were probably deposited in a low-energy, dysaerobic, shelf sea environment. Farther west, black mudstones and sandstones of Caradoc (late Ordovician) age are preserved in the core of the Tywi Anticline (Figure 2b). In contrast to the older Ordovician sequences of the Builth Inlier, these rocks are dominantly graptolitic, and were deposited in a deeper-water, basinal environment; extrusive volcanism was expressed as thin bentonite layers. High-level dolerite sills and discordant bodies were intruded into the Llanvirn, Llandeilo and Caradoc sequences.

Bioturbated early Ashgill (late Ordovician) mudstones of the Tywi Anticline, as elsewhere in the central and southern parts of the Welsh Basin, record the entry of oxygenated, turbiditic slope-apron facies. Turbidite conglomerate and sandstone bodies are locally present. During late Ashgill times a worldwide (eustatic) lowering of sea level, attributed to the growth of ice sheets on Gondwana, led to a marked increase in the rate of sediment supply to the slope-apron. A thick wedge of silt-laminated turbidite mudstones prograded into the basin, west of the Garth Fault. Sediment instability resulted in mass movement and the formation of destratified and slump-folded strata. Coarse-grained turbidites were deposited as lobes and in channels sourced from nearshore environments to the east. Late in the Ashgill, when sea level was at its lowest during the acme of glaciation, shallow-water shelly sandstones were locally deposited across the Tywi Lineament.

The latest Ashgill to early Llandovery (early Silurian) postglacial rise in sea level saw an abrupt decline in the volume of sediment supplied to the slope-apron and the coincidental introduction of anoxic bottom conditions. The hemipelagic mudstones which now accumulated between turbidites were undisturbed by burrowers and preserve depositional lamination and graptolites. A mid-Llandovery regression reintroduced oxygenated bottom waters allowing soft-bodied organisms to live and feed on the slope-apron, resulting in bioturbation and the destruction of graptolite remains. The subsequent alternation of anoxic and oxic slope-apron facies, during the mid-Llandovery, may reflect further oscillations in sea level; certainly a widespread unit of anoxic facies deposited in latest mid-Llandovery times coincided with a major transgression recognised in adjacent shelf successions.

Figure 2 The Llanilar and Rhayader district (facing page).

a Physical features.
b Simplified geological map.

During the late Llandovery, large volumes of sandy turbidites, sourced from uplifting tracts to the south, entered the west of the district and encroached eastwards across the former slope-apron as a series of turbidite lobes. Their geometry was largely controlled by contemporary faulting along the Glandyfi, Central Wales and Tywi lineaments, with eastward migration of the lobes reflecting the sequential reactivation of underlying basement structures. Farther east, contemporaneous faulting and uplift along the Tywi Lineament resulted in submarine mass-wasting of the earlier Llandovery and Ashgill slope-apron sequence and produced west-facing scarps. Localised accumulations of slumps, debrites and coarse-grained turbidites were the products of this erosional episode.

A temporary cessation of fault activity, late in the Llandovery, saw easterly sourced, mudstone slope-apron deposition resume across the Tywi Lineament and then spread westwards as the earlier sandy turbidite lobes were abandoned. This phase of slope-apron development persisted into the early Wenlock (mid Silurian), but was abandoned as renewed movements on the major faults of the Tywi Lineament coincided with the reinvasion of southerly derived sandy lobes. The succeeding late Wenlock sequence of turbidite mudstones and laminated hemipelagites records the re-establishment and westward progradation of anoxic slope-apron facies along the eastern margin of the basin. A eustatic lowering of sea level in late Wenlock times allowed a bottom fauna of shelly organisms and burrowers of shelf affinity to colonise the easternmost part of this slope-apron.

Although shelf facies of early and mid Llandovery age were deposited east of the Garth Fault, none are exposed in the district. Mid and late Llandovery eustatic transgressions drove shelly, shallow-water sandstones across these concealed sequences and onto the long emergent and denuded early Ordovician rocks of the Builth Inlier. The overlying late Llandovery, burrow-mottled mudstones which crop out in the south-eastern sector of the district formed in an oxygenated distal shelf setting. In contrast, anoxic bottom conditions introduced during the Wenlock, in common with many other areas throughout the Iapetus province, are reflected in laminated distal shelf mudstone facies. Disturbed strata within these laminites indicate sediment instability on a west-facing slope.

Evidence from the district suggests that tectonic activity took place at intervals from Ordovician times until the Carboniferous. During the Ordovician, fault movements along the Pontesford Lineament possibly controlled the location of the basin margin and focused volcanism. The patterns of late Silurian deposition and erosion, in the basin and at its margin, demonstrate that there were contemporary movements along the Glandyfi, Central Wales and Tywi lineaments. Although these movements can be shown to have had a dip-slip component, significant amounts of strike-slip displacement cannot be discounted. During the Acadian Orogeny (late early to mid-Devonian) the Welsh Basin underwent transpressive deformation, metamorphism and uplift. South-east-directed shortening across the central part of the basin resulted in the formation of regional fold structures and the development of cleavage. The intensity of the Acadian deformation decreases rapidly to the east of the Tywi Lineament, against the edge of the stable Midland Platform. The grade of regional Acadian metamorphism ranged from diagenetic zone to low epizone, and was largely controlled by the depth of burial, although some areas of epizone grade may be the result of localised high strain. Late Acadian transverse faulting was associated with the emplacement of mineralised veins which form part of the Central Wales Mining Field. Subsequent movement on some transverse faults occurred in early Carboniferous times, and the last movement on the Ystwyth Fault may be Hercynian.

No sediments deposited between late Wenlock times (about 420 million years ago) and the last glaciation of the Quaternary (about 20 000 years ago) have been preserved in the district. However, the grade of metamorphism of the Ordovician and Silurian mudstones suggests that a thick cover sequence of late Silurian to early Devonian rocks was formerly present.

The Late Devensian glacial deposits of the district are mostly tills and morainic deposits derived from local sources, with minor amounts of glaciofluvial sands and gravels. Along the coast there are small patches of till deposited from the Irish Sea ice sheet, which covered Cardigan Bay at that time. Following the retreat of the ice from the valleys and upland areas, periglacial solifluction deposits formed on the valley sides. In the last 10 000 years alluvial sediments, peat and marine deposits have accumulated.

PREVIOUS RESEARCH

The earliest reference to the district was made by Murchison (1839) in his account of the Silurian rocks of Wales and the Borders. The first geological maps (Geological Survey of Great Britian, 1844, 1850) established the broad distribution of the sedimentary rocks in the district. The surveys also distinguished the volcanic rocks and dolerites of the Builth Ordovician Inlier and the Tywi Anticline. However, all the rocks in the district were assigned to the Silurian System, as the maps predate the naming of the Ordovician System by Lapworth (1879).

Graptolites from the Builth Ordovician Inlier were described by Elles and Wood (1908) and R A Hughes (1989), and the stratigraphy and faunal assemblages of the Llanvirn and Llandeilo successions of the inlier by Elles (1940) and Jones and Pugh (1941). The trilobite fauna was described by Hughes (1969, 1971, 1979) and certain populations were analysed in greater detail by Sheldon (1988). A map of the inlier has been published at 1:25 000 scale by the Institute of Geological Sciences (1977). The Caradoc and Ashgill rocks of the Tywi Anticline were first studied in detail by Roberts (1929). Descriptions of the igneous rocks associated with the Ordovician sequences were published by Jones and Pugh (1948a and b; 1949) and Roberts (1927).

Studies of the basinal Silurian succession of the district include those of Lapworth (1900) between Rhayader and the Elan valley, O T Jones (1909) between Devil's Bridge and Cwmystwyth, Davies (1926, 1928) and Davies and Platt (1933) between the Elan valley and the southern edge of the district, Williams (1927) around Pont-rhyd-y-groes, Roberts (1929) around Abbeycwmhir and W D V Jones (1945) between Cwmystwyth and Waun Marteg. The account of the Silurian strata of the adjacent Aberystwyth district (Cave and Hains, 1986) is relevant to the present work. Accounts of the Llandovery and Wenlock shelf sequences of the district and adjacent areas include Andrew (1925), Jones (1947), Ziegler et al. (1968) and Harris (1987).

Coarse-grained turbidites along the basin margin near Rhayader were first described by Kelling (1964) and have been documented in detail by Kelling and Woollands (1969), Woollands (1970), Holroyd (1978) and Kelling and Holroyd (1978). Within the basin, work on the turbidites of the Aberystwyth Grits by Wood and Smith (1959) has been extended by Lovell (1970), Anketel and Lovell (1976), Crimes and Crossley (1980, 1991), Smith and Anketel (1992) and Wilson et al. (1992). Studies of Llandovery turbidite facies within the *greistoniensis* Biozone have been made by Smith (1987a; 1988), and for parts of the *turriculatus* Biozone by Clayton (1992). Similar studies of the Wenlock turbidites have been made by Dimberline (1987), Dimberline and Woodcock (1987) and Dimberline et al. (1990). The diagenetic aspects of Llandovery basinal facies have been described by Smith (1987c) and Milodowski and Zalasiewicz (1991a and b). The geochemistry of the turbidite mudstones (Ball et al., 1992) and the provenance of the heavy mineral suites (Morton et al., 1992) have also been studied.

Accounts of the interaction of sedimentary facies and structure in the southern part of the Welsh Basin during the Ordovician are provided by Cave (1965), James (1983), Woodcock (1984a), Smallwood (1986), and Woodcock and Gibbons (1988); and during the Silurian by James and James (1969), Smith (1987b), Woodcock (1990), James (1983, 1991) and Waters et al. (1992). Regional assessments of the Llandovery sequences have been made by Smith and Long (1969) and Cave (1979), and of the Wenlock sequences by Bassett (1974) and Hurst et al. (1978). Evolutionary models for the Welsh Basin are given by Jones (1938; 1956), George (1963), Bassett (1963) and Woodcock (1984a).

The identification of the major tectonic structures of the district was first made by O T Jones (1912). Regional structural reviews have been published by O T Jones (1922), Bassett (1969), George (1970), Woodcock (1984a) and Woodcock et al. (1988). Detailed structural descriptions in the district are confined to the hinge zone of the Tywi Anticline (Roberts, 1929), and the Glandyfi Lineament (Wilson et al., 1992). The deep structure of this part of the Welsh Basin has been interpreted from the regional gravity and aeromagnetic data by Carruthers et al. (1992) and McDonald et al. (1992).

The mineralised vein systems of the central part of district were described in detail by O T Jones (1922). The form, mineralogy and genesis of the veins have also been discussed by Finlayson (1910), Phillips (1972, 1986) and Raybould (1974). A 1:100 000 scale geological map of the Central Wales Mining Field was published by the Institute of Geological Sciences (1974) and was also included in a report on the Mineral Reconnaissance Programme undertaken across the central and eastern parts of the district (Ball and Nutt, 1976). Isotopic studies on the mineralisation have included those by Ineson and Mitchell (1975) and Fletcher et al. (1993). Historical accounts of the mines in the district are given by Bick (1974), Foster-Smith (1978, 1979), Hall (1971, 1989), Jones and Moreton (1977) and Hughes (1981). The results of a geochemical survey across the stratiform lead mineralisation in the Builth Inlier has been given in a report by Marshall et al. (1987). The hydrology of the mineral springs at Llandrindod Wells has been published by Edmunds and Robins (1991).

The glacial and periglacial deposits of the coastal section have been described by Bowen (1973a, 1974), Watson and Watson (1967), Watson (1970, 1977, 1982), Campbell and Bowen (1989) and Vincent (1976). In the central part of the district, details of the periglacial nivation cirques have been given by Watson (1966), Potts (1971), and Campbell and Bowen (1989). The form and the pollen biozonation of the peat bogs in the Elan valley (Moore and Chater, 1969; Moore, 1970) and near Tregaron (Godwin and Conway, 1939; Godwin and Mitchell, 1938; Hibbert and Switsur, 1976; Campbell and Bowen, 1989) have been extensively studied.

Field itineraries to parts of the district have been published by Bates (1982), Siveter et al. (1989) and, based to some extent on the results of the recent survey, by Smith et al. (1991), Waters et al. (1993) and Wilson et al. (1993).

TWO

Ordovician

Ordovician rocks, ranging in age from Llanvirn to Ashgill (Table 1), crop out in the eastern part of the district within the Rhayader sheet area (Figure 3). Rocks of Llanvirn and Llandeilo age are confined to a small area immediately east of Llandrindod Wells [SO 060 609], occupying the north-western part of the Builth Ordovician Inlier. Rocks of late Caradoc and Ashgill age form the core of the Tywi Anticline (Plate 1), extending from Cwmcringlyn Bank [SO 073 725] in the north-east to Bryn Moel [SN 9300 5975] in the south-west; Ashgill strata also crop out in the core of the Rhiwnant Anticline, south-west of the Caban-coch reservoir, and in the vicinity of Pentre [SO 0882 6790] (Figure 3).

The Ordovician strata of the Builth Inlier comprise a succession of dark grey mudstones with intercalated reworked tuffs, lavas and a suite of high-level basic intrusions. They range in age from the *Didymograptus artus* Biozone of the Lower Llanvirn to the *Nemagraptus gracilis* Biozone, which extends into the lowermost Caradoc (Williams et al., 1972; Hughes, 1989; Rushton, 1990). The mudstones were deposited in a low-energy, marine shelf environment and locally contain abundant trilobites and graptolites. Two mudstone formations, the Camnant Mudstones and the Llanfawr Mudstones, have been recognised, separated by a sequence composed mainly of volcanic rocks (Figure 4).

The volcanic rocks were emplaced during a period of widespread igneous activity that affected the Welsh Basin from Arenig to Caradoc times. The volcanism was the result of south-east-directed subduction on the south-east side of the Iapetus Ocean (Phillips et al., 1976) and is considered to have occurred within a faulted 'ensialic' back-arc basin, associated with this subduction (Dunkeley, 1979; Kokelaar et al., 1984). The main phase of volcanism in the Builth Inlier is represented in the area east of Llandrindod Wells by a sequence of reworked tuffs and subordinate pyroclastic deposits and lavas, here termed the Builth Volcanic Formation. It is of late Llanvirn (*Didymograptus murchisoni* Biozone) age and correlates with the volcanic succession of Jones and Pugh (1941, 1949) up to and including their Cwm-amliw Series. The latter is accorded member status here as the Cwm-amliw Tuff, as is the basal ash-flow tuff, the Llandrindod Tuff (equivalent to the Llandrindod Series of Jones and Pugh, 1941). A higher sequence of volcanic breccias and agglomerates, here termed the Trelowgoed Volcanic Formation, overlies the Llanfawr Mudstones of *gracilis* Biozone age, north-east of Llandrindod Wells.

The Caradoc and Ashgill rocks of the Tywi Anticline are predominantly mudstones with subordinate sandstones, of marine shelf and slope environments. Debrites and slumped sequences are common in the upper part of the Ashgill sequence. Beds of bentonite occur at

Table 1 Biostratigraphical zonation of the Ordovician system.

~ Ma	SERIES	STAGES	GRAPTOLITE BIOZONES	
435	ASHGILL	HIRNANTIAN	*persculptus*	
			extraordinarius	
		RAWTHEYAN	*anceps*	*pacificus*
		CAUTLEYAN		*complexus*
				?
		PUSGILLIAN	*complanatus*	
	CARADOC	ONNIAN	*linearis*	
			?	
		ACTONIAN	*clingani*	
		MARSHBROOKIAN		
		WOOLSTONIAN	?	
		LONGVILLIAN		
		SOUDLEYAN	*multidens*	
		HARNAGIAN		
		COSTONIAN		
455	LLANDEILO	U	*gracilis*	
		M		
		L		
	LLANVIRN	U	*teretiusculus*	
			murchisoni	
		L	*artus* ('*bifidus*')	
470	ARENIG (PART)	FENNIAN	*hirundo*	

intervals throughout the Caradoc sequence, but are absent from the Ashgill.

The Caradoc strata are assigned to the St Cynllo's Church Formation, which is largely synonymous with the

Plate 1   View along the core of the Tywi Anticline, looking south-west from Camlo Hill
[SO 039 686].

Bioturbated mudstones of the Nantmel Mudstones underlie the low drift-covered ground in the centre of the
photograph. Blocks of sandstone and conglomerate belonging to Nantmel Mudstones are scattered over the
slope in the foreground; similar lithologies form a vertical, south-east-younging, sequence across the distant
high ground on the left. The scarp features on the distant high ground on the right are formed by units of
disturbed beds in the Yr Allt Formation, dipping gently to the north-west.

Carmel Group of Roberts (1929). The formation is composed of black graptolitic mudstones, and has correlatives along the Tywi Lineament (Davies, 1933; Jones,
1949; Strahan et al., 1909; Smallwood, 1986); but it contrasts with the mixed shelly and graptolitic facies of the
Builth Inlier and the shelly Caradoc facies of the Welsh
Borderland (C P Hughes, 1969, 1971, 1979; Williams et
al., 1972; Whittard, 1979; R A Hughes, 1989). The St
Cynllo's Church Formation spans the *Diplograptus multidens* to *Dicranograptus clingani* biozones of the Caradoc.
Faunas of the *gracilis* Biozone have not been identified,
and therefore the formation is probably younger than
the Ordovician strata of the Builth Inlier (Hughes,
1989). *Pleurograptus linearis* Biozone faunas have not been
recorded; this may reflect current uncertainties in Caradoc biozonation (Rushton, 1990) or the faulted nature
of many contacts between the Caradoc and Ashgill strata.

A major lithological change, from black graptolitic
mudstones to paler, burrow-mottled mudstones, is recog-

nised throughout Wales and is thought to coincide with
the Caradoc–Ashgill boundary. This boundary was
regarded by Price (1984) as a stratigraphical break corresponding to, at least, the highest part of the Onnian
(latest Caradoc) and the lower Pusgillian (earliest Ashgill)
stages. Although the Pusgillian Stage has not been recognised in the district, the contact between the Ashgill and
Caradoc strata appears conformable. It is doubtful,
therefore, that the stratigraphic break seen elsewhere in
Wales has the basinwide significance that was suggested by
Price (1984), Woodcock (1990b) and Toghill (1992). In
the present district the lithological change marks a marine
regression. The oldest Ashgill strata are assigned to the
Nantmel Mudstones, which equate with the lower part of
the Camlo Hill Group of Roberts (1929). The earliest
faunas, of mid- to late Cautleyan age, occur in the Cefnnantmel Member, a distinctive sequence of shelly sandstones within the lower part of the formation. Three units
of laminated hemipelagite at higher levels within the

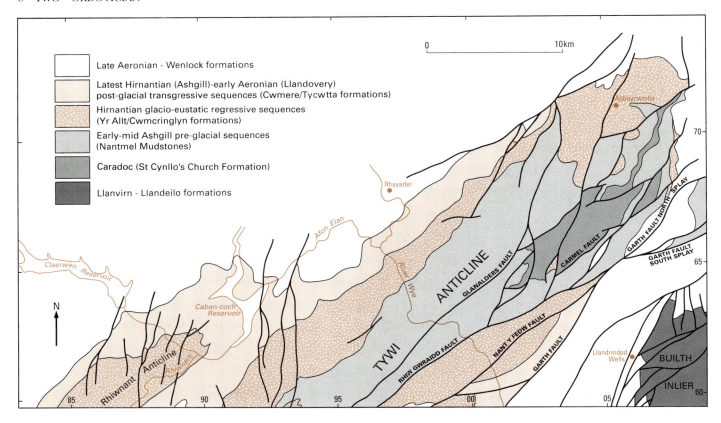

Figure 3 Distribution of Ordovician and earliest Silurian rocks in the district.

formation have yielded graptolites probably indicative of the *Dicellograptus anceps* Biozone.

Throughout the Welsh Basin the late Ashgill is characterised by regression, attributed to a glacioeustatic fall in sea level and reflected by a widespread change in the pattern of sedimentation (Brenchley and Cullen, 1984; Brenchley and Newall, 1984; Brenchley, 1988). In the district it is marked in the Ashgill strata of the Tywi Anticline by an upward change from predominantly burrow-mottled mudstones to thinly laminated silty mudstones and slumped sequences of the Yr Allt Formation, which record an increase in the rate of sedimentation and an associated reduction of the burrowing benthos. In the north-east of the Rhayader sheet area the Yr Allt Formation is overlain by the shallow-water Cwmcringlyn Formation, which contains a cool-water, shelly Hirnantian fauna and represents the acme of the glacioeustatic regression in the district. The Cwmcringlyn Formation compares with the shallow subtidal Scrach Formation (Woodcock and Smallwood, 1987) of the Llandovery area.

On both limbs of the Tywi Anticline the Yr Allt Formation is conformably overlain by sequences related to the widespread late Hirnantian transgression that followed the glacioeustatic low sea level. On the south-eastern limb, the Tycwtta Mudstones, with graptolites of latest Hirnantian to Rhuddanian age, succeed the Yr Allt Formation. On the north-western limb, the Yr Allt Formation is overlain by the Mottled Mudstone Member of

the Cwmere Formation (Cave and Hains, 1986) which includes, near its base, laminated hemipelagites containing graptolites of the *Normalograptus persculptus* Biozone. The Mottled Mudstone Member is a widespread marker band throughout the Welsh Basin (Temple, 1988); it is locally overlain by the Caban Conglomerate Formation, of mainly Llandovery age. The base of the *Parakidographtus acuminatus* Biozone (base Silurian System) has been proved about 40 m above the top of the Cerig Gwynion Grits (see p.77). The latest Ordovician to early Silurian sequences, which postdate the glacioeustatic regression, are described in Chapter 3.

Ordovician rocks near Pentre [SO 0882 6790] consist mainly of sandy mudstones and have yielded a poorly preserved, shallow-water assemblage of trilobites and bivalves indicative of an Ashgill age. These rocks constitute the Pentre Formation; their correlation with the Ashgill sequences of the Tywi Anticline is uncertain as the two areas are separated by faulting.

LLANVIRN–LLANDEILO ROCKS

Rocks of Llanvirn to Llandeilo age crop out in the north-western part of the Builth Inlier (Figure 5), across the hilly ground to the east of Llandrindod Wells [SO 060 609]. They comprise a sequence of dark grey mudstones interbedded with reworked tuffs, ash-flow tuffs, lavas and volcanic breccias, which are intruded by

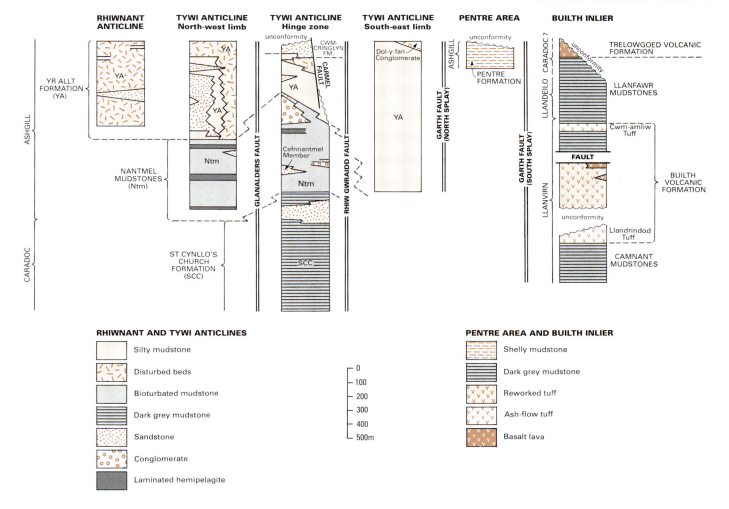

Figure 4 Ordovician sequences in the district.

a series of dolerite sills. The age of the exposed strata ranges from the *artus* Biozone (lower Llanvirn) to the *gracilis* Biozone (upper Llandeilo–basal Caradoc). At present it is uncertain whether any rocks of Caradoc age are present in the inlier. The maximum exposed thickness of the Llanvirn to Llandeilo sequence in the district is 1600 m. Local unconformities have, however, given rise to marked thickness changes in some of the formations.

The nomenclature used here for the north-eastern part of the Builth Inlier (Figure 4) largely replaces the biostratigraphical divisions of previous researchers (Elles, 1940; Jones and Pugh, 1941; Hughes, 1969; Institute of Geological Sciences, 1977). However, formal definitions of new formations, except the Llanfawr Mudstones, await the geological survey of the more extensively exposed ground to the south and east of the district. The Camnant Mudstones consist of dark grey, rusty weathering mudstones intercalated with sandstone and tuff beds. The succeeding Builth Volcanic Formation consists of reworked tuffs, lavas and mudstones bounded by two distinctive units of ash-flow tuff, the Llandrindod Tuff at the base and the Cwm-amliw Tuff at

the top. The overlying Llanfawr Mudstones comprise richly fossiliferous dark grey mudstones with a few tuff layers. The volcanic breccias of the Trelowgoed Volcanic Formation are the youngest Ordovician rocks exposed in the inlier.

Biostratigraphy

The faunas of the Builth Inlier, known since the time of Murchison (1839), are mainly mixed shelly and graptolitic assemblages of Llanvirn, Llandeilo and possibly earliest Caradoc age. They were listed and discussed by Elles (1940) who proposed a succession of local graptolite-trilobite biozones; though these zones have played a part in the development of mid-Ordovician biostratigraphy in the Welsh Basin, the paucity of extensive sections has limited their precision. Table 2 shows the suggested correlations with the standard graptolite zones.

The *Didymograptus artus* Biozone (Fortey and Owens, 1987, p.90) replaces the misnomer *Didymograptus bifidus* Zone of earlier British authors: the base of the zone defines the base of the Llanvirn Series (Table 1). The problems of distinguishing the *artus* Biozone from the

Figure 5
Llanvirn–Llandeilo rocks
of the north-western part
of the Builth Ordovician
Inlier.

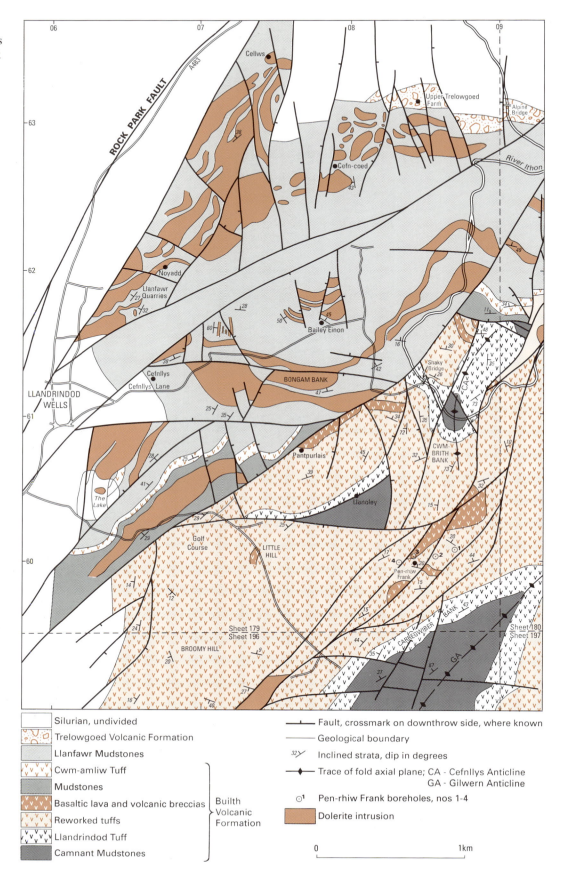

Silurian, undivided

Trelowgoed Volcanic Formation

Llanfawr Mudstones

Cwm-amliw Tuff

Mudstones

Basaltic lava and volcanic breccias — Builth Volcanic Formation

Reworked tuffs

Llandrindod Tuff

Camnant Mudstones

Fault, crossmark on downthrow side, where known

Geological boundary

Inclined strata, dip in degrees

Trace of fold axial plane; CA - Cefnllys Anticline
GA - Gilwern Anticline

Pen-rhiw Frank boreholes, nos 1-4

Dolerite intrusion

0   1km

Table 2 Correlation of the graptolite zones from the Builth Ordovician Inlier.

Elles' (1940) zones for the Builth area		Graptolite biozones	British series
			Caradoc (part)
			--------- ? ---------
Zone of *Nemagraptus gracilis* and *Ogygiocaris buchi*		*Nemagraptus gracilis* (part)	Llandeilo
Zone of *Glyptograptus teretiusculus* and *Ogygiocaris buchi*		*'Glyptograptus' teretiusculus*	--------- ? ---------
Zone of *Didymograptus murchisoni*		*Didymograptus murchisoni*	Llanvirn
[Main Volcanic Series]			
Zone of *Didymograptus bifidus* and *Ogyginus corndensis*	Subzone of *D. speciosus* and *Cryptolithus gibbosus*	*Didymograptus artus*	

overlying *Didymograptus murchisoni* Biozone by means of pendent didymograptids have been discussed by Bulman (1958), Skevington (1973) and Kennedy (1989). The use of diplograptids may prove more useful biostratigraphically, but in the Builth Inlier such graptolites are relatively few.

Elles (1940) recorded *'Amplexograptus' confertus* and *Didymograptus artus*, indicative of the *artus* Biozone, in the oldest fossiliferous strata of the inlier. The overlying beds contain *Didymograptus geminus* var. *latus*, *D. speciosus*, *Thysanograptus retusus* and the trilobite *Cryptolithus gibbosus*, and have been assigned to the *speciosus* Subzone (*artus* Biozone). However, Jenkins (1987) referred all the material on which *D. geminus latus* and *D. speciosus* were established to *D. murchisoni*, and these beds may therefore belong with the *murchisoni* Biozone. According to Hughes (1989), *Lasiograptus [Thysanograptus] retusus* is confined to the *Glyptograptus teretiusculus* Biozone in the Builth Inlier, but Strachan (1986) has stated that the 'broad form' of *L. retusus*, figured from the Llanvirn of Cwm-brith Bank on the River Ithon by Elles and Wood (1908, p.34, fig. 3c), is closely comparable with *Gymnograptus*(?) sp. from the *murchisoni* Biozone of the Shelve area. *Cryptolithus gibbosus* from the *speciosus* Subzone of the Builth area and *Bettonia frontalis* from the Betton Formation (*murchisoni* Biozone) of the Shelve area have been referred by Hughes (1971) to *Bettonia* (now *Bettonolithus*) *chamberlaini*. Recent revision of the supposedly diagnostic taxa therefore indicates that those in the *speciosus* Subzone cannot be distinguished from those at the base of the *murchisoni* Biozone. The recorded appearances of the various species are in closer accord if they are treated as part of the *murchisoni* Biozone rather than the *artus* Zone.

Didymograptus murchisoni is abundant at some localities in the *murchisoni* Biozone and is accompanied by a few biserial graptolites (Hughes, 1989). Shelly faunas, especially brachiopods, are relatively abundant where the buildup of volcanic piles created facies suitable for local colonisation (Lockley and Williams, 1981; Williams, Lockley and Hurst, 1981).

Hughes (1989) identified the base of the *Glyptograptus teretiusculus* Biozone in the stream section west of Pantpurlais (= Bach y graig section). Although the graptolites of the biozone are relatively well represented in the Builth Inlier, both they and the shelly faunas are of limited use for correlation with the calcareous facies of the type Llandeilo Series. It has been suggested that the base of the *teretiusculus* Biozone may lie at a slightly lower horizon than the base of the Llandeilo Series (Kennedy, 1989). The trilobite fauna was described by Hughes (1969, 1971, 1979) and certain populations were analysed in greater detail by Sheldon (1988). The nature of the trilobite fauna suggests that it was adapted to muddy substrates in deeper shelf environments of the Welsh Basin (Fortey and Owens, 1987).

The base of the *gracilis* Biozone is recognised by the incoming of *Nemagraptus gracilis* (Hughes, 1989, p.22). Although the graptolite fauna is fairly well represented in the Builth Inlier the base of the zone has not been recognised there. The trilobite fauna is akin to that of the *teretiusculus* Biozone, but differences have been noted at the specific level. The base of the Caradoc Series, as understood at the time of writing, lies high in the *gracilis* Biozone (Table 1); it is not known whether the Builth succession extends to the base of the Caradoc, but to date no fossils diagnostic of the Caradoc have been recorded in the area.

Camnant Mudstones

The formation crops out in the cores of the Cefnllys [SO 089 596] and Gilwern [SO 087 610] anticlines (Figure 5), and its presence is inferred from the topographic depression near Llanoley [SO 081 604]. The Camnant Mudstones equate with the mudstones of the *bifidus (artus)* Biozone of Elles (1940), Jones and Pugh (1949), Hughes (1969), and Institute of Geological Sciences (1977). The formation takes its name from the Camnant Brook section, situated some 3 km to the south of the district and described by Elles (1940, p.391). Its base is not exposed and the formation is sharply overlain by ash-flow tuffs of the Builth Volcanic Formation. The exposed thickness of the formation in the district ranges from about 100 m in the Cefnllys Anticline to 300 m in the Gilwern Anticline.

The Camnant Mudstones comprise a monotonous sequence of dark to medium grey fossiliferous mudstones intercalated in the upper part with sandstone beds, up to 1.5 m thick, and thin tuffs beds. The tuffs are moderately to poorly sorted and contain angular to subangular clasts of highly altered microporphyritic volcanic rock.

The principal fossil localities in the Camnant Mudstones lie to the south-east of the district. Fossils of the *artus* Biozone, including *Didymograptus (D.) artus, D. (sensu lato) acutidens* and *'Amplexograptus' confertus* were recorded from beds low in the succession at Nant, and fossils from low in the *murchisoni* Biozone (*speciosus* Subzone of Elles, as discussed above) are reported from Camnant Brook and other localities (Elles, 1940, pp.394, 398).

DETAILS

Outcrops of the Camnant Mudstones in the district are confined to a section [SO 0862 6104 to SO 0877 6097] across the core of the Cefnllys Anticline, along the south bank of the River Ithon and on the northern slopes of Cwm-brith Bank. Homogeneous, dark grey graptolitic mudstones are well exposed at river level, and coarse-grained sandstones, ranging up to 1.5 m thick, occur near the top of the formation at the eastern end of the section [SO 0877 6097].

Builth Volcanic Formation

The formation includes the volcanic and associated sedimentary rocks which lie between the Camnant Mudstones and the Llanfawr Mudstones. The formation crops out in a broad north-east-trending tract, up to 2 km wide, between the Llandrindod golf course [SO 070 597] and Castle Bank [SO 088 616].

Rhyolitic ash-flow tuffs occur at the base and top of the formation, and constitute the Llandrindod Tuff (Member) and Cwm-amliw Tuff (Member) respectively. Reworked tuffs form the bulk of the intervening sequence, but in the upper part they are intercalated with basalt lavas and volcanic breccias. The reworked tuffs are overlain by purplish grey to dark grey mudstones with thin interbeds of tuff.

LLANDRINDOD TUFF

The Llandrindod Tuff was previously referred to as the 'Rhyolitic Ash' of the 'Llandrindod Series' by Jones and Pugh (1949) and as 'Rhyolitic tuffs' at the base of the 'Main tuff group' by the Institute of Geological Sciences (1977). It forms a well-defined scarp feature on Carregwiber Bank [SO 0815 5935 to SO 0900 5980] and on Castle Bank [SO 0880 6115 to SO 0910 6160], and crops out along the hill [SO 0730 6032 to SO 0825 6067] behind Upper Llanoley. On Carregwiber Bank the Llandrindod Tuff is about 75 m thick, but on Castle Bank, only 1.8 km to the north, it is at least 150 m thick.

The Llandrindod Tuff consists of a basal pyroclastic breccia which grades upwards into lapilli-tuff. A rhyolitic composition is inferred on the basis of petrographic evidence and geochemical analyses of rocks collected from near Carregwiber and Cwm-brith banks (Furness, 1978). The basal pyroclastic breccia is at least 10 m thick; it commonly contains numerous very pale brown 'rhyolite' blocks, which locally form highly elongate slabs up to 1.5 m in length. Typically the lapilli-tuff is massive and poorly sorted, but in the higher parts of the unit it is commonly thinly planar bedded, with rare diffuse cross-lamination. The tuff consists of abundant pumice lapilli, glass shards, numerous feldspar crystals and few lithic fragments. The pumice and shards have largely been replaced by a microcrystalline mosaic of quartz and feldspar, with minor chlorite. The feldspar crystals have been altered in varying degrees to white mica, but twin lamellae are commonly visible. In samples from Carregwiber Bank pumice lapilli appear to be most abundant in the upper part of the tuff. The lithic fragments are angular to subangular and consist of mainly cryptocrystalline volcanic rocks, but in places there are a few feldspathic microcrystalline lithologies. The bedding, where present, is defined by alternating shard-rich and clast-rich beds.

DETAILS

On Carregwiber Bank the basal pyroclastic breccia is exposed to the east [SO 0832 5940] of Carregwiber farm, and to the west [SO 0875 5965] of Bwlchyfedwen. The grey-green breccia consists of 'rhyolite' fragments between pebble and granule size, but rarely up to 1.5 m in length, set in a fine-grained massive matrix. Outcrops of the overlying, very hard, lapilli tuff occur along the crest of the escarpment and behind Carregwiber farm. The tuff forms finely bedded units, up to 1 m thick, separated by thin beds of fissile, very pale grey, fractured tuff [SO 0816 5938]. In places [SO 0872 5970] low-angled cross-stratification is visible. Small isolated exposures [SO 0809 6055 and SO 0777 6038] of hard pale greenish grey tuffs with granule-sized, dark greyish green fragments occur along the south-east-facing slopes behind Upper Llanoley. Similar lithologies crop out [between SO 088 612 and SO 089 613] across the southern end of Castle Bank. On the summit area of Castle Bank [SO 0885 6137] the highest parts of the Llandrindod Tuff display thin planar bedding. Nearby exposures [SO 0863 6118] to the north of the River Ithon, upstream from Shaky Bridge, consist of purplish grey, very hard, splintery tuff beds up to 1 m thick, intercalated with friable units of coarse-grained tuff with abundant flattened dark grey pyroclasts. Similar bedded tuffs are also exposed in the road section [SO 0860 6104] to the south the river.

STRATA BETWEEN THE LLANDRINDOD AND CWM-AMLIW TUFFS

The strata between the Llandrindod and Cwm-amliw tuffs are dominated by a sequence of reworked tuffs (volcaniclastic sandstones) intercalated with a few basalt lava flows and associated volcanic breccias. The mudstone unit at the top of the sequence is conformably overlain by the Cwm-amliw Tuff, but its lower contact with the reworked tuffs is faulted. The reworked tuffs unconformably overlie the Llandrindod Tuff and are equivalent to the lower part of the Builth Volcanic Series of Jones and Pugh (1949) and the bulk of the 'Main tuff group' of the Institute of Geological Sciences (1977). They crop out in a broad tract between Carregwiber Bank [SO 085 596] and the north-east-trending fault which runs through Pantpurlais [SO 0767 0682]. Reworked tuffs and basalt lava flows were intersected in boreholes drilled near Pen-rhiw Frank (Marshall et al., 1987) (Figure 5); the borehole logs are presented in Appendix 1.

The reworked tuff sequence ranges in thickness from about 200 m in the area between Llanoley [SO 0803 6046] and Pantpurlais [SO 0768 6079], to about 250 m between Carregwiber Bank and Pen-rhiw Frank [SO 0851 6005], and to at least 300 m on the western side of Castle Bank [SO 087 612]. However, at all localities the upper contact is faulted.

Coarse-grained reworked tuffs dominate the lower part of the sequence. They consist of poorly sorted, often angular, clasts of volcanic rock and, very rarely, mudstone and sandstone, set in fine-grained chloritic matrix. The size of the clasts ranges up to 7 cm. Among the finest-grained components are discrete feldspar crystals and glass shards. The tuffs are mainly structureless but locally grain-size variations or concentrations of volcanic shards define planar beds which range in thickness from 5 to 10 cm. A 20 m-thick bedded reworked tuff is present in Pen-rhiw Frank Borehole 3 (Appendix 1).

The volcanic clasts consist predominantly of feldsparphyric and vesicular basalts with a chloritised groundmass that may originally have been glassy. Some of the highly vesicular clasts appear scoriaceous. Several of the volcanic clasts are lithologically similar to the basalt lavas and volcanic breccias that occur in the upper part of the sequence (see below). Other clasts include pumice, which has a fibrous appearance under the microscope, and very pale brown 'rhyolite,' which is similar to the pyroclasts in the Llandrindod Tuff. Rare, nonvolcanic clasts consist of dark grey mudstone and silty sandstone.

Fine-grained reworked tuffs dominate the upper part of the sequence. Locally they display parallel bedding, and in Pen-rhiw Frank Borehole 1 (Appendix 1) fine-grained tuffs, 5 m thick, are interbedded with coarser tuffs. The fine-grained tuffs are better sorted and the clasts more rounded than those in the coarse-grained tuffs.

Basalt lava flows and volcanic breccias intercalated with the reworked tuff sequence were described by the Institute of Geological Sciences (1977) as intrusive 'keratophyre' and 'andesite,' but have been reinterpreted here as being of extrusive origin. They crop out

in the vicinity of Pen-rhiw Frank [SO 0851 6005], between Pantpurlais [SO 0768 6078] and Shaky Bridge [SO 0848 6125], and on either side of Castle Bank [at SO 0883 6164; SO 0873 6161; SO 0893 6120]. Basalt lava flows also occur in Pen-rhiw Frank Boreholes 2 [SO 0856 6002] and 3 [SO 0843 6006] (Appendix 1).

Two basalt lava flows and associated volcanic breccias are present in the Pen-rhiw Frank area, and on the western side of Castle Bank. The lower flow is up to 30 m thick, and has volcanic breccia at the base (thin) and top. The breccia is composed of angular to subangular blocks of highly vesicular basalt, up to 25 cm across, in a matrix of vesicular basalt fragments, shards and feldspar crystals less than 1 mm in diameter. However, there are both clast- and matrix-supported volcanic breccias. The upper flow consists of feldsparphyric basalt with small vesicles. The contacts with underlying and overlying reworked tuffs are sharp, with no apparent baking. In thin section the basalt lavas contain up to 40 per cent feldspar phenocrysts and a maximum of 10 per cent mafic microphenocrysts, which are usually contained within a vesicular microcrystalline groundmass. All the mafic minerals have been replaced by chlorite, calcite and opaque oxide, but a few olivine crystals can still be distinguished on the basis of crystal form. The feldspar phenocrysts are commonly zoned, and have sieve-textured cores. Twin-lamellae extinction angles indicate an overall An_{30} to An_{40} composition. The groundmass consists of abundant feldspar laths together with grains of sphene and leucoxene which have replaced former mafic minerals and ilmenite. The vesicles have been filled mainly with quartz or calcite, but there is often some marginal opaque oxide and chlorite.

Mudstones form the uppermost of the strata between the Llandrindod and Cwm-amliw Tuffs, and are equivalent to the 'Lower *D. murchisoni* Shales' of Jones and Pugh (1949), Hughes (1969) and Institute of Geological Sciences (1977). They crop out in a narrow tract between Pantpurlais [SO 0767 6079] and Llandrindod Hall [SO 0642 6010], and over a small area between Neuadd [SO 0906 6178] and Castle Bank [SO 089 6150]. The contact with the underlying reworked tuffs is faulted, but a few thin mudstone and siltstone interbeds in the highest parts of the reworked tuff sequence suggest that the contact may be transitional. The mudstones are about 150 m thick, and are sharply overlain by the Cwm-amliw Tuff. They are poorly exposed, except close to the contacts with dolerite intrusions, and are purplish grey to dark grey with faint bedding traces. In places they are interbedded with thin, pale grey, fine- to medium-grained tuffs up to 2 cm thick.

DETAILS

Coarse-grained reworked tuffs crop out [between SO 8010 5944 and SO 0805 5952] to the north of Carregwiber farm. They consist of abundant pale yellow-brown clasts of rhyolite set in a grey-green matrix of sand- to granule-sized fragments. The clasts are generally pebble-sized, but rare angular blocks up to 30 cm across are also present. Similar lithologies above the Llandrindod Tuff near Upper Llanoley [SO 0812 6073] contain vesicular rhyolite clasts. Similar tuffs occur in the Pen-rhiw Frank boreholes (Appendix 1). In the deepest of these, Borehole 3 [SO 0842 6006], 112 m of greenish to brownish grey coarse-grained tuff is present, of which the upper part is bedded. The angular to subangular clasts range up to 2.5 cm across, and consist of scoriaceous and feldsparphyric tuffs and rare lithic fragments. Partial analyses of these breccias suggest 'dacitic' to andesitic' whole-rock compositions (Marshal et al., 1987).

Basalt lava is exposed in a small scarp [SO 0881 6051] on the southern side of Cwm-brith Bank. The greenish grey to pale green basalts contain abundant calcite-filled vesicles up to 1.5 cm in diameter. Approximately 15 m of weathered vesicular basalt are present in the top of the Pen-rhiw Frank Borehole 3 [SO 0843 6006]. The lava flow appears fairly homogeneous, with little brecciation. The basal contact with the underlying reworked tuffs is obscured. In Borehole 2 [SO 0856 6002] there are two basalt lava flows separated by reworked tuffs. The lower interval is about 1.5 m thick and composed of a homogeneous nonvesicular basalt. However, the upper flow has a basal brecciated zone, 25 cm thick, overlain by a nonvesicular basalt, 2.75 m thick, with a highly vesicular and partly brecciated top, 7 m thick. Both flows have irregular, and in the case of the thin lower flow, sharply defined contacts with the overlying reworked tuffs.

Blocks from the volcanic breccia exposed on the north-western slopes [SO 0873 6161 to SO 0880 6154] of Castle Bank are petrographically similar to the lava. In the breccia exposed to the north of Carregwiber Bank [SO 0835 5987], the blocks display variations in abundance of vesicles and feldspar phenocrysts.

Purplish grey mudstones crop out near Llandrindod Hall [at SO 0651 6026], along the north-western contact of a dolerite intrusion. Similar mudstones [SO 0669 6023], to the south-east of the intrusion near the Llandrindod Club House, are interbedded with thin, fine- to medium-grained tuffs, between 1 and 2 cm thick. To the east of Neuadd grey mudstones are locally exposed alongside a track [SO 0893 6172]. Laminated mudstones and siltstones are interbedded with the uppermost parts of the reworked tuff sequence near Pantpurlais [SO 0794 6097] and in a stream section [SO 0828 6096] between Llanoley and Cwm-brith Bank.

CWM-AMLIW TUFF

The Cwm-amliw Tuff is an acid ash-flow tuff which was previously referred to as the Cwm-amliw Series by Jones and Pugh (1949), the Cwm-amliw Ash by Hughes (1969) and the Cwm-amliw Formation by Furness (1978). It was identified only as 'tuffaceous beds at or near the base of the Upper *Didymograptus murchisoni* Shales' by the Institute of Geological Sciences (1977). The Cwm-amliw Tuff is exposed along a narrow faulted crop between Llandrindod Hall [SO 062 601] and Pantpurlais [SO 076 609], and over a small area [SO 089 619 to SO 093 619] in the vicinity of Neuadd.

A sharp contact between the Cwm-amliw Tuff and underlying mudstone is exposed at the roadside [SO 0643 6058] adjacent to the north-eastern edge of The Lake, but the contact with the overlying Llanfawr Mudstones is nowhere exposed.

The Cwm-amliw Tuff is a pale grey, predominantly fine-grained, very hard rock consisting of shards, together with discrete crystals of feldspar and quartz. It has a maximum thickness of 50 m near Llandrindod Wells. In thin section, although the original shardic texture is still distinguishable, recrystallisation has formed a widespread finely

microcrystalline mosaic of quartz and feldspar. Other secondary minerals include calcite and opaque oxide.

DETAILS

A small quarry [SO 0696 6071] north-east of The Lake exposes 2 m of very fine-grained grey ash-flow tuff with patches of coarser-grained material. At this locality the Cwm-amliw Tuff forms a low north-east-trending topographic feature and similar lithologies crop out on the hillside to the south-west. Fine-grained ash-flow tuffs are exposed on the south side of a small ridge [SO 0885 6189], to the west of Neuadd farm.

Llanfawr Mudstones

The Llanfawr Mudstones incorporate the upper *Didymograptus murchisoni*, *Glyptograptus teretiusculus* and *Nemagraptus gracilis* shales of Elles (1940), Jones and Pugh (1949), Hughes (1969) and Institute of Geological Sciences (1977). The formation crops out entirely to the north of the Builth Volcanic Formation, in a broad tract between Llandrindod Wells and Trelowgoed [SO 0900 6311]. The basal contact with the Cwm-amliw Tuff is not exposed; in the extreme north of the Builth Inlier the formation is overlain by the Trelowgoed Volcanic Formation. North-east of Pantpurlais [SO 0867 6079] the Llanfawr Mudstones are faulted against reworked tuffs of the Builth Volcanic Formation. The thickness of the Llanfawr Mudstones is estimated to be 500 m, but this includes numerous dolerite sills.

The Llanfawr Mudstones consist of a monotonous sequence of dark grey, locally finely laminated, fossiliferous mudstones interbedded with a few tuff beds up to 90 cm thick. The type section is located along a stream section [SO 0707 6103 to SO 0723 6097] to the west of Pantpurlais. This richly fossiliferous succession which is 135 m thick and dips north-west, includes several tuff beds, and has been previously described as the Bach y graig section (Elles, 1940, fig. 5, p.404).

The lower part of the formation contains abundant pendent didymograptids (Figure 6a), and the transition from the *murchisoni* Biozone to the *teretiusculus* Biozone is recorded in the type section, to the west of Pantpurlais (Elles, 1940, p.404; Hughes, 1989, p.16). There, *Diplograptus? decoratus*, *Glyptograptus* sp. and *Pseudoclimacograptus angulatus sebyensis* persist above the level at which *D. murchisoni* dies out. Trilobites from the type section (Sheldon, 1988) include species of the genera *Barrandia*, *Cnemidopyge*, *Ogyginus*, *Ogygiocarella*, *Platycalymene*, *Protolloydolithus*, and *Rorringtonia*, together with *Segmentagnostus mccoyii* and rare examples of the cyclopygids *Microparia lusca* (Plate 2l) and *Emmrichops? extensus*. The brachiopod fauna is composed of inarticulate forms including *Lingulella* cf. *displosa* (Lockley and Williams, 1981), and *Metaconularia?* (Plate 2g) has also been collected.

The fauna of the higher beds of the Llanfawr Mudstones belong to the *gracilis* Biozone. Llanfawr quarries [SO 0650 6170 and SO 0655 6185] have yielded a rich and diverse fauna. The graptolites recorded by Elles (1940, p.418), and revised by Hughes (1989, p.13) include: *Normalograptus brevis brevis*, *Dicellograptus cam-*

briensis, D. salopiensis, Dicranograptus brevicaulis, Nemagraptus gracilis, N. cf. *subtilis* (Figure 6d,e) and species of *Cryptograptus* and *Glossograptus*. The most important trilobites are *Cnemidopyge bisecta* (Plate 2j,k) *Homalopteon murchisoni, Ogygiocarella angustissima* (Plate 2m,n), *Nobiliasaphus powysensis, Platycalymene duplicata, Telaeomarrolithus intermedius* and *Trinucleus fimbriatus* (Hughes, 1969, 1971, 1979). The brachiopods are mainly inarticulate forms, but the dalmanellid *Tissintia* sp. has also been recorded (Lockley and Williams, 1981).

Details

The Llanfawr Mudstones are poorly exposed except along the stream bed in the type section [SO 0707 6103 to SO 0723 6097] and in Llanfawr quarries [SO 0650 6170 and 0655 6185]. In the most northerly of the quarries 20 m of medium grey to dark grey mudstones with a pale grey siltstone bed 10 cm thick are exposed between two dolerite sills. In the neighbouring quarry to the south a similar thickness of mudstone occurs beneath a dolerite sill. Fossiliferous grey and purplish mudstones interbedded with a few pale grey tuffs, between 2 and 3 cm thick, are exposed in a trackside cutting [SO 0685 6139] north of Cefnllys Lane. In a few places, hard, baked mudstones are present along the contacts with the dolerite sills, for example in the Llanfawr quarries. At Bongam Bank [SO 0787 6121] faintly laminated, dark grey mudstone, containing indeterminate trilobite fragments, grades into hard, pale mudstones adjacent to an overlying dolerite sill. A similar transition is seen in Llanfawr quarries and in a track [SO 0815 6140] 350 m to the north-east.

Trelowgoed Volcanic Formation

This formation equates with the volcanic rocks exposed to the north of the '*Nemagraptus gracilis* shales' and previously referred to as the 'tuff and agglomerate of Trelowgoed' (Institute of Geological Sciences, 1977). It crops out over a small area in the vicinity of Upper Trelowgoed farm [SO 0845 6313], but is better exposed just to the east of the district along the River Ithon [SO 0905 6312], near Alpine Bridge. The basal contact with the underlying Llanfawr Mudstones is nowhere exposed, and the formation is unconformably overlain by late Llandovery mudstones. The approximate present thickness of the Trelowgoed Volcanic Formation is 100 m.

The formation consists of massive purplish grey volcanic breccia, made up of subangular pebble- to cobble-sized clasts of green to greyish green or purplish grey feldsparphyric basalt, set in a finer-grained matrix of basaltic fragments and feldspar crystals. Feldspar laths are also abundant in the groundmass of the breccia, together with granular (mafic?) microlites and possible altered glass. Secondary alteration of the feldspar to sericite and chlorite is common, and many of the vesicles are filled with chlorite and calcite. The only exposure [SO 0846 6314] of the formation in the district occurs in the yard of Upper Trelowgoed farm.

Environments of deposition

All the Ordovician mudstones of the Builth Inlier are comparable in appearance and fauna. It is probable, there-fore, that they were deposited in similar depositional conditions; however, only the Llanfawr Mudstones are assessed in detail here. According to Sheldon (1987a, p.563) the lower part of the Llanfawr Mudstones were deposited in 'persistent low-energy, dysaerobic conditions, probably in a silled marine basin several hundred metres deep'. Sheldon's (1987b) conclusion was based on factors that included the absence of high-energy features such as current structures and winnowing, and the presence of articulated trilobites, some in moulted configuration and many of them blind. Low oxygen levels were suggested by the persistence of fine sedimentary lamination, the lack of bioturbation and the high organic content. A connection with the open ocean was inferred from the presence of widely distributed species of graptolites throughout the mudstone sequence (Hughes, 1989), but a submarine ridge or sill was invoked to account for the stagnation of the shelf areas and the general exclusion of mesopelagic cyclopygid trilobites (Fortey and Owens, 1987, p.105). However, the scarcity of cyclopygids, which have a preference for deep-water sites in the late Ordovician (Price and Magor, 1984), may alternatively indicate a relatively shallower depositional environment. The benthic trilobites of the Llanfawr Mudstones are more similar to those of the 'Raphiophorid community' than the 'atheloptic association' of Fortey and Owens (1987, p.106); this suggests that they inhabited a muddy shelf setting, offshore from the shallow shelf environment, but in water less deep than that occupied by the atheloptic trilobites. The persistence of low-energy, dysaerobic conditions in the mudstones indicates that the sea floor lay at depths below which tidal scour and storm surges were generally effective. Although the interbedded volcanic rocks (discussed below) include material erupted subaerially, this was introduced onto the shelf by gravity-driven processes from edifices that were probably being submerged by a eustatic sea-level rise during *gracilis* Biozone times (Fortey and Cocks, 1986). It is concluded that the Ordovician mudstones of the Builth Inlier accumulated in a persistently low-energy, dysaerobic shelf environment. The depth is uncertain but generally lay below storm-generated current activity.

The tuffs in the upper part of the Camnant Mudstones are the earliest products of extrusive volcanism in the district. This volcanism was a precursor to that of late Llanvirn times, which commenced with the subaqueous emplacement of a rhyolitic ash-flow tuff, the Llandrindod Tuff. The source of this tuff probably lay outside the district. The overlying reworked tuffs of the Builth Volcanic Formation comprise the products of contemporaneous basaltic volcanism, which, along with material eroded from the ash-flow tuff, were deposited as a series of sediment gravity flows. Effusive basaltic volcanism produced the lavas which occur interbedded with the reworked tuffs. The brecciated bases and tops of some lavas are similar to those erupted subaerially, but the reworked tuffs immediately beneath display no evidence of emergence; a subaqueous emplacement is therefore considered more likely. The Cwm-amliw Tuff was also emplaced subaqueously but its fine grain size, compared with that of the Llandrindod Tuff, indicates a

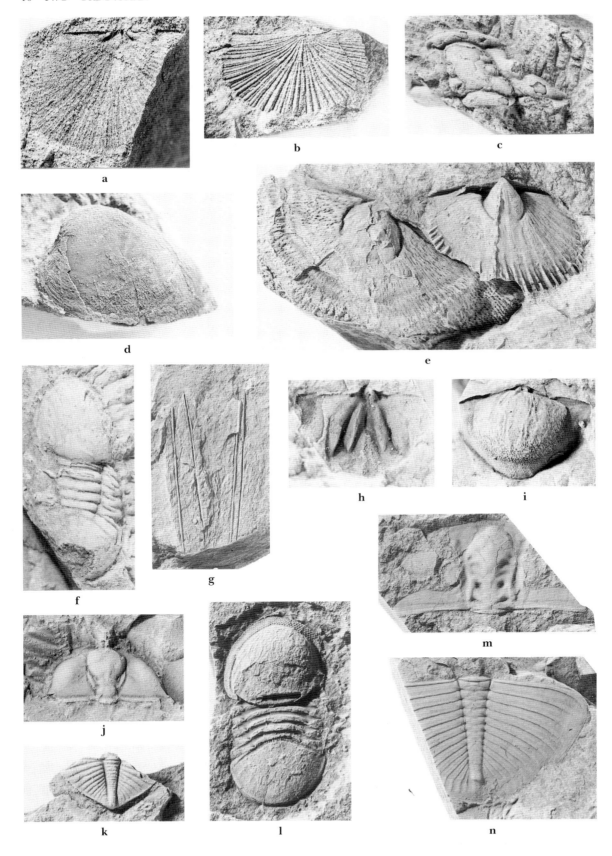

more distal facies. The reddish colouration along with the highly vesicular nature of the basalt blocks that comprise the volcanic breccias of the Trelowgoed Volcanic Formation suggests fragmentation by explosive subaerial volcanism, but the poor sorting could indicate that they were redeposited as submarine sediment gravity flows.

CARADOC ROCKS

St Cynllo's Church Formation

The Caradoc rocks of the Tywi Anticline, the St Cynllo's Church Formation, crop out between Treflyn farm [SO 0152 6435] in the south-west, and Bachell Brook [SO 0755 7115] in the north-east, occupying a 1 to 2 km-wide tract in the central part of their crop, but thinning north-eastwards from Lower Rhymney farm [SO 0525 6776] and south-westwards from Gilfach farm [SO 0307 6558]. They form the high ground on either side of the River Dulas valley south-east of Nantmel [SO 0242 6638], and occur on the prominent ridge of Baxter's Bank [SO 0580 6729] where they are intruded by dolerite sills; scattered small exposures also occur in the largely drift-covered ground immediately east of Baxter's Bank. Roberts (1929) suggested that the Caradoc rocks were bounded to the east by the Carmel Thrust and that the Caradoc and Ashgill strata formed a conformable, westward-younging succession in the western limb of the Tywi Anticline. This survey has shown that between Nantmel and Baxter's Bank the overall younging direction in Caradoc strata is to the south-east and that, locally, strike-faults delimit both its western and eastern boundaries.

The St Cynllo's Church Formation is largely equivalent to the Carmel Group of Roberts (1929). The base of the formation is nowhere seen and the higher parts are mostly faulted; it is estimated to be 850 to 950 m thick south-east of Nantmel. Three informal divisions have been recognised, based mainly on sections along the River Dulas valley and between Carmel [SO 0542 6658] and Baxter's Bank. These comprise a lower unit, dominantly mudstones, an intermediate sequence of interbedded sandstones and mudstones, and an upper unit of black, locally graptolitic, mudstones.

The lower unit of the St Cynllo's Church Formation is intermittently exposed in a 800 m-wide crop from Gilfach farm to the vicinity of Bwlch Mawr farm [SO 0430 6714]. It comprises up to 650 m of soft, fissile mudstones with subordinate sandstones and bentonites (Plate 3). The mudstones are dark bluish grey and black, locally colour banded, and weather rusty brown. Thin (1 to 3 mm) silty laminae occur locally and some micaceous partings are present. Sandstone beds occur throughout, locally in bundles of 3 or 4 beds, but they rarely form more than 5 per cent of the sequence. They are mostly dark grey, fine grained, parallel bedded and 1 to 5 cm thick, in places ranging up to 25 cm. The sandstones are usually structureless, but a few of the thicker beds are planar laminated in their lower part. Bioturbation is rare in both sandstones and mudstones.

Bentonites represent a relatively small proportion of the lower mudstone unit. They form conspicuous beds of deeply weathered, white, buff and orange-brown material, mostly 1 to 2 cm thick but locally up to 40 cm thick (Plate 3). Some of the thicker horizons are made up of interlaminations of bentonite and mudstone. Individual beds contain variable amounts of fine- to coarse-grained, tuffaceous material, scattered throughout or forming a graded interval at the base. The highly ductile bentonites were a focus for bedding-parallel slip during folding, and reveal numerous quartz-fibre slickensided surfaces.

These mudstones pass upwards into a coarsening-upward sequence of interbedded sandstones and mudstones up to 250 m thick, which represents the middle unit of the St Cynllo's Church Formation. These beds crop out mainly between Lower Rhymney and Gilfach, forming the high ground of the Dulas valley in the vicinity of Penllan [SO 0358 6560] and The Grange [SO 0416 6583]; the outcrop is largely fault bounded in the south-west. In the lower part, the sandstones are subordi-

Plate 2 Ordovician trilobites and brachiopods from the Rhayader sheet area (facing page).

a, b *Eostropheodonta hirnantensis*. Internal mould of brachial valve and external mould showing fascicostellate ribbing. Cwmcringlyn Formation, forestry quarry [SO 0585 7219] north of Abbeycwmhir. (JZ 4037, JZ 4044), both ×2. **c** *Gravicalymene* cf. *pontilis*, cranidium. Pentre Formation, Pentre farm [SO 0886 6808]. (JZ 5449), ×2. **d** A burrowing bivalve similar to the Upper Palaeozoic genus *Edmondia* (the dentition is not preserved and the genus is accordingly uncertain). Pentre Formation, Pentre farm [SO 0886 6808]. (JZ 4083), ×2. **e** *Leptaena* cf. *rugosa* (left) and *Dolerorthis* cf. *intercostata*. Nantmel Mudstones, near [SO 077 710] about 400 m north-west of Fron-rhyd-newydd. (R O Roberts collection, no. ROR 1950), ×2. **f** *Microparia* cf. *caliginosa*. This specimen differs from typical examples of *M. caliginosa* because the pygidium shows a well-marked second pleural furrow. St Cynllo's Church Formation, quarry [SO 0356 6627] west of Maesygelli farm. (JZ 5232), ×4. **g** *Metaconularia?* sp. Distal part of a flattened exoskeleton. Llanfawr Mudstones, Bach y graig stream-section [SO 073 609]. (Zv 4049, presented T Keenan), ×2. **h, i** *Eoplectodonta (Kozlowskites) nuntius*, internal moulds of brachial and pedicle valves. Cefnnantmel Member of Nantmel Mudstones, Cefnnantmel [SO 0355 6697]. (RX 3798), ×6, and (RX 3791), ×3. **j, k** *Cnemidopyge bisecta* (Elles), cranidium and pygidium. Llanfawr Mudstones, 'Llandrindod' (Llanfawr quarry?). (Zw 291, 292, Crosfield collection), ×2. **l** *Microparia lusca*. The compound eyes are united in front, giving a single visual surface. Llanfawr Mudstones, Bach y graig stream-section [SO 071 611], Llandrindod Wells. (GSM 104586, plaster cast of specimen in T Keenan collection), ×6. **m, n** *Ogygiocarella angustissima*, cranidium and pygidium. Llanfawr Mudstones, 'Llandrindod' (Llanfawr quarry?). (Zw 289, 290, Crosfield collection), ×1.5.

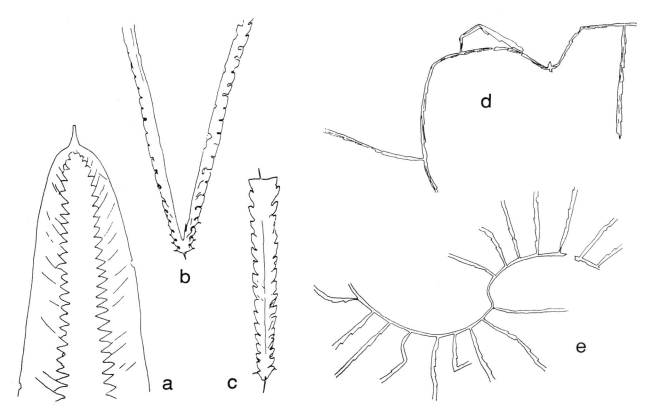

Figure 6 Selected graptolites from the Llanfawr Mudstones.

a *Didymograptus (D.) murchisoni*, the morph commonly referred to the subspecies *geminus*. Llanvirn Series, *murchisoni* Biozone, outcrop [SO 065 606] north of The Lake, Llandrindod Wells. (RX 3285, partly restored from the counterpart RX 3284) ×4. **b** *Dicellograptus intortus* Lapworth. Llandeilo Series, *gracilis* Biozone, "Llandrindod" [Llanfawr quarry?]. (Zv 4509, Crosfield collection), ×4.
c *"Glyptograptus" teretiusculus* of Elles & Wood, possibly now referable to *Hustedograptus* Mitchell. Llandeilo Series, *gracilis* Biozone, "Llandrindod" [Llanfawr quarry?]. (Zv 4510, Crosfield collection), ×4.
d *Nemagraptus* cf. *subtilis* Hadding, showing the sparse cladia. Llandeilo Series, *gracilis* Biozone, "Llandrindod" [Llanfawr quarry?]. (Zw 288, Crosfield collection), ×4. **e** *Nemagraptus gracilis*. Llandeilo Series, *gracilis* Biozone, "Llandrindod" [Llanfawr quarry?]. (Zw 287, Crosfield collection), ×3.

nate to mudstones and occur mainly as thin (1 to 3 cm) flaggy beds in bundles 2 to 3 m thick; in the upper part of the unit the sandstones become predominant, locally forming beds up to 1.5 m thick. They are generally grey, weathering to brown, fine to medium grained, and quartzose with micaceous partings and laminae. Beds are commonly parallel-laminated throughout but, in places, are diffusely ripple cross-laminated in their upper part; a few surfaces display straight, undulatory or linguoid ripple marks. The bases of several of the sandstone beds reveal biogenic sole markings and groove casts. The mudstones are dark grey, weathering bluish green, red-brown and buff, locally with laminae and thin ribs of silt-stone or fine-grained sandstone. In the upper part of the unit, the mudstones are locally faintly burrow mottled and in places near the top they are coarsely micaceous. Porcellanous, pale grey, tuffaceous siltstones and associated bentonitic mudstones occur as distinctive, but subordinate, lithologies throughout the sequence.

The highest mudstone unit of the St Cynllo's Church Formation sharply overlies the sandstone unit and comprises up to 50 m of soft, black, fissile, pyritous mudstones, which weather to a rusty brown colour. The mudstones are mostly homogeneous, but weak burrow mottling occurs at a few horizons and, locally, faint silty or pyritic laminations and thin wispy-bedded sandstones are present. The mudstones occupy a narrow, sinuous crop between Bachell Brook and Lower Rhymney, and also crop out on Baxter's Bank and in small fault-bounded wedges between Carmel [SO 0550 6655] and Cefn-draenog [SO 0333 6470]. The higher parts of the unit are intermittently exposed in Bachell Brook and the area between Gwerncynydd [SO 0270 6490] and Treflyn [SO 0152 6435] farms, where they include graptolite-rich horizons containing faunas of the *clingani* Biozone. They are therefore, in part at least, the equivalent of the Nod Glas of the Corris and Welshpool areas (Pugh, 1923; Cave, 1965).

Plate 3 Overturned, eastward-younging sequence of black mudstones and thin white bentonite beds in the St Cynllo's Church Formation. Cutting [SO 0374 6614] behind Maesygelli farm.

BIOSTRATIGRAPHY

The St Cynllo's Church Formation is dominantly in a graptolitic facies. Graptolites were collected from several localities during the present survey (Figure 7), and use was made of collections by R O Roberts, P Clarke and Dr R A Hughes from the Sedgwick Museum, Cambridge.

The distribution of species from the St Cynllo's Church Formation (Table 3) confirms the presence of the *multidens* and *clingani* biozones. However, the zonal boundary has not been established due to the structural complexities in the area. The *gracilis* Biozone has not been positively identified from the formation, and thus it is apparently younger than the strata of the Builth Inlier (Hughes, 1989) and the Llanwrtyd area (Stamp and Wooldridge, 1923).

Localities yielding specimens compared with *Amplexograptus arctus* and *Normalograptus [Climacograptus] brevis* (Localities 4 to 8; Table 3) are regarded as representing the lower part of the *multidens* Biozone, from comparison with the succession in the Dicranograptus Shales of

South Wales. *Amplexograptus leptotheca* occurs in the *multidens* Biozone in Shropshire (Hughes, 1989), and a comparable species is taken to indicate this biozone in the St Cynllo's Church Formation (Localities 4 and 5, Table 3). A fauna containing *D. foliaceus* (Locality 9, Table 3) is also referred to the *multidens* Biozone, as the species is not known to range higher, although some of its associates occur in proved *clingani* Biozone assemblages at Gwerncynydd farm [SO 0270 6490]. *Amplexograptus* cf. *perexcavatus* and *Lasiograptus harknessi* (Locality 11, Table 3) occur in a sequence within the lowermost part of the formation; both forms are known to range up into the *clingani* Biozone, although the presence there of a single cyclopygid trilobite *Microparia* cf. *caliginosa* (Plate 2f) is more suggestive of the *multidens* Biozone.

Localities [SO 0270 6490] at Gwerncynydd farm have yielded *Dicranograptus clingani*, characteristic of the lower part of the *clingani* Biozone, associated with *Climacograptus antiquus*, *C. spiniferus*, and two forms, *Orthograptus quadrimucronatus* s.l. and *O.* cf. *quadrimucronatus spinigerus*, that elsewhere range up into the *linearis* Biozone. No species diagnostic of the *linearis* Biozone has been recorded in the district. An unnamed species of *Normalograptus*, characterised by a bulbous virgellar structure (Figure 7i), was recorded at Gwerncynydd farm. This form is also present in Bachell Brook [SO 0746 7104] (Locality 15, Table 3) and suggests the presence of the *clingani* Biozone there. The sporadic distribution of *clingani* Biozone faunas in the district probably reflects the many faulted contacts within the formation. Other strata attributable to the biozone may be present elsewhere in the crop. Numerous poorly preserved *Pseudoclimacograptus modestus* associated with a slender *Orthograptus* aff. *apiculatus*, akin to *O. uplandicus* (Strachan, 1986) occur [SO 0273 6498] to the north of Gwerncynydd farm (Locality 1; Table 3). Both forms resemble species hitherto recorded from the *gracilis* Biozone in the Shelve area, where they range up to the base of the *multidens* Biozone, but here appear to lie in the uppermost division of the St Cynllo's Church Formation, in close proximity to localities containing graptolites of the *clingani* Biozone.

Diverse acritarch floras obtained from near Carmel [SO 0542 6659] and near Esgairwy [SO 0644 6898] are listed in Appendix 2a. They include species of *Actinotodissus*, *Arkonia*, *Baltisphaerosum*, *Coryphidium*, *Diexallophasis*, *Multiplicisphaeridium*, *Ordovicidium*, *Stellechinatum*, *Striatotheca* and *Villosacapsula*. The floras are similar to those recorded by Turner (1984) from the type Caradoc area in Shropshire, but one taxon, *Diexallophasis sanpetrensis*, has not previously been recorded from strata older than the *D. complanatus* Biozone of early Ashgill age (Jacobson and Achab, 1985). One locality has yielded recycled acritarchs of mid-Cambrian to Ordovician age (Appendix 2a).

DETAILS

The lower unit of the St Cynllo's Church Formation is intermittently exposed around the village of Nantmel [SO 0342 6638]. Sections at Maesygelli farm [SO 0374 6614] and along the

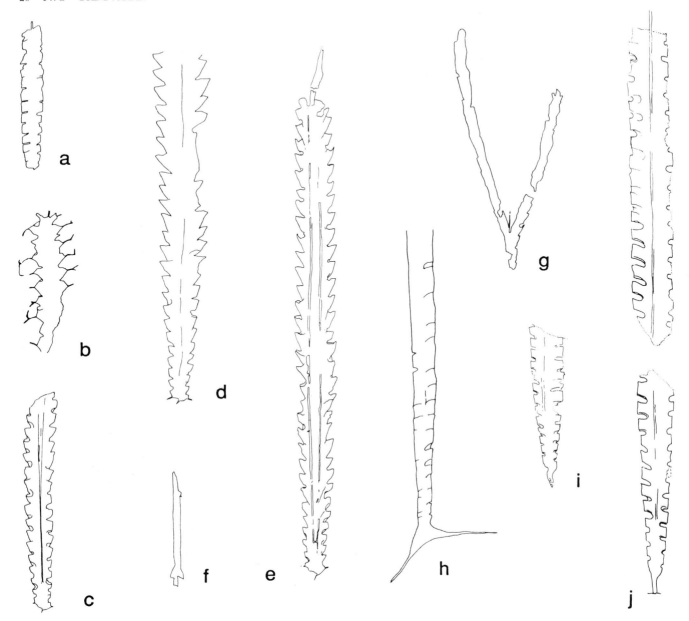

Figure 7   Selected graptolites from the St Cynllo's Church Formation.

a   *Pseudoclimacograptus modestus.* (JZ 6800), from trackside exposure [SO 0273 6498] near Gwerncynydd farm. The horizon is not certain, but *P. modestus* is usually found in the *gracilis* to lowest *multidens* biozones.   **b**   *Lasiograptus harknessi.* (JZ 5292), stream section [SO 0367 6648] west of Maesygelli farm; *multidens* Biozone.   **c**   *Amplexograptus arctus.* (JZ 3082), Pen-y-bank quarry [SO 0561 6670]; *multidens* Biozone (JZ 4813).   **d**   *Orthograptus* aff. *apiculatus.* (ROR 1868, R O Roberts collection). Pen-y-banc quarry [SO 0561 6670]; *multidens* Biozone. It differs from *O.* cf. *calcaratus* because one of the proximal spines arises from the antivirgellar side of the sicula rather than from theca 1². **e**   *Orthograptus* cf. *calcaratus.* (MW-80-A17, P Clarke collection) Gwerncynydd farm [about SO 027 649]; *clingani* Biozone. The specimen is one of many from this locality that show splitting of the virgula.   **f**   *Corynoides* aff. *curtus.* (JZ 5292), stream section [SO 0367 6648] west of Maesygelli farm; *multidens* Biozone. The specimen differs from typical *C. curtus* in having a vestigial third theca near the apex.   **g**   *Dicranograptus clingani.* (JZ 6740), Gwerncynydd farm [SO 0271 6490]; *clingani* Biozone.   **h**   *Climacograptus spiniferus* (now referred to *Diplacanthograptus*). (JZ 6693), Gwerncynydd farm [SO 0271 6490]; *clingani* Biozone.   **i**   *Normalograptus* sp. nov. (JZ 6773). Gwerncynydd farm [SO 0271 6490]; *clingani* Biozone. This rare form, which has a thickened, bulbous virgellar prolongation, appears to be restricted to the *clingani* Biozone.   **j**   *Ensigraptus* cf. *caudatus.* (JZ 6750), Gwerncynydd Farm [SO 0271 6490]; *clingani* Biozone. *E. caudatus* is typical of the lower part of the *clingani* Biozone in Scotland but is otherwise unknown in Wales.

All specimens magnified ×5. Numbers prefixed JZ are housed in the palaeontological collections of the BGS, Keyworth.

Table 3   Distribution of graptolites in the St Cynllo's Church Formation.

	Localities														
	1	2	3	4	5	6	7	8	9	10	11	12	13	14	15
Amplexograptus arctus	·	·	·	●	cf.	aff.	?	·	·	·	·	·	·	·	·
A. aff. *leptotheca* [broad]	·	·	·	●	●	·	·	·	·	·	·	·	·	·	·
A. cf. *perexcavatus*	·	·	·	·	·	·	·	·	·	·	●	·	·	·	·
A. sp.	·	·	?	·	·	·	·	·	·	·	·	·	?	·	·
Climacograptus antiquus antiquus	·	·	·	·	·	·	cf.	?	·	·	·	●	·	·	·
C. aff. *antiquus* [broad]	·	·	·	·	·	·	·	·	·	·	·	●	·	·	·
C. spiniferus	·	·	·	·	·	·	·	·	·	·	·	●	·	·	·
C. sp.	·	·	·	·	·	·	·	·	●	·	·	·	·	·	·
Corynoides curtus	·	·	·	●	cf.	·	·	●	·	aff.	·	·	·	·	·
Cryptograptus sp.	·	·	·	·	·	·	·	●	·	·	·	·	·	·	·
Dicranograptus clingani	·	·	·	·	·	·	·	·	·	·	·	●	●	·	·
D. cf. *ramosus spinifer*	·	●	·	·	·	·	·	·	·	·	·	·	·	·	·
Diplograptus foliaceus	·	·	·	·	·	·	?	·	●	cf.	·	·	·	·	·
Diplograptus sp.	·	·	·	·	?	·	·	●	·	●	●	·	·	·	?
"Glyptograptus" sp.	·	·	·	·	·	·	●	·	·	·	·	·	·	·	·
Lasiograptus costatus	·	·	●	·	·	·	·	·	·	·	·	·	·	·	·
L. harknessi	·	·	·	●	·	·	·	·	·	●	●	·	·	·	·
Normalograptus cf. *brevis*	·	·	·	·	·	·	●	·	·	●	·	·	·	·	·
N. mohawkensis	·	·	·	·	·	·	?	·	·	·	·	·	·	cf.	·
N. sp. [bulbous virgella]	·	·	·	·	·	·	·	·	·	·	·	·	●	·	●
N. spp.	·	·	·	·	·	·	·	·	·	·	·	·	·	·	●
Orthograptus aff. *apiculatus* [narrow form]	●	·	·	·	·	·	·	·	●	●	·	●	●	·	·
O. cf. *calcaratus* (group)	·	·	·	·	·	·	·	●	·	·	·	·	·	·	sp.
O. quadrimucronatus s.l.	·	·	·	·	·	·	·	·	·	·	·	●	·	·	·
O. cf. *quadrimucronatus spinigerus*	·	·	·	·	·	·	·	·	·	·	·	●	·	·	·
Pseudoclimacograptus modestus	●	·	·	·	·	·	·	·	·	·	·	·	·	·	·
Microparia cf. *caliginosa* [Trilobita]	·	·	·	·	·	·	·	·	·	·	●	·	·	·	·
	multidens Biozone											*clingani* Biozone			

Localities

1 trackside exposure [SO 0273 6498] 90 m north-north-east of Gwerncynydd farm; 2 west of Carmel Chapel (R O Roberts collection); 3 west of Keeper's Lodge, Fronrhydnewydd (R O Roberts collection); 4 Pen-y-bank quarry [SO 0561 6670], Baxter's Bank and R O Roberts, P Clarke and R A Hughes collections; 5 field west of Bwlchbryndinam (R O Roberts collection); 6 road cutting [SO 0444 6563] on A44, east of Penllan farm; 7 south end of quarry [SO 0622 6863], near Esgairwy farm; 8 north end of quarry [SO 0624 6865], near Esgairwy farm; 9 quarry [SO 0426 6585] at Grange, 1250 m east of Gilfach, Nantmel; 10 stream [SO 0367 6648] west of Maesygelli farm, Nantmel; 11 stream [SO 0356 6627] west of Maesygelli farm, Nantmel; 12 Gwerncynydd farm [SO 0271 6490], east of farmhouse and P Clarke and R A Hughes collections; 13 stream [SO 0280 6500] 150 m north-east of Gwerncynydd farm; 14 stream [SO 0281 6500] 8 m upstream of 13; 15 Bachell Brook [SO 0746 7104], 320 m at 100° from Dyfanner.

For key to range chart symbols see Appendix 3.

farm track [SO 0353 6625] to the aqueduct show a strongly folded, locally inverted and mainly eastward-younging, sequence of rusty weathering, dark bluish grey mudstone with beds of dark grey, fine-grained sandstone up to 5 cm thick, and packets of thin (1–2 cm) white bentonites. A small quarry [SO 0356 6627] (Locality 11, Table 3) near the aqueduct, exposes a conspicuous 40 cm-thick bentonite; strata about 3 m above this horizon have yielded cf. *Amplexograptus perexcavatus*, *Lasiograptus harknessi*, and a single specimen of the cyclopygid trilobite *Microparia* cf. *caliginosa*, suggestive of the *multidens* Biozone.

Dark grey mudstones of the lower part of the formation, exposed in the stream [SO 0367 6648] north-north-east from the aqueduct, have also yielded a fauna indicative of the *multidens* Biozone (Locality 10, Table 3), including *L. harknessi*, cf. *Diplograptus foliaceus*, *Corynoides* aff. *curtus* and *Normalograptus* cf. *brevis*.

A track section between [SO 0350 6575] and [SO 0406 6554] in the forestry plantations on the south-west side of the Dulas valley affords almost continuous exposure through most of the sandstone-dominated middle unit of the formation. Dark greenish grey mudstone, with subordinate thin beds of sandstone and bentonite, similar to those at Maesygelli farm, are intermittently exposed at the western end of the section; they are succeeded eastwards by a sequence of fine-grained, thinly bedded sandstones. The contact is gradational and poorly exposed, but the sandstones crop out in a quarry [SO 0365 6562] where they occupy the core of a large syncline. They are generally 2 to 3 cm thick, in bundles 2 to 3 m thick, interbedded with dark, greenish weathering, silty mudstone. The sandstones are planar bedded and diffusely cross-laminated with micaceous partings; biogenic structures occur on the bases of a few units. The sequence is repeated by folding along the track section to the east, with the underlying mudstones reappearing in the cores of anticlinal closures. Along the section [between SO 0387 6530 and SO 0407 6534] the sandstone unit is overturned, eastward-younging and, at the eastern end of the section, is faulted against brownish grey, faintly bioturbated, silty Nantmel Mudstones.

Although the sequence in the forestry plantations is undated, nearby strata in a small quarry [SO 0426 6585] (Locality 9, Table 3) on the north-east side of the Dulas valley have yielded a fauna indicative of the *multidens* Biozone. The quarry exposes rusty weathering, dark grey mudstones with packets, up to 40 cm thick, of fine-grained, planar and cross-laminated sandstones, sheared and thrust in a large anticline, and probably representing a horizon close to the gradational base of the sandstone unit. *D. foliaceus* and *Orthograptus* aff. *apiculatus*, together with *Climacograptus* sp. and *Dicellograptus* or *Dicranograptus* sp., have been obtained from the mudstones near the base of the section.

The upper part of the sandstone unit is exposed in a farm track [SO 0462 6589 to SO 0475 6604]. The sandstones are up to 20 cm thick, interbedded with subordinate mudstones and, locally, with tuffaceous units of grey, splintery, siliceous mudstone, bentonites and ash bands up to 90 cm thick. Ripple-marked bedding surfaces on a few sandstones indicate that the sequence is overturned to the south-east. The highest beds are well exposed 50 m east of the track, in a small quarry [SO 0481 6600] behind Hirfron Cottage. The section shows up to 15 m of fine- to medium-grained sandstone with subordinate, dark grey, coarsely micaceous mudstone, overturned south-eastwards. The sandstones are mostly thinly bedded, but are locally up to 50 cm thick and planar laminated in their lower part with well-defined, linguoid-ripple-marked tops. Grooves, trails and other indeterminate biogenic structures occur on the bases of a few beds. The sandstone sequence is succeeded by black mudstone, which is poorly exposed in the trackside 40 m east of the cottage.

The highest sandstone beds, up to 75 cm thick and overturned south-eastwards, are again exposed in Camlo Brook [SO 0541 6658] adjacent to Carmel Chapel and in a tributary stream [SO 0529 6648] 150 m to the south-west. A continuous sequence from these sandstones into the upper mudstone unit of the St Cynllo's Church Formation is exposed in Camlo Brook downstream from the chapel. In this section, thin horizons of dark grey, burrow-mottled mudstone are interbedded with the sandstones. About 100 m downstream from the chapel the sandstones are overlain by mudstones of the upper unit. The contact is sharp and appears conformable, although the sequence at this point may be repeated by faulting. The overlying mudstones are black and dark grey, faintly colour laminated with a strong primary bedding-parallel fabric which has probably been enhanced during deformation; they are locally graptolitic. They are thrown [SO 0553 6653] against grey, burrow-mottled, silty mudstones of the Nantmel Mudstones by a steep westward-inclined reverse fault, the Carmel Fault (Carmel Thrust; Roberts, 1929). Although there is little evidence of disturbance along the contact, the throw on the fault may be large; the stratigraphical evidence suggests that a significant thickness of Ashgill strata is missing in Camlo Brook.

Black, graptolitic mudstones of the upper unit, intruded by a dolerite sill in Pen-y-Banc Quarry [SO 0561 6670] (Locality 4, Table 3), have yielded a low-diversity fauna of probable *multidens* Biozone age, comprising abundant *Amplexograptus arctus*, together with *A.* aff. *leptotheca*, *Corynoides curtus* and *L. harknessi* (the latter from a collection of Dr P Clarke in the Sedgwick Museum, Cambridge).

Similar black mudstones are also exposed in a small quarry [SO 0622 6863] 550 m south-west of Esgairwy farm (Localities 7 and 8, Table 3), where they have yielded *Orthograptus* ex.gr. *calcaratus*, *Climacograptus* ?*antiquus antiquus*, *Normalograptus* cf. *brevis*, ?*N. mohawkensis*, *Cryptograptus* sp. and *Corynoides curtus*. The fauna, although poorly preserved, is generally indicative of the *multidens* Biozone. However, the possible presence of *C. mohawkensis* may indicate the *clingani* Biozone, and *C. antiquus* may be synonymous with *Diplograptus multidens compactus* Elles and Wood (Hughes, 1989) which also ranges into the *clingani* Biozone.

A stream section [SO 0707 7031] west of Keepers Lodge (Locality 3, Table 3) has yielded *Lasiograptus costatus* (from the R O Roberts collection of the Sedgwick Museum, Cambridge), which probably ranges no higher than the *multidens* Biozone. What may be the highest beds of this mudstone succession are exposed in Bachell Brook [SO 0746 7104] (Locality 12, Table 3) where they are overlain, with apparent conformity, by the Nantmel Mudstones. There the graptolites are poorly preserved, but some *Normalograptus* spp. resemble forms associated with *clingani* Biozone faunas at Gwerncynydd farm (see below).

Exposures of black, graptolitic mudstone in the vicinity of Gwerncynydd farm [SO 0270 6490] have yielded a well-preserved fauna attributable to levels within the *clingani* Biozone (BGS and Sedgwick Museum collections). The presence of *Dicranograptus clingani* confirms the biozonal age; *Climacograptus spiniferus* is common. A section [SO 0271 6490] behind the old farmhouse (Locality 12, Table 3) has yielded *D. clingani*, *C. spiniferus*, *C.* ?*antiquus antiquus*, *Orthograptus* aff. *apiculatus*, *O. quadrimucronatus* s.l. and *Ensigraptus* cf. *caudatus*, *O.* cf. *quadrimucronatus spinigerus*, and in the adjacent stream section [SO 0280 6500] (Locality 13, Table 3) examples of *D. clingani*, *O.* aff. *apiculatus* and *Normalograptus* spp. are present. Previous collections from the stream, although less precisely located, have yielded in addition *C.* cf. *antiquus lineatus*?, *C.* cf. *antiquus antiquus* and *N. mohawkensis*. *Pseudoclimacograptus modestus* and *O.* aff. *apiculatus*, obtained from a track section [SO 0273 6498] 90 m north of the farm (Locality 1, Table 3) appear anomalously old (basal *multidens* Biozone).

ENVIRONMENTS OF DEPOSITION

The almost total absence of trilobites in the St Cynllo's Church Formation suggests deposition at greater depths, and in a more distal setting, than is inferred for the lithologically similar Llanvirn–Llandeilo sequences of the Builth Inlier, which contain mixed graptolite-trilobite assemblages. The sequence is representative of the euxinic facies that was widely developed throughout the Welsh Basin from the Llanvirn to Caradoc epochs (Price, 1984; Woodcock, 1984a), and is considered to record high global sea level (Leggett, 1978). Levels of anoxicity may have been enhanced by an increase in plankton production, resulting from flooding of cratonic platforms to create extensive shelf areas (Leggett, 1980).

The coarsening-upward motif of the lower and middle parts of the St Cynllo's Church Formation, with the gradual change from mud- to sand-dominated deposition, increasing thickness of sandstone beds and greater degree of bioturbation in the sandstones and mudstones, is indicative of a major regression within the *multidens* Biozone. Individual sandstones are interpreted as event beds, each deposited by a sediment density flow (turbiditic and/or storm-induced). Comparable regressive sequences have been recorded in the Welshpool area (Cave and Price, 1978), and the same event may be represented by slumped deposits in the Ceiswyn Formation

(Pratt, 1990; Pratt et al., 1995) of the Corris area. The regression appears to correspond with the early Caradoc (post-*gracilis*) eustatic fall in sea level proposed by Fortey and Cocks (1986).

The dark grey, pyritous and generally unbioturbated mudstones which sharply overlie the sandstones, indicate deposition in an oxygen-deficient environment, attributable to a major transgressive pulse (cf. McKerrow, 1979). The fossiliferous (*clingani* Biozone) levels at the top of the sequence correspond with the acme of this event and correlate with the widespread deposition of black, locally phosphatic mudstones across the Welsh Basin (the Nod Glas; Cave, 1965). There is no evidence for a break beneath this stratigraphical unit in the Rhayader sheet area, and elsewhere within the Welsh Basin the base of the coeval Nod Glas is transitional and conformable with underlying dark grey mudstone sequences (Lockley, 1980; Pratt, 1990; Temple and Cave 1992). However, in the north-eastern part of the basin around Welshpool and Bwlch-y-Groes, the base of the Nod Glas is associated with non-sequence (Cave, 1965; Cave and Price, 1978; Temple and Cave, 1992). The evidence suggests, therefore, that deposition of these transgressive late Caradoc facies was probably diachronous, starting in the *multidens* Biozone in the south-west of the basin (cf. Lockley, 1980), but only inundating exposed north-eastern parts in *clingani* Biozone times.

The possibility that there was periodic tectonic activity along the Tywi Lineament during the Ordovician has been discussed by George (1963), Woodcock (1984) and Woodcock and Gibbons (1988). Cave (1965) proposed that a bathymetric high along the line of the lineament, during the Caradoc, restricted clastic input from the east and allowed the phosphatic black mudstones of the Nod Glas to accumulate at relatively shallow depths. Localised 'periclinal' highs along the lineament throughout the Ordovician were inferred by Smallwood (1986) to explain the distribution of shallow-water facies. In the Rhayader sheet area, however, the Caradoc of the Tywi Lineament comprises a thick sequence in graptolitic facies, suggesting an active depocentre and indicating an area of bathymetric depression.

Deposition of the upper mudstone unit of the St Cynllo's Church Formation appears to coincide broadly with the cessation of volcanism within the Welsh Basin, for it contains only a few bentonites and high-level basic sills. Caradoc volcanism was associated with rifting in a back-arc setting (Howells et al., 1991), and the complex patterns of sedimentation (Cave, 1965; Smallwood, 1986) within the Welsh Basin at this time undoubtedly reflect the complexities of this rifted terrain. In the context of basin development, the postulated extensive late Caradoc transgression can be attributed to regional subsidence following the cessation of rifting and volcanism.

ASHGILL ROCKS

The Ashgill sequences of the district can be divided into two parts, the formations deposited prior to and during the Hirnantian regression and those deposited during the subsequent latest Hirnantian transgression. The latter are described in Chapter 3 with the Silurian rocks with which they are naturally grouped. The former crop out within a broad belt, up to 6 km wide, along the Tywi Anticline (Figure 3), from Cwmcringlyn Bank [SO 073 725] in the north-east to Rhos Saith Maen [SN 945 600] in the south-west. They are also exposed over smaller areas within the core of the Rhiwnant Anticline, between Afon Claerwen [SN 892 620] and Pant Glas [SN 865 597], and in the vicinity of Pentre [SO 0886 6808].

The Ashgill succession described here attains its maximum exposed thickness (perhaps 2200 m) along the south-eastern limb of the Tywi Anticline to the east of the Glanalders Fault (Figures 3 and 4). Four formations have been recognised within the district (Figure 4). The Nantmel Mudstones consist of a monotonous sequence of pale to medium grey, burrow-mottled mudstone intercalated with three distinct units of laminated hemipelagite. To the south-east of the Glanalders Fault shelly sandstones of the Cefnnantmel Member occur in the lower part of the formation, and conglomerate and sandstone units are present in the higher parts. The overlying Yr Allt Formation records the late Ashgill glacioeustatic regression and comprises a thick sequence of laminated silty mudstones, sandstones and conglomerates. Along the north-western limb and the closure of the Tywi Anticline, and within the Rhiwnant Anticline, the bedding of the Yr Allt Formation is strongly disturbed by synsedimentary slumping and in-situ destratification. The Cwmcringlyn Formation overlies the Yr Allt Formation across the closure of the Tywi Anticline, between Cwmcringlyn Bank [SO 073 725] and Y Glog [SO 048 718]. It consists of a thin sequence of shallow-water, cross-laminated sandstones. Fossiliferous silty mudstones and thin sandstones of the Pentre Formation are exposed on the eastern border of the district near Pentre [SO 0886 6808]. The formation is bounded by splays of the Garth Fault, and correlations with the Ashgill sequences to the west are uncertain.

Nantmel Mudstones

The Nantmel Mudstones comprise the lower part of the Camlo Hill Group of Roberts (1929) and correlate with parts of the Cefn Cynllaith and Craig Las formations described by Mackie and Smallwood (1987) to the south-west of the Rhayader sheet area, and mapped by James (1991) in the vicinity of Llanwrthwl [SN 975 637]. The Nantmel Mudstones crop out within the core of the Tywi Anticline (Figure 8), from Cwmbedw [SO 056 699] in the north-east to Rhos Saith-maen [SN 945 600] in the south-west. The formation also occurs in a series of fault slices along the eastern flank of the Tywi Anticline, from Bachell Brook [SN 0745 7105] to Drum Ddu [SO 960 596].

At most localities the contact between the St Cynllo's Church Formation and the overlying Nantmel Mudstones is faulted, but west of Broad Oak and Bryn-Camlo [between SO 0695 7062 and SO 0545 6809], the contact is apparently conformable. The contact is exposed in

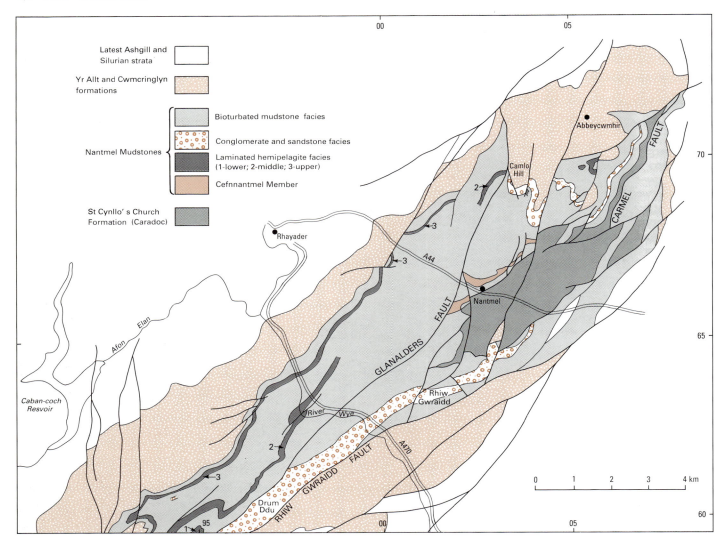

Figure 8 Facies of the Nantmel Mudstones.

a stream section [SO 0747 7103] in Bachell Brook and north-west of Lower Rhymney [SO 0509 6793]. At both localities the junction is sharp.

The Nantmel Mudstones are overlain by the Yr Allt Formation along the north-western limb of the Tywi Anticline between Y Gamriw [SN 944 612] and Castle Hill [SO 010 685], and around its closure on the southern slopes of Camlo Hill [SO 035 691] and at Lanwen [SO 063 699]. The junction is poorly exposed but crops out in the Cwm Pistyll stream section [SN 9418 6083]. On the south-eastern limb the higher parts of the formation are faulted out. The Nantmel Mudstones have an estimated thickness of 1100 m in the core of the Tywi Anticline, but thin to less than 800 m towards the north-east in the closure of the anticline.

Four sedimentary facies are present in the Nantmel Mudstones: bioturbated mudstone facies; laminated hemipelagite facies; conglomerate and sandstone facies; and shelly sandstone facies (Cefnnantmel Member).

A mid- to late Cautleyan trilobite faunal assemblage has been collected from the shelly sandstones at Cefnnantmel [SO 0355 6697]. The Pusgillian Stage has not yet been recognised in the area, but the lower part of the mudstone sequence is poorly fossiliferous and the contact between the Ashgill and Caradoc strata is commonly faulted. The laminated hemipelagites have yielded a restricted graptolite fauna of probable *anceps* Biozone age.

BIOTURBATED MUDSTONE FACIES

This facies, which constitutes over 70 per cent of the Nantmel Mudstones, crops out across the low ground within the core of the Tywi Anticline (Plate 1) to the south-east of Abbeycwmhir [SO 056 712] as far as Rhos Saith Maen [SN 945 600], on the southern boundary of the district (Figure 8). Intermittent exposures occur in fault slices along the south-eastern flank of the Tywi Anticline from Bachell Brook [SO 0745 7105] to the Wye

River near Doldowlod [SO 000 625]. The facies consists predominantly of blocky, pale to medium grey mudstones with common burrow mottling picked out by dark grey mudstone (Plate 4a). Bedding is defined by very thin, sandstone beds and siltstone laminae, local colour banding, rare lenses and thin beds of dark grey, phosphatic mudstone, and diagenetic, calcareous nodules. The mudstone beds range in thickness from 5 to 20 cm. The burrows generally form branching unorientated networks, but some lie preferentially along the bedding or define small *Chondrites* clusters. The diameters of the burrows range from less than 1 mm to 10 mm. The intensity of burrow mottling is variable across the outcrop of the formation; in general the

mottling is less apparent on the south-eastern limb of the Tywi Anticline.

The bioturbated mudstones contain few macrofossils, but have yielded diverse acritarch assemblages, which include fourteen taxa (Appendix 2a) indicative of the Caradoc–Ashgill time interval. Additional taxa include *Arkonia* cf. *tenuata*, *Baltisphaeridium longispinosum* spp. *delicatum*, *Cheleutochroa meionia*, *C.* cf. *homoia*, *Coryphidium bohemicum*, *Dicrodiacrodium normale*, *Frankea breviuscula*, *Navifusa similis*, *Striatotheca frequens*, *S. principalis parva* and *S.* cf. *principalis*, which are known from pre-Ashgill strata but for which there is no previous Ashgill record. Most of these taxa are assumed to be indigenous. Some assemblages contain *Elektoriskos williereae*, *Multiplici-*

Plate 4 Facies of the Nantmel Mudstones.

a Bioturbated mudstone facies. Well-bedded, colour-banded mudstone with large dark grey burrow mottles. Road cutting [SO 0245 6630] on the A44 near Nantmel. **b** Conglomerate and sandstone facies. Vertical, eastward-younging, sequence of graded conglomeratic sandstone, sandstone and pebbly siltstone. Road cutting [SN 9974 6258] on A470 near Doldowlod House.

a

b

sphaeridium wrightii, Tylotopalla cf. *caelamenicutis* and *Very-hachium hamii,* none of which have been recorded from pre-Ashgill strata. Recycled acritarchs of both late Cambrian and early Ordovician age occur in some samples (Appendix 2a).

DETAILS

The large road cutting [SO 0243 6633 to SO 0257 6623] on the A44 near Nantmel has been taken as the type section of the Nantmel Mudstones. It comprises a gently dipping sequence of light to medium grey, burrow-mottled mudstones, colour banded in places and with scattered fine-grained sandstone laminae and thin beds (Plate 4a). The base of the formation is exposed 350 m east of Dyfaenor in Bachell Brook [SO 0747 7103]. There, black graptolitic mudstones of the St Cynllo's Church Formation pass upwards into a 2 to 3 m-thick unit of interbedded blocky black mudstone and paler burrow-mottled mudstone, which is overlain by medium to pale grey mudstone with abundant large dark grey burrow mottles. The upper contact of the formation is exposed in Cwm Pistyll [SN 9418 6083]. Medium to pale grey, burrow-mottled Nantmel Mudstones, which crop out in the steeply incised gully at the top of the stream, are sharply overlain by laminated silty mudstones of the Yr Allt Formation.

Folded bioturbated mudstones, with silt laminae and rare beds of silty sandstone up to 4 cm thick, are exposed along the banks of the Wye River between Dolcreiglyn [SN 9766 6300] and Doldowlod Bridge [SN 9779 6375]. A more continuous track section [SO 005 635] along the east side of Rhiw Graidd exposes over 500 m of bioturbated mudstones beneath a thick conglomerate unit. To the south-west in a roadcutting [SN 9941 6292] on the A470 west of Argoed Mill vertical bedding in highly friable bioturbated mudstones is identified by differential weathering of the tops of beds. Bedding in the bioturbated mudstone facies is also well displayed in a small quarry [SN 9633 6255] at Gamrhiw Uchaf, and in the streams which incise the north-east slopes of Drum Ddu [SN 965 610], where the bioturbated mudstones are intercalated with thin siltstone and hard phosphatised mudstone beds.

Grey, silty mudstones with diffuse dark grey burrow mottling and scattered thin sandstone beds are exposed to the east of the Carmel Fault in Camlo Brook [SO 0555 6651] near Bailey-Walter. A small quarry [SO 0587 6699] to the south-west of Cae-meiriol exposes micaceous silty mudstones, locally bioturbated, with scattered argillaceous, planar- and cross-laminated, sandstones in beds up to 8 mm thick. Silty mudstones with diffuse burrow mottling occur in stream sections [SO 0725 6828 and SO 0722 6860] to the east of Bryn-Camlo. To the south-west of Rhiw Graidd at Upper Cilgee [SN 0154 6307], medium to dark grey silty mudstones with scattered sandstone laminae display faint burrow mottling, whereas exposures along strike at Lower Cilgee [SO 0107 6288] are unmottled.

LAMINATED HEMIPELAGITE FACIES

This facies forms three distinct units, on average 25 m thick, along the western limb of the Tywi Anticline (Figure 8), between Rhos Saith-maen [SN 950 597] and Gwern-hesgog [SO 0318 6962]. To the east of the Glanalders Fault the facies crops out in a few isolated exposures around the hinge zone of the Tywi Anticline, south of Abbeycwmhir. Correlation of these hemipelagites with the three units to the west of the Glanalders Fault is uncertain.

The facies consists of thin alternating laminae of dark grey, organic-rich material, which contains poorly preserved graptolite fragments, and medium grey mudstone. The laminations are on a millimetre scale. They compare closely with the intervals laminated hemipelagite within the Silurian succession of the district (p.47). The preservation of organic carbon and the lack of bioturbation indicate that the facies accumulated under anoxic bottom conditions. Thin, medium grey, massive mudstones also form part of the facies locally. The hemipelagites have yielded a sparse, low diversity, graptolite fauna which includes rare examples of *Orthograptus abbreviatus,* in places accompanied by *Normalograptus miserabilis,* together suggestive of the *anceps* Biozone. Sparse floras of spore dyads and tetrads and a few acritarchs have also been collected from the hemipelagite facies (Appendix 2a).

DETAILS

The lower unit of laminated hemipelagite is exposed only in a small section [SN 9502 5960] in Nant Hafen, close to the southern boundary of the Rhayader sheet area. There, rusty weathering, laminated hemipelagites are interbedded with a few thin sandstones. The unit overlies a succession of bioclastic sandstones and burrow-mottled mudstones that are exposed to the south of the district.

The middle unit of laminated hemipelagite can be traced from Afon Chwerfri [SN 9438 5973] along the western slopes of Drum Ddu to the River Wye near Llanwrthwl [SN 9776 6171]. It is well exposed in a narrow gorge [SN 9431 5980 and SN 9443 5963] at the headwaters of Afon Chwerfri. This section contains a few thin beds of medium grey massive mudstone and rare, thin sandstone lenses. The laminated hemipelagites have yielded *Orthograptus abbreviatus,* chiefly characteristic of the *anceps* Biozone. This species together with *Normalograptus miserabilis* has also been collected from the middle unit in a stream [SN 965 608] which drains Drum Ddu. A comparable section in a farm track to the north-east contains an abundant monospecific fauna of *Normalograptus* cf. *normalis.* The unit is also exposed in the rock platform and along the southern bank [SN 9790 6273 to SN 9767 6295] of the River Wye near Dolycreiglyn. There, laminated hemipelagites are interbedded with burrow-mottled mudstones in the lower part of the unit and with rare, silty mudstone beds, up to 2 cm thick, in the upper part. Farther north, a thick sequence of laminated hemipelagites crops out [between SO 0278 6914 and SO 0292 6919] in Cwm Scwlws stream and to the east [SO 0318 6962] of Gwern-hesgog. These exposures are tentatively correlated with the middle hemipelagite unit.

The upper unit of laminated hemipelagite parallels the contact of the Nantmel Mudstones with the Yr Allt Formation from Bryn Maen [SN 937 597] in the south to as far north as Ty-isaf [SO 020 688]. It overlies a distinctive sequence of thin sandstones interbedded with bioturbated mudstones. In Cwm Pistyll [SN 9422 6071] over 16 m of finely laminated, rusty weathering mudstones include scattered interbeds of light grey, homogeneous mudstones up to 1 cm thick. Graptolites collected from this locality include *Orthograptus abbreviatus* and *Normalograptus* cf. *miserabilis.* To the north of the River Wye, laminated hemipelagic mudstones of the uppermost unit are only exposed in a cut [SO 0083 6792] behind Neauddllwyd farm.

To the east of the Glanalders Fault the laminated hemipelagites within the Nantmel Mudstones are restricted to a few outcrops [SO 0555 6966; SO 0539 6965; SO 0555 6982] along forestry tracks near Cwmbedw, and a small stream section [SO 0391 6900] on the southern slopes of Camlo Hill.

CONGLOMERATE AND SANDSTONE FACIES

This facies forms at least two main units within the Nantmel Mudstones; they are best preserved in the hinge zone and along the south-eastern limb of the Tywi Anticline (Figure 8). A major unit crops out along the high ground between Drum Ddu [SN 970 604] and Rhiw Gwraidd [SO 023 636], with a similar conglomerate to the north-east, between the high ground to the north of Rhiw-goch [SO 055 682] and Lan-wen [SO 066 699]; a fault-bounded conglomerate also occurs to the south-east of Black Bank [SO 046 693 to SO 052 686]. The highest sandstone and conglomerate is exposed across the hinge zone of the Tywi Anticline on the southern slopes of Camlo Hill [SO 034 693 to SO 042 681], where they are overlain by silty mudstones of the Yr Allt Formation (Figure 9). Thin sandstones belonging to this facies are also present, locally, beneath the uppermost unit of laminated hemipelagite facies on the north-western limb of the Tywi Anticline.

The facies consists of fine- to coarse-grained sandstones and pebbly mudstones, locally conglomeratic, interbedded with pale to medium grey mudstones and darker grey silty mudstones (Plate 4b). These lithologies consist of both turbidite and debrite deposits. The turbidite sandstones, which range in thickness up to 90 cm, are predominantly parallel laminated, but beds also display normal grading and cross-laminated tops. The bases are sharp and erosive and some display load structures and rare flutes. The thicker sandstones may contain small scattered pebbles or pass into pebbly sandstones with conglomeratic bases (Plate 4b). The sub-rounded pebbles are generally less than 2 cm across, and consist of intraformational mudstone and sandstone, vein quartz and acid volcanic rocks, together with minor amounts of crinoid, brachiopod and coral debris. The pebbly mudstones (debrites) form discrete lenticular beds up to 1 m thick consisting of an unsorted matrix-supported pebble assemblage, similar to that of the pebbly sandstones, together with rare tabular rafts of sandstone, set in a dark grey silty mudstone matrix. Some beds grade laterally into clast-supported pebble and cobble conglomerates. In places the thick-bedded sequences fine upwards into packets of interbedded thin sandstones and silty mudstones. These sandstones are generally less than 10 cm thick and some display rippled tops.

The shell debris within the sandstone facies is very poorly preserved. Material from previous Geological Survey collections, probably from the lowermost beds of the upper unit of the facies recorded from the south-east part of Camlo Hill [about SO 052 685], include the brachiopods *Chonetoidea radiatula* and the trilobite *Tretaspis* sp.. The assemblage is similar to that collected from the shelly sandstones of the Cefnnantmel Member; it is probably of Cautleyan or Rawtheyan age. The facies has also yielded acritarch floras of low diversity, but includes recycled taxa of mid-Cambrian to early Ordovician age (Appendix 2a).

DETAILS

A major conglomeratic unit is well exposed to the north-east of Banc Ystrad-wen [SN 9899 6226 to SN 9917 6172], along the steep valley slope of the River Wye, where approximately 450 m of interbedded sandstones, pebbly sandstones, conglomerates, pebbly siltstones and light grey silty mudstones are exposed in a vertical to overturned south-east-younging section. At the western end of the section [SN 9899 6226], above Carreg-yn-fol wood, folded tabular-bedded, fine- to medium-grained sandstones, between 10 and 70 cm thick, are interbedded with pebbly siltstones and silty mudstones. The sandstones, which make up over 80 per cent of this part of the section, have parallel-laminated bases and cross-laminated tops. Rare flutes on the bases of some turbidite sandstone beds suggest that the transport direction was from the north-east. Farther south-east [SN 9894 6205], beds of pebbly sandstones and siltstones, up to 2 m thick, are interbedded with fine-grained sandstones and silty mudstones. An erosional channel, nearly 2 m deep, in the top of a pebbly siltstone is filled with thinly bedded sandstones and silty mudstones [SN 9874 6205]. Farther south along the spur [SN 9893 6201], fine-grained sandstones with rippled tops, up to 10 cm thick, are interbedded with grey mudstones with siltstone laminae and thin sandstones. The upper part of the section [between SN 9905 6177 and SN 9910 6174] consists of mica-rich, argillaceous sandstones up to 40 cm thick, pebbly mudstones, locally conglomeratic, and thin calcareous sandstones interbedded with siltstones. A similar sequence is exposed in the road cutting [SN 9970 6263 to 9984 6254] on the A470 near Doldowlod House. There, interbedded sandstones, conglomerates, pebbly siltstones and sandstones (Plate 4b) thin and fine upwards into a sequence of interbedded grey silty mudstones and fine- to medium-grained sandstones, some of which preserve isoclinal slump folds. To the north-east of the A470 a quarry [SN 9997 6283] behind Laundry Cottage exposes an interbedded sequence of medium-grained, grey sandstones, up to 60 cm thick, and grey silty mudstones. The sandstones are generally parallel-laminated, but some have cross-laminated tops and trace fossils on their bases. At the top of the quarry a pebbly mudstone infills a channel in the underlying sandstones.

A narrow ridge [SO 055 681 to SO 061 691] to the north of Rhiw-goch reveals scattered blocks of fine- and medium-grained sandstones. The quarry [SO 0608 6908] west of Esgairwy exposes a sequence of medium- to coarse-grained, locally pebbly, sandstones up to 20 cm thick, interbedded with pale to medium grey, faintly colour-banded mudstones. Isolated outcrops of pebbly sandstone and medium-grained sandstone occur on the southern slopes of Lan-wen [SO 0644 6972]. Further exposures to the west [between SO 046 693 and SO 054 688] consist of fine- to medium-grained sandstones, pebbly sandstones, up to 75 cm thick, pebbly mudstones containing large blocks of coarse and pebbly sandstone, and dark grey silty mudstones. This sequence is deformed into a series of tight east-facing folds which plunge steeply towards the north.

The highest conglomerate and sandstone is exposed in a series of asymmetrical folds with overturned south-east limbs, on the southern slopes [SO 034 693 to SO 042 682] of Camlo Hill (Figure 9). The unit consists of thick-bedded conglomerates, argillaceous and pebbly quartzose sandstones and pebbly siltstones interbedded with dark grey silty mudstones. Clasts within the conglomeratic deposits include intraformational sandstones and mudstones, together with vein quartz and igneous rock types. The sequence fines upwards into thinly interbedded sandstones and silty mudstones. The complete section through the unit is well exposed along the

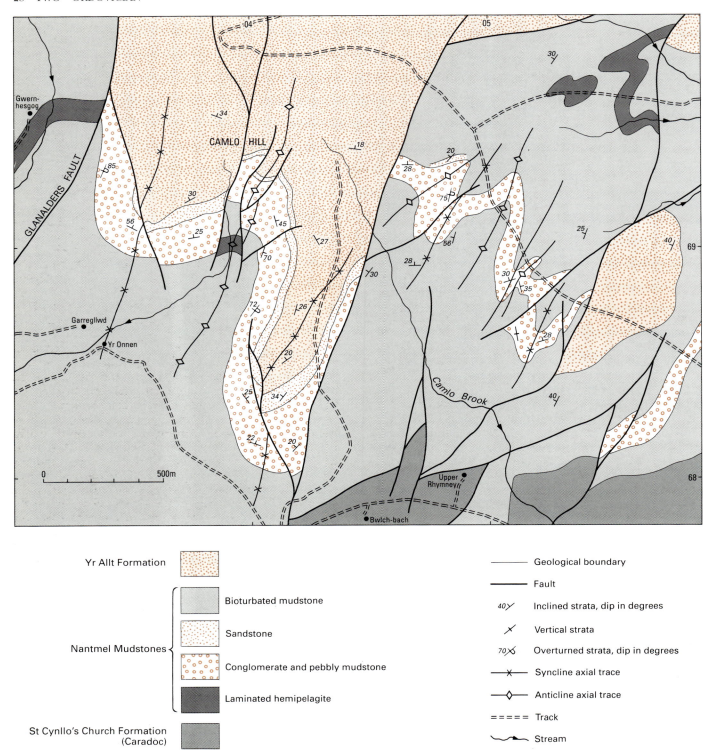

Figure 9   Geological map of the Camlo Hill area.

Legend:

Yr Allt Formation

Nantmel Mudstones
- Bioturbated mudstone
- Sandstone
- Conglomerate and pebbly mudstone
- Laminated hemipelagite

St Cynllo's Church Formation (Caradoc)

——— Geological boundary
——— Fault
40⟋ Inclined strata, dip in degrees
⟋ Vertical strata
70⤬ Overturned strata, dip in degrees
—⤬— Syncline axial trace
—◇— Anticline axial trace
===== Track
⌇⌇ Stream

overturned limb of the main anticline [between SO 040 685 and SO 041 690].

On the north-western limb of the Tywi Anticline the sandstone facies is exposed directly beneath the uppermost unit of laminated hemipelagite. In the Cwm Pistyll section [SN 9425 6061] discontinuous fine-grained sandstones and siltstones, up to 15 cm thick, are interbedded with dark grey laminated siltstones and bioturbated silty mudstones. The sandstones constitute up to 15 per cent of the succession. Similar sandstones crop out at the same stratigraphic level at Neauddllwyd [SO

0082 6790], and in a small quarry [SO 0260 6963] and farm track [SO 0270 6972] to the east of Little Vaynor.

SHELLY SANDSTONE FACIES (CEFNNANTMEL MEMBER)

Within the lower part of the Nantmel Mudstones a distinctive facies of interbedded shelly sandstones and weakly bioturbated mudstones has been named the Cefnnantmel Member (Figure 8). The member lies within burrow-mottled mudstones, and crops out along the south-eastern slopes of Cefnnantmel [between SO 0270 6636 and SO 0375 6719]. At least 75 m of the bioturbated mudstone facies underlie the Cefnnantmel Member as far as the faulted contact with Caradoc rocks. The facies consists of rusty weathering, largely decalcified, sharp-based, normally graded, planar- and cross-laminated shelly sandstones interbedded with medium to dark grey slightly burrow-mottled mudstones. The sandstones, which constitute between 5 and 40 per cent of the member, are generally fine grained and range in thickness from 1 to 2 cm, and locally up to 10 cm. Some of the lower sandstone beds have coarse shelly bases which contain abundant trilobite and brachiopod fragments. The sandstones decrease in frequency upwards, and the bioturbation in the interbedded mudstones becomes more prominent.

A diverse fauna has been collected from outcrops [SO 0355 6697 and SO 0279 6646] on Cefnnantmel. Twelve trilobite genera have been recorded including *Acidaspis magnospina*?, *Gravicalymene* cf. *deani*, *Kloucekia* cf. *extensa* and *Tretaspis* cf. *radialis*. A diverse brachiopod fauna includes *Chonetoidea radiatula*, *Eoplectodonta (Kozlowskites) nuntius* (Plate 2h,i), dalmanellids and *Platystrophia anomala*. The fauna suggests a Cautleyan, probably mid-Cautleyan, age. The Cefnnantmel Member has also yielded a poorly preserved acritarch assemblage of late Ordovician aspect (Appendix 2a).

ENVIRONMENTS OF DEPOSITION

The bioturbated mudstones of the Ashgill sequence contrast sharply with the underlying black graptolitic shales of Caradoc age, and mark the establishment of predominantly oxygenated bottom conditions over most of the southern Welsh Basin. Although this change in sedimentary environment postdated a period of non-sequence or minimal sediment input in other parts of Wales (Price, 1984; Cave, 1965), there appears to have been continuous and sustained sedimentation in the district across the Tywi Anticline. The lithological change probably reflects, therefore, a lowering of sea level, which culminated with the deposition of the shelly sandstones of the Cefnnantmel Member, and the westward progradation of shallower-water mudstones.

Bioturbation, colour banding, and phosphatic layers in the bioturbated mudstone facies indicate deposition under oxic bottom conditions and suggest comparision with the mud-dominated turbidites of the Silurian slope-apron facies (see Chapter 3). The three units of laminated hemipelagite record periods when anoxic bottom conditions were established across large parts of the Welsh Basin (Waters et al., 1992), such that the alternation of mud and organic debris settling from suspension was undisturbed by bioturbation (Cave, 1979; Dimberline et al., 1990). As with comparable Silurian facies, these may record a series of transgressive events.

The Cefnnantmel Member occurs near the base of the Nantmel Mudstones and consists of a shelly sandstone facies. The sharp-based sandstone beds with accumulations of disarticulated, commonly broken shells at their bases indicate sudden incursions of high-energy sand-laden currents. The presence of fauna typical of several muddy outer-shelf habitats suggests mixing and emplacement of the shells by storm-generated currents beyond the limits of colonisation by contemporary shelly benthos.

The conglomerate and sandstone facies consists of debrites and turbidites containing reworked clasts and shells derived from the adjacent shelf and beyond. The lower unit of the facies can be traced for over 10 km just to the north-west of the Rhiw Gwraidd Fault. Although faulted in places, the unit appears to have a relatively constant thickness over much of the length of the outcrop. The flutes on the bases of some sandstones indicate that the transport direction was to the south-west, parallel to the Rhiw Gwraidd Fault. Apart from a thin sandstone sequence beneath the uppermost unit of laminated hemipelagite, there is no indication that the conglomerate and sandstone facies is present within the Nantmel Mudstones on the north-west limb of the Tywi Anticline; it certainly does not correlate with the sandstone facies in the overlying Yr Allt Formation as proposed by James (1991). Although contemporary activity on the Rhiw Gwraidd Fault may have confined the sandstone and conglomerate facies to a trough trending north-east to south-west, the lack of correlation may be an expression of early Silurian (Telychian) strike-slip faulting, which juxtaposed dissimilar sequences across the Tywi Lineament. The upper unit of conglomerate and sandstone facies is confined between the Glanalders Fault and a splay of the Rhiw Gwraidd Fault. Again, the absence of similar facies on the north-western limb of the Tywi Anticline may reflect later strike-slip faulting.

Yr Allt Formation

The Yr Allt Formation overlies the Nantmel Mudstones and is equivalent to the upper part of the Camlo Hill Group of Roberts (1929). It also partly encompasses the Cefn Ystradffin Formation (Mackie and Smallwood, 1987) and the Chwefru Formation (Lockley, 1978), south-west of the district. In the Rhayader sheet area the Yr Allt Formation crops out around the Tywi Anticline (Figure 10): along the north-western limb in the belt of high ground between Bryn Moel [SN 930 597] and Vaynor [SO 018 693], across the closure in the hills surrounding Abbeycwmhir [SO 056 711], along the south-eastern limb in a narrow tract to the east of Henfryn [between SO 077 683 and SO 076 704], and between the Rhiw Gwraidd and Garth faults from Llanfihangel-helygen [SO 048 645] to Blaenglynolwyn [SO 988 598]. It also occupies the core of the Rhiwnant Anticline between Pant Glas [SN 865 598] and Craig y Bwlch [SN 896 626].

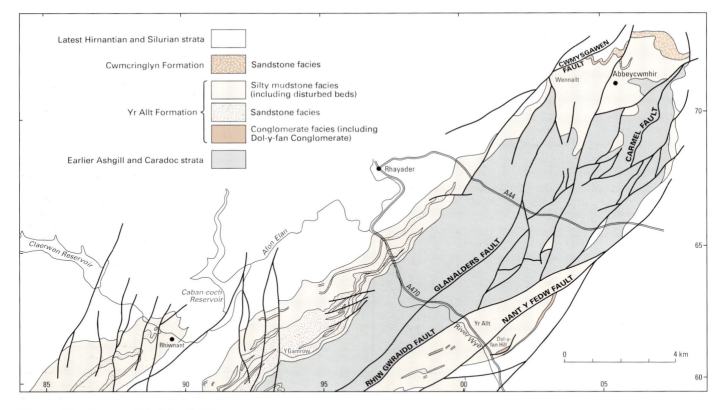

Figure 10   Facies of the Yr Allt Formation.

Throughout most of the district the basal contact of the Yr Allt Formation is poorly exposed, and on the south-eastern limb of the Tywi Anticline it is generally faulted. However, at the top of Cwm Pistyll stream section [SN 9418 6083] the contact with the underlying Nantmel Mudstones is sharp. The nature of the upper contact of the Yr Allt Formation is highly variable (Figure 4). Along the north-western limb of the Tywi Anticline and within the Rhiwnant Anticline the formation is overlain conformably by the Mottled Mudstone Member (Cwmere Formation) of late Hirnantian age (*persculptus* Biozone). Between the Cwmysgawen and Carmel faults, the formation passes upwards into Hirnantian sandstones of the Cwmcringlyn Formation, whereas over the area to the east of the Carmel and Rhiw Gwraidd faults and to the west of the Nant y Fedw Fault, it is disconformably overlain by late Telychian mudstones. Farther east, between the Nant y Fedw and Garth faults, the formation is conformably overlain by the Tycwtta Mudstones of late Hirnantian age. The Yr Allt Formation has an estimated thickness of approximately 1000 m east of the Rhiw Gwraidd Fault, but thins to less than 650 m across the hinge and on the north-west limb of the Tywi Anticline.

The Yr Allt Formation comprises three facies (Figure 10): silty mudstone facies, including slumped and destratified strata (disturbed beds); sandstone facies; and conglomerate facies (including the Dol-y-fan Conglomerate Member).

In general, the formation contains few macrofossils but a collection made during the earliest Geological Survey (about 1856) on the north end of Camlo Hill [near SO 041 695] contains a rich reworked fauna of shallow-water aspect. It includes decalcified corals, brachiopods (*Meifodia?*), molluscs (*Similodonta* and *Lyrodesma*), the trilobite *Stenopareia* sp., and crinoid columnals. Acritarch floras of moderate diversity are found within the formation, including the genera *Actinotodissus*, *Arkonia*, *Baltisphaeridium*, *Baltisphaerosum*, *Cheleutochroa*, *Diexallophasis*, *Eupoikilofusa*, *Multiplicisphaeridium*, *Ordovicidium*, *Peteinosphaeridium*, *Priscotheca*, *Stellechinatum*, *Striatotheca*, *Tunisphaeridium*, *Tylotopalla* and *Villosacapsula* (Appendix 2a). Recycled acritarchs include *Acanthodiacrodium* spp., *Stelliferidium* cf. *fimbrium*, *S. stelligerum*, and *Vulcanisphaera africana* from late Cambrian to early Ordovician, and *Arbusculidium filamentosum* from the Arenig to early Llanvirn (Appendix 2a).

SILTY MUDSTONE FACIES (INCLUDING DISTURBED BEDS)

This facies is characteristic of much of the Yr Allt Formation and consists of grey, micaceous silty mudstone intercalated with thin laminae of dark grey siltstone or fine-grained sandstone (Plate 5a). The siltstone laminae vary in thickness from 1 to 3 mm and are spaced at intervals of 5 to 20 mm (Plate 5a). Locally, the silty mudstones are interbedded with a few thin, locally shelly, sandstones which range in thickness up to a few centimetres. These sandstone packets are generally too thin and discontinuous to map. At a few localities quartzite pebbles form discontinuous layers or randomly scattered clasts within the silty mudstones.

The effect of slumping in the Yr Allt Formation varies from slight disruption of the bedding fabric, through syndepositional folding and faulting to the total destruction of the silty laminations and the break-up of the thicker sandstone beds into scattered randomly orientated blocks. The slumped silty mudstones display an irregular anastomosing cleavage, which contrasts with the pervasive planar cleavage seen in the undisturbed lithologies. All these features are characteristic of the disturbed beds, which form mappable units within the silty mudstone facies. In the Rhayader sheet area the disturbed beds become generally more frequent westwards, and in the Rhiwnant Anticline undisturbed laminated silty mudstones are restricted to thin units less than 15 m thick. The units of disturbed beds are generally less than 20 m thick, but rarely up to 120 m, and along the north-western limb of the Tywi Anticline, to the west of Llanwrthwl [SN 976 637], they form distinct scarp features (Plate 1). Where the lamination has been preserved, or the sandstone beds have retained some cohesion, slump folds can be recognised within the disturbed beds. Detached, small-scale, tight to isoclinal, slump folds are the most common, but close recumbent folds with axial plane separations of over 8 m are present locally (Plate 5b). The average trend of the axes of the slump folds is approximately north-east, which would suggest a palaeoslope striking north-east to south-west.

DETAILS

The road cuttings on the A470 between Garden Cottage [SO 0022 6218] and Holmes Wood [SO 0083 6136] constitute the type section of the Yr Allt Formation. This well-exposed section consists of dark grey, thinly laminated, silty mudstones interbedded locally with rusty weathering, coarse-grained sandstones. The closely spaced laminae are generally of siltstone or fine-grained sandstone and are commonly micaceous. The sandstones, a few centimetres thick, appear decalcified and some contain indeterminate bioclastic debris. The sequence is cut by a north-west-dipping cleavage which, in places, is partitioned into anastomosing domains, giving an 'augened' appearance to the weathered surface.

Around the closure of the Tywi Anticline the laminated silty mudstones display local destratification and are interbedded with thick slump units. Laminated mudstones occur on Wennallt [SO 037 712] where the small quarries [SO 0429 7115 and SO 0332 7131], and the crags [SO 0417 7055] near Lower Cwm-Hir expose moderately dipping, thinly laminated, silty mudstones interbedded with a few thin, rusty sandstones, some displaying trace fossils on their bases.

On the north-western limb of the Tywi Anticline, to the south-west [SN 985 656] of Ty-mawr, laminated silty mudstones are interbedded with slump units and sandstones. They are generally poorly exposed and form the slack features on the south-east-facing escarpment. In the River Wye, dark grey laminated silty mudstones interbedded with a few fine-grained sandstones are exposed [SN 9750 6449]. Contiguous exposures on the slopes [SN 9701 6397] to the west of Hill Cottage display similar laminations, with slightly disrupted sandstone beds up to 1 cm thick. At Coed-Blaen-y-cwm [SN 9522 6335] laminated silty mudstones are interbedded with units of pebbly mudstone. The subrounded pebbles range up to 2 cm in diameter, and consist of white quartz and subordinate acid volcanic rocks.

In the Rhiwnant Anticline a thin unit of undisturbed silty mudstone facies is exposed in a small quarry [SN 8947 6150] to the south of Rhiwnant farm. There the facies comprises well-bedded, pale to medium grey mudstones and silty mudstones. The bedding thickness varies from less than 0.2 cm to approximately 2 cm. To the north of Pant-y-gwartheg [SN 8979 6200] and to the south of Cwm-clyd [SN 8918 6207] the laminated mudstones and silty mudstones are interbedded with thin sandstones up to 3 cm thick.

The crags [SN 974 652] near Glaslyn consist of dark grey, silty, micaceous mudstones with randomly orientated balls and wisps of fine-grained sandstone up to 15 cm across. The rock is totally destratified and cut by irregular shiny cleavage surfaces which are partitioned into small domains. Small isolated slump folds, with steep axial surfaces, are present locally. The outcrops [SN 9632 6423] to the north-west of Tan-yr-allt have retained their original silty lamination, but the sequence has been deformed into a series of disharmonic slump folds which are associated with several generations of dislocation. Slump folds in laminated silty mudstones are also well exposed in a quarry [SN 9642 6459] near Cefn.

On Craig Ddu [SN 9594 6365] recumbent slump folds occur in a sequence of finely laminated siltstones and fine-grained sandstones (Plate 5b). The folds have axial plane separations of up to 8 m, and are associated with syndepositional slide planes. Numerous small-scale isoclinal slump folds are present [SN 9588 6370] in the overlying disrupted sandstone beds. The average orientation of the axes of the slump folds in this vicinity is north-east to south-west.

A sequence of slump sheets is well displayed across the axial zone of the Rhiwnant Anticline. In the crags of Craig y Bwlch [SN 8951 6223 to SN 8985 6200], north of Nant Claerwen, slump sheets, between 10 and 30 m thick, dip shallowly towards the north-east. The destratified silty mudstones contain rare balls and wisps of sandstone, and are cut by an anastomosing network of intersecting cleavage fabrics, which gives the rock a shiny appearance. The slump sheets are separated by thin beds of laminated silty mudstones which form the slack features on the escarpment.

The lowest beds of the Yr Allt Formation in the core of the Rhiwnant Anticline are exposed in the Rhiwnant stream section [between SN 877 605 and SN 888 609]. The water-worn surfaces display the destratified bedding fabrics, with streaks of light and dark grey silty mudstones, balls of disaggregated sandstone beds, and rare slump folds. At one place [SN 8799 6080], angular white quartz clasts, up to 5 mm across, are randomly distributed through the rock. On the crags of Craig y Llysiau [between SN 886 611 and SN 890 612] disrupted sandstone balls and lenses, up to 15 cm across, are set in a dark grey silty mudstone matrix.

SANDSTONE FACIES

This facies is best developed to the west of the Glanalders Fault, and crops out along the north-western limb of the Tywi Anticline and within the Rhiwnant Anticline (Figure 10). Small lenticular units of the sandstone facies are also present across the closure and along south-eastern limb of the Tywi Anticline. The facies varies from scattered fine-grained turbidite sandstones, 2 to 5 cm thick, to massive turbidite sandstone beds, up to 2 m thick, interbedded with thin silty mudstones. The thicker sandstone beds are commonly amalgamated and can form up to 90 per cent of some sections. The thick tur-

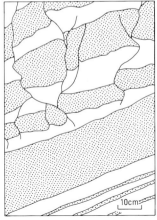

bidites are medium to coarse grained, locally graded, with sharp bases and parallel-laminated and cross-laminated tops. The basal parts of some units contain angular to rounded clasts of black mudstone, quartz and acid volcanic rock up to 2 cm across. Dewatering and load structures have formed at the contact with the underlying silty mudstones, and synsedimentary faulting has affected some of the thinner-bedded sequences (Plate 5d). The thinner turbidite sandstones display parallel lamination or, more rarely, cross-lamination.

The thickest development of the sandstone facies occurs within the lower part of the Yr Allt Formation along the north-western limb of the Tywi Anticline (Figure 10). Approximately 400 m of interbedded thick turbidite sandstones and silty mudstones are present over the well-featured, but poorly exposed, area around Carnau [SN 940 612]. Farther to the north-east and south-west the sandstones become thinner, and separate into discrete packets interbedded with sequences of laminated and slumped silty mudstones. These sandstones die out to the north-east of the River Wye near Ty Mawr [SN 986 657]. At least two thin units of the sandstone facies are present in the upper part of the Yr Allt Formation (Figure 10). The most extensive crops out between Coed-Blaen-y-cwm [SN 950 633] and Upper Esgair-rhiw [SN 998 673].

DETAILS

Outcrops near Carnau [SN 9387 6135 and SN 9460 6147] consist of light grey, medium- to fine-grained, feldspathic sandstones up to 2 m thick. The ungraded sandstones display a weak bedding-parallel lamination and in places have cross-laminated tops. On Bryn Melys [SN 9302 6050] discontinuous conglomeratic layers are present within the sandstones. The subrounded clasts average 1 cm across and consist of pink and white vein quartz, acid volcanic rock, and subordinate dark grey mudstone. Four kilometres to the north-east, on the hill [SN 972 641] behind Dolgai, the sandstone facies is reduced to two units, each less than 30 m thick. At one place [SN 9701 6399] the lower unit is composed of massive, medium-grained, feldspathic sandstones, between 0.2 and 1.2 m thick, interbedded with flaggy, finer-grained, parallel-laminated sandstones. The bases of many of the sandstones display trace fossils. Exposures in the lower sandstone unit also occur in the road cut [SN 9751 6454] on the A470. There, the sandstones are medium to coarse grained, in beds up to 40 cm thick, locally graded, with parallel- and cross-laminated tops in places. The

Plate 5   Facies of the Yr Allt and Cwmcringlyn formations (facing page).

a   Silty mudstone facies. Forestry track on Y Glog [SO 047 717].   **b**   Large-scale recumbent slump fold in disturbed beds of the silty mudstone facies. Outcrop [SN 9594 6365] on Graig Ddu.   **c**   Dol-y-fan Conglomerate. Outcrop [SO 013 610] on Dol-y-fan Hill.   **d**   Sandstone facies. Synsedimentary faulting in sandstones and interbedded silty mudstones. Sandstones are stippled in sketch. Roadside outcrop [SN 9751 6454] on A470.   **e**   Wave-rippled sandstone in the Cwmcringlyn Formation. The sandstone bed is 4 cm thick. Quarry [SO 0585 7219] near Cwm-poeth Cottage.

bases of a number of the sandstones exhibit load structures, and parts of the sequence are cut by synsedimentary faults (Figure 5d). The sandstones of the upper unit, exposed in the River Wye section [SN 9739 6464], are between 5 and 60 cm thick and interbedded with well-laminated, silty mudstones. Several of the thicker sandstone beds contain angular clasts of mudstone up to 2 cm across at their bases, and have parallel- and cross-laminated tops. Contiguous exposures on the hillside to the east display massive beds, possibly amalgamated, of medium- to coarse-grained pebbly sandstones, up to 2.5 m thick, interbedded with slumped, thinly laminated, silty mudstones. Cuspate dewatering structures are present along the bases to several of the sandstones.

Sandstones in the upper part of the Yr Allt Formation are exposed in the track [SN 9671 6436] through Cefn Wood, where a sequence of light grey, medium-grained sandstones, between 1 and 50 cm thick, is interbedded with thin flaggy sandstones and laminated silty mudstones. The turbidite sandstones are generally parallel-laminated and some exhibit convolute bedding. To the north-east of the River Wye [SN 9745 6530], the same unit consists of massive sandstone beds with coarse-grained, feldspathic bases containing mudstone clasts. Parallel lamination and load structures are evident in some beds. These sandstones are interbedded with fine-grained, parallel-laminated sandstones and silty mudstones.

A prominent unit of the sandstone lithofacies, between 26 and 60 m thick, crops out in the core of the Rhiwnant Anticline. It consists of massive, commonly amalgamated, sandstone turbidites up to 3 m thick, interbedded with thinly laminated and destratified silty mudstones. Two good stream sections are present along the steeply dipping south-eastern limb of the anticline. In the Rhiwnant valley [SN 8832 6083] the unit is approximately 28 m thick and consists of amalgamated sandstone beds, between 0.3 and 1.0 m thick, which display parallel-laminated bases, poor grading and cross-laminated tops. The basal part of one sandstone contains abundant dark grey mudstone flakes. The thicker sandstones, which make up over 95 per cent of the sequence, are overlain by thinner sandstones interbedded with streaky, silty mudstones. In Pant Glas stream [SN 8781 6035] the unit consists of fine-grained sandstones, up to 8 cm thick, interbedded with dark grey silty mudstones overlain by massive amalgamated sandstones up to 2 m thick. These are succeeded by thinner and less frequent sandstones displaying parallel-laminated bases and cross-laminated tops. Well-featured massive sandstones of the same unit crop out on the shallow north-west-dipping limb of the Rhiwnant Anticline along Craig Rhiwnant [SN 874 609 to SN 884 612], and in Nant Carw [SN 8725 6075].

The thin undisturbed laminated and silty mudstone units of the Yr Allt Formation in the Rhiwnant Anticline are in places interbedded with fine-grained sandstones, which make up less than 15 per cent of the sequence. The turbidite sandstones commonly display cross-laminated tops and vary in thickness from 1 to 3 cm, and rarely up to 8 cm. Units of this facies are exposed [at SN 8786 6030] in Pant Glas, in the crags [SN 8979 6200] below Craig y Bwlch and [at SN 8918 6207] near Cwm-clyd.

The sandstone facies of the Yr Allt Formation across the hinge zone of the Tywi Anticline near Abbeycwmhir is restricted to a thin unit of dark grey massive and pebbly sandstones exposed around a hill in Great Park [SO 059 717]. At Fowler's Cave [SO 0582 7154] the top of the sandstone unit is a scoured surface overlain by a muddy debrite, which contains disrupted rafts and small clasts of sandstone.

The sandstone facies between the Nant y Fedw and Rhiw Gwraidd faults, on the south-eastern limb of the Tywi Anticline, is confined to discontinuous sandstones up to 2 m thick within a monotonous sequence of relatively undisturbed laminated silty mudstones. The quarry [SN 9940 6116] at Bryn Wood displays a channelised medium-grained sandstone with a massive conglomeratic base and parallel-laminated top. The broken outcrops [SN 9522 6339] to the east of Garth consist of a vertical, eastward-younging, sequence of coarse-grained sandstones and conglomerates, which display grading and scour structures.

CONGLOMERATE FACIES (INCLUDING THE DOL-Y-FAN CONGLOMERATE)

The main unit of this facies is the Dol-y-fan Conglomerate (Member) which forms a lenticular unit at the top of the Yr Allt Formation on the south-eastern limb of the Tywi Anticline. It crops out between the Wye valley and Werngronllwyd where it is truncated by the Nant y Fedw Fault. The type locality of the member is the prominent crags [SO 0130 6095] on Dol-y-fan Hill. The Dol-y-fan Conglomerate comprises turbidite conglomerates and sandstones with some pebbly mudstone debrites. The member attains its maximum thickness of 85 m in the north and, in general, thins south-westwards. However, on Dol-y-fan Hill, fault-repeated crops reveal a marked north-westwards thickening (from 7 to 50 m). It overlies the silty mudstone facies with a sharp and commonly erosional contact.

The coarse-grained turbidites mainly comprise clast supported, pebble-cobble conglomerates with some pebbly coarse-grained sandstone and sandstone. Individual turbidites that grade upwards from conglomerate through pebbly sandstone to sandstone are rare. Most comprise thick units of massive conglomerate, several metres thick. Normal and reverse grading and pebble imbrication is locally developed and amalgamation of individual turbidites is common. The predominantly extrabasinal clasts are mainly rounded and largely of quartz. The bases of conglomerate units are commonly erosional and in some cases, notably at the base of the member, deeply channelised.

The debrites comprise massive, matrix-supported conglomerates in which the clasts are mainly close-packed in a muddy sandstone matrix. The clasts largely comprise rounded, extrabasinal pebbles and cobbles but scattered angular intrabasinal clasts of mudstone and sandstone, up to boulder size, are locally present. The moulds of leached crinoid ossicles have been noted in crags [SO 0232 4168] at the north-east end of Dol-y-fan Hill.

A few thin units of silty mudstone facies up to 2 m thick are locally interbedded in the member and in some cases exhibit diapiric injection structures.

The member exhibits an overall fining-upwards sequence with sandstones generally becoming dominant in the uppermost part. A lateral change into a sandstone-dominated sequence, where the member is very thin, has been noted locally, as in Cwm-nant [SO 0255 6183]. Pebble imbrication in the turbidite conglomerates suggests transport towards the west-south-west.

Debrites of the conglomerate facies also form a lenticular unit approximately 30 m thick at Coed Tydabuan [SN 9905 6105]. The unit is contained within laminated silty mudstones. The debrites consist of rounded clasts of mudstone, sandstone, siltstone, quartz and volcanic rock,

up to 10 cm in diameter, set in a siltstone matrix containing finer-grained clasts. Towards the south-west they pass into a sequence of pebbly siltstones and lenses of light grey, fine-grained sandstone.

The Dol-y-fan Conglomerate has yielded microfloras of variable diversity, dominated by acritarchs, although all but one have also included sporomorphs (Appendix 2a). The assemblages are similar to those found within the silty mudstone facies. However, the notable difference is the abundance of the enigmatic taxon *Moyeria cabotti*, together with sporomorphs, in mudstones within the conglomerate.

DETAILS

The most south-westerly exposure of the Dol-y-fan Conglomerate is in the east bank [SO 0094 6071] of the River Wye, where up to 40 cm of coarse-grained pebbly sandstone occurs in an isolated 3 m-wide lens, overlying silty mudstone facies with a sharp contact.

As a result of faulting, the member crops out twice on Dol-y-fan Hill [SO 015 611]. The southern crop is 7 to 20 m thick and well exposed as a line of crags. It mainly comprises clast-supported, pebble-cobble conglomerates passing up into coarse-grained sandstones in the uppermost part. The northern crop is a faulted syncline in which the member overlies the silty mudstone facies with a sharp and commonly channelled contact. At least 50 m thick, the member comprises a basal 5 m thick debrite, with boulder-sized rafts of intrabasinal (? Yr Allt Formation) mudstone and sandstone, overlain by clast-supported, turbidite conglomerates and debrites. Higher in the sequence, a metre-thick lenticular unit of silty mudstone facies has suffered diapiric injection due to loading by the overlying conglomerate.

In Cwm-nant [SO 0255 6183] the member overlies silty mudstone facies with a sharp contact and comprises 12 m of coarse-grained sandstones with scattered beds of pebble conglomerate up to 10 cm thick. A few thin beds of silty mudstone facies up to 40 cm thick occur in the top 5 m. The contact with the overlying Tycwtta Mudstones is not exposed.

ENVIRONMENTS OF DEPOSITION

The Yr Allt Formation and its correlatives to the south-west (Mackie and Smallwood, 1987) are considered to represent a westward-prograding, mud-dominated sequence which developed across the Tywi Lineament in Hirnantian times (James, 1991). The lithological change from the predominantly bioturbated mudstone facies of the Nantmel Mudstones to the silty mudstone facies of the Yr Allt Formation is taken to mark the onset of the Hirnantian glacioeustatic regression (Waters et al., 1992; cf. Brenchley, 1988). The lowering of sea level was associated with increased rates of sediment supply from the adjacent shelf areas to the east, and reduced levels of bioturbation. The silty mudstones are interpreted as turbidites deposited in a slope-apron setting.

The presence of thick units of disturbed beds on the north-western limb of the Tywi Anticline and across the Rhiwnant Anticline indicate frequent slope failure and slumping on the oversteepened front of the prograding muddy wedge. The orientation of the slump folds would suggest a palaeoslope striking north-east to south-west.

The conglomerate and sandstone facies are thought to have been deposited in submarine channel systems and sandy lobes that were sourced from the shelf. On the south-eastern limb of the Tywi Anticline the lenticular conglomerate and coarse-grained sandstone bodies display abundant scour features and are typical of channelised deposits. Current vectors, together with the apparent geometry of the Dol-y-fan Conglomerate, suggest that it was deposited in a north-east- to south-west-oriented channel by high-concentration turbidity currents and debris flows directed towards the south-west.

Along the north-western limb of the Tywi Anticline the sandstone facies is mainly confined to a large-scale lenticular body within a variably slumped sequence. The sandstones are interpreted as turbidites deposited on the slope apron as a series of small sandy lobes. The sandstones of the Rhiwnant Anticline form a more distal part of the same lobe system. Sediment transport direction, based on the geometry of the sandstone bodies, was from an easterly quadrant. Cross-laminations in the sandstones of the Rhiwnant Anticline suggest transport from both the south-east and south-west, the variation being attributed to the spreading out of the lobes (James, 1986).

The lack of direct correlation of the channel and lobe systems, both in relative size and stratigraphic position, is probably due to postdepositional strike-slip faulting along the Tywi Lineament.

Cwmcringlyn Formation

The Cwmcringlyn Formation equates with the highest part of the Camlo Hill Group of Roberts (1929). It is exposed only between the Cwmysgawen and Carmel faults, from Y Glog [SO 048 718] to Cwmcringlyn Bank [SO 073 725] (Figure 10). The formation has a gradational contact with the underlying Yr Allt Formation, but is unconformably overlain by the late Telychian Dolgau Mudstones (Figures 35 and 39). The thickness of the formation is about 200 m on Cwmgringlyn Bank, and thins to less than 50 m at Y Glog. It is unclear whether this thinning is due mainly to lateral facies change or to erosion prior to the deposition of the Dolgau Mudstones.

The lower part of the formation consists of cross-laminated medium- to fine-grained sandstones, up to 70 cm thick, interbedded with packets of thinner, flaggy, flaser-bedded sandstones with mudstone partings. In places, the thicker sandstones contain mudstone flakes and rare quartz pebbles. The upper surfaces of the sandstones exhibit low-amplitude, symmetrical ripples, with straight crests spaced at between 6 and 10 cm. The internal structures of the ripples are undulatory and bidirectional (Plate 5e). Higher in the formation, planar, possibly massive sandstones up to 3 m thick enter the succession.

A restricted Hirnantian brachiopod fauna is present within the lower sandstones exposed in a small quarry [SO 0585 7219] to the north-east of Cwm-poeth Cottage. The fauna is dominated by *Eostropheodonta hirnantensis* (Plate 2a,b), together with fragmentary *Kinnella kielanae?* and *Rhynchotrema?* sp. The localities reported by Roberts (1929) appear to have yielded similar restricted faunas that are thought to be referable to the Hirnantian Stage. Acritarch assemblages include 16 indigenous taxa (Appendix 2a) and recycled acritarchs of late Cambrian to early Ordovician age.

DETAILS

The type section of the Cwmcringlyn Formation is defined by a series of scattered outcrops on Cwmcringlyn Bank, including a quarry section [SO 0764 7244] which exposes a 3 m-thick, medium grey, fine- to medium-grained, massive sandstone with a sparse scatter of small quartz pebbles and rounded flakes of mudstone, and an outcrop [SO 0705 7276] west of Bank Style in medium- to fine-grained, parallel-laminated sandstones, up to 7 cm thick, finely interbedded with silty mudstone. The upward transition from the laminated silty mudstones of the Yr Allt Formation is well exposed along a forestry track on Y Glog; scattered thin ripple-laminated sandstones, up to 3 cm thick, are seen interbedded in Yr Allt Formation silt-laminated mudstones in a quarry [SO 0474 7175] at the base of the section and increase in abundance upwards towards the base of the Cwmcringlyn Formation [at SO 0469 7195]. The base of the latter is taken where the thin sandstones account for more than 25 per cent of the sequence. The sedimentary structures distinctive of the Cwmcringlyn Formation are well preserved in the quarry [SO 0585 7219] north-east of Cwm-poeth Cottage. This section consists of medium grey, fine- to medium-grained, cross-bedded sandstones up to 50 cm thick, which contain occasional mudstone flakes. The tops of the sandstones display north-north-west-trending straight-crested, symmetrical ripples with wavelengths of about 8 cm. The basal sandstones are overlain by thinly bedded, bioturbated sandstones, up to 3 cm thick, with dark grey mudstone partings. The sandstones are commonly cross-laminated and lenticular-bedded. This locality has yielded a Hirnantian brachiopod fauna dominated by *Eostropheodonta hirnantensis*.

ENVIRONMENTS OF DEPOSITION

The Cwmcringlyn Formation gradationally succeeds the regressive Yr Allt Formation, and records deposition during the acme of Hirnantian glacioeustatic regression in the district. Its fauna is indicative of cool to temperate marine conditions. The sedimentary bedforms of the sandstones are characteristic of a subtidal, shallow-water environment, and the symmetrical, straight-crested ripples were formed by wave action (Raaf et al., 1977) on a west-facing shoreface. The presence of rip-up clasts along the bases to some of the thicker sandstones record episodic high energy, probably storm-generated, events.

Pentre Formation

The Pentre Formation crops out between fault splays of the Garth Fault (Figure 3) and the type section is in a small quarry [SO 0886 6808] near Pentre farm (Figure 38). The formation has no obvious lithological correlative in the main Ashgill strata to the west, but is similar to parts of the Ashgill succession in the Garth area (Williams and Wright, 1981), to the south of the district. It is unconformably overlain by the late Telychian Henfryn Formation. The base of the formation is not exposed; an estimated 200 m of strata are exposed between Bryn-Nicholas [SO 076 658] and Pentre [SO 089 682].

The Pentre Formation consists of well-bedded, dark grey, micaceous silty mudstones with scattered siltstone laminae and thin sandstone beds. The mudstones have a characteristic nodular and rusty weathered appearance, and are variably burrow mottled. The fine-grained argillaceous sandstones range up to 6 cm in thickness and contain scattered mudstone clasts and shell debris.

The sandstones at Pentre farm [SO 0886 6808] have yielded a poorly preserved shallow-water fauna including burrowing bivalves (Plate 2d), orthocones, the trilobites *Brongniartella* cf. *sedgwicki*, *Gravicalymene* cf. *pontilis* (Plate 2c) and *Mucronaspis?* sp., and echinoderm columnals and plates. Acritarchs and other palynomorphs have also been recovered, including *Peteinosphaeridium trifurcatum* subspp., *?Helosphaeridium citrinipeltatum* and spore tetrads (Appendix 2a). An Ashgill age is inferred, but precise dating is not possible.

DETAILS

The type section of the Pentre Formation is exposed in a small quarry [SO 0886 6808] near Pentre farm (Figure 38). The section consists of a 6 m sequence of buff-weathered, fine-grained, shelly, argillaceous sandstone, overlain by nodular weathered silty mudstone. The details of the fauna collected at this locality are recorded above. The nearby track [SO 0878 6795] at Pentre farm exposes a 15 m section of well-bedded, buff silty mudstones interbedded with dark grey, fine-grained sandstones up to 6 cm thick, and rare thin beds of white, possibly bentonitic, clay. The stream bed and banks of Clywedog Brook [between SO 0813 6747 and SO 0838 6648] provide intermittent exposures of nodular weathering silty mudstones and scattered argillaceous sandstones with shell debris.

ENVIRONMENTS OF DEPOSITION

The sedimentary and faunal character of the Pentre Formation is indicative of deposition in a shallow-water shelf environment, dominated by mud and silt deposition and the periodic input of sand from the foreshore to the east. The trilobites do not show the great diversity that characterises the outer-shelf environments of the Ashgill, but include rare *Gravicalymene*, homolanotid and dalmanitacean fragments suggestive of shallower-water conditions (cf. Thomas, 1979).

INTRUSIVE ROCKS

Dolerite intrusions occur within the Llanvirn and Llandeilo mudstones of the Builth Inlier to the east of Llandrindod Wells, and within the Caradoc black shales of the Tywi Anticline in the vicinity of Baxter's Bank. Jones and Pugh (1948a; 1948b) described the overall form, distribution and mode of emplacement of the dolerite intrusions throughout the Builth Inlier, and the details of the intrusions near Llandrindod Wells. They concluded that the dolerites formed mainly concordant lenticular bodies intruded at several levels within the sequence of mudstones. A series of strongly discordant dolerites was also recognised, which they interpreted as part of a feeder system. The thin dolerite sills of the Tywi Anticline were described by Roberts (1927). The sills are exposed along the south-eastern limb of the Tywi Anticline in folds developed immediately to the north-west of the Carmel Fault.

Dolerites of the Builth Inlier

In the Rhayader sheet area the dolerite intrusions of the Builth Inlier are confined to the mudstones in the Builth Volcanic Formation and the Llanfawr Mudstones (Figure 5). Farther to the south-east they also crop out in the Camnant Mudstones (Jones and Pugh, 1948a; Institute of Geological Sciences, 1977). They intrude, therefore, rocks of both Llanvirn and Llandeilo age, including the *murchisoni*, *teretiusculus* and *gracilis* biozones.

The majority of the dolerite intrusions are sills (Plate 6), locally lenticular, interleaved with the mudstones or slightly transgressive to them. The sills range in thickness up to approximately 70 m, and some can be traced for over 2 km along strike. In the upper part of the Llanfawr Mudstones the dolerites crop out in a series of ovoid-shaped intrusions, ranging in length from less than

Plate 6 Concordant contact between a dolerite sill and mudstone of the Llanfawr Mudstones. Small quarry [SO 0787 6121] on Bongam Bank.

100 m up to 550 m. These were interpreted by Jones and Pugh (1948b) to represent a laccolithic body with some of the smaller masses acting as feeders. A few dolerite intrusions have markedly discordant relationships with the bedding of the mudstones and are thought to have the form of dykes or small plug-like bodies (Jones and Pugh, 1948a). The margins of the dolerites are generally sharp and planar, and no mudstone xenoliths have been observed within the intrusions. The mudstones adjacent to the dolerites have been baked to a distance ranging from less than 10 cm to over 15 m.

Most of the dolerites are greenish grey, medium- to coarse-grained, equigranular rocks with the local development of white or dark green vesicles. At Bailey Einon quarry [SO 0758 6170] the vesicles are dispersed throughout the dolerite, whereas at Llanfawr quarry [SO 0647 6176] they are concentrated along the margins of the intrusions. The dolerites display a primary igneous texture of intergranular subhedral albitic feldspar, pseudomorphed mafic minerals and opaque oxide, but in places this is totally obscured by pervasive alteration. There are only very rare occurrences of unaltered clinopyroxene. In most cases the former mafic minerals have been completely replaced by an assemblage of calcite, chlorite and oxide, but some retain subhedral or euhedral crystal forms indicative of former pyroxene and, more rarely, olivine.

DETAILS

The dolerites are well exposed in the Llanfawr quarries, and have been described in detail by Jones and Pugh (1948a). The dolerites are generally concordant with the bedding in the Llanfawr Mudstones. Baked mudstones adjacent to the dolerites range in thickness from less than 10 cm to approximately 1.5 m. Discordant contact relationships are less common, but are exposed in the central Llanfawr quarry [SO 0648 6183]. On the western side there is an apparent abrupt termination of one of the dolerites, and on the eastern face a dolerite has vertical contacts with north-west-dipping mudstones. Concordant contacts of other dolerite intrusions are exposed in a quarry [SO 0787 6121] on Bongam Bank (Plate 6) and in a quarry [SO 0758 6171] west of Bailey Einon farm. A contact of a transgressive dolerite crops out in a small quarry [SO 0650 6024] to the south-east of The Lake at Llandrindod Wells. The steep south-south-west dip of the contact contrasts markedly with the moderate north-west dip of the Llanfawr Mudstones to the east and south-east of The Lake. Similarly, a series of north-trending dolerite dykes is exposed on low ridges [SO 0713 6160] to the north-east of Cefnllys. A more complex transgressive contact relationship is displayed in a roadside exposure [SO 0668 6199] near Noyadd farm, where a narrow vertical wedge of mudstone, at least 1 m high, is flanked by dolerite. Zones of brecciation have developed along some of the dolerite contacts. In the quarry [SO 0801 6261] to the east of Cefn-coed farm the dolerite is separated from mudstone by a 1 m-wide breccia containing angular dolerite and baked mudstone clasts. This brecciation may be the result of post-intrusion faulting.

Dolerites of the Tywi Anticline

Slightly transgressive dolerite sills are intruded into black mudstones of the St Cynllo's Church Formation along Baxter's Bank, between Pen-y-banc quarry [SO 0560 6665] and 400 m north-north-west of Bwlchbryndinam [SO 0626 6794]. They lie just to the north-west of the Carmel Fault and have been deformed into a tight, overturned syncline and anticline pair (Figure 11). At Pen-y-banc quarry the contacts with the mudstone are locally irregular, sheared and veined with quartz, and in places truncate the bedding. The mudstones show slight baking within 1 m of the contact with the dolerite. Approximately 1.5 km south-west of the Baxter's Bank intrusion a

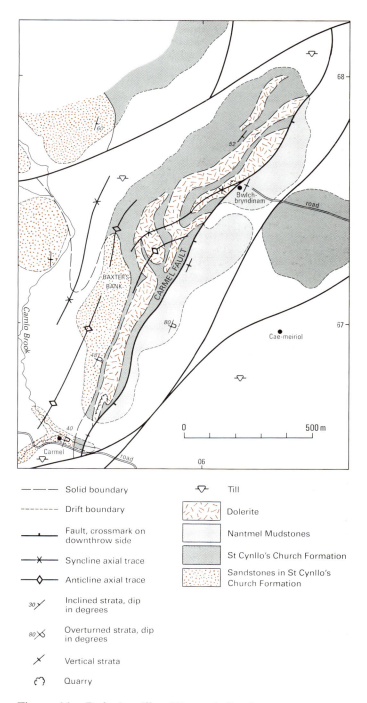

Figure 11   Dolerite sills of Baxter's Bank.

thin dolerite crops out [at SO 0447 6562] by the A44 road near Caerfagu Mill.

The dolerite of Baxter's Bank is uniformly fine to medium grained, but highly altered (Roberts, 1927). Some unaltered feldspar crystals remain, and their optical properties indicate a labradorite composition, but mostly the feldspar has been altered either to 'saussurite' or to an assemblage of albite, calcite, epidote and quartz. There are no primary mafic minerals but pseudomorphs after olivine are distinguishable.

Chilled marginal zones of the dolerite intrusions consist of feldspar microphenocrysts and spherulites of radiating acicular feldspar microlites within a mostly chloritised groundmass. The spherulites are indicative of devitrified glass. These zones rarely exceed 10 cm in width.

Age relations and mechanism of intrusion

The dolerites of the Builth Inlier and Tywi Anticline are interpreted as high-level intrusions associated with several episodes of basaltic magmatism from Llanvirn to late Caradoc times. There is no evidence that the sediments were unlithified at the time of intrusion; the dolerite contacts are sharp and planar, and no mudstone was incorporated into the magma. However, the discontinuous form of some of the dolerites, in particular those in the upper parts of the Llanfawr Mudstones, could have resulted from the disruption of the magma as it intruded into ductile, partially lithified, mudstone. The discordant dykes and plug-like dolerites are thought to represent the feeder systems for the sills and concordant lenticular bodies (Jones and Pugh, 1948b).

THREE

Silurian: introduction and biostratigraphy

INTRODUCTION

The base of the Silurian System in the district, and throughout the Welsh Basin, falls within formations which, though predominantly Llandovery in age, include late Ordovician strata. The major lithological changes that define the bases of these formations occur within the late Ashgill *persculptus* Biozone (Temple, 1988) and reflect the transgressive rise in sea level which followed the late Ordovician glacioeustatic lowstand (Brenchley and Cullen, 1984; see Chapter 2). The latest (postglacial) Ashgill strata are therefore related lithostratigraphically and sedimentologically to the early Silurian deposits and for convenience are here described with them. A schematic cross-section showing the relationships of the main Silurian lithostratigraphical divisions in the district is shown in Figure 12.

Although the outcrop of Silurian rocks covers more than three-quarters of the district, only the Llandovery and Wenlock series are present. The Llandovery sequence that crops out west of the Tywi Lineament is more than 4 km thick and is the thickest recorded in southern Britain. Along and to the east of the lineament, attenuated and incomplete Llandovery sequences occur between rocks of Ordovician and Wenlock age. Outcrops of Wenlock strata are restricted to the north-eastern and south-eastern corners of the district; they are conformably overlain by younger Silurian rocks to the south and east of the district. The stages and the graptolite biozones of the Llandovery and Wenlock series, as used in this memoir, are shown in Table 4.

The Silurian rocks of the district are divisible into 'shelf' and 'basinal' successions identified on the basis of facies and biota. However, during much of the Silurian there appears to have been a ramp-like continuum of facies (cf. Hancock et al., 1974) between the shallower and deeper depositional settings of the district, with no evidence to suggest a 'shelf break'. However, there is evidence that the Tywi Lineament in the Telychian and early Wenlock formed a belt of relatively steep, tectonically generated gradients. Today, gross facies and thickness differences occur across the major fractures of the lineament, notably the Garth Fault and its splays. This fault therefore serves as a convenient feature to delineate the shelf succession to the east, from basinal successions to the west.

SHELF SUCCESSION

This is limited to the east of the Garth Fault, where it comprises two depositional systems (cf. Galloway, 1989), initiated in the late Hirnantian and persisting through to the late Ludlow (Figure 13). Strata of the lower system (late Hirnantian to Aeronian) crop out immediately to the south of the district in the Garth Inlier (Figure 42), but are also thought to occur at depth in the district. The upper system (latest Aeronian and younger), although dominated by unfossiliferous, burrow-mottled mudstones and laminated graptolitic mudstones of distal shelf aspect, contains thin, transgressive, shelly sandstones overlying a basal unconformity, demonstrating that this region of the district was subject to periodic shoaling and subaerial exposure. In-situ shelly benthos mainly belongs to the deep-water *Clorinda* and *Visbyella trewerna* communities (Ziegler et al., 1968; Hurst et al., 1978).

BASINAL SUCCESSION

Characterised by episodes of rapid subsidence and high rates of sediment accumulation, the basinal succession contains no evidence of subaerial exposure. Soft-bodied, burrowing animals dominated the indigenous bottom-dwelling (benthic) fauna and gave rise to trace fossil assemblages belonging to the *Nereites* ichnofacies (Smith, 1987a). Bottom-dwelling shelly forms, confined to the eastern margin of the basin, were restricted to rare trilobites. Shelly, free-swimming animals (nekton), orthocones and nowakiids, are present in the east, but the most common macrofaunal remains preserved throughout the basin are the free-floating (planktonic) graptolites. The basinal succession comprises several depositional systems constructed exclusively of resedimented and hemipelagic deposits, with turbidites forming the dominant component.

Unlike shelf systems, which are bounded by disconformities (Galloway, 1989), turbidite systems (Mutti and Normark, 1987) are bounded by significant facies changes. In the district, such changes can commonly be correlated with disconformities in the shelf systems of the Midland Platform, an expression of the increased clastic input to the basin during periods of eustatic sea-level fall or local tectonic uplift.

Two principal types of turbidite system have been recognised, slope-apron and sandstone-lobe (Figure 13). The slope-apron systems comprise wedge-shaped accumulations of turbiditic and hemipelagic mudstone, that thin westwards from the shelf to the western edge of the district. They locally contain small lenticular bodies of coarse-grained turbidites. Both the fine- and coarse-grained turbidites of these systems were supplied from the shelf to the east. Slope-apron deposition was active over much of the basin throughout the Llandovery and Wenlock, except for periods during the Telychian and early

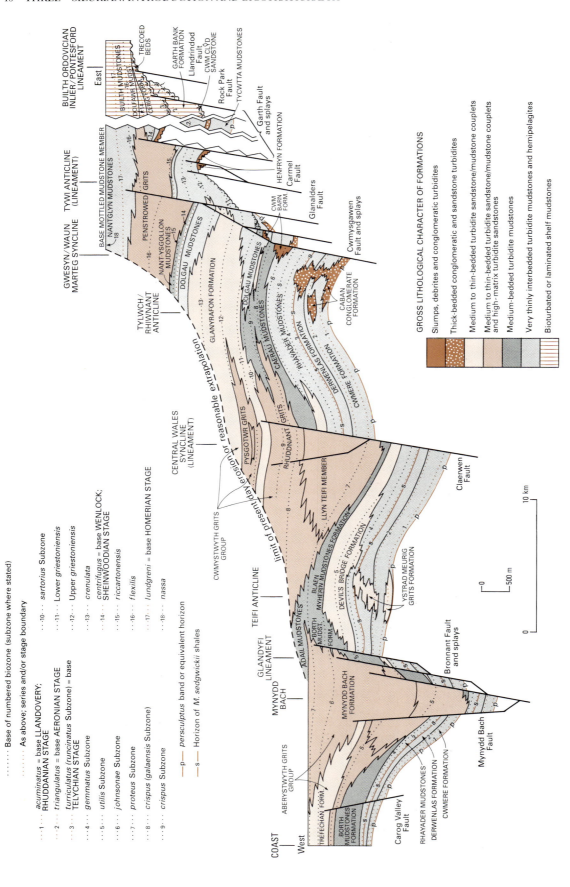

Figure 12 Schematic pre-Acadian lithostratigraphical cross-section of the latest Hirnantian (Ashgill) to Homerian (Wenlock) succession of the district.

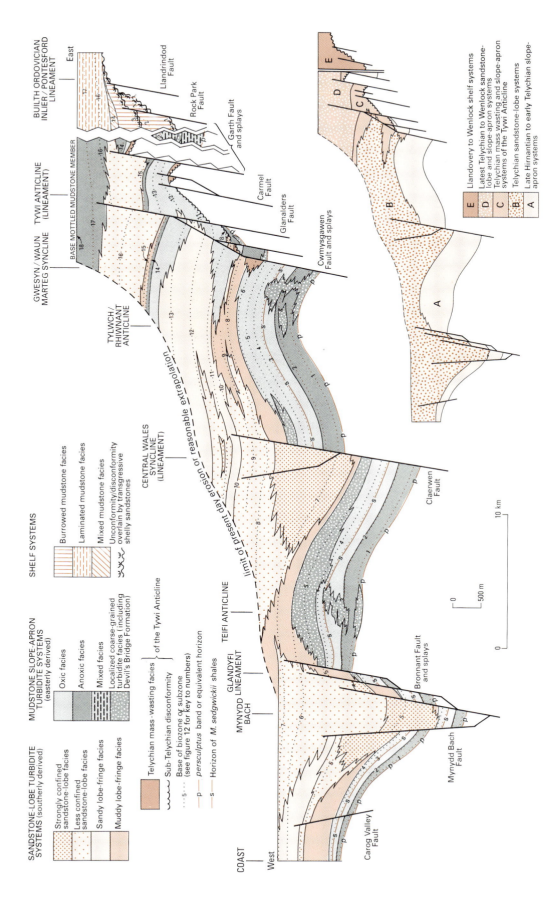

SANDSTONE-LOBE TURBIDITE
SYSTEMS (southerly derived)

- Strongly confined
  sandstone-lobe facies
- Less confined
  sandstone-lobe facies
- Sandy lobe-fringe facies
- Muddy lobe-fringe facies

MUDSTONE SLOPE-APRON
TURBIDITE SYSTEMS
(easterly derived)

- Oxic facies
- Anoxic facies
- Mixed facies
- Localized coarse-grained
  turbidite facies (including
  Devil's Bridge Formation)

Telychian mass-wasting facies } of the Tywi Anticline

Sub-Telychian disconformity

......5..... Base of biozone or subzone
(see figure 12 for key to numbers)

p ..... perisculptus band or equivalent horizon

s ..... Horizon of M. sedgwickii shales

SHELF SYSTEMS

- Burrowed mudstone facies
- Laminated mudstone facies
- Mixed mudstone facies
- Unconformity/disconformity
  overlain by transgressive
  shelly sandstones

COAST

West

Carog Valley
Fault

Mynydd Bach
Fault

Bronnant Fault
and splays

GLANDYFI
MYNYDD LINEAMENT
BACH

TEIFI ANTICLINE

limit of present day erosion or reasonable extrapolation

CENTRAL WALES
SYNCLINE
(LINEAMENT)

Claerwen
Fault

TYLWCH/
RHIWNANT ANTICLINE

Cwmysgawen
Fault and splays

GWESYN / WAUN
MARTEG SYNCLINE

Glanalders Fault

Carmel
Fault

TYWI ANTICLINE
(LINEAMENT)

BASE MOTTLED MUDSTONE MEMBER

Garth Fault
and splays

Rock Park
Fault

Llandrindod
Fault

BUILTH ORDOVICIAN
INLIER / PONTESFORD
LINEAMENT

East

10 km

0

500 m

0

E   Llandovery to Wenlock shelf systems

D   Latest Telychian to Wenlock sandstone-
    lobe and slope-apron systems

C   Telychian mass-wasting and slope-apron
    systems of the Tywi Anticline

B   Telychian sandstone-lobe systems

A   Late Hirnantian to early Telychian slope-
    apron systems

Figure 13 Schematic pre-Acadian facies and system architecture of the latest Hirnantian (Ashgill) to Homerian (Wenlock) succession of the district. Inset shows simplified architectural elements A–E described in text.

Table 4   The stages and graptolite biozones of the late Ashgill, Llandovery and Wenlock series: left-hand column after Rickards (1976); right-hand column shows modifications, subzones and informal subdivisions used in this memoir.

System	Series	Stage	Biozones (Rickards, 1976)
Silurian	Wenlock	Homerian	ludensis
			nassa
			lundgreni
		Sheinwoodian	ellesae
			flexilis
			rigidus
			riccartonensis
			murchisoni
			centrifugus
	Llandovery	Telychian	crenulata
			griestoniensis
			crispus
			turriculatus
		Aeronian	sedgwickii
			convolutus
			leptotheca
			magnus
			triangulatus
		Rhuddanian	cyphus
			acinaces
			atavus
			acuminatus
Ordovician	Ashgill	Hirnantian	persculptus

Biozonal modifications used in this memoir	Subzones and informal subdivisions
spiralis/centrifugus interregnum	
spiralis	
crenulata	
griestoniensis	upper
	lower
crispus	sartorius
	crispus
	galaensis
turriculatus	carnicus
	proteus
	johnsonae
	utilis
	renaudi
	gemmatus
	runcinatus
halli	
sedgwickii	
convolutus	upper
	lower

Biozonal modifications used in this memoir	Subzones and informal subdivisions
acinaces	upper
	lower
atavus	
acuminatus	upper
	lower
persculptus	

After Loydell, 1991

Wenlock, when a series of sandstone-lobe systems, supplied from the south, invaded the basin. The arrival of these sandy turbidites coincided with a prolonged period of intrabasinal growth faulting; the distribution of the turbidites was controlled by the location of the faults.

The constituent facies of the turbidite systems are constructed of a limited range of depositional components, none of which is uniquely characteristic of any particular facies, formation or system. They can be broadly classified into two genetic components, resedimented and hemipelagic, as described below.

Resedimented components

The resedimented components of the slope-apron and sandstone-lobe systems were emplaced by a variety of mass-flow processes. Individual resedimented units were deposited very rapidly (probably in hours or days), yet they are the dominant component, comprising over 95 per cent of the sandstone-lobe systems. Although less abundant in slope-apron systems (30 to 70 per cent), some types of mass-flow deposit are largely confined to, and therefore diagnostic of, such systems.

Resedimented lithologies range from boulder conglomerates, through sandstones and siltstones, to mudstones. Individual units may comprise a single lithology, but commonly display a combination of two or more. Many of the units occur in stacked sequences; others occur singly or in clusters of two or three, intercalated with units of a different character.

The resedimented components of the district are divisible into three categories: disturbed beds (slumped and destratified strata), debrites, and turbidites (Figure 14a).

DISTURBED BEDS (slumped and destratified strata)

Resedimented units which display complex slump folds and contortions, internal slide planes and disrupted bedding (Plates 7a and 19d), commonly include, or pass into, deposits in which the primary layering has been lost, or is only preserved in isolated and contorted wisps or balls. During the present work it was found to be impracticable to differentiate slump-folded strata from intimately associated, variably destratified mass-movement deposits and they are, therefore, all grouped together as 'disturbed beds'.

Loss of the primary depositional fabric in disturbed beds appears to have inhibited the subsequent imposition of the regional Acadian cleavage, and they are therefore commonly distinguished in the field by their poorly or irregularly cleaved character (cf. Warren et al., 1984). They range up to tens of metres in thickness. The disturbed beds commonly have sharp basal contacts which cross-cut the underlying, undisturbed deposits (Plate 7a). The downslope movement of some units may have been considerable; travel distances of over 100 km are reported for modern examples (Nardin et al., 1979). However, in others, folding and destratification may record earthquake-triggered diapirism and liquefaction, with negligible lateral movement.

DEBRITES

Debrites typically comprise mudstones, with dispersed larger clasts ranging up to cobble size (pebbly mudstones). They occur in highly lenticular, thin to very thick beds. Clast concentration varies from widely spaced to closely packed, so that they grade into matrix-supported conglomerates. Internally they appear largely structureless, although clast-rich varieties commonly exhibit crude normal grading. Clast types include well-rounded, extrabasinal igneous and metamorphic lithologies, and angular blocks and rafts of intra- and peribasinal sedimentary rocks.

Debrites are regarded as the product of cohesive debris flows (Lowe, 1982; Pickering et al., 1986), highly viscous flows in which the larger clasts are supported by a plastic, cohesive mud matrix. Such flows are initiated and sustained on slopes in excess of 1° (Nardin et al., 1979); they are relatively slow moving and freeze rapidly where the sustaining gradients shallow.

TURBIDITES

Turbidites are the deposits of sediment-laden density flows (turbidity currents), which flow down slopes of less than 1°. At the present day, they transport gravel- to mud-grade material for great distances across ocean floors from adjacent continental margins. Although fluid turbulence is generally cited as the principal sediment-support mechanism in such flows, it is now known that long-distance transport of coarse sand and gravel relies on additional factors such as hindered settling, matrix buoyant lift and dispersive pressure operating within flows with very high sediment concentrations (see Lowe, 1982). In contrast, very fine-grained sediments can be carried in turbulent suspension at very low concentrations and at very low flow velocities (Stow and Piper, 1984). Thus, although many of the turbidites recognised in the Silurian of the Welsh Basin belong to the proximal-to-distal suite of Bouma (classical) turbidites (Bouma, 1962; Walker, 1967), other 'coarse-grained' and 'fine-grained' varieties were emplaced by flows of widely differing character. For a comprehensive review of turbidite deposits and processes the reader is referred to Pickering et al. (1986). The classification of turbidites used in this account is presented in Table 5 and Figure 14.

COARSE-GRAINED TURBIDITES

Resedimented conglomerates and coarse-grained, mud-depleted sandstones are not adequately described by the Bouma (1962) scheme. The processes active in their deposition have been assessed by Lowe (1982), and his ideal sequence of divisions in a single resedimented unit is adopted here (Figure 14b).

Type A: conglomerate/sandstone couplets and conglomerate/ sandstone/mudstone triplets These include a range of conglomeratic turbidites subdivided on thickness and the presence of mudstone.

Type Ai: medium- to very thick-bedded conglomerate/sandstone couplets Individual Ai turbidites are 2 to 6 m thick and comprise a medium- to very thick-bedded, clast-supported conglomerate gradationally overlain by pebbly sandstone and sandstone

a

Very low

Low

High

sediment* mechanisms
concentration

sediment support*

Fluid turbulence

Hindered settling

Matrix bouyant lift

Dispersive pressure

Spectrum of (Bii) turbidites

0.2m

0.2m

1.0m

1.5m

0.1m

0.1m

0.1m

0.1-
0.01
m

0.1-
0.01
m

0.05
m

(E)

T7-8
T4-6
T3
T1-2
T0

(Dii)

(Di)

***

(Cii)

(Ci)

Te
Td
Tc
Tb
Ta

(Bi)

S3

S2

(Ai)

S3
S1-2
R3
R2

(Aii)**

Te
Td
Tc
Tb
Ta

conglomerate
sandstone
siltstone
mudstone

COARSE-GRAINED   BOUMA   FINE-GRAINED

TURBIDITES

10
m

2.0m

DISTURBED BEDS
(slumped &
destratified units)

DEBRITES

Cross lamination

Parallel lamination

Convolute lamination

Silt laminae

Fluid-escape structures

Pseudonodules

Rip-up clasts

High levels of mud
matrix (matrix support)

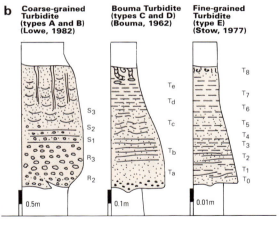

b

Coarse-grained Turbidite (types A and B) (Lowe, 1982)	Bouma Turbidite (types C and D) (Bouma, 1962)	Fine-grained Turbidite (type E) (Stow, 1977)

S3
S2
S1
R3
R2

Te
Td
Tc
Tb
Ta

T8
T7
T6
T5
T4
T3
T2
T1
T0

0.5m

0.1m

0.01m

Figure 14 Classification of the resedimented deposits of the district.

a Types and inter-relationships of resedimented deposits of the district. Solid arrows connect deposits which may be part of an evolutionary continuum for individual flows. Broken arrows connect end members of a spectrum of deposits without necessarily implying an evolutionary linkage. Turbidite types A–E are described in Table 5. (* applies to turbidites only; ** combines features of both coarse-grained and Bouma turbidites; *** many very thin-bedded varieties may represent 'base-cut-out' type E turbidites and properly lie within the field of fine-grained turbidites).

b Idealised internal divisions for coarse-grained, Bouma and fine-grained turbidites, after the authors indicated.

(Figure 14b; Plates 7b, 13a and d). The conglomerates are typically poorly sorted with clasts ranging from boulders over 50 cm in diameter, to medium-grained sand. Individual conglomerate beds have erosive, locally channelled bases (Plate 13a) and are typically highly lenticular. They may be structureless, or exhibit both inverse grading (R_2 division) and normal grading (R_3 division)(Plates 7b and 13d). Clasts may be of both intra- and extrabasinal type, and include penecontemporaneous, shelf-derived, shelly benthos. Although the overlying sandstones are commonly massive with pebbles dispersed in a medium or coarse-grained sandstone matrix (S_3 division), others show pebbly layers with basal shear laminations (S_2 division) and, more rarely, cross-stratification (S_1 division). Seldom, however, is

the ideal vertical sequence S_1, S_2, S_3 observed. Multiple stacking and amalgamation of units commonly make recognition of individual Ai turbidites difficult (Plate 13a).

These deposits are interpreted as resedimented gravels and gravelly sands transported by high-density turbidity currents. The physical processes operating in such flows have been assessed by Lowe (1976; 1982). Structureless units in the conglomerates may record rapid frictional freezing of the flow; inversely graded divisions record deposition in traction carpets, and normally graded divisions record deposition from suspension (Lowe, 1982; Pickering et al., 1986). In the overlying pebbly sandstones, the S_1, S_2 and S_3 divisions record deposition under traction, from a traction carpet and from suspension respectively.

Table 5 Turbidite classification used in this memoir.

COARSE-GRAINED TURBIDITES	type Ai	medium- to very thick-bedded conglomerate/sandstone couplets
	type Aii	thin- to thick-bedded, shelly conglomerate/sandstone/mudstone triplets
	type Bi	medium- to thick-bedded sandstones (<15% matrix)
	type Bii	medium- to thick-bedded 'high matrix' sandstones (>15% mud matrix)
BOUMA TURBIDITES	type Ci	thin- to medium-bedded sandstone/mudstone couplets
	type Cii	very thin- to thin-bedded sandstone/mudstone couplets
	type Di	siltstone/mudstone couplets with medium to thick mudstone divisions
	type Dii	very thin- to thick-bedded mudstones
FINE-GRAINED TURBIDITES	type E	very thin- to thin-bedded silt-laminated mudstones

Type Aii: thin- to thick-bedded, shelly conglomerate/sandstone/mudstone triplets   Individual turbidites of this type are 0.15 to 1.3 m thick and comprise three divisions: a basal, typically shell-rich conglomerate resting on a sharp erosion surface; a middle calcareous sandstone; and a capping mudstone (Figure 14a; Plates 7c and 19a–d). The conglomerates are clast supported and predominantly of pebble and granule grade with a coarse- to fine-grained sandstone matrix. Cobbles and rare boulders may also be present. Clast assemblages, in addition to a variety of both rounded and angular lithic fragments and vein quartz, characteristically include abundant and varied, derived, shelly benthos and black, phosphatised mudstone pebbles (Plate 19a).

The conglomeratic division is sharply overlain by coarse- to fine-grained, shelly-graded sandstone, exhibiting pervasive parallel-and/or cross-lamination. At the top it grades rapidly, commonly via a laminated siltstone, into the capping mudstone. In stacked sequences of these turbidites, any of the three divisions may be dominant; loss of the mudstone division and amalgamation of the conglomeratic and sandstone divisions is common.

Type Aii turbidites can be compared to Bouma turbidites (types C and D of this account). The basal conglomeratic portions may be viewed as very coarse-grained T_a divisions, whilst the overlying sandstone and mudstone divisions can be compared with the T_{b-e} divisions. In contrast with the much thicker Ai turbidites, therefore, Aii turbidites may record deposition from high-concentration turbidity currents, which carried proportionally greater amounts of sand and mud, or from currents in which a basal, high-concentration gravel transporting part did not de-couple from the overlying lower-concentration part that carried sand and mud.

Type B: medium- to thick-bedded sandstones and muddy sandstones   These include a range of sandy turbidite lithologies, which are subdivided into two types dependant on the proportion of mud matrix (less than 15 per cent and 15 to 50 per cent).

Type Bi: medium- to thick-bedded sandstones (less than 15 per cent mud matrix)   This type comprises medium- to coarse-grained, locally pebbly, grain-supported sandstones, that mainly occur in thick to very thick beds up to 3.5 m thick (Figure 20; Plates 7d and 12a). Sharp bases commonly exhibit bulbous or rope-like load structures, and are locally channelled. Mudstone rip-up clasts are common. The basal, laminated or cross-bedded division (S_1) of Lowe (1982) (Figure 14) has not been recognised, but thin pebbly layers with basal shear laminations (S_2 division) occur in the basal parts of some sandstones. The main part of the sandstone is either structureless or normally graded (S_3 division) but may locally display dish-structures or, more rarely, diffuse parallel-lamination. Scattered pebbles and granules are sometimes present in the lower part of this division. The sandstone beds are laterally persistent, although commonly lenticular (Plate 12a). Amalgamation of two or more beds is widespread.

Type Bi turbidites were deposited from high-concentration turbidity currents. Such sand transporting flows are sustained by general fluid turbulence and 'hindered settling' (Lowe 1982) and, in contrast to flows which carry gravel, rely less on grain-to-grain dispersive pressure. The locally developed S_2 divisions record the development of thin, gravel-transporting traction carpets at the base of some flows. The dominant, structureless or normally graded sandstone (S_3 division) records rapid sediment fall-out from high-density suspensions. The resultant loosely packed, cohesionless sediments were particularly prone to postdepositional liquefaction; dish-structures record the subsequent upward movement and expulsion of fluids from these beds.

Type Bii: medium- to thick-bedded 'high matrix' sandstones (more than 15 per cent mud matrix)   Type Bii turbidites (Figure 14) comprise medium to thick, typically parallel-sided sandstone beds up to 1.5 m thick with sharp, locally gutter-casted and loaded bases (Plate 17a and b). Grain sizes range from coarse sand to silt, although granules, scattered throughout or concentrated in basal lags, are common. Large contorted rafts of subjacent strata, over 1 m long and up to 20 cm thick, also occur and commonly float within the beds (Plate 7f). Clots, wisps and pseudonodules of silt or fine-grained sandstone may also be present (Plate 7e). The proportion of mud matrix varies; units impoverished in coarser grain sizes grade into sandy mudstones or muddy siltstones (Plate 7e) and some have matrix-support textures. Apart from crude, normal or inverse grading in their basal portion, most Bii turbidites are structureless, although rare examples (less than 10 per cent) exhibit diffuse cross-lamination in the top 1 or 2 cm. In others a thin, coarser, graded and less muddy basal division is well segregated from an overlying, thicker, finer-grained, mud-dominated division (Plate 7e).

Variations in mud content account for marked differences in field appearance of Bii turbidites. Less muddy varieties form hard, massive, tabular beds resistant to weathering. In contrast, those with a greater mud content are softer, more friable and typically display the regional Acadian cleavage (Plates 17a and b). Stacking or amalgamation of Bii turbidites is uncommon; they mainly occur as individual beds separated by sequences of Bouma turbidites. Such sequences display a marked bimodality in sandstone thickness, a direct function of the two types of resedimented sandstone present (Clayton, 1992).

Although Bii turbidites compare with the T_a division of Bouma turbidites (see Smith, 1988), the general absence of tractional structures and the local presence of well-defined, coarse-grained basal divisions, floating clasts and inversely graded bases suggest that they cannot be explained by a Bouma turbidite model (Clayton, 1992).

The apparent gradation between Bi and Bii turbidites suggests that the latter also record deposition from high-concentration turbidity currents. In the currents that deposited Bii turbidites, however, the buoyant lift and lubrication provided by the dense, interstitial mud/water mixtures were significant factors in grain transportation, enabling such flows to carry sand further than their mud-depleted relatives (type Bi). The character of such flows varied, according to the cohesiveness and proportion of mud, between turbulent suspensions (which deposit Bouma turbidites) and cohesive debris flows. Type Bii turbidites may have been generated in two ways: by high-concentration, sandy turbidity currents entraining mud and becoming converted into highly fluid, weakly cohesive debris flows, or by cohesive debris flows entraining water and evolving into turbulent suspensions.

Normal grading in these otherwise structureless turbidites suggests rapid fall-out from muddy suspensions. Matrix-support, and floating contorted rip-up clasts suggest that many units (or parts of units) formed by cohesive freezing and were emplaced as liquified, mud-silt-sand slurries with little internal organisation. Rip-up clasts were deformed by internal hydro-plastic shear as the flow froze. The well-segregated, coarser, graded basal division present in some Bii turbidites may be explained by the action of the dense, mud-rich, interstitial fluid, which caused a decrease in turbulence as the flow decelerated. As a result, the coarser grains fell from suspension to form a basal graded unit. Capping cross-laminated divisions record tractional reworking by, and sedimentation from, later low-concentration turbidity currents. Pseudonodules record late-stage liquefaction and the foundering of overlying sand or silt beds deposited from penecontemporaneous low-concentration turbidity currents.

BOUMA TURBIDITES

Type C: sandstone/mudstone couplets   This class embraces a range of turbidites which display the characteristics described by Bouma (1962) (Figure 14b). Subdivision is on the basis of internal structure and sandstone bed thickness.

Type Ci: thin- to medium-bedded sandstone/mudstone couplets
The coarse- to fine-grained basal sandstones of these turbidites form medium to thin, parallel-sided, normally graded beds up

to 30 cm thick. They pass upwards via thin siltstone intervals into mudstone divisions which are commonly thicker. Together these lithological components may exhibit the complete sequence of Bouma (1962) divisions (Figure 14). Sharp sandstone bases are commonly fluted or grooved and may display a variety of trace fossils. The basal sandstones rarely exhibit the massive graded T_a division, being dominated by parallel-laminated (T_b) and cross- or convolute-laminated (T_c) divisions. Overlying siltstone–mudstone units constitute the upper T_d and T_e divisions, respectively (Plates 8a, 16d and e).

Type Ci turbidites may form stacked sequences, but commonly occur singly, or in groups of two or three, intercalated within sequences of type Cii turbidites.

Type Cii: very thin- to thin-bedded sandstone/mudstone couplets
This type comprises a medium- to fine-grained basal sandstone up to 10 cm thick overlain by much thicker mudstone (up to 30 cm). The sandstones rarely exhibit fluted bases; bottom structures are confined to small tool marks, grooves and trace fossils. Cross-lamination (T_c division) predominates internally with thin parallel-laminated siltstones (T_d division) commonly intervening below the capping mudstone (T_e division) (Figure 14a; Plates 8a, 16d and e). Type Cii turbidites mainly occur in stacked sequences and commonly form a background facies to other turbidite types.

Apart from rare T_a divisions, which record rapid suspension deposition from flows (or parts of flows) with initial high coarse-fraction concentrations, type C turbidites display features characteristic of deposition from lower-concentration turbidity currents. In contrast to types A and B, the sequence of sedimentary structures in type C sandstones reflects a more gradual deceleration of the flow, which allowed time for material derived from suspension to be reworked within tractional bed loads. The hydrodynamic interpretation of the sedimentary structures of each Bouma division is well established (Bouma, 1962; Walker, 1965).

Type D: siltstone/mudstone couplets and mudstones   Turbidites of this type exhibit the T_d and T_e divisions of Bouma turbidites and equate directly with the siltstone and mudstone intervals observed in type C turbidites.

Type Di: siltstone/mudstone couplets with medium to thick mudstone divisions   These comprise a thin (up to about a centimetre), commonly laminated, basal siltstone (or fine-grained sandstone) (T_d division) overlain by a typically much thicker (up to 30 cm), graded to massive mudstone (T_e division) (Figure 14a; Plates 8a and 15b).

Type Dii: very thin- to thick-bedded mudstones   In this type the basal siltstone division observed in type Di is poorly developed or absent. Graded or massive mudstone (T_e division) comprises the whole of the turbidite which may range in thickness from 1 to 40 cm (Figure 14a; Plate 8b). Slope-apron turbidite systems in the district are characterised by very thin-bedded Dii turbidites; thicker-bedded varieties typify mudstone facies of the sandstone-lobe turbidite systems.

Type D turbidites are the products of low-concentration turbidity currents depleted in sand-grade material, either from source or following its up-current deposition. However, many very thin-bedded type Dii turbidites may form a continuum with type E turbidites (Figure 14a).

FINE-GRAINED TURBIDITES

Type E: very thin- to thin-bedded silt-laminated mudstones   These are not true Bouma turbidites, although they may broadly be compared to the T_d and T_e divisions of such deposits. They have recently been assessed by Stow and Piper (1984; and references therein), whose scheme of internal divisions (based on Stow, 1977) (Figure 14b) is adopted here.

Complete turbidites of this type (Plates 8c and 15c) consist of a thin basal siltstone division (T_o), up to 1 cm thick, commonly displaying low-amplitude fading-ripple structures, overlain by a thicker (up to 5 cm), variably silt-laminated mudstone. In the latter, divisions displaying well-defined siltstone laminae (T_{1-3}) pass up into diffusely laminated, silt-depleted mudstone (T_{4-5}). Mudstone, comprising a lower graded division (T_6) and an upper ungraded division (T_7), caps this 'ideal' sequence. Type E turbidites mainly occur in stacked sequences in which either the lower or upper parts of the ideal Stow and Piper sequence are poorly developed ('base-cut-out' and 'top-cut-out' sequences of Stow et al., 1984).

Studies of modern analogues (Stow and Bowen, 1980) suggest that such turbidites are deposited by very low-concentration turbidity currents, which are much thicker and slower moving than those transporting sand. As with Bouma turbidites, the succession of internal divisions reflects the progressive waning of individual flows (see Stow and Piper, 1984).

Hemipelagic components

Hemipelagic deposits (or hemipelagites) represent the muddy background sediment of the slope-apron and sandstone-lobe systems. Individual hemipelagites are rarely more than 3 cm thick and record the slow, but constant rain of sediment out of suspension on to the sea bed. They differ from true pelagic sediments in being composed principally of terrigenous mud and silt (Stow and Piper, 1984). Hemipelagic sediments are a minor component of sandstone-lobe systems but locally comprise over 50 per cent of slope-apron systems. Hemipelagites occur as two distinct types, laminated and burrowed (Cave, 1979), thought to reflect differences in the oxygen content of the contemporary bottom waters.

Laminated hemipelagites comprise delicate and persistent, regular alternations of dark grey, carbonaceous laminae and pale grey, silt-rich laminae, all less than a millimetre thick (Plate 8d). Each pair of laminae is thought to represent an annual varve (Dimberline et al., 1990). Contacts with underlying turbidite mudstones are sharp. Ovoid aggregates of silt and mud, common in the silty laminae, are interpreted as the compacted faecal pellets of pelagic fauna. The absence of bioturbation and the preservation of organic debris demonstrate that such hemipelagites accumulated under anoxic (anaerobic) bottom conditions (Cave, 1979). They are commonly graptolitic.

Burrowed hemipelagites mainly comprise pale grey or green mudstones, in which darker mottles represent the infills of burrow systems, principally of *Chondrites* type (Plate 8E). Although devoid of carbonaceous material, a streaky and diffuse colour lamination is commonly present. Distinctive black, but white-weathering, phosphate nodules or beds up to several millimetres thick, commonly occur at or near the base of these hemipelagites.

The presence of bioturbation in these hemipelagites reflects deposition under relatively oxic (dysaerobic) bottom conditions. Their pale colour is the result of con-

temporaneous diagenesis, during which a predominantly downward-migrating oxidation front consumed the organic carbon in the accumulating hemipelagite, and in some cases the topmost part of the underlying turbidite mudstone (Cave, 1979; Smith, 1987c). Where burrowed hemipelagites overlie thick (over 4 cm), medium grey, turbidite mudstones (types C and D), they are easily distinguished as fairly sharp-based, paler coloured bands, 1 to 2 cm thick. However, their recognition, when interbedded with very thin (less than 2 cm) mudstone turbidites (type E), is in some cases difficult. In the simplest cases, medium grey or green, type E turbidites interbedded with paler, burrowed hemipelagites, give rise to sequences that are colour banded on a centimetre scale (Plate 10c). However, in many cases the oxidation front advanced well below the hemipelagite/turbidite interface, so that much of the turbidite mudstone is also pale and burrowed. Where successive oxidation fronts have merged, the result is a totally oxidised, pale coloured sequence, locally many metres thick (Plate 8c). In such cases, burrow mottling is pervasive and it is only the presence of relict tractional siltstone laminae that enables the turbiditic components to be identified (Plate 10d).

The associated phosphate nodules record the early diagenetic precipitation of apatite cements and are a byproduct of the oxidation front process (Cave, 1979; Smith, 1987c).

SILURIAN DEPOSITIONAL SYSTEMS

The Silurian succession of the district can be divided into five architectural elements (Figure 13, inset): A) late Hirnantian to early Telychian slope-apron systems; B) Telychian sandstone-lobe systems; C) Telychian mass wasting and slope-apron systems of the Tywi Anticline; D) latest Telychian to Wenlock sandstone-lobe and slope-apron systems; and E) Llandovery to Wenlock shelf systems. A separate chapter is devoted to each element (see chapters 4 to 8).

As shown in Figures 12 and 13 each element comprises one or more depositional systems, which in turn consist of several lithostratigraphical units. The present survey has enabled the many local lithostratigraphical terms in the district to be rationalised. Previous terminology has been used where practicable; new names have been introduced where no suitable existing term was available. The nomenclature used on the maps, together with pre-existing classifications for the district and adjacent areas, is shown in Table 6.

BIOSTRATIGRAPHY

Correlation within the relatively uniform lithologies of the Silurian sequence have relied heavily on biostratigraphical evidence. Indigenous shelly benthos (shelly biofacies) are restricted to the shelf facies. Pelagic macrofauna, principally graptolites, and microflora, principally acritarchs, together constitute the graptolitic biofacies,

Table 6 Previous and current lithostratigraphical classifications for the district and adjacent areas.

SYSTEMS	SERIES		Area / Biostratigraphy	WEST OF BRONNANT FAULT — THIS MEMOIR	BRONNANT FAULT TO CENTRAL WALES LINEAMENT				
					Jones 1909 (Ponterwyd)	Jones 1922 (Mining field Memoir)	Jones & Pugh 1916 (Machynlleth)	Cave & Hains 1986 (Aberystwyth)	THIS MEMOIR
SILURIAN	LUDLOW	GORSTIAN	nilssoni						
	WENLOCK	HOMERIAN	ludensis						
			nassa						
			lundgreni						
		SHEINWOODIAN	ellesae						
			flexilis						
			rigidus						
			riccartonensis						
			murchisoni						
			centrifugus						
	LLANDOVERY	TELYCHIAN	(interregnum)						Glanyrafon Formation″
			spiralis						
			crenulata						Pysgotwr Grits
			griestoniensis — upper			Cwmystwyth Grits			Glanyrafon Formation′
			griestoniensis — lower						Rhuddnant Grits
		crispus	sartorius		Rhuddnant Grits			Blaen Myherin Mudstones Formation	
			crispus		Rhuddnant Shales			Borth Mudstones Formation	Blaen Myherin Mudstones Formation
			galaensis						
			carnicus	Trefechan Formation					
			proteus	Adail Mudstones					
		turriculatus	johnsonae	ABERYSTWYTH GRITS GROUP / Mynydd Bach Formation	Blaen Myherin Mudstones	Frongoch Formation		Dolwen Mudstones Formation	Borth Mudstones Formation
			utilis					Devil's Bridge Formation / Cwmsymlog Fm	Devil's Bridge Formation
			renaudi	Borth Mudstones Formation	Dolwen Mudstones				Rhayader Mudstones
			gemmatus		Devil's Bridge Group				
			runcinatus				YSTWYTH STAGE		
		AERONIAN	sedgwickii/halli		Castell Group			Dark Grey Mudstones Member	M. sedgwickii shales
			convolutus			M. sedgwickii shales			Derwenlas Formation
			leptotheca			Derwen Group		Derwenlas Formation	Ystrad Meurig Grits Formation
			magnus		Rheidol Group		PONTERWYD STAGE		
			triangulatus						Cwmere Formation
	RHUDDANIAN		cyphus			Gwestyn Formation	Cwmere Group	Cwmere Formation	
			acinaces						Ystrad Meurig Grits Formation
			atavus						
			acuminatus		Eisteddfa Group			Mottled Mudstone Member	
ORDOVICIAN	ASHGILL	HIRNANTIAN	persculptus		Bryn Glas Group	Van Formation	PLYNLIMON STAGE	Brynglas Formation	

which though characteristic of the basinal facies also typifies some shelf facies deposited under anoxic bottom conditions.

Graptolites have provided the prime method of correlation. Most of the 22 British graptolite biozones that span the Llandovery and Wenlock series have been widely recognised in the district. The average duration of graptolite biozones in the Silurian has been estimated to be about 0.8 Ma, that of subzones and informal subdivisions less than 0.5 Ma (Zalasiewicz, 1990). The shelly

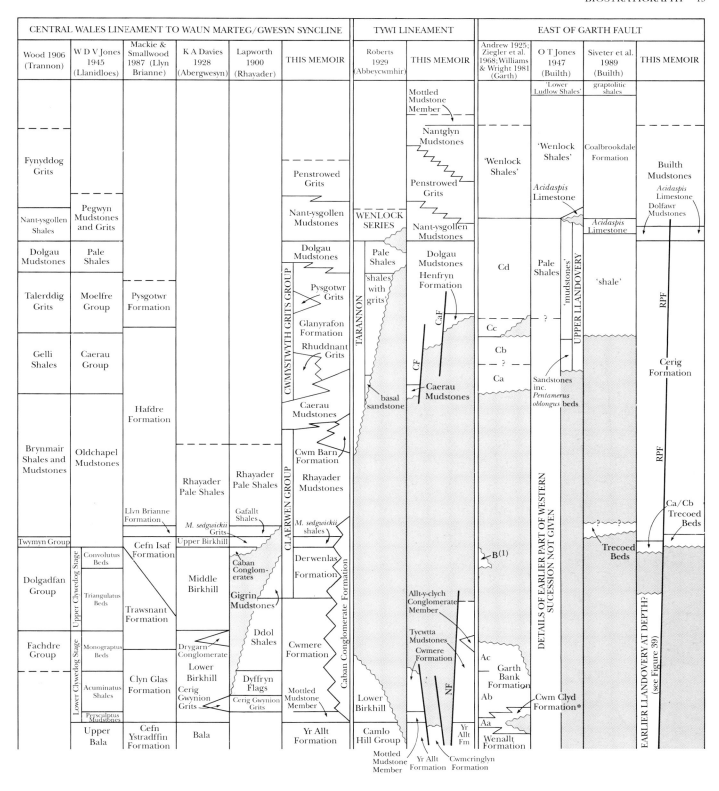

faunas, mostly brachiopods, have proved to be of only limited biostratigraphical value. Derived brachiopod faunas obtained from the resedimented deposits of the basin have, in general, proved too fragmentary to be of use. Acritarchs, the most biostratigraphically useful paly-

nomorphs in the Silurian of the Midland Platform, have not proved so useful in the Llandovery basinal successions of the district, where the assemblages are more sparse and less diverse than those of the shelf. However, the presence of diagnostic acritarch taxa at some levels has improved correlation between the graptolite and acritarch biozonations.

Graptolites

In the district, graptolites tend to be preserved in partial to full relief. Fragmentary periderm is frequently preserved, and usually surrounds an internal mould of pyrite or limonite. Graptolites preserved as hollow internal moulds or as periderm infilled by sediment are more rare. Graptolites recovered from sandstones are commonly flattened. The graptolites are variably deformed, depending on mode of preservation, grade of metamorphism and enclosing lithology. Brittle fracture of pyritised graptolites is typical, the graptolites being broken into segments at their weakest points; chlorite fills the fractures and strain-shadow voids which formed around the graptolite during deformation. The graptolites often show well-preserved thecal structures, though commonly they were obtained only as partial rhabdosomes because the rock fractured along cleavage and not bedding during collection. This has made the identification of some graptolite taxa, for example the spiraliform monograptids, difficult.

Although graptolites are the predominant macrofossils in the Llandovery and Wenlock basinal and distal shelf successions, they are not equally distributed. They are largely confined to laminated hemipelagites, referred to in the past as graptolitic shales. Rare examples also occur in the turbidite mudstones and sandstones; of these the basal portions of thin laminated fine-grained sandstones have proved the most productive. The degree of resolution provided by graptolite biostratigraphy depends in practice, therefore, on the frequency and proportion of laminated hemipelagites within the local sequence. Between the *persculptus* Biozone and the lower part of the *magnus* Biozone they occur at regular intervals and provide an almost continuous graptolitic sequence. In the remainder of the Llandovery, laminated hemipelagites occur at variably spaced intervals and represent a very small proportion of the succession. However, nearly all the graptolite zones have been identified. In the Wenlock, laminated hemipelagites provide a fairly continuous graptolitic record in both the shelf and basinal successions.

For the most part, the established British zonal system (summarised in Rickards, 1976) has been used (Table 4). The subdivisions of the *turriculatus* Biozone, recently developed by Loydell (1989; 1991a) in western mid-Wales are followed, but with some modification. A subdivision of the *crispus* Biozone into three subzones and the *greistoniensis* Biozone into two informal subdivisions (Zalasiewicz, 1990), is also employed. The interval occupied by the *crenulata* Biozone (as defined by Rickards, 1976) has been subdivided into the *crenulata* and *spiralis* biozones (*sensu* Boucek, 1953; Bjerreskov, 1975) with an interregnum separating the latter from the basal Wenlock *centrifugus* Biozone. A summary of the ranges of the latest Ordovician and Silurian graptolite taxa recorded from the district is given in Appendix 3. The details of the individual zones, including that of the

latest Ordovician, *persculptus* Biozone, are described below. Examples of typical Llandovery and Wenlock graptolites are shown in Figure 15.

ASHGILL (HIRNANTIAN)

***persculptus* Biozone** This biozone is typified by the presence of the zone fossil *Normalograptus? persculptus* and the abundance of the narrow, related species *Normalograptus? parvulus* ('*Glyptograptus*' cf. *persculptus* sensu Williams, 1983), together with long-ranging species of *Normalograptus*. The biozone is well exposed in the Wye valley, south-east of Rhayader (Table 9, locality A).

LLANDOVERY BIOZONES

***acuminatus* Biozone** In the district the biozone can be divided into two parts, corresponding to those noted elsewhere by Rickards (1976). *Akidograptus ascensus, Normalograptus? parvulus* and *N? persculptus,* the last as a late form with no median septum on the reverse side (cf. Davies, 1929), characterise the lower part. Upper levels of the biozone are marked by the appearance of *Parakidograptus acuminatus* and the rarity of *N? persculptus* and *N? parvulus*. The biozone is well exposed in the Wye valley (Table 9, localities A and B), where the majority of the *acuminatus* Biozone faunas belong to the lower part but a possible *P. acuminatus* from higher levels suggests the upper part.

***atavus* Biozone** First recognised in the Wye valley by H Lapworth (1900, pp.78–79), who termed it the *Monograptus tenuis* Zone due to a misidentification of the zone fossil, it was renamed the *Monograptus* (now *Atavograptus*) *atavus* Zone by Jones (1909). Re-examination of Lapworth's type locality [SN 9799 6738] (Table 9, locality C), however, suggests that only 5 m of strata, referable to the upper part of the *atavus* Biozone, are currently exposed. The recovered assemblage includes *A. gracilis* and species of *Normalograptus* in addition to the zone fossil *Atavograptus atavus*.

***acinaces* Biozone** This biozone, as employed here, can be divided into two parts, recognised in the Wye valley (Table 9, localities C and D). In the lower part *Lagarograptus acinaces* occurs together with normalograptids of lower Rhuddanian aspect (e.g. *N. normalis, N. rectangularis*). In the upper part *L. acinaces* is accompanied by a variety of monograptids and diplograptids including *Pribylograptus incommodus, Pr. sandersoni, Pristiograptus fragilis pristinus, Coronograptus cyphus cyphus, Dimorphograptus confertus, Metaclimacograptus hughesi* sensu Bulman and Rickards, *Rhaphidograptus toernquisti* and *Glyptograptus tamariscus*.

***cyphus* Biozone** This was originally defined by H Lapworth (1900, p.79) in the Wye valley. Subsequent work, however, has shown that the zone fossil, *Coronograptus cyphus cyphus*, ranges down into the *acinaces* Biozone, and this has led to different authors using different criteria to determine the *acinaces–cyphus* zonal boundary. Following the re-examination of the type section (Table 9, locality D), the base of the *cyphus* Biozone is here taken at the incoming of graptolites of the *revolutus/austerus* groups (cf. Rickards, 1976). Other graptolites in the biozone include *Atavograptus strachani* and subspecies of *Coronograptus gregarius*.

***triangulatus* Biozone** *M. triangulatus triangulatus* is the characteristic fossil, associated with other triangulate monograptids and *Coronograptus gregarius gregarius*. In the Wye valley, where the biozone is well exposed (Table 9, locality D), faunas of the middle part of the biozone (as defined by Rickards, 1976) were found above a sparsely fossiliferous interval that overlies *cyphus* Biozone faunas. This interval presumably occupies the lower part of the biozone at this locality.

***magnus* Biozone** An association of the zone fossil *Normalograptus magnus* with *Monograptus triangulatus fimbriatus* (= *M. pectinatus* Richter according to Bjerreskov 1975) is characteristic, whilst *Pseudoglyptograptus barriei* is also restricted to this biozone in the district. Localities providing diverse biozonal assemblages include the Tynygraig railway cutting [SN 6859 7012] (Table 10), the Wye valley (Table 9, locality D), and a road section [SN 7324 7203], near Ysbyty Ystwyth (Table 13, locality 6).

***leptotheca* Biozone** This was recorded from the Derwenlas Formation by H Lapworth (1900, p.81) in the Wye valley, on the basis of the abundance of the zone fossil, *Pribylograptus leptotheca*. Although the locality (Table 9, locality E) is no longer exposed, Lapworth's recorded assemblage seems generally comparable to the *leptotheca* Biozone assemblage in the Tynygraig railway cutting [SN 6857 7017] (Table 10), where the zone fossil is common. Somewhat different assemblages are associated with the Ystrad Meurig Grits Formation at Hendre quarry [SN 7200 6987 and SN 7216 6952] (Table 13, localities 7 and 8). There the zone fossil is absent, and the faunas are dominated by *Monograptus argenteus*, with subsidiary *Monograptus denticulatus* sensu Sudbury, *M. lobiferus*, *M. imago* and '*Glyptograptus sinuatus*' *sinuatus*.

***convolutus* Biozone** Assemblages in which a narrow form of *M. convolutus* (*M.* aff. *convolutus* of Storch, 1980) is associated with *M. clingani*, *M. denticulatus* Törnquist and, more rarely, *M. urceolinus* appear to characterise the lowest levels of the biozone within the Teifi Anticline (Table 13, localities 10 and 11). More typical assemblages are seen in the Tynygraig railway cutting [SN 6867 7007 to SN 6864 7010 and SN 6858 7022 to SN 6842 7047] (Table 10), where the biozone is recorded from three separate units of anoxic facies. *M. lobiferus* and related forms are abundant, together with *M. convolutus* sensu stricto, *M. capillaris*, *M. decipiens* and *Rastrites hybridus hybridus*. *Cephalograptus cometa extrema*, an indicator of the upper part of the biozone (cf. Hutt, 1974–75) occurs in the topmost unit of anoxic facies in the Tynygraig railway cutting (Table 10) and locally elsewhere in the Teifi Anticline (Table 13, localities 25–28).

***sedgwickii* and *halli* biozones** At most localities in the district the *sedgwickii* Biozone cannot be distinguished from the overlying *halli* Biozone. The *sedgwickii* Biozone is characterised by an abundance of the zone fossil, locally with *Lagarograptus tenuis* (Rickards, 1976), as in the Tynygraig railway cutting [SN 6860 7011] (Table 10) and in a nearby forestry track section (Table 13, locality 33). This assemblage is confined to the *M. sedgwickii* shales at the base of the Rhayader Mudstones. However, in the upper levels of the *M. sedgwickii* shales and succeeding parts of the Rhayader Mudstones, several species, most notably *Monograptus halli*, appear for the first time and *L. tenuis* is absent. This fauna constitutes the *halli* Biozone of Jones and Pugh (1916), first defined in the Machynlleth district. However, the *halli* Biozone has not been widely recognised and was therefore omitted from Rickards' (1976) summary of Silurian graptolite biostratigraphy. Recent work in the Aberystwyth district by Loydell (1989; 1991a; see also Loydell, 1991b) has essentially confirmed the findings of Jones and Pugh (1916), and led him to reinstate the *halli* Biozone. However, the difficulty of distinguishing *M. halli* from *M. sedgwickii* using incomplete specimens has precluded its systematic recognition during this survey. The biozone has only been identified in a few places, largely on the basis of *M.*

halli, (e.g. roadside exposure [SN 6815 6570] near Swyddffynnon). Other species common in the *sedgwickii–halli* Biozone interval include *M. involutus*, *M. contortus* and *Metaclimacograptus undulatus*.

***turriculatus* Biozone** This has recently been subdivided into six subzones, in western mid-Wales by Loydell (1989; 1991a) (Table 4). Since the completion of the survey, Loydell (1992) has subdivided the *turriculatus* Biozone into the *guerichi* and *turriculatus* biozones, based on the division of material formerly attributed to *M. turriculatus*, into two temporally distinct species. An early slender form, *M. guerichi* (termed *Spirograptus guerichi* in Loydell, 1992), ranges from the *runcinatus* Subzone into the lower part of the *utilis* Subzone, while *M. turriculatus* sensu stricto (termed *Spirograptus turriculatus* by Loydell, 1992) does not appear until the upper part of the *utilis* Subzone. This account employs the earlier concept of a *turriculatus* sensu lato Biozone; references to the 'zonal' fossil are also sensu lato.

The lower part of Loydell's (1991a) subzonal scheme (*runcinatus–gemmatus* subzones) is based on an examination of sections that contain the subzonal boundaries. In contrast, the upper part relies on the placing of assemblages from isolated localities into a biostratigraphic sequence using both local, lithostratigraphical relationships (*gemmatus–renaudi* subzones) and the gross sense of younging in the folded and faulted, incompletely exposed, coastal section (*renaudi* to *proteus* subzones). The *maximus* Subzone, previously recognised as the lower division of the *turriculatus* Biozone (e.g. Cave and Hains, 1986) has been rejected by Loydell (1991b) as invalid.

Loydell's (1991a) subdivision of the *turriculatus* Biozone has been employed during the present survey, although not all the subzones could be recognised; in particular, it was commonly impossible to distinguish between the lowest three subzones. An additional subzone, the *carnicus* Subzone, has been erected to cover the interval at the top of the *turriculatus* Biozone, not represented by fossiliferous strata in the sections studied by Loydell.

runcinatus Subzone The base of this subzone was defined by Loydell (1989; 1991a) at Nant Fuches-wen quarry [SN 7716 8065], near Ponterwyd, in the Aberystwyth district. Succeeding the *halli* Biozone, it is characterised by a low-diversity fauna dominated by *Monograptus runcinatus* and *M. turriculatus*. Although no *runcinatus* Subzone faunas were found during the survey, many localities yielding low-diversity assemblages, assigned to the *runcinatus* to *renaudi* subzone interval, may belong to this subzone. It is only known from the Devil's Bridge Formation.

gemmatus Subzone The base of this subzone was defined by Loydell (1989; 1991a) in the Glan-fred Borehole [SN 6305 8812], near Borth, at the first occurrence of a characteristic and diverse assemblage including *Monograptus gemmatus*, *Glyptograptus fastigatus*, *Rastrites fugax*, *M. runcinatus* and *Pristiograptus renaudi* above that of the low-diversity *runcinatus* Subzone assemblage. The *gemmatus* Subzone has been recognised at several localities in the district, including a quarry (Table 13, locality 36) east of Pont-rhyd-y-groes, and Grogwynion Mine (Table 13, locality 37), east of Llanafan.

renaudi Subzone The position of this subzone relative to the *gemmatus* Subzone has not been established in a continuous section. However, because in the Aberystwyth district the lower part of the Borth Mudstones in the Glan-fred Borehole has yielded *gemmatus* Subzone faunas and the basal Aberystwyth Grits, that overlie them on the coast, contain *renaudi* Subzone faunas, Loydell (1991a) concluded that the latter subzone was the younger. The *renaudi* Subzone is characterised by an abundance of the zone fossil in association with *Monograptus*

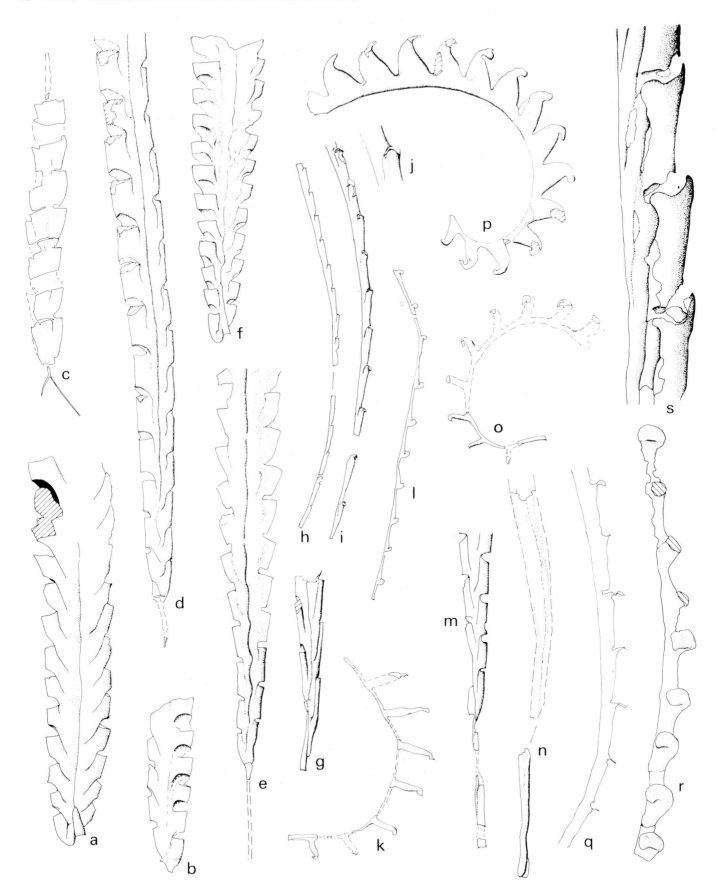

Figure 15 Illustrations of selected late Ordovician and Silurian graptolites from the district.

a Hirnantian, Rhuddanian and Aeronian graptolites (facing page)

a) *Normalograptus? persculptus*, Prysg stream section [SN 9807 6718], *acuminatus* Biozone (JZ 417); b) *Normalograptus? parvulus*, quarry [SN 8855 6268] below Claerwen reservoir, *persculptus* Biozone (JZ 8687); c) *Glyptograptus? avitus*, Cerrig Gwynion quarry [SN 9722 6577], uppermost *persculptus* Biozone (JZ 4813); d) *Normalograptus normalis*, Nant Paradwys [SN 8919 6702], *acuminatus* Biozone (JZ 2306); e) '*Glyptograptus' sinuatus sinuatus*, Hendre quarry [SN 7200 6987], *leptotheca* Biozone (JZ 9259); f) *Normalograptus? magnus*, River Wye [SN 9762 6745], *magnus* Biozone (DJ 7768, latex cast of external mould); g) *Akidograptus ascensus*, Cerrig Gwynion quarry [SN 9723 6597], *acuminatus* Biozone (JZ 4829); h) *Atavograptus gracilis*, [SN 9799 6738], *atavus* Biozone (JZ 9431); i–j) *Monograptus* cf. *sudburiae*, River Wye [SN 9775 6738], *cyphus* Biozone (JZB 846); k) *Rastrites hybridus hybridus*; Nant Cwm-nel [SN 6854 6957], *convolutus* Biozone (JZ 535); l) *Monograptus capillaris capillaris*; Nant Cwm-nel [SN 6854 6957], *convolutus* Biozone (JZ 565); m) *Rhaphidograptus toernquisti*, Tynygraig railway cutting [SN 6859 7011], *triangulatus* or *magnus* Biozone (JZ 8065a); n) *Cephalographus cometa extrema*, Tynygraig railway cutting [SN 6842 7047], upper part of *convolutus* Biozone (JZB 605); o) *Monograptus decipiens*, Nant Cwm-nel [SN 6854 6957], *convolutus* Biozone (JZ 543); p) *M. triangulatus fimbriatus*, Tynygraig railway cutting [SN 6859 7012], *magnus* Biozone (JZ 8457); q) *Lagarograptus tenuis*, forestry track [SN 7675 7294] east of Pont-rhyd-y-groes, *sedgwickii* Biozone (JZ 9077); r) *Monograptus involutus*, road cutting [SN 7675 7294] near Swyddffynnon, *halli* Biozone (JZ 886); s) *Pribylograptus leptotheca*, Tynygraig railway cutting [SN 6856 7015], *leptotheca* Biozone (JZB 181). All ×10 except h ×5, and j and s ×25. Numbers in round brackets prefixed by JZ, DJ or JZB refer to specimen numbers in BGS collections.

b Telychian graptolites (p.54)

a) *Streptograptus storchi*, A44 road cutting [SN 8580 9198], *turriculatus* Biozone, *carnicus* Subzone (DJ 8297); b) *Glyptograptus fastigatus*, Grogwynion Mine [SN 7131 7252] west-north-west of Ysbyty Ystwyth, *turriculatus* Biozone, *gemmatus* Subzone (JZ 7216); c) *Monograptus gemmatus* (proximal fragment), quarry [SN 6744 6705] at Gwenhafdre farm, *turriculatus* Biozone, *gemmatus* Subzone (JZ 610); d) *Monograptus gemmatus* (distal fragment), locality and horizon as for (c); e) *Monograptus runcinatus*, quarry north-east of Tregaron [SN 7130 6157], *turriculatus* Biozone, *gemmatus* Subzone (JZ 7143); f) *Pristiograptus renaudi*, quarry [SN 7109 6309] south-east of Pontrhydfendigaid, *turriculatus* Biozone, *rucinatus–renaudi* subzones (JZ 6149); g) *Monograptus utilis*, Nant Rhuddnant [SN 7979 7837], *turriculatus* Biozone, *utilis* Subzone (DJ 6508); h) *Monograptus carnicus*, quarry [SN 9445 7650] at Cwmgwary farm, *turriculatus* Biozone, *carnicus* Subzone (JZ 1034); i) *Monograptus turriculatus*, quarry [SN 6400 7015] near Lledrod, *turriculatus* Biozone, *utilis* Subzone (JZ 7484); j) *Monograptus cavei*, road cutting [SN 6400 7015] near Pont-rhyd-y-groes, *turriculatus* Biozone, *? utilis* Subzone (JZ 7621); k) *Monograptus crispus*, Afon Ystwyth [SN 83889 7545], *crispus* Biozone, middle part (DJ 7270); l) *Streptograptus pseudobecki*, cliff section [SN 5572 7494] south of Aberystwyth, *turriculatus* Biozone, *proteus* Subzone; m) *Monograptus clintonensis*, Pandy Brook [SO 0843 6847], *griestoniensis* Biozone, lower part (JZ 2710); n) *Monograptus pragensis pragensis*, Pandy Brook [SO 0843 6847], *griestoniensis* Biozone, lower part (JZ 2683); o) *Streptograptus exiguus*, track section [SN 8187 6035] Twyi Forest, *turriculatus* Biozone, *carnicus* Subzone (JZ 2173); p) *Streptograptus sartorius*, quarry [SN 9975 7477] near Pant-y-dwr, *crispus* Biozone, upper part (SPT 173); q) *Monoclimacis* cf. *griestoniensis* sensu Elles and Wood, Pandy Brook [SO 0843 6847], *griestoniensis* Biozone, lower part (CAV 1853); r) *Monoclimacis griestoniensis*, quarry [SN 9258 7994] east of Llangurig, *griestoniensis* Biozone, upper part (DJ 7517); s) *Monoclimacis* cf. *crenulata*, sensu Elles and Wood, trackside quarry [SO 0817 7417] near Crychell Cottage, east of Bwlch-y-sarnau, probably *crenulata* Biozone (DJ 4191); t) *Monoclimacis linnarssoni*, quarry [SO 0092 7707] north-west of Bwlch-y-sarnau, probably *spiralis* Biozone (DJ 9996). All ×10 except i and n ×5, and p ×20. Numbers in round brackets prefixed by JZ, DJ or SPT refer to specimen numbers in BGS collections.

c Wenlock graptolites (p.55)

a) *Monoclimacis vomerina* cf. *gracilis*, west bank [SO 0362 5949] of River Ithon, *spiralis-centrifugus* interregnum (JZ 3759); b) *Monograptus* aff. *minimus cautleyensis*, scree from base of Nant-ysgollon Mudstones, Waun Marteg quarry [SO 0092 7707], *spiralis-centrifugus* interregnum (note long interthecal septae) (SPT 78); c) *Monoclimacis flumendosae flumendosae*, quarry [SO 0656 7553] south-east of Mynydd-llys farm, lower *linnarssoni* Biozone (JZ 3860); d) *Monograptus radotinensis inclinatus*, quarry [SO 0161 7739] north-west of Bwlch-y-sarnau, upper *riccartonensis* Biozone (SPT 1439); e) *Monograptus flexilis*, quarry [SO 0656 7553] south-east of Mynydd-llys farm, lower *linnarssoni* Biozone (JZ 3895); f) *Cyrtograptus rigidus cautleyensis*, quarry [SO 0656 7553] south-east of Mynydd-llys farm, lower *linnarssoni* Biozone (JZ 3913); g) *Monograptus flemingii flemingii*, quarry [SO 0922 6575] north-north-east of Crossgates, *?ellesae* Biozone (JZ 3241); h) *Plectograptus? bouceki*, track [SO 0901 7166] east of Abbeycwmhir, *rigidus–linnarssoni* biozones (SPT 1072); i) *Pristiograptus pseudodubius*, road section [SO 0860 7894] north of Llanbadarn Fynydd, *nassa* Biozone (JZ 6111). a and c–g ×5, b and h–i ×10. Numbers in round brackets prefixed by JZ or SPT refer to specimen numbers in BGS collections.

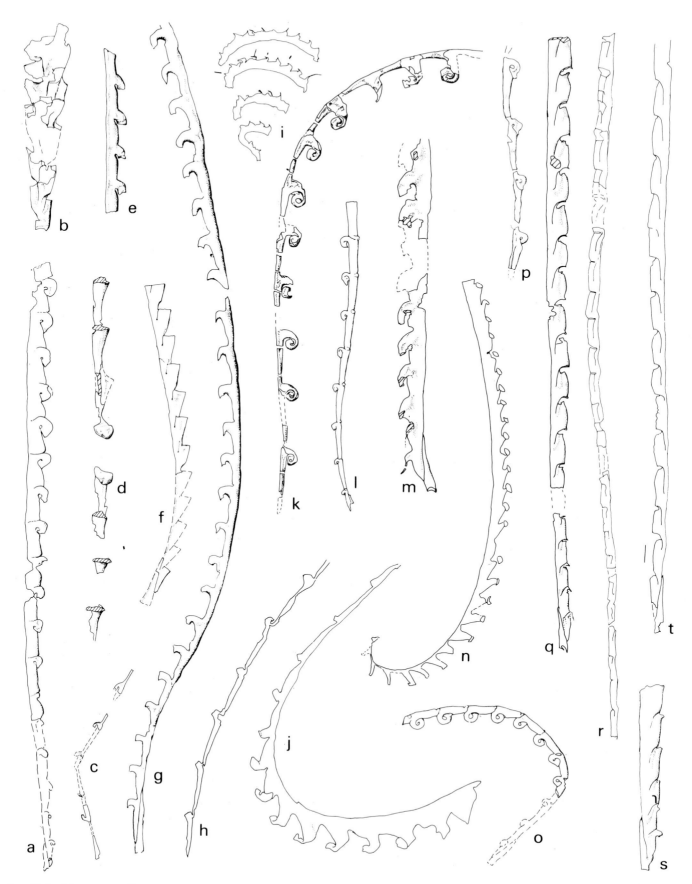

b Telychian graptolites

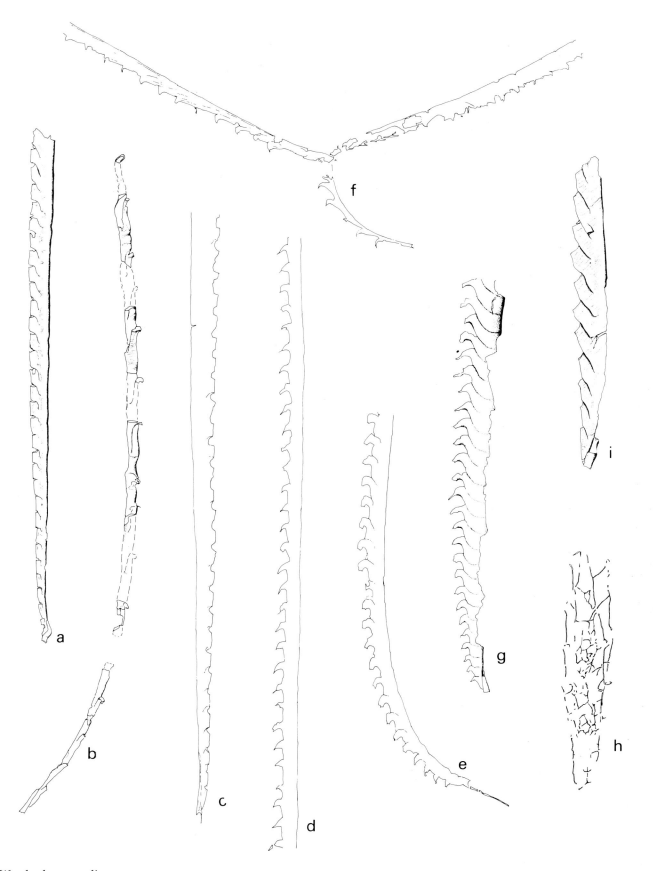

c   Wenlock graptolites

planus and *M. cavei*, the latter also being present in the overlying *utilis* Subzone, but not in the underlying *gemmatus* Subzone. The only fauna in the district unambiguously referable to the *renaudi* Subzone was found in a forestry track section [SN 6584 7110], near Lledrod. The *renaudi* Subzone, in common with the earlier *runcinatus* Subzone, may represent a low-diversity interval of relatively short duration.

utilis Subzone   Although this fauna notably includes several species that are found in lower subzones, such as *M. cavei* and *Streptograptus plumosus*, the incoming of taxa such as *Monograptus utilis*, *Pristiograptus bjerringus* and *Streptograptus barrandei* make this a readily identifiable interval in the district. It is present in quarries [SN 9644 7597 and SN 9544 7523] west of Pant-yr-dwr, and a road section [SN 7483 7309] at Pont-rhyd-y-groes. The assemblages, however, are commonly not uniform in composition; work subsequent to the survey has revealed temporally restricted, stratigraphically useful faunal associations (Loydell, 1992).

johnsonae Subzone   This subzone has been rarely, but consistently recognised as a low-diversity interval separating *utilis* and *proteus* subzonal assemblages. It is characterised by an abundance of *Streptograptus johnsonae* and *M. marri*. The subzone appears to occupy a narrow interval of strata, possibly reflecting a short-lived biostratigraphic event. It is present [SN 9540 7621 and SN 9523 7621] west of Cwmyrychen, and in a track section [SN 9430 7417], west of Pant-y-dwr.

proteus Subzone   The subzone is marked by the incoming of *Monograptus proteus*, *M. rickardsi* and *Monoclimacis? galaensis*, but the index fossil itself is not common. Assemblages with *M. proteus* occur in crags [SN 7662 6577] on the north side of the Teifi valley, east of Pontrhydfendigaid and the coast section [SN 5573 7492], south-west of Pantyrallad (Loydell, 1989; 1991a). Many localities assigned tentatively to this subzone are low-diversity assemblages dominated by hooked monograptids, in particular *Monograptus marri*.

carnicus Subzone   Stratigraphic levels immediately beneath strata assigned to the *crispus* Biozone (Tables 14 and 15) contain an assemblage which includes the distinctive slender 'spirograptid' *Monograptus carnicus* and the gently ventrally curved streptograptid *Streptograptus whitei*, together with common *S. storchi* and *Monoclimacis? galaensis*. This assemblage has not been recognised by Loydell (1989; 1991) and probably represents the highest recognisable biostratigraphic level within the *turriculatus* Biozone (Table 4). It is named herein after its most characteristic element. *S. nodifer* sensu Haberfelner and *S. storchi* have not been found in the overlying *crispus* Biozone, though, in one locality (in the A44 road section [SN 8580 8198] west of Llangurig, just north of the district), *S. storchi* has been found associated with a rare taxon resembling *M. crispus*, but having narrower and closer-set distal thecae. *M. exiguus*, a species most characteristic of the *crispus* and lower *griestoniensis* biozones, has also been found in strata referred to the *carnicus* Subzone in a track section [SN 8187 6035] in Tywi Forest.

crispus Biozone   This is marked by the incoming of *M. crispus* and *M. discus*. As neither are common in the lowermost part of the zone the recognition of the base is generally difficult; it is not known which of the two species appears first in Wales. Three successive assemblages can be recognised and these are here given subzonal status (Table 4).

galaensis Subzone   This is a partial range subzone, its base defined by the overlap of *M. crispus* and/or *M. discus* and *Monoclimacis? galaensis*. *M. carnicus* ranges no higher than this subzone and *M. clintonensis* and *Streptograptus exiguus* are also common.

M. crispus is not common, however. Localities assigned to this subzone include the upper part of the Afon Dulas section (Table 14) and a quarry [SN 9833 7517] near Pant-y-dwr.

crispus Subzone   This is a partial range subzone defined by the overlap of *M. crispus* and *S. loydelli*. Both taxa are abundant. The subzone is well displayed in the Cwm Ystwyth section (Table 15).

sartorius Subzone   This is an interregnum, comprising the interval between the disappearance of *M. crispus* and the first appearance of monoclimacid monograptids sensu stricto. *M. pragensis* and *S. sartorius* appear for the first time. Many collections from this assemblage are dominated by relatively non-diagnostic hooked monograptids related to *M. marri*, as in the upper levels of the *turriculatus* Biozone. This interval is well displayed in the Cwm Ystwyth section (Table 15).

griestoniensis Biozone   This is recognised by the presence of the narrow monoclimacids *Monoclimacis griestoniensis* sensu stricto and the distinctive *Mcl.* cf. *griestoniensis* sensu Elles and Wood. The zone can be broadly divided into two parts based largely on the relative abundance of these two taxa. *Mcl.* cf. *griestoniensis* appears earlier (Table 4) and dominates the lower part of the biozone, along with species common to the upper part of the *crispus* Biozone, including *M. pragensis* and *S. sartorius*. Such an assemblage has been recognised in the Cwm Ystwyth section (Table 15) and in Pandy Brook [SO 0843 6847] near Henfryn, Abbeycwmhir. *Mcl. griestoniensis* sensu stricto is rare until the upper part of the zone, where it typically dominates most faunas, as in a forestry track [SO 0560 7290] near Abbeycwmhir.

crenulata Biozone   The *crenulata* Biozone in Britain has traditionally encompassed all strata between the top of the *griestoniensis* Biozone and the base of the early Wenlock *centrifugus* Biozone (Rickards, 1976). More zones have been recognised in this interval in Europe (e.g. Boucek, 1953, Bjerreskov, 1975) and elements of this more precise biostratigraphy have been recognised locally in Britain (Loydell and Cave, 1993). In the district, two of the these intervening zones have been recognised, the *crenulata* and *spiralis* biozones, overlain by a low-diversity *spiralis–centrifugus* interregnum (Table 4).

The incoming of broad vomerinids, particularly *Mcl. vomerina vomerina*, is used to denote the lower boundary of the *crenulata* Biozone (sensu Boucek, 1953 and Bjerreskov, 1975). The zonal index fossil, *Mcl. crenulata* sensu Elles and Wood (*Mcl. crenulata* sensu stricto has been synonomised with *Mcl. vomerina vomerina*) has not been recognised with certainty in the district, although a taxon resembling this species has been recovered from a quarry [SO 0817 7410] east of Bwlch-y-sarnau. Other species common in this biozone are *M. discus* and *M. tullbergi*, present at the above locality, and in a quarry, [SO 0296 7456] at Bwlch-y-sarnau.

spiralis Biozone   An apparently post-*crenulata* Biozone assemblage has been recognised in the uppermost unit of laminated hemipelagites in the Dolgau Mudstones, in Waun Marteg quarry [SO 0093 7706] (Table 17), near Bwlch-y-sarnau. This assemblage includes newcomer species such as *Monograptus parapriodon* and *M.* cf. *falx*, together with common *Monoclimacis linnarssoni*. It probably correlates broadly with the *spiralis* Biozone faunas of Europe (e.g. Boucek, 1953, Bjerreskov, 1975). The occurrence of *M. spiralis* from the Cerig Formation in Pentre Brook [SO 0486 6398] also suggests the presence of this zone.

'spiralis/centrifugus interregnum'   Graptolite faunas, indicative neither of the *spiralis* Biozone nor the *centrifugus* Biozone, characterise the lower parts of the Nant-ysgollon Mudstones and the laterally equivalent Dolfawr Mudstones. These assemblages are dominated by broad vomerinids referable to *Monoclimacis vomerina vomerina*, narrower forms here referred to *Mcl. vomerina gracilis*, hooked monograptids of the *priodon* group and *Retiolites geinitzianus* together with a taxon seemingly related to, but not conspecific with, the early Wenlock species *M. minimus cautleyensis*. These interregnum faunas are typically found on the west bank of the River Ithon [SO 0356 5935 to SO 0368 5955], near Llandrindod Wells, immediately south of the district.

WENLOCK BIOZONES

centrifugus **Biozone**   The base of the *centrifugus* Biozone is taken at the first appearance of cyrtograptids. These occur together with vomerinid, hooked monograptid and retiolitid graptolites, broadly similar to those occurring below. The biozone is well exposed in a track section [SO 0420 7490] near Hendy, east of Bwlch-y-sarnau.

murchisoni **Biozone**   This Biozone has been recognised on the basis of the zone fossil occurring together with hooked monograptids and slender vomerinids; *Retiolites geinitzianus* is absent. It has been recognised in Arlair brook [SO 0568 6092], Llandrindod Wells (Table 19).

riccartonensis **Biozone**   *Monograptus riccartonensis* dominates this biozone, commonly to the virtual exclusion of other taxa. The biozone is well exposed in Arlair brook, Llandrindod Wells (Table 19). The upper part of the biozone also includes the related, but more slender *M. radotinensis inclinatus*, together with the first appearance of *Monoclimacis flumendosae flumendosae*, as in a forestry quarry [SO 0161 7739] at Waun Marteg.

rigidus **Biozone**   This zone has not been recognised definitely in the district, but faunas including *Mcl. flumendosae flumendosae*, slender pristiograptids and rare *Plectograptus bouceki* may belong to this interval. Such assemblages have been found in a track section [SO 0901 7166] east of Abbeycwmhir.

linnarssoni (= *flexilis*) **Biozone**   A rich fauna belonging to the lower part of the biozone was collected from a quarry [SO 0656 7553] near Mynydd-Llys, north-east of Bwlch-y-sarnau. Both zone fossils are present, together with *Cyrtograptus rigidus cautleyensis*.

ellesae **Biozone**   This zone has not been recognised in the district.

lundgreni **Biozone**   Faunas of the *lundgreni* Biozone have been proved in the district only in the A483 road section [SO 0879 7872 to SO 0869 7887] north-east of Bwlch-y-sarnau, where they are characterised by subspecies of the *flemingii* group of monograptids, together with *Cyrtograptus hamatus* and *Pristiograptus pseudodubius*.

nassa **Biozone**   A *nassa* Biozone fauna dominated by *Gothograptus nassa* associated with a broader gothograptid and pristiograptids of the *dubius* group, is present in the Mottled Mudstone Member of the Nantglyn Mudstones in the A483 road section [SO 0860 7894], north-east of Bwlch-y-sarnau.

Brachiopods

In the shelly shelf facies of the Midland Platform and its margins, correlations have been made by means of brachiopods, mainly using lineages within *Borealis–Pentamerus*, *Stricklandia–Costistricklandia* and *Eocoelia* (see Bassett, 1989 for summary). The presence of sparse graptolite faunas within the shelly biofacies has enabled limited correlation to be made between the two schemes (Cocks et al., 1984).

Indigenous brachiopod faunas are restricted in the district to thin beds in the Cerig Formation, namely the Trecoed Beds and the *Acidaspis* Limestone and their correlatives, and the Mottled Mudstone Member of the Nantglyn Mudstones. Pentamerids and stricklandiids are present only in the Trecoed Beds. The faunas from the *Acidaspis* Limestone and Mottled Mudstone Member comprise long-ranging taxa of the deep-water *Clorinda* and *Visbyella trewerna* benthic communities respectively.

Acritarchs

Prior to the present survey, there were no published data available on Llandovery acritarch distribution in the Welsh Basin. Previous studies had concentrated on the shelf sediments of the Midland Platform and its margins (Hill, 1974; Dorning, 1981a and b; Hill and Dorning, 1984; Mabillard and Aldridge, 1985; Dorning and Bell, 1987). During the survey, four palynological traverses were undertaken across the basinal and slope-apron successions (Appendix 2b). Only spot samples were taken from the distal shelf facies due to a lack of continuous sections. In most samples, no distinction was made between turbidite mudstone and hemipelagite. Wenlock strata were not sampled.

Assemblages were found to be sparse and of low diversity compared to the Llandovery shelf successions of the Midland Platform and its margins. In the basinal succession the most abundant and diverse floras were recovered from the Dolgau Mudstones (Appendix 2biii); the most sparse came from the Cwmere Formation in Nant Paradwys and the lower part of the Rhayader Mudstones in the Claerwen valley (Appendix 2bii).

The assemblages are generally poorly preserved, the processes on acanthomorph acritarchs are commonly damaged or missing, and surface sculpture is difficult to interpret, making taxonomic identification problematical. The nature of the bottom conditions during early diagenesis had a major effect on acritarch preservation. Oxic bottom conditions during deposition of the Cerig Formation, most of the Rhayader Mudstones, and parts of the Dolgau Mudstones resulted in the oxidation of much of the contained organic matter, thus accounting for their sparse and very poorly preserved assemblages.

West of the Tywi Lineament, thermal maturation levels are high, as indicated by dark brown to grey/black acritarch body colours, and the preservation is poor. In contrast, across and to the south-east of the lineament, maturation levels are lower, as shown by yellow to dark brown body colours.

There is a general trend towards increasing diversity and abundance of acritarchs through the Llandovery in Britain and Ireland (Dorning *in* Aldridge et al., 1979) and acritarch assemblages recovered from the district confirm this (Table 7). Rhuddanian floras from the basinal succession in the district are impoverished, with a total of only 12 taxa recorded. However, a sample of late Rhuddanian (*acinaces* or *cyphus* Biozone) age from shelf

Table 7 Llandovery acritarch taxonomic diversity.

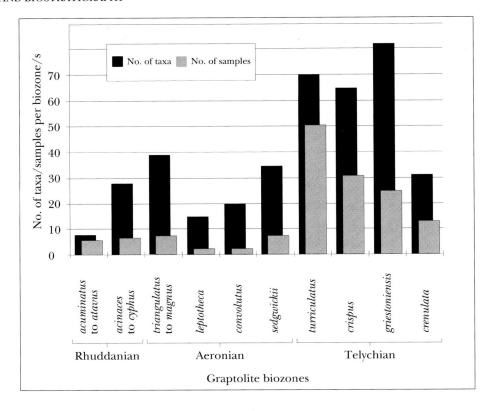

facies at Comin Coch [SN 9577 5169], Garth, just to the south of the district, has yielded 23 taxa. Although early Aeronian assemblages are more diverse, those from the mid Aeronian appear to be impoverished, though this may in part reflect the low number of samples studied. Acritarch abundance and diversity show a marked increase in the Telychian, but there is a return to impoverished assemblages in the *crenulata* Biozone. The diversity maximum in the *griestoniensis* Biozone is also recorded in the Welsh Borderland (K J Dorning, personal communication, 1991).

A modified form of the Llandovery acritarch zonation of Dorning and Bell (1987) has been applied in the district (Table 8). Dorning and Bell's scheme is based on earlier work by Hill (1974), and Hill and Dorning (1984) on the type Llandovery succession. The good graptolite evidence in many of the sections sampled in the district has enabled the ranges of several acritarch taxa to be revised. As a result, Dorning and Bell's (1987) scheme has been modified by the deletion of the *D. monospinosa* Biozone and the introduction of two zones, the *Ammonidium.* sp. B and *Gracilisphaeridium encantador* biozones. The details of the revised scheme are described below.

***Helosphaeridium citrinipeltatum* Biozone**   The base of the biozone is defined by the first appearance of *Helosphaeridium citrinipeltatum* near the base of the Rhuddanian sequence in the Llandovery district (Hill and Dorning, 1984). The biozone has not been recognised in the district.

***Tylotopalla robustispinosa* Biozone**   The base of the biozone is defined by the first appearance of *Tylotopalla robustispinosa*

sensu lato (Hill and Dorning, 1984). In the district this species first occurs within the *acuminatus* Biozone. It has been recognised in Prysg brook [SN 9809 6716] in the Wye valley (Appendix 2biii).

***Multiplicisphaeridium fisheri* Biozone**   The base of the biozone is defined by the first appearance of *Multiplicisphaeridium fisheri* (Hill and Dorning, 1984). This lies within the range of the *acinaces* to *cyphus* biozones at Comin Coch [SN 9577 5169], near Garth, just south of the district. *Multiplicisphaeridium paraguaferum* also has a range base within this interval.

***Oppilatala eoplanktonica* Biozone**   The base of this biozone is defined by the first appearance of *Oppilatala eoplanktonica* and is correlated with the base of the *triangulatus* Biozone by Cocks et al. (1984). No samples were collected from strata of *triangulatus* Biozone age during the survey, but *O. eoplanktonica* was present in samples taken from within the overlying *magnus* Biozone in the Wye valley [SN 9761 6747], Rhayader (Appendix 2biii).

***Ammonidium* sp. B Biozone**   The base of this new biozone is defined here as the first appearance of *Ammonidium* sp. B in the *convolutus* Biozone in the Wye valley section [SN 9744 6770] (Appendix 2biii). *Lophosphaeridium* sp. A and *Domasia limaciformis* also have range bases at this horizon.

***Ammonidium microcladum* Biozone**   The base of this biozone is defined (Hill and Dorning, 1984) by the first appearance of *Ammonidium microcladum*, and is correlated with the base of the *sedgwickii* Biozone by Cocks et al. (1984). In the district this species is not recorded below the base of the *turriculatus* Biozone. However, the biozone can be recognised by the appearance of *Salopidium granuliferum, A. listeri* and *A.* sp. A, which occur in the *sedgwickii* Biozone.

***Dactylofusa estillis* Biozone**   The base of this biozone is defined by Hill and Dorning (1984) on the first appearance of

Table 8   Llandovery acritarch biozonation adopted in this memoir, showing the range of selected taxa within the district. The ranges are derived from charts in Appendix 2b and spot samples MPA 29285, MPA 29286, MPA 33365 and MPA 33516.

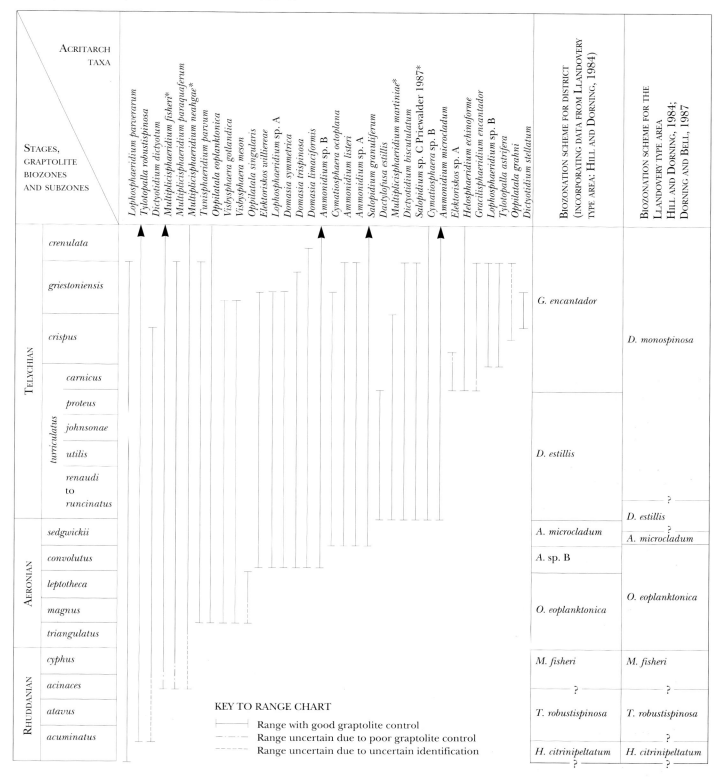

Dactylofusa estillis, and is correlated (Cocks et al., 1984; Dorning and Bell, 1987) with a level in the late Aeronian *sedgwickii* Biozone. In the district, however, this species first appears near the base of the *turriculatus* Biozone. K J Dorning (personal communication, 1991) has indicated that the lithological correlation and graptolite control in the southern part of the Llandovery type area (Cocks et al., 1984) is so poor that the first appearance in the present district may be more reliable. The biozone is present at Gwenhadfre Farm [SN 6744 6705].

Gracilisphaeridium encantador Biozone   The base of this new biozone is defined on the first appearance of *Gracilisphaeridium encantador* within the *carnicus* Subzone of the *turriculatus* Biozone at Blaenycwm [SN 829 754], in the eastern part of Cwm Ystwyth (Appendix 2bi). *Elektoriskos* sp. A and *Helosphaeridium echiniforme* also have range bases at this horizon.

The *Deunffia monospinosa* Biozone, which occurs between the *Dactylofusa estillis* Biozone and the overlying Wenlock *Deunffia brevispinosa* Biozone in Dorning and Bell's (1987) scheme, is not recognised here. *Deunffia monospinosa*, the species used to define the base of the zone (Hill and Dorning, 1984), is only recorded from the *crenulata* Biozone in the district, whilst in Shropshire it enters above the base of the range of *G. encantador* (K J Dorning, personal communication, 1991). This suggests that the correlation of the base of the *monospinosa* Biozone with the base of the *turriculatus* Biozone in the southern part of the Llandovery type area (Cocks et al., 1984, fig. 69) may be in doubt.

Other palynomorphs

Other palynomorphs recovered from the district include chitinozoans, scolecodonts, and sporomorphs. Chitinozoans are less widely distributed than acritarchs, but are nevertheless common at many localities. They are more prone to thermal degradation than acritarchs and are often only identifiable at generic level. Distribution of scolecodonts appears to be similar to that of chitinozoa, but they are of little biostratigraphical use at present as their taxonomy and biostratigraphy are poorly documented. Sporomorphs (mainly trilete spores of the *Ambitisporites* type) are rare in the shelf facies and very rare in the basinal facies.

Plate 7   Disturbed beds and coarse-grained turbidites (facing page).

a   Slump-folded disturbed beds. Note sharp, cross-cutting basal contact (dashed). Nantglyn Mudstones, quarry [SO 0500 7730]. **b**   Type Ai coarse-grained turbidites. Arrows point to conglomeratic bases of four separate turbidites (1 to 4); note inverse grading at base of first unit and channelised base (dashed) to third. Caban Conglomerate Formation, Cabancoch quarry [SN 9240 9460].   **c**   Type Aii coarse-grained turbidite, Cwm Barn Formation, Cwm Barn quarry [SO 0082 7081].   **d**   Type Bi coarse-grained turbidite. Cerig Gwynion Grits, Afon Claerwen [SN 8844 6272].   **e**   Silty mudstone (type Bii) turbidite/debrite. Note the cleaved character of the unit and presence of siltstone pseudonodules (ps). Rhuddnant Grits, cutting [SN 8446 8113].   **f**   Muddy sandstone (type Bii) turbidite/debrite with large, contorted rip-up clasts. Mynydd Bach Formation, cliff [SN 485 645] north of Aberarth.

Plate 8   Bouma and fine-grained turbidites, and hemipelagites (p.62).

a   Sequence of Bouma turbidites, including (see arrows) sandstone/mudstone couplets (types Ci and Cii), siltstone/mudstone couplets (type Di) and mudstone turbidites (type Dii); cross-marks indicate position of intercalated hemipelagites (laminated). Trefechan Formation, cliff [SN 576 796], Allt-wen. **b**   Medium-bedded type Dii Bouma turbidites interbedded with laminated hemipelagites (dark bands). Borth Mudstones, cliff [SN 603 887], Borth.   **c**   Type E fine-grained turbidites (silt-laminated mudstones). Note lenticular basal siltstone divisions (dark-coloured), diffuse but pervasive undulatory and parallel silt-lamination, burrow mottling (pale-coloured infills) and syndepositional microfaulting. Cut block of Rhayader Mudstones, Claerwen reservoir [SN 8639 6521]. **d**   Laminated hemipelagite. Note sharp upper and lower contacts with interbedded turbidite mudstones. Cut block of Caerau Mudstones, Claerwen reservoir [SN 8577 6516]. **e**   Burrow-mottled (dark-coloured infills) hemipelagite. Note relict lamination, sharp contact with succeeding siltstone-based turbidite, but diffuse contact with underlying turbidite mudstone. Cut block of Caerau Mudstones, Wye valley northeast of Dolhelfa [SN 9341 7416]. Scale bars in centimetres.

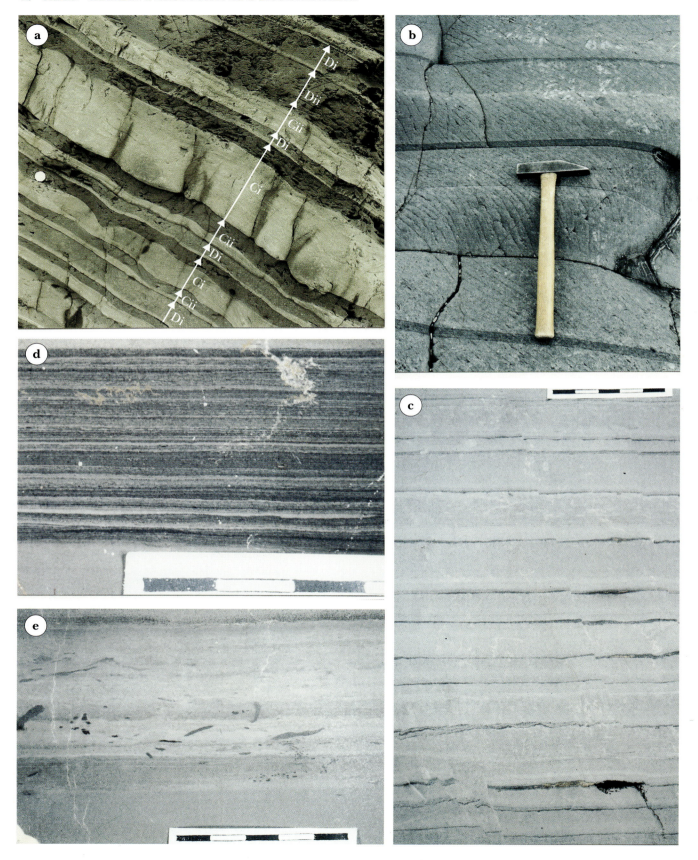

Plate 8   Bouma and fine-grained turbidites, and hemipelagites.

FOUR

Silurian: late Hirnantian to early Telychian slope-apron systems

The transgression that followed the late Ashgill glacio-eustatic lowstand (see Chapter 2) was responsible for a major change in sedimentation patterns in the district. West of the Garth Fault, the late Ashgill regressive sequences of the Tywi and Rhiwnant anticlines were abruptly replaced by easterly supplied, mud-dominated, slope-apron facies. Successive systems of these facies, active throughout latest Hirnantian (*persculptus* Biozone) to early Telychian times, aggraded across the district to build a westward-thinning wedge of sediment. These systems comprise three interdigitating types of formation: mudstone formations (Cwmere Formation, Tycwtta Mudstones, and Claerwen Group); localised coarse-grained turbidite formations, disposed as a series of easterly sourced channel fills and lobes (Allt-y-clych Conglomerate, Caban Conglomerate Formation, and Ystrad Meurig Grits Formation); and early Telychian, easterly supplied, sandy turbidite facies (Devil's Bridge Formation).

MUDSTONE FORMATIONS

The mudstone formations (Plate 9) are predominantly constructed of mudstone turbidites and hemipelagites. Three facies are recognised, anoxic, oxic and mixed, based on the type and proportion of hemipelagite present (Figure 13). Anoxic and oxic facies comprise sequences in which the hemipelagites are respectively laminated or burrowed. Mixed facies comprise sequences

in which the two types of hemipelagite occur in roughly equal proportions. The Cwmere Formation and Tycwtta Mudstones, two laterally equivalent formations, overlie the Yr Allt Formation with sharp lithological contact. The Cwmere Formation crops out to the west of the Tywi Lineament and consists predominantly of anoxic facies; the Tycwtta Mudstones crop out to the east, between the Nant-y-fedw and Garth faults, and consist of mixed facies.

The Claerwen Group only crops out to the west of the Tywi Lineament, where it overlies the Cwmere Formation. It comprises the Derwenlas Formation and Rhayader Mudstones, both of which are predominantly constructed of oxic facies.

Cwmere Formation

The Cwmere Formation, first defined in the Aberystwyth and Machynlleth district (Jones and Pugh, 1916; Cave and Hains, 1986), predominantly comprises thinly interbedded turbidite mudstones and laminated hemipelagites (anoxic facies) (Plate 10a and b). Very thin turbidite sandstones and siltstones are commonly present, but rarely exceed 10 per cent in any section. Weathered exposures of the formation are characteristically rust-stained, a result of the oxidation of the pyrite in the hemipelagites. The formation includes the Ddol Shales of Lapworth (1900). The Mottled Mudstone Member, a unit up to 10 m thick in which the hemipelagites are pre-

Plate 9   Hirnantian to Aeronian succession of the Claerwen valley, looking north-westwards [from SN 8843 6275]. Thick-bedded sandstones of the Cerig Gwynion Grits (foreground) are succeeded by mudstones of the Cwmere Formation (middle distance) and Derwenlas Formation (feature at skyline). (A14710).

dominantly burrowed (oxic facies), is developed at the base of the formation.

In the Rhayader sheet area the formation crops out along the north-western limb of the Tywi Anticline west of the Glanalders Fault, and along the limbs of the adjacent Gwesyn Syncline and Rhiwnant Anticline. It interdigitates with the Caban Conglomerate Formation (Figures 12, 18 and 19).

In the Llanilar sheet area the formation mainly crops out in a series of anticlines in the axial zone of the Teifi Anticline between Swyddffynon and just east of Pont-rhyd-y-groes.

In the Rhayader sheet area, south-west of Cerrig Llwyd y Rhestr, on the north-western limb of the Rhiwnant Anticline, where there are no intercalations of Caban Conglomerate Formation, the Cwmere Formation is about 400 m thick. Where tongues of Caban Conglomerate Formation are present, the base and the top of the Cwmere Formation may be separated by up to 450 m of strata, though, as to the south of Carn Gafallt [SN 942 647], the bulk of the interval may be occupied by the intertonguing, coarse-grained formation. North-east of Carn Gafallt, where the upper tongues of the Caban Conglomerate Formation die out, up to 180 m of Cwmere Formation overlie the remaining basal tongue (Cerig Gwynion Grits and Dyffryn Flags).

In the Llanilar sheet area, only the upper part of the Cwmere Formation is exposed. There, up to 160 m of strata overlie facies of the Ystrad Meurig Grits Formation, the base of which is not seen. The Cwmere Formation thins to 70 m in the adjacent Aberystwyth district (Cave and Hains, 1986).

MOTTLED MUDSTONE MEMBER

This member, first named in the Aberystwyth and Machynlleth districts (Jones and Pugh, 1916; Cave and Hains, 1986) occurs at the base of the Cwmere Formation and comprises 3 to 10 m of thinly interbedded pale and medium grey burrowed hemipelagite with subordinate beds of medium grey turbidite mudstone. Thin turbidite sandstones and siltstones make up less than 5 per cent of the unit. In the basal 2 m, two thin beds of dark grey laminated hemipelagite, separated by burrowed hemipelagite, comprise the *persculptus* Band of Jones and Pugh (1916). The Band is 52 cm thick and contains *persculptus* Biozone graptolites. In the district, the Mottled Mudstone crops out only in the Rhayader sheet area, where it overlies the Yr Allt Formation and for much of its extent is overlain by Cerig Gwynion Grits facies of the Caban Conglomerate Formation. Channelling at the base of the Cerig Gwynion Grits has locally resulted in an attenuated Mottled Mudstone Member sequence, as on the north-east side of the Wye valley, in the higher levels of Cerrig Gwynion quarry [SN 9723 6564 to SN 9766 6585]. In many places the crop defines a marked hollow between these two more resistant units, such as on the summit and north-west slopes of Allt Goch [SN 940 630], 2 km south of Elan village.

The base of the member is marked by the abrupt appearance of well-bedded, pale, commonly burrow-mottled mudstones above the commonly disturbed, dark grey, silty mudstones of the Yr Allt Formation. A change in cleavage style, from widely spaced and irregular in the Yr Allt Formation, to closely spaced and regular in the Mottled Mudstone Member, commonly occurs at the contact.

The Mottled Mudstone Member mainly comprises pale and medium grey, colour-banded burrowed hemipelagites, some of which exhibit a streaky, diffuse lamination. Silt-laminated turbidite mudstones (type E) with laminae up to 5 mm thick are present, interbedded with hemipelagites in the lowermost and topmost parts of the member. Rare thin, turbidite sandstone/mudstone and siltstone/mudstone couplets (types Cii and Di), in which the fine-grained sandstones range up to 2 cm thick, are restricted to the uppermost part of the member; they are most common where the member is overlain by Dyffryn Flags. The laminated hemipelagites of the *persculptus* Band contain abundant pyritised graptolites and irregularly shaped pyrite nodules up to 1 cm in diameter.

STRATA ABOVE THE MOTTLED MUDSTONE MEMBER

The Cwmere Formation mudstones above the Mottled Mudstone Member only directly overlie it on the western limb of the Rhiwnant Anticline. Elsewhere in the

Plate 10 Lithologies of the Cwmere and Derwenlas formations (facing page).

a Cwmere Formation, interbedded pale-coloured turbidite mudstones (very thin type Dii) and darker laminated hemipelagites, Afon Arban [SN 8688 6329]. (A14676).
b Cwmere Formation, lithologies as a. Note minor burrow mottling in laminated hemipelagites. Cut block, Nant Paradwys [SN 8918 6088]. **c** Derwenlas Formation, colour-banded subfacies. Cut block, Claerwen valley [SN 8719 6343].
d Derwenlas Formation, bioturbated oxic facies (arrowheads point to silty bases of relict type E turbidites). Cut block, Claerwen dam [SN 8709 6362]. **e** Derwenlas Formation, silt-laminated and colour-banded oxic subfacies. Note dark phosphatic nodules (p) at or near base of paler oxidised layers. Fallen block, Hendre quarry [SN 7215 6965]. **f** Derwenlas Formation, colour-banded oxic subfacies displaying synsedimentary faulting and bedding disruption, Afon Arban [SN 8684 6330]. Scale bars in centimetres.

Plate 11 Lithologies of the Rhayader Mudstones (p.66).

a Strongly silt-laminated, pale grey-green oxic subfacies; road cutting, Rhayader [SN 9694 6835]. (A14716). **b** *M. segwickii* shales (anoxic facies), interbedded turbidite mudstones (very thin type Dii) and laminated hemipelagites. Cut block, Claerwen reservoir [SN 8707 646404]. **c** Anoxic facies. Note synsedimentary faulting and bedding convolution, and difficulty in distinguishing individual laminated type E turbidites from associated hemipelagic intervals. Claerwen reservoir [SN 8703 6445]. **d** Colour-banded oxic subfacies, Claerwen reservoir [SN 8690 6465]. **e** Packet of sandstone/mudstone couplets (type Cii turbidites) (sandstone-rich oxic subfacies), Claerwen reservoir [SN 8626 6503]. (A14690). **f** Interbedded sandstone/mudstone couplets (type Cii turbidites) and diffusely laminated and burrow-mottled hemipelagites (sandstone-rich oxic subfacies). Cut block, Pyllau Clais [SN 9670 7422]. Scale bars in centimetres.

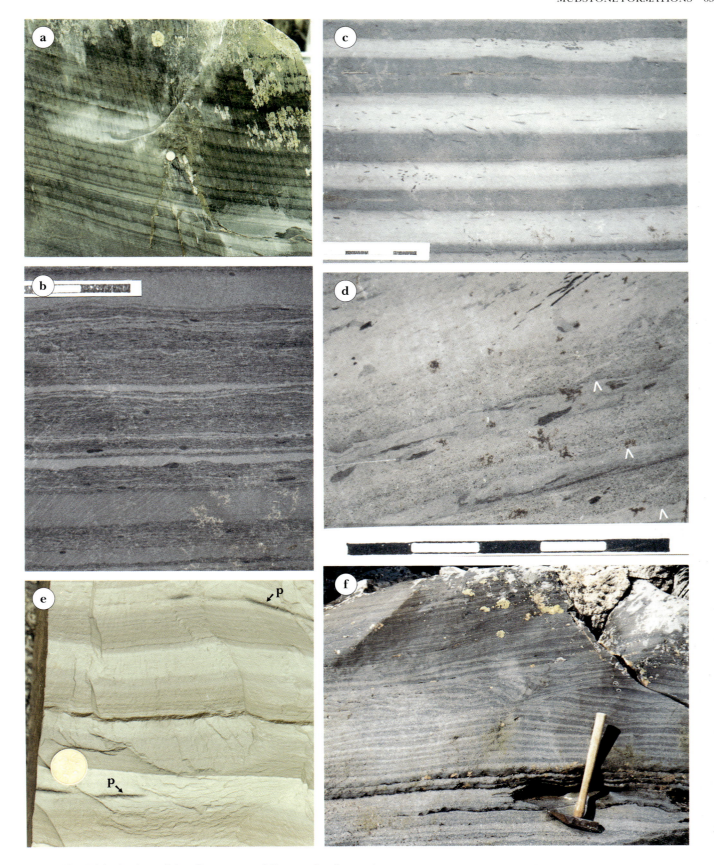

Plate 10   Lithologies of the Cwmere and Derwenlas formations.

Plate 11   Lithologies of the Rhayader Mudstones.

Plate 12 Caban Conglomerate Formation.

a Lower face of Cerrig Gwynion quarry [SN 9709 6563] exposing about 55 m of Cerig Gwynion Grits, overlying (bottom right) the Mottled Mudstone Member of the Cwmere Formation. Note lateral wedging of massive sandstone beds (type Bi turbidites), packets of thinner-bedded mudstones and sandstone/mudstone couplets (types C and D turbidites).

b Outcrop of the Caban Conglomerate Formation on the northern side of the Elan valley as viewed from Cnwch [SN 9270 6420]: 1 — second sequence of Caban Conglomerates facies ('Lower Conglomerate' of Lapworth); 2 — Dyffryn Flags ('Intermediate Shales' of Lapworth); 3 — third sequence of Caban Conglomerates facies ('Upper Conglomerate' of Lapworth); 4 — Sedgwickii Grits; 5 — Gafallt Shales. The prominent gully (top left) marks the position of the Foel Fault. (A14697).

a

b

Rhayader sheet area, facies of the Caban Conglomerate Formation intervene. In the Llanilar sheet area, Cwmere Formation mudstones overlie facies of the Ystrad Meurig Grits Formation.

The junction with the Mottled Mudstone Member is taken at an upward change from burrowed to predominantly laminated hemipelagites and the accompanying replacement of type E by very thin-bedded type D mudstone turbidites. Contacts with facies of the Caban Conglomerate and Ystrad Meurig Grits formations are typically gradational. They are drawn where very thin-bedded mudstone turbidites (type D) become predominant, and sandstones (parts of type C turbidites) that typify facies of

these coarser formations, are reduced to less than 10 per cent.

The Cwmere Formation above the Mottled Mudstone Member predominantly consists of thinly interbedded, grey turbidite mudstones (types Di and Dii) and dark grey, laminated hemipelagites with very subordinate turbidite sandstone/mudstone couplets (mainly type Cii but with rare Ci).

Individual mudstone turbidites (types Di and Dii) are generally up to 4 cm but locally up to 20 cm thick. Although the two types occur in variable proportions in outcrops, type Dii are generally more abundant. The basal siltstone or fine-grained sandstone laminae of the type Di

Plate 13   Lithologies of the Caban Conglomerate Formation.

Plate 13   Lithologies of the Caban Conglomerate Formation (facing page).

a   Caban Conglomerates facies (second sequence). Note channelised bases (dashed), amalgamation and lenticularity of component Ai turbidites. Caban-coch quarry [SN 9240 9460]. **b**   Close-up of conglomerate in a. **c**   Sedgwickii Grits, thin section of sandstone (shelly quartz arenite) (left side — ppl, right side — Xpl, scale bar 0.5 mm). Elan valley [SN 9285 6476]. (E65458).   **d**   Locality as a; note inverse, followed by normal, grading within conglomerate division. (A14709).   **e**   Sedgwickii Grits, medium- and thin-bedded sandstone/mudstone couplets (type Ci and Cii turbidites), Cnwch [SN 9270 6416]. (A14700).

Plate 14   Lithologies of the Ystrad Meurig Grits Formation (p.70).

a   Chwarel Goch facies, thin section of sandstone (fine-grained quartz arenite, Xpl, scale bar 0.5 mm). Forestry cutting [SN 7153 7196], Coed Craigyrogof. (E65464).   **b**   Ystrad Meurig Grits facies (upper part of first sequence). Thicker, commonly amalgamated sandstones include structureless type Bi and 'top-cut-out' type Ci turbidites; thinner-bedded sandstone/mudstone couplets (type Cii turbidites) locally intervene. Ceredigion Council quarry [SN 7011 6802].   **c**   locality as b, showing mudstone lined channel and composite nature of many of the thick sandstone beds.   **d**   Henblas facies associated with the first sequence of Ystrad Meurig Grits facies. Fallen block, Hendre quarry [SN 7215 6965].   **e**   locality as d, loaded flute-casts on base of turbidite sandstone, fallen block.   **f**   Ystrad Meurig Grits facies, locality as d. Thin section of quartz granule conglomerate (top — Xpl, bottom — ppl, scale bar 0.5 mm. (E65465).

Plate 15   Lithologies of the Devil's Bridge Formation, Blaen Myherin Mudstones Formation and Caerau Mudstones (p.71).

a   Devil's Bridge Formation, sandstone/mudstone couplets (type Cii turbidites). Quarry [SN 7223 6228] north of Tregaron.   **b**   Blaen Myherin Mudstones, mainly thin- to medium-bedded type Dii turbidites. Arrow points to parallel-laminated basal siltstone of a type Di turbidite. Nant Rhuddnant [SN 8010 7850].   **c**   Caerau Mudstones, colour-banded oxic facies, showing internal divisions of type E (T_{0-7}) and type D (T_{d-e}) turbidite mudstones and diffusely laminated hemipelagites (hp). Note that the bases of strongly oxidised layers (pale bands) and the lower limits of burrowing commonly lie within turbidites. Claerwen reservoir [SN 8574 6521].   **d**   Caerau Mudstones, mixed facies. Lower oxic portions comprise variably laminated and *Chondrites* burrow-mottled (b) type E turbidites intercalated with diffusely laminated, burrowed hemipelagites and associated early diagenetic phosphate nodules (ph); upper anoxic parts comprise structureless, thin-bedded type Dii turbidites interbedded with dark grey, well-laminated hemipelagites. Claerwen reservoir [SN 8657 6493].

Plate 16   Lithologies of the Aberystwyth Grits Group (p.72).

a   Mynydd Bach Formation, packet of commonly structureless, thick-bedded and locally amalgamated turbidite sandstones (type B) overlain by thinner-bedded turbidite sandstone/mudstone couplets (type C). Craig Fawr quarry [SN 5895 6220], Penuwch.   **b**   Mynydd Bach Formation, cleaved muddy sandstone (type Bii turbidite/debrite), Trefenter [SN 605 677].   **c**   Trefechan Formation, stacked sequence of Bouma turbidites, mainly sandstone/mudstone couplets (type C). Cliffs [SN 583 825] north of Aberystwyth.   **d** and **e** Trefechan Formation, turbidite sandstone/mudstone couplets (mainly type Cii), Allt-wen [SN 576 796].   **f**   Mynydd Bach Formation, thin section of Bii turbidite sandstone (lithic greywacke, Xpl, scale bar 0.5 mm). Roadside quarry [SN 6483 7169] near Gaer fawr. (E65467).   **g**   Trefechan Formation, thin section of turbidite sandstone (quartz arenite, Xpl, scale bar 0.5 mm). Track cutting [SN 5663 7517] near Blaenplwyf. (E65469).

Plate 17   Lithologies of the Cwmystwyth Grits Group (p.73).

a   Rhuddnant Grits (Llyn Teifi Member), stacked sequence of cleaved, thick-bedded, structureless muddy sandstones (type Bii turbidites). Nant Egnant [SN 777 660].   **b**   Rhuddnant Grits (Llyn Teifi Member), cleaved muddy sandstone (type Bii turbidite) resting on basal sandstone of a type Cii turbidite. Road [SN 783 684] near Llyn Teifi.   **c**   Rhuddnant Grits (Lyn Teifi Member), type Bii turbidite with folded rip-up clasts of laminated sandstone. Graig Ddu [SN 8147 7398].   **d**   Glanyrafon Formation, stacked sequence of thinly bedded turbidite sandstone/mudstone couplets (mainly type Cii). Afon Ystwyth [SN 8442 7565]. (A14728).   **e**   Pysgotwr Grits, packet of pebble and granule-rich type Bii turbidites. Crags [SN 8363 6886] near Llyn Cerrigllwydion Uchaf.

Plate 18   Lithologies of the Pysgotwr Grits, Penstrowed Grits, Nantglyn Mudstones and Cerig Formation (p.74).

a   Pysgotwr Grits, thin section of type Bii turbidite sandstone (lithic greywacke, top — ppl, bottom —Xpl, scale bar 0.5 mm). Road [SN 8508 7569], Cwm Ystwyth. (E65472).   **b**   Penstrowed Grits, medium- to thick-bedded type Bii turbidite sandstones intercalated with thin-bedded turbidite sandstone/mudstone couplets and mudstones (types C and D). Roadside quarry [SO 0364 7503] near Bwlch-y-sarnau.   **c**   As b, thinly interbedded turbidite mudstones (type Dii) and laminated hemipelagites.   **d**   Nantglyn Mudstones, thinly interbedded turbidite mudstones and laminated hemipelagites with thin slump-folded and destratified disturbed bed (centre). Road cutting [SO 0871 7884] near Llanbadarn Fynydd.   **e**   Cerig Formation, block of basal sandstone with coquinas of disarticulated *Pentamerus oblongus*. Commin-coch [SN 9580 5173], Garth.

Plate 14   Lithologies of the Ystrad Meurig Grits Formation.

Plate 15   Lithologies of the Devil's Bridge Formation, Blaen Myherin Mudstones Formation and Caerau Mudstones.

Plate 16   Lithologies of the Aberystwyth Grits Group.

73

Plate 17 Lithologies of the Cwmystwyth Grits Group.

Plate 18 Lithologies of the Pysgotwr Grits, Penstrowed Grits, Nantglyn Mudstones and Cerig Formation.

---

turbidites are up to 5 mm thick, but are commonly discontinuous and lenticular, occurring as a series of low-amplitude, starved ripples.

The proportion of hemipelagite in the formation varies from 30 to 70 per cent. Individual beds range up to 5 cm thick. Although predominantly of the laminated variety, rare burrowed hemipelagites are locally present. They are most common in the top few metres of the formation, beneath the Derwenlas Formation.

Turbidite sandstone/mudstone couplets (type C), occurring either scattered or as thin bundles, are largely restricted to regions of interdigitation with the Caban Conglomerate and Ystrad Meurig Grits formations. Individual sandstones, in type Ci and Cii turbidites, are up to 20 cm and 5 cm thick, respectively.

BIOSTRATIGRAPHY

Graptolites from the Rhayader sheet area indicate that the Cwmere Formation ranges in age from the *persculptus* Biozone to the *magnus* Biozone. The Mottled Mudstone Member is entirely of *persculptus* Biozone age. The distribution and approximate thicknesses of individual biozones are depicted in Figure 19. Faunal lists from localities (Figures 18a and 21) in both the Cwmere Formation and the intercalations of Caban Conglomerate Formation are given in Tables 9 and 12.

On the north-western limb of the Tywi Anticline, north-east of the Afon Elan, the mudstones overlying facies of the Caban Conglomerate Formation range from possibly as low as upper *acuminatus* Biozone through to the *magnus* Biozone. There, in the Wye valley (Figure 19; Table 9), *atavus* Biozone graptolites have been found 25 m above the base of these mudstones in the lowest part of the A470 road section [SN 9799 6738 to SN 9794 6749]. Graptolites from the remainder of the A470 section and from a stratigraphically higher section in the River Wye at Ddole Farm [SN 9776 6737 to SN 9762 6745], have demonstrated that the *acinaces*, *cyphus* and *triangulatus* biozones are 75 m, 47 m and 20 m thick respectively (Table 9). The top 14 m of the mudstones contain *magnus* Biozone graptolites.

In the Llanilar sheet area, *magnus* Biozone graptolites also occur in the uppermost part of the Cwmere Formation, whilst upper *acuminatus* Biozone faunas have been proved in the uppermost part of the underlying tongue of Ystrad Meurig Grits Formation (Chwarel Goch facies) (Table 13). Although graptolite faunas have been found in the remainder of the formation (Table 13) none of the intervening biozones has been definitely proved.

The only shelly taxon from the Cwmere Formation in the district is a probable nowakiid tentaculitoid, *Nowakia gwynensis*, from the Wye valley [SN 9799 6741] east of Rhayader. The locality is reassigned here to the *acinaces* Biozone rather than the *atavus* Biozone reported by Tunnicliff (1989).

The mudstones in Nant Paradwys [SN 8918 6088 to SN 8933 6002] have yielded sphaeromorph acritarchs and indeterminate acanthomorphs of no biostratigraphical value (Appendix 2bii). More diverse assemblages including *Dicyotidium dictyotum*, *Lophosphaeridium parverarum*, *Multiplicisphaeridium fisheri* and *Tylotopalla caelameni-*

cutis were obtained from *magnus* Biozone mudstones at Ddole Farm [SN 9761 6747] (Appendix 2biii). They are referable either to the *M. fisherii* or to the *O. eoplanktonica* Biozone. Recycled acritarchs within the mudstones from both localities include *Acanthodiacrodium* spp. and *Vulcanisphaera* sp. from the early Ordovician and *Peteinosphaeridium nanofurcatum*, *Stellechinatum* cf. *brachyscolum* and *Villosacapsula irrorata*? from the mid- to late Ordovician.

DETAILS

Mottled Mudstone Member

On the north-western limb of the Rhiwnant Anticline part of the Mottled Mudstone is exposed in Ffos y Rhest east of Llyn Carw (Figure 21b). Scattered sections in the member are common around the closure and on the south-eastern limb of the Rhiwnant Anticline. On Craig Cwm clyd [SN 8850 6250 to SN 8890 6248], on the north side of the Claerwen valley, the sharp basal contact with the Yr Allt Formation and the gradational contact with the overlying Dyffryn Flags are well exposed. Rare turbidite sandstone/mudstone couplets (types Di and Cii) with thin basal sandstones up to 2 cm thick are present in the top 1.5 m. A kilometre to the south-east on Craig y Bwlch [SN 8991 6205] (Figure 18a, locality 1), *Normalograptus* spp. were obtained 60 cm above the base of the member in an incomplete section (Table 12). On the south-eastern limb of the Rhiwnant Anticline, the member is about 10 m thick in Nant Paradwys [SN 8917 6097] (Figure 21c) and 6 m thick in Nant yr Ych [SN 8779 5993]. Both sections exhibit a gradational passage up into Dyffryn Flags.

Sections through the Mottled Mudstone Member on the north-western limb of the Tywi Anticline, where it is overlain by Cerig Gwynion Grits, include a stream section [SN 9533 6450] north of Cwm and Cerrig Gwynion quarry [SN 9710 6559] (Figure 20; Table 9; Plate 12a), both south-west of Rhayader. At the latter locality the following section is seen:

	Thickness m
CABAN CONGLOMERATE FORMATION, CERIG GWYNION GRITS FACIES	
Sandstone, medium-grained with sharp base	seen
CWMERE FORMATION, MOTTLED MUDSTONE MEMBER	
Hemipelagite, pale and medium grey, colour banded, diffusely laminated and burrowed. Scattered mudstone turbidites with siltstone laminae up to 5 mm thick (type E) in upper 1.3 m. Turbidite sandstone/mudstone couplet (type Di) with basal 2 cm sandstone defines base	2.0
Hemipelagite as above	1.9
Upper leaf of *persculptus* Band: hemipelagite, dark grey, laminated with abundant irregular pyrite nodules up to 1 cm in diameter. Abundant *persculptus* Biozone graptolites are preserved in pyrite (Table 9)	0.08
Hemipelagite, pale and medium grey, colour-banded, burrowed and diffusely laminated	0.27
Lower leaf of *persculptus* Band: hemipelagite as in upper leaf	0.17
Hemipelagite, pale and medium grey, colour-banded, diffusely burrowed and laminated. Scattered pyrite nodules in upper part. Abundant interbedded turbidite mudstones with siltstone laminae up to 1 mm thick (type E) in basal 35 cm.	

N. persculptus sensu stricto? near base indicative of *persculptus* Biozone. Sharp base   1.38

YR ALLT FORMATION
Mudstone, dark grey, silty, micaceous, massive and
  destratified   seen

The most north-easterly exposure of the member is in a gully [SO 0166 6983], on the eastern side of the Dulas valley.

Strata above the Mottled Mudstone Member

Rhayader sheet area

Rhiwnant Anticline   Deeply weathered mudstones, immediately beneath the Caban Conglomerates facies, are exposed at Llyn Carw [SN 8554 6120]. They contain an *acuminatus* Biozone fauna (Figure 21b, locality 12 and Table 12). In Ffos y Rhest [SN 8588 6165], 500 m to the north-east, an *acinaces* or *cyphus* Biozone fauna is present in the mudstones also lying beneath the Caban Conglomerates facies (Figure 21b, locality 22 and Table 12).

Crags [SN 8865 6271 to SN 8722 6341] on the northern side of the Claerwen valley on the north-western limb of the anticline, provide a semi-continuous section through the Cwmere Formation mudstones between the Caban Conglomerate Formation and the Derwenlas Formation. At the base, a gradational contact with the Dyffryn Flags is exposed in scattered crags [SN 8865 6271 to SN 8852 6295] east of Ciloerwynt. Scattered turbidite sandstone/mudstone couplets (type Cii), with basal sandstones up to 4 cm thick, occur in the lower 50 m of mudstones; in some packets the sandstone content is up to 15 per cent. An *acuminatus* Biozone fauna is present about 40 m above the Cerig Gwynion Grits in a small gully [SN 8703 6200] 1.5 km to the south-east (Figure 18a, locality 11 and Table 12). Between Ciloerwynt and the prominent scarp formed by the Derwenlas Formation on the north side of the Claerwen valley, turbidite mudstones and hemipelagites predominate. The junction with the Derwenlas Formation is best seen on the south side of the valley in Afon Arban, immediately downstream of the footbridge [SN 8686 6329]. There, the top few metres of the Cwmere Formation contain thin packets with dark grey, burrowed hemipelagites.

The upper part of the Cwmere Formation mudstones on the north-western limb of the Rhiwnant Anticline are intermittently exposed in Nant Methan [SN 9080 6469 to SN 9011 6533]. They mainly comprise very thin-bedded turbidite mudstones

Plate 19   Lithologies of the Cwm Barn and Henfryn formations (facing page).

a   Cwm Barn Formation, basal conglomerate of Aii turbidite. Note polymict clast assemblage, unaltered (grey) and phosphatised (black) mudstone clasts, and clast imbrication. Cut block, Cwm Barn quarry [SO 0082 7081] (scale bar in centimetres).   **b**   As a; note clasts of oxic-facies mudstone with burrow mottling and anoxic (laminated) hemipelagite (hp).   **c**   Thin section of conglomerate with abundant shell debris (ppl, scale bar 0.5 mm), locality as a. (E65461).
d   Lower portion of Aii turbidite, polymict cobble conglomerate with moulds of leached shelly benthos capped by parallel- and convolute-laminated calcareous sandstone. Locality as a.   **e**   Henfryn Formation, thin section of pebbly sandstone (quartz arenite; left side — ppl, right side — Xpl, scale bar 0.5 mm). Quarry [SO 0774 7029] near Keeper's Lodge. (E65462).

and hemipelagites in which type Di turbidites are rare. The distribution of the *cyphus*, *triangulatus* and *magnus* Biozones in the upper part of Nant Methan reported by Davies (1928, fig. 17) has not been confirmed, but his thickness of 30 m for the top two zones is in accord with that in the Wye valley (Table 9).

On the south-eastern limb of the Rhiwnant Anticline the transition from the Cerig Gwynion Grits into the overlying Cwmere Formation mudstones is well exposed in crags [SN 9021 6178] on the northern side of the Claerwen valley. The transition zone comprises 4 m of Dyffryn Flags. To the south-west, Nant Paradwys [SN 8918 6088 to SN 8933 6002] provides a semicontinuous, graptolitic section above the Cerig Gwynion Grits (Figure 21c; Table 12). At the base, a thin sequence of Dyffryn Flags passes up into very thinly interbedded mudstone turbidites (type Dii) and laminated hemipelagites; silt-based, mudstone turbidites (type Di) and burrow-mottled hemipelagites are rare. Long-ranging normalograptids, including *N. normalis*, typify the lower part of the sequence, though a fauna of the *acuminatus* Biozone has been recorded [SN 8919 6072] (Figure 21c, locality 9) about 155 m above the base. The *atavus* Biozone or younger is suggested by a fauna [SN 8916 6031] (Figure 21c, locality 21), 70 m below a tongue of Caban Conglomerates, while graptolites from immediately beneath this tongue [SN 8917 6023] (Figure 21c, locality 23), suggest either the *acinaces* or *cyphus* Biozone. Despite an allowance for strike faulting, a minimum thickness of 395 m has been estimated for the mudstone sequence below the tongue of Caban Conglomerates facies. Above the conglomerate, 10 m of Dyffryn Flags pass up gradationally into 90 m of Cwmere Formation mudstones, intermittently exposed up to the junction with the Derwenlas Formation. Graptolites from 20 m above the top of the Dyffryn Flags [SN 8925 6011] (Figure 21c, locality 25), are indicative of either the upper *cyphus* or lower *triangulatus* Biozone. Graptolites from 20 m higher in the section [SN 8926 6009] (Figure 21c, locality 26), suggest the *triangulatus* Biozone.

The gradational junction with the overlying Dyffryn Flags, immediately west of the Foel Fault, is exposed in a forestry track [SN 9192 6263 to SN 9196 6310] (Figure 18a, locality 13) at Graig Fawr. Near the contact, type Ci turbidites with basal sandstones up to 20 cm thick appear. Although the mudstones have yielded only long-ranging taxa (Table 12) an *acuminatus* Biozone age is likely, for the upper part of this biozone occurs in the overlying Dyffryn Flags. A comparative section can be traced between the road section [SN 9110 6400 to SN 9150 6383] and the crags [SN 9152 6387 to SN 9160 6402] on the slopes above the western side of the Caban-coch reservoir.

East of the Foel Fault, scattered thin packets of Dyffryn Flags facies with fine- to coarse-grained sandstones up to 10 cm thick and forming up to 15 per cent of the lithology, are present in the upper part of the Cwmere Formation mudstones between the tongues of Caban Conglomerate Formation. They are well displayed on the eastern shore of the Caban-coch reservoir [SN 9215 6362 to SN 9223 6391]. At the northern end of the section, graptolites from such a packet are indicative of the *acuminatus* Biozone (Figure 21d, locality 10 and Table 12).

Western limb of the Tywi Anticline   South-west of Carn Gafallt, the lower part of the Cwmere Formation mudstones and the gradational contact with the underlying Dyffryn Flags are exposed in Marchnant [SN 9058 6055 to SN 9056 6110] and Nant Rhyd-goch [SN 9153 6170 to SN 9140 6179].

North-east of Carn Gafallt, Cwmere Formation mudstones overlying the basal tongue of the Caban Conglomerate Formation are exposed in the Wye valley (Table 9). The lower part of the sequence is seen on the eastern side of the A470

Table 9 Distribution of graptolites in the latest Ordovician to Aeronian succession of the Wye valley.

Sections: see map on facing page

A localities in Cerrig Gwynion quarry [SN 9710 6559, SN 9718 6564, SN 9732 6579, SN 9729 6575, SN 9722 6577 and SN 9723 6597] in ascending stratigraphic order; **B** Prysg stream [SN 9825 6695, SN 9814 6711 and SN 9806 6719] in ascending stratigraphic order; **C** A470 road cutting [SN 9799 6738 to SN 9794 6749]; **D** west bank of River Wye [SN 9776 6737 to SN 9763 6744]; **E** excavation [about SN 9758 6757] in River Wye recorded by H Lapworth (1900); **F** reef [SN 9744 6770] in River Wye; **G** Rhyd-hir brook [SN 9746 6781].

For key to range chart symbols see Appendix 3; horizontal dashed lines indicate that position of base of biozone is uncertain due to the absence of a continuous section. ∗ — species recorded by H Lapworth (1900) but not verified during survey. ∗1 — graptolites from the H Lapworth collection examined in the course of this survey.

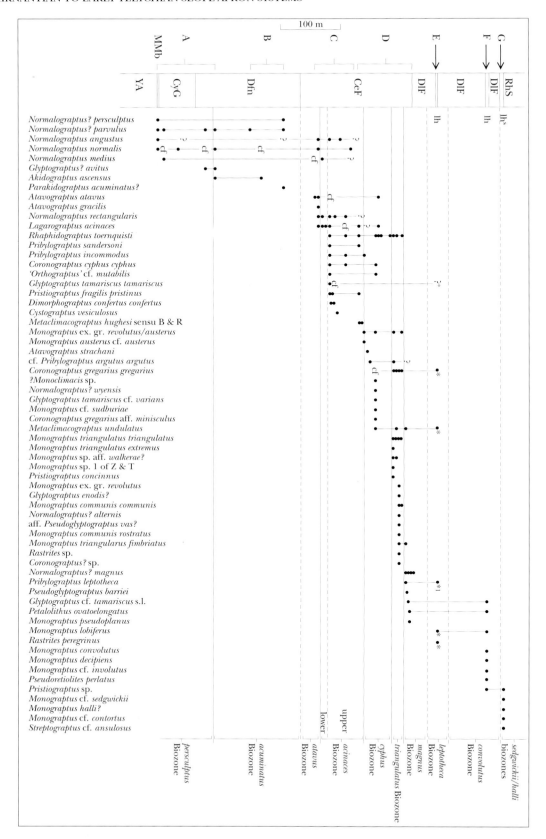

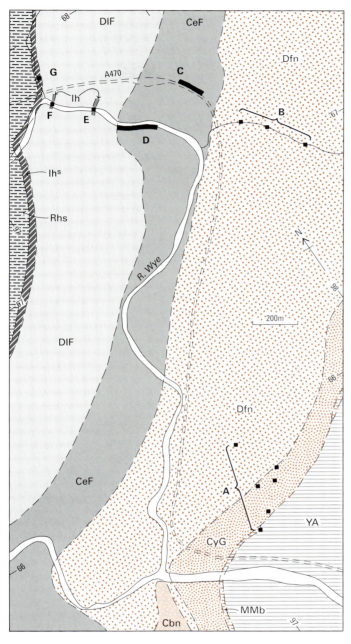

road [SN 9799 6738 to SN 9794 6749], but is separated from the underlying Dyffryn Flags by an exposure gap of about 25 m. The turbidite mudstones and hemipelagites contain scattered turbidite sandstone/mudstone couplets (type Cii), with basal sandstones up to 4 cm thick, that locally form up to 10 per cent of the lithology. Graptolites from the section demonstrate that the lower 7 m belongs to the *atavus* Biozone and the overlying 48 m to the *acinaces* Biozone. The road section is the type locality for the *atavus* Biozone. An exposure gap of about 15 m separates the section from the upper part of the mudstone sequence exposed in the west bank of the River Wye [SN 9776 6737 to SN 9762 6745] at Ddole Farm (Table 9, locality D). There, 90 m of mudstones spanning the *acinaces* to *magnus* biozones occur beneath the Derwenlas Formation. The locality is the type section for the *cyphus* Biozone.

A forestry track section [SO 0295 7079], west of Lower Cwmhir, in turbidite mudstones and hemipelagites provides one of the most north-easterly sections. Beyond this, the formation is absent due to Telychian overstep. Together with long-ranging acritarch taxa the section has yielded the chitinozoa *Belonechitina aspera* and *B. postrobusta*, both known to range from the *persculptus* to the *acinaces* Biozone.

Llanilar sheet area

Teifi Anticline   The Tynygraig railway cutting [SN 6859 7015 to SN 6861 7012] exposes the topmost part of the Cwmere Formation thrust over the *M. sedgwickii* shales at the base of the Rhayader Mudstones (Figure 22). The mudstones lack type Di turbidites and contain *magnus* and possibly *triangulatus* Biozone graptolites (Table 10).

In the Hendre Anticline, Afon Ystwyth [SN 7287 7198 to SN 7327 7208] exposes a nearly complete section through the local mudstone sequence from its gradational contact with the Chwarel Goch facies (Ystrad Meurig Grits Formation) to the Derwenlas Formation. Graptolites from the topmost metre of the mudstones from the road section [SN 7324 7203] above the river are indicative of the *magnus* Biozone (Figure 22, locality 6 and Table 13).

In a tributary of Afon Ystwyth [SN 7613 7292 to SN 7613 7274] (Figure 22, localities 3 to 5) in Coed Bwlchgwallter, east of Pont-rhyd-y-groes, graptolites from about 60 m [SN 7616 7287] (Table 13, locality 4) and 30 m [SN 7612 7275] (Table 13, locality 3) below the top of the Cwmere Formation belong either to the *acinaces* or *cyphus* Biozone. A triangulate monograptid from the top of the formation [SN 7614 7273] in this section (Table 13, locality 5) suggests either the *triangulatus* or the *magnus* Biozone.

Tycwtta Mudstones

The Tycwtta Mudstones comprise very thinly interbedded turbidite mudstones and both laminated and burrowed hemipelagites (mixed facies). The formation takes its name from the type section on the banks and bed of the River Wye [SO 0094 6062 to SO 0123 6000], south-east of Tycwtta farm. Thin beds and laminae of turbidite siltstone and sandstone occur scattered throughout, but do not exceed 10 per cent of the sequence. Up to half of the succession is locally slumped or destratified. A thin unit of interbedded conglomerates, debrites and sandstones, the Allt-y-clych Conglomerate Member, is locally developed in the middle part of the formation as exposed.

Previous authors (Andrew, 1925; Davies, 1980; Lockley, 1978; 1983; James, 1983; 1991) have included

Table 10   Distribution of graptolites in the Aeronian succession of the Tynygraig railway cutting [SN 6866 7007 to SN 6842 7047].

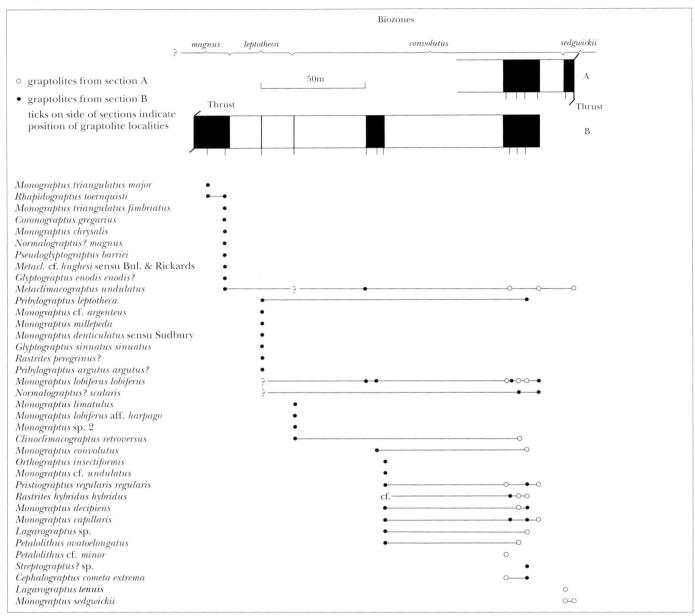

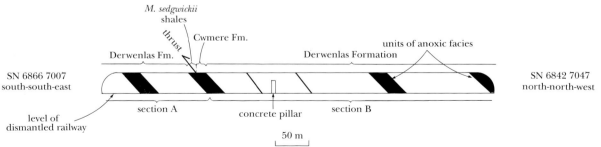

Sketch of south-western side of Tynygraig railway cutting. Section is oblique to dip.

the formation with the late Ashgill regressive mudstones, the Yr Allt Formation of this account. Between the Wye valley and the Chwerfri valley, south of the district, the Tycwtta Mudstones comprise the upper part of Andrew's (1925) 'Bala rocks', and the uppermost part of Lockley's (1978; 1983) Chwerfri Mudstone Formation.

The formation crops out on the eastern limb of the Tywi Anticline between Llanfihangel-helygen and the southern edge of the district. It is bounded to the south-east by the Garth Fault, which cuts out the highest part of the formation. About 740 m of the formation are exposed in the district.

Between the southern edge of the district and the Wye valley, the Tycwtta Mudstones rest on Yr Allt Formation mudstones but between the Wye valley and Werngronllwyd, the formation overlies the Dol-y-fan Conglomerate Member of the latter formation. In both cases the base of the Tycwtta Mudstones is taken at the appearance of pale and medium grey, thinly interbedded turbidite mudstones and burrowed hemipelagites. The contact is gradational.

The anoxic facies comprises very thinly interbedded, medium grey, turbidite mudstones (types Di and Dii) and dark grey laminated hemipelagites. The latter are up to a centimetre thick and form up to 50 per cent of the facies. The oxic facies is mainly colour banded on a milli-metre to centimetre scale, reflecting incomplete very early diagenetic oxidation (see p.83). It comprises very thinly interbedded, medium grey, variably silt-laminated turbidite mudstones (type E) and pale grey, burrowed hemipelagites. Throughout the formation, the two facies occur in alternating packets up to tens of metres thick. In the type section, burrow-mottled oxic facies forms the basal 4 m and compares with the Mottled Mudstone Member of the Cwmere Formation.

Turbidite sandstone/mudstone couplets (mainly type Cii but with rare type Ci) occur throughout the Tycwtta Mudstones. The basal sandstones of these turbidites are up to 6 cm and 20 cm thick respectively. They are most prominently developed in the gradational sequences overlying the Dol-y-fan and Allt-y-clych conglomerates.

Mudstones exhibiting slump folding and destratifica-tion occur throughout the succession, in packets up to tens of metres thick. These disturbed beds vary from homogeneous mudstones with randomly orientated dia-genetic nodules and balls and wisps of sandstone to units in which as much as 50 per cent of the bedding fabric is preserved but grossly disturbed by ramifying listric surfaces. All the disturbed levels exhibit an irregular curved cleavage.

The Allt-y-clych Conglomerate, first named by Andrew (1925) in the Chwerfri valley south of the district, extends north-eastwards to the Wye valley, where its out-crop is terminated by a splay of the Garth Fault. It is up to 20 m thick, and lies about 400 m above the base of the formation. The base is sharp and commonly erosional but is locally deformed by large-scale loading; its contact with the overlying mudstones is gradational. Marked vari-ations in thickness reflect channelling at the base. It comprises debrites and turbidite pebble-cobble conglom-erates, pebbly sandstones and sandstones. The debrites

form units several metres thick. They comprise rounded pebbles and cobbles, scattered or close-packed in a silty mudstone or very muddy fine-grained sandstone matrix. The conglomerates and pebbly sandstones comprise type Ai turbidites, commonly amalgamated. Massive turbidite sandstones (type Bi) are limited to the uppermost part of the member. Clasts in the debrites and conglomerates are mainly of vein quartz and sandstone with scattered igneous rocks. Derived brachiopods and crinoid ossicles occur locally.

BIOSTRATIGRAPHY

The Tycwtta Mudstones range in age from the late Ash-gill (Hirnantian) to the Aeronian. Graptolite faunas are considerably more sparse in the formation than in the coeval Cwmere Formation. Graptolites of probable *per-sculptus* Biozone age have been obtained about 60 m above the base of the formation in the Wye valley; the biozone has been unequivocally proved at a comparable level near Brochen [SN 9933 5851], south of the district. A fauna from about 400 m above the base of the formation in the River Wye, has a range within the *atavus* to *cyphus* biozones. Graptolites from about 100 m below and about 70 m above the Allt-y-clych Conglomerate, in quarries just south of the district at Allt-y-clych [SN 9796 5724] and Ty Mawr [SN 9859 5749] respectively, sug-gest that the conglomerate is probably of either *acinaces* or more likely of *cyphus* Biozone age.

Derived Rhuddanian brachiopods have been recorded from the Allt-y-clych Conglomerate at Pen y Rhiw [SN 994 582], just south of the district, by Lockley (1978; 1983) and Temple (1987).

Six samples collected at regular intervals from the type section in the River Wye, between the base of the forma-tion and just above the Allt-y-clych Conglomerate have yielded only impoverished floras of long-ranging Ordovi-cian to Silurian aspect. Taxa present include *Diexallopha-sis denticulata*, *D. sanpetrensis*, *Michrystridium* spp., *Multi-plicisphaeridium* spp. and *Veryhachium trispinosum*. How-ever, a sample from the river bank [SO 0117 6004], about 80 m above the base of the Allt-y-clych Conglom-erate has yielded the acritarchs *Ammonidium listeri*, *Oppi-latala eoplanktonica* and *Salopidium* cf. *granuliferum*. The ranges of these taxa in the district fall within the *A. microcladum* to *G. encantador* biozones interval. However, *A. listeri* and *S. granuliferum* have been recognised beneath levels with *convolutus* Biozone graptolites in the Church Stretton district (H Barron, personal communi-cation, 1991), and appear therefore to range down well into the Aeronian. The presence of *O. eoplanktonica* sug-gests that the flora is no older than the earliest Aeronian (*triangulatus* Biozone). This assessment is supported by the existence of late Rhuddanian graptolites, recovered from 70 m above the Allt-y-clych Conglomerate (see above).

DETAILS

The type section in the banks and bed of the River Wye [SO 0094 6062 to SO 0123 6000] is cut by a splay of the Garth Fault. About 400 m of mudstones are exposed between the base of the formation [SO 0094 6062] and the fault. At the base,

4 m of colour-banded and burrow-mottled, thinly interbedded turbidite mudstones (type E) and hemipelagites, gradationally overlie Yr Allt Formation mudstones on the west bank and an isolated lens of Dol-y-fan Conglomerate, protruding from the river bed, on the east bank. Throughout the remainder of the section, similar oxic-facies mudstones alternate in roughly equal proportions with sequences of anoxic-facies mudstones composed of very thin-bedded turbidite mudstones (type D) and laminated hemipelagites. Both facies occur in packets tens of metres thick. Abundant thick units of destratified and slump folded mudstones are also present. The irregular curving cleavage surfaces, characteristic of the destratified mudstones, are well displayed in the rock platform that forms the river bed. A 2 m-thick lens of medium-grained, micaceous, structureless, high-matrix sandstone crops out in variably destratified mudstones in the west bank [SO 0092 6063]. There is no evidence to suggest whether it is a large clast or a turbidite sandstone (type Bii) that has pillowed in situ. In the east bank [SO 0095 6065], 60 m above the base of the formation, *Normalograptus? persculptus* and *N. angustus?* indicate a probable *persculptus* Biozone age. Immediately north of the fault, about 400 m above the base of the formation on the east bank [SO 0106 6027], cf. *Atavograptus atavus* and *Normalograptus* spp. suggest an age within the *atavus* to *cyphus* Biozone interval. The Allt-y-clych Conglomerate is exposed in the River Wye immediately south of the splay of the Garth Fault. It is 8 m thick with a sharp base. At the base of the overlying mudstones, a 4 m-thick sequence with abundant turbidite sandstone/mudstone couplets (type Cii) includes sandstones up to 5 cm thick. To the south, exposure in the mudstones is limited to submerged reefs and isolated sections in the river banks.

The gradational junction of the Tycwtta Mudstones and Dol-y-fan Conglomerate is exposed in track sections [SO 0297 6243 and SO 0312 0254] north of Fronhir. There, turbidite sandstone/mudstone couplets (types Ci and Cii) form the basal part of the Tycwtta Mudstones and decrease in abundance upwards.

CLAERWEN GROUP

The Claerwen Group correlates with the Derwen Group of Jones and Pugh (1916). Cave and Hains (1986) introduced the name Derwenlas Formation for the lower part of the group in the Aberystwyth district. Together with the overlying Rhayader Mudstones, the Derwenlas Formation forms a coherent lithological group of predominantly oxic facies, which is well exposed on the shore of the Claerwen reservoir [SN 873 634 to SN 650 858] in the Rhayader sheet area. To avoid the confusion that would arise from the resurrection of the Derwen Group, the new name Claerwen Group is proposed for the two formations.

Derwenlas Formation

The Derwenlas Formation broadly correlates with the lower part of the Castell Group of Jones (1909), the lower part of the Derwen Group of Jones and Pugh (1916) and the Gigrin Mudstones of Lapworth (1900) (Table 8). The formation is largely composed of mudstones of oxic facies, which comprise very thinly interbedded, variably silt-laminated, turbidite mudstones and burrowed hemipelagites (Plate 10c–f). A few thin packets of anoxic facies, composed of thinly interbedded turbidite mudstones and laminated hemipelagites, punctuate the formation. The proportion of hemipelagite observed in the formation varies between 30 and 40 per cent.

In the Rhayader sheet area, the Derwenlas Formation has two principal north-east- to south-west-trending outcrops, separated by the intertonguing Caban Conglomerate Formation (Figure 19). The northern one, situated east of Rhayader, lies along the north-western flank of the Tywi Lineament and is limited to the north by the Cwmysgawen Fault. The southern one occurs on the western limb of the Rhiwnant Anticline, south-west of the Carreg-ddu reservoir. A minor outcrop of the formation occurs in the core of the Gwesyn Syncline, south-west of Llanerch Cawr. The main crop of the formation in the Llanilar sheet area forms much of the core of the Teifi Anticline, south of the Ystwyth Fault, where apart from a small area north-west of Tynygraig, it interdigitates with tongues of the Ystrad Meurig Grits Formation (Figure 23). Two small periclinal inliers occur north of the Ystwyth Fault, north-east of Devil's Bridge.

Throughout the district, the formation normally succeeds the Cwmere Formation. Its lower contact is taken at the base of a thin transitional sequence, less than 5 m thick, in which the dark grey laminated hemipelagites of the underlying formation are replaced by pale, burrow-mottled ones. As a consequence of this change, the Derwenlas Formation is more resistant to weathering and typically forms a prominent, commonly craggy feature (Plate 9), above the gentler topography of the Cwmere Formation. This feature is commonly obscured or lost completely where the formation passes into the contemporaneous Caban Conglomerate or Ystrad Meurig Grits formations.

The Derwenlas Formation in the Rhayader sheet area, outside the zone of intertonguing with the Caban Conglomerate Formation, is about 165 m thick. In the northern crop, the formation thins rapidly southwards towards Carn Gafallt, as the Caban Conglomerate Formation progressively replaces it. The Carreg-ddu reservoir provides a convenient divide between the southern crop of the Derwenlas Formation and the Gafallt Shales of the Caban Conglomerate Formation, though in detail the junction is a lateral transition, which involves much small-scale intertonguing of the two units.

In the Llanilar sheet area the thickness of the Derwenlas Formation, uninterrupted by tongues of Ystrad Meurig Grits Formation, is estimated to range from 170 to 190 m. These thicknesses, which greatly exceed the 35 to 100 m reported from the Aberystwyth district (Cave and Hains, 1986), reflect proximity to the depositional site of the Ystrad Meurig Grits Formation. Where the Derwenlas Formation contains tongues of Ystrad Meurig Grits Formation, the combined thickness can be up to about 400 m, as in the area east of Pont-rhyd-y-groes (Figure 23).

Away from the zones of interface with the Caban Conglomerate and Ystrad Meurig Grits formations, the oxic facies of the Derwenlas Formation is composed almost exclusively of turbidite mudstones (type E) thinly interbedded with pale coloured, burrowed hemipelagites.

The type E turbidite mudstones are principally base-cut-out units (T_{4-7}) with diffuse and irregular siltstone wisps and laminae in their lower parts. However, scattered units with well-defined, silt-laminated (T_3) divisions are common. Where clearly defined, type E turbidites and intercalated burrowed hemipelagites rarely exceed 3 cm and 2 cm in thickness respectively.

A silt-rich subfacies (Plate 10e) of the oxic facies containing many top-cut-out (T_{0-3}), strongly silt-laminated, type E turbidite mudstones, predominates over wide areas peripheral to the crops of the Caban Conglomerate and Ystrad Meurig Grits formations. It occurs throughout the formation in the Teifi Anticline and dominates the upper half of the formation in the Rhayader sheet area. Near the contacts, scattered thin turbidite sandstone/mudstone couplets (type Cii) are also present within the silt-rich subfacies.

Striking colour variations in the oxic-facies mudstones are a reflection of the differing degrees of penecontemporaneous oxidation. Two gradational, diagenetic subfacies, one colour banded and the other pale grey-green, are recognised. In the colour-banded subfacies (Plate 10c), medium grey or green bands alternate with paler, burrow-mottled bands on a 1 to 2 cm scale. Depending on the degree of colour contrast, the banding is either strongly or weakly defined. Though the banding crudely reflects the alternation of turbidite mudstone and hemipelagites, in detail the bases of the paler bands commonly do not coincide with the bases of hemipelagites, but fall within turbidite mudstones. Decimetre-thick units of pale grey, pervasively burrow-mottled units locally occur within the subfacies and record the merging of oxidation fronts affecting several consecutive turbidite/hemipelagite couplets (Plate 10d). The pale grey-green subfacies is unbanded, and is considered to represent the most advanced stage of this oxidation continuum, where all trace of banding and dark burrow mottling has been lost. Both subfacies occur in sequences up to tens of metres thick.

The banded subfacies typically occurs adjacent to units of anoxic-facies mudstones and commonly passes upwards and/or downwards into the pale grey-green subfacies. This diagenetic zonation appears to mirror the changes that would be expected in bottom-water chemistry prior to and/or following periods of anoxic deposition.

The anoxic facies comprises very thinly interbedded medium grey turbidite mudstones (type D) and dark grey laminated hemipelagites. The former rarely exceed 3 cm and the latter 2 cm. Units of such facies form the graptolite bands of earlier workers (Lapworth, 1900; Jones, 1909; Jones and Pugh, 1916; Cave and Hains, 1986). Six such units are recognised in the Rheidol Gorge, close to the southern margin of the Aberystwyth district (Jones, 1909; Cave and Hains, 1986). In the Llanilar sheet area, four have been proved in a semi-continuous section in the Tynygraig railway cutting (Table 10), where, in upward order, they range up to 1.3 m, 2 m, 10 m and 20 m in thickness. A fifth, identified as palaeontologically distinct from the others, is known from an isolated section [SN 7174 71520]. In the

Rhayader sheet area only three have been proved. The distribution of the units in the Derwenlas Formation in the district is shown in Tables 9 and 10 and Figures 16, 22 and 23.

The Derwenlas Formation displays abundant small-scale, soft-sediment deformation phenomena. These include synsedimentary faults, with either normal or low-angle thrust displacements, convolute bedding affecting several turbidite mudstone/hemipelagite couplets, and bedding discordances (Plate 10f). Such features collectively record the effects of post-depositional, intrastratal slippage and accommodation (creep).

BIOSTRATIGRAPHY

As the topmost part of the Cwmere Formation belongs to the *magnus* Biozone and the lowest unit of anoxic facies in the Derwenlas Formation contains *leptotheca* Biozone graptolites (Figure 22 and Tables 9, 10 and 13), the base of the formation in the district, by convention, lies within the *magnus* Biozone. This contrasts with the Aberystwyth district, where *triangulatus* Biozone faunas have been recorded (Cave and Hains, 1986) from the lowest part of the Derwenlas Formation. Graptolite assemblages from the remaining four anoxic-facies units in the district are indicative of the *convolutus* Biozone (Figure 22 and Tables 9, 10 and 13). In the Teifi Anticline a possible early *convolutus* Biozone fauna from an anoxic-facies unit near Ysbyty Ystwyth [SN 7174 7152] (Table 13, locality 10), probably immediately postdates or predates the lowest *convolutus* Biozone anoxic-facies unit in the Tynygraig railway cutting [SN 6855 7022] (Table 10). The topmost anoxic-facies unit in the Llanilar sheet area locally contains *Cephalograptus cometa extrema*, which indicates the upper part of the biozone.

The Derwenlas Formation has yielded the oldest, diverse Llandovery acritarch assemblages recorded from the district (Appendix 2bi, ii and iii). Referable to the *O. eoplanktonica* Biozone, they commonly include *Diexallophasis sanpetrensis*, *Multiplicisphaeridium fisheri*, *Oppilatala singularis* and *Tylotopalla robustispinosa*. Also present are *Lophosphaeridium* sp. A, *Multiplicisphaeridium neahgae*, *M. paraguaferum*, *Tunisphaeridium parvum* and *Oppilatala eoplanktonica*. In the upper part of the formation *Ammonidium* sp. B, *Domasia limaciformis*, *D. symmetrica* and *D. trispinosa* first appear in the *convolutus* Biozone. The richest assemblages are recorded from the River Wye section at Ddole Farm. Recycled acritarchs from the formation include *Acanthodiacrodium* spp. and *Vulcanisphaera pila* from the early Ordovician, and *Stellechinatum* cf. *.brachyscolum* from the mid- to late Ordovician.

DETAILS

Rhayader sheet area

Northern crop   The most north-easterly sections of the Derwenlas Formation are in Rhyd-hir Brook [SO 0003 6942 and SN 9989 6924 to SN 9967 6919], north-west of Berthabley, where colour-banded, oxic-facies mudstones form the basal part of the formation. Scattered sections in similar mudstones occur in the lower half of the formation around Cefnceido [SN 9840 6825], and include both a road [SN 9849 6832] and stream section [SN 9825 6818], in which a thin unit of anoxic-facies mudstones has yielded poorly preserved biserial grapto-lites. Colour-banded oxic-facies mudstones succeed the Cwmere Formation in the River Wye [SN 9762 6746 to SN 9760 6756] at Ddole Farm. The upper part of the formation com-prising grey-green subfacies mudstones is intermittently exposed upriver. Lapworth (1900) obtained *leptotheca* Biozone graptolites from a thin unit of anoxic-facies mudstones in a temporary section [SN 9762 6746] in the east bank of the river (Table 9). A higher unit of anoxic-facies mudstones exposed in reefs [SN 9744 6769], near the confluence with Rhyd-hir Brook, lies near the top of the formation and has yielded *convo-lutus* Biozone graptolites (Table 9).

A disused quarry [SN 9712 6724], west of Ddole Farm, exposes 3 m of silt-rich subfacies. A 6 m-thick sequence in a quarry [SN 9641 6640], near Glan-Elan, comprises an upward transition from pale grey-green into colour-banded subfacies mudstones that underlie a unit of anoxic-facies mudstones. Extensive exposures along Nant y Feddail [SN 9535 6524 to SN 9534 6541], south-west of Dolifor Farm, are predominantly in pale grey-green subfacies.

Southern crop   Nant Methan [SN 9012 6534 to SN 8993 6554], east of the Carreg Ddu reservoir, exposes an incomplete section through the Derwenlas Formation. Although largely in oxic facies, units of anoxic-facies mudstones are developed at intervals. One such unit, exposed upstream of the footbridge [SN 8999 6549] and in an adjacent tributary [SN 9004 6547], was assigned to the *sedgwickii* Biozone by Davies (1928). However, it contains *Monograptus lobiferus* and *Pristiograptus* cf. *regularis* sensu lato suggesting, instead, the *convolutus* Biozone. The 'leptotheca band' cited by Davies (1928) from this vicinity was not located.

The road [SN 8720 6341 to SN 8705 6366] leading to the top of the Claerwen dam exposes the lower half of the for-mation, which is entirely in oxic-facies mudstones (Figure 16). The gradational base is poorly exposed, but the succeeding colour-banded subfacies, passing upwards into pale grey-green subfacies, is well displayed south of the public toilets. Abun-dant silt laminae up to 5 mm thick (silt-rich subfacies), appear in the grey-green subfacies just north-west of the toilets and define a coarsening-upwards sequence. They persist through-out the remainder of the formation's oxic-facies mudstones, intermittently exposed north of the dam, along the reservoir shore [SN 8705 6366 to SN 8706 6402]. Two units of anoxic-facies mudstones, both giving rise to marked linear depres-sions on the moorland to the east, occur in the section. The lower unit [SN 8705 6380] has yielded a solitary *Petalolithus ovatoelongatus* (Table 11, locality 1), whilst the upper one [SN 8704 6387] contains a *convolutus* Biozone fauna (Table 11, locality 2).

South of the Claerwen dam, the base of the formation is well exposed in Afon Arban [SN 8686 6329], downstream of the footbridge. The gradational junction with the Cwmere For-mation occurs over an interval of about 5 m and is reflected not only by the entry and rapid increase in the proportion of burrowed hemipelagites, but in a gradual increase in the degree of bioturbation within successive hemipelagic units. The lower part of the Derwenlas Formation is in colour-banded sub-facies mudstones, and is well displayed upstream of the foot-

Figure 16 Graphic log of the Claerwen Group in the Rhayader sheet area, based principally on the Claerwen reservoir section.

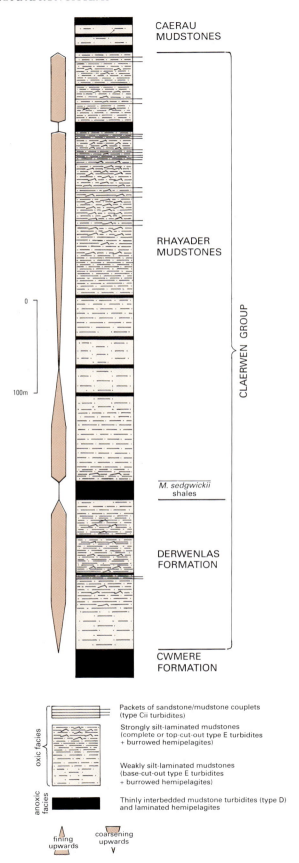

bridge. The effects of early post-depositional intrastratal disturbance (small-scale faulting, sliding and convolution) are well developed. Upstream of the Llyn Carw Fault [SN 8671 6325], the upper part of the Derwenlas Formation is in pale grey-green, silt-rich subfacies mudstones. Two units of anoxic-facies mudstones are present and the transitions from pale grey-green to colour-banded subfacies are well displayed on either side of them.

Llanilar sheet area

Sections in the Derwenlas Formation are confined to the Teifi Anticline. Figure 22 illustrates the distribution of the formation in the Teifi Anticline and its intimate relationship with the Ystrad Meurig Grits Formation. The figure also shows the main graptolite localities in the formation in this area, whilst Table 13 gives the faunal lists. The gradational base of the formation is well seen on Allt Dihanog in Nant Bwlchgwallter [SN 7614 7273], in roadside crags [SN 7324 7220] west of Pont-rhyd-y-groes and in the nearby gorge of Afon Ystwyth [SN 7322 7208].

The most extensive sections through the formation are the disued railway cuttings west of Tynygraig. Silt-rich, oxic-facies mudstones prevail and are well seen in the two southern cuttings [SN 6913 6950 to SN 6899 6967 and SN 6889 6981 to SN 6881 6992]. The northern series of cuttings [SN 6867 7003 to SN 6847 7044] exposes a complete section through the Derwenlas Formation but the sequence is disrupted by thrust faulting (Table 10). At the south-eastern end the uppermost part of the Cwmere Formation (*triangulatus* Biozone) is thrust over the Derwenlas Formation and basal part of the Rhayader Mudstones. Below the thrust, two units of anoxic-facies mudstones have yielded, successively, graptolites of the *convolutus* and *sedgwickii* biozones; those of the latter were recovered from blocks within the fault zone. Above the thrust, west of the Cwmere Formation, the cuttings expose a near-continuous section through the lowest 150 m of the 180 m-thick Derwenlas Formation. The dominantly silt-rich, oxic-facies mudstones contain four units of anoxic-facies mudstones, the lowest referable to the *leptotheca* Biozone and the upper three to the *convolutus* Biozone (Table 10). In contrast to other sections within the northern part of the Teifi Anticline (Figure 23), the cutting contains no facies of the Ystrad Meurig Grits Formation. The section therefore provides an important constraint on the geographical distribution of that coarse-grained turbidite system.

North of the Ystwyth Fault the upper part of the formation was encountered in Cwm Rheidol No. 1 borehole [SN 7302 7833], beneath the fourth sequence of Ystrad Meurig Grits facies. A 12 m-thick unit of anoxic-facies mudstones in the formation yielded upper *convolutus* Biozone graptolites (Appendix 1; locality 28 on Table 13 and Figure 23).

Rhayader Mudstones

The Rhayader Mudstones, the Rhayader Pale Shales of Lapworth (1900), are well exposed in the district. Lapworth's name takes stratigraphic precedence over the equivalent Cwmsymlog Formation of Cave and Hains (1986) and the Oldchapel Mudstones of Jones (1945). The formation is lithologically similar to the Derwenlas Formation, largely comprising an oxic facies of very thinly interbedded, variably silt-laminated, turbidite mudstones and burrowed hemipelagites (Plates 8c and 11). A few thin units of anoxic facies consisting of very thinly interbedded turbidite mudstones and laminated hemipelagites are also present. The base of the formation is marked by a widely recognised unit of anoxic-facies mudstones, the *M. sedgwickii* shales of Jones and Pugh (1916).

The Rhayader Mudstones have extensive outcrops in the Rhayader sheet area where they give rise to well-exposed, craggy terrain. They form a broad belt occupying the core of the Tylwch Anticline between the Wenallt, in the north, and the Elan valley in the south. A contiguous outcrop trends south-westwards through the Claerwen reservoir to the southern edge of the district and lies along the western limb of the Rhiwnant Anticline.

The Rhayader Mudstones are less well exposed in the Llanilar sheet area, where they crop out principally along the limbs of the Teifi Anticline. Outcrops along the eastern limb of the anticline form a linear belt extending southwards from Coed Bwlchgwallter [SN 755 725] via Ffair-Rhos to the east of Pontrhydfendigaid. To the south, the Rhos Gelli-gron Anticline and other mesoscale folds account for the complex local outcrop patterns displayed by the Rhayader Mudstones to the north-east of Tregaron. West of the Tregaron–Pontrhydfendigaid road, the formation passes beneath the thick drift of Cors Caron. On the western limb of the Teifi Anticline the formation occupies a north–south trending crop terminated by the Ystwyth Fault system at Pen-y-bont. To the north, additional outcrops occur in faulted anticlines around Llanafan and, to the east, at Grogwynion Mine. The formation also crops out in narrow periclinal inliers in the vicinity of Devil's Bridge.

Except in the vicinity of Elan village, where the formation succeeds and passes laterally into the upper part of the Caban Conglomerate Formation (Figure 19), and a narrow tract east of Pont-rhyd-y-groes, where it overlies the Hafod Member of the Ystrad Meurig Grits Formation, the base of the Rhayader Mudstones everywhere rests on the Derwenlas Formation and is defined by the base of the *M. sedgwickii* shales. Ranging up to 30 m thick in the Rhayader sheet area and 15 to 20 m in the Llanilar sheet area, the *M. sedgwickii* shales give rise to a prominent slack in drift-free areas. The *M. sedgwickii* shales have been mapped in two areas within the crop of the Caban Conglomerate Formation (Figures 18, 19 and 21d). In the intervening ground, laterally equivalent sequences of turbidite mudstones with sparse very thin laminated hemipelagites are included in facies of the latter formation (pp.106–107). East of the Central Wales Syncline the formation is succeeded by the Caerau Mudstones, but to the west by the Devil's Bridge Formation. In the Rhayader sheet area the formation is generally about 450 m thick, but is less than 300 m thick where the basal parts are replaced by facies of the Caban Conglomerate Formation. In the Llanilar sheet area, the formation ranges in thickness up to a maximum of 200 m. Across the north-eastern part of this area it exhibits marked attenuation and is locally reduced to less than 20 m thick. This belt of attenuation is comparable and contiguous with that in the adjacent Aberystwyth district (Cave and Hains, 1986) and reflects the lateral passage of the Rhayader Mudstones into the Devil's Bridge Formation (Figures 12 and 25).

Table 11 Distribution of graptolites in the early Telychian sequence of the Claerwen reservoir section.

Biozones:	convolutus (C)		sedgwickii/halli (S)				turriculatus (T)							crispus (C)							
Ages of localities	C	C	S?	S	S-Tren		Tru-u	Tu	Tu?	Tu?	Tu-p	Tp?	Tc	Cg-c							
Locality No:	1	2	3	4	5	6	7	8	9	10	11	12	13	14	15	16	17	18	19	20	21
Pe. ovatoelongatus	•	.	.	.	.	.	.	.	.	.	.	.	.	.	.	.	.	.	.	.	.
M. lobiferus s.l.	.	•	.	.	.	.	.	.	.	.	.	.	.	.	.	.	.	.	.	.	.
P. regularis	.	cf.	.	cf.	.	.	.	.	.	.	.	.	.	.	.	.	.	.	.	.	.
N? scalaris	.	cf.	.	.	.	.	.	.	.	.	.	.	.	.	.	.	.	.	.	.	.
Ra. hybridus	.	•	.	.	.	.	.	.	.	.	.	.	.	.	.	.	.	.	.	.	.
Lagarograptus sp.	.	•	.	.	.	.	.	.	.	.	.	.	.	.	.	.	.	.	.	.	.
M. sp.	.	•	.	.	.	.	.	.	.	.	.	.	.	.	.	.	.	.	.	.	.
Me. undulatus	.	.	cf.	cf.	.	.	.	.	.	.	.	.	.	.	.	.	.	.	.	.	.
M. sedgwickii	.	.	.	•	.	.	.	.	.	.	.	.	.	.	.	.	.	.	.	.	.
M. contortus	.	.	.	.	cf.	.	.	.	.	.	.	.	.	.	.	.	.	.	.	.	.
M. sedgwickii/halli	.	.	.	.	.	cf.	.	.	??	.	.	.	.	.	.	.	.	.	.	.	.
M. bjerreskovae	.	.	.	.	.	.	?	.	.	.	.	.	.	.	.	.	.	.	.	.	.
Str. pseudoruncinatus	.	.	.	.	.	.	cf.	cf.	.	.	.	.	.	.	.	.	.	.	.	.	.
M. marri	.	.	.	.	.	.	•	.	??	•	.	cf.	.	cf.	.	.	??	?	•	?	•
P. bjerringus	.	.	.	.	.	.	.	•	.	.	.	.	.	.	•	.	.	.	.	.	.
P. nudus s.l.	.	.	.	.	.	.	.	.	??	.	??	.	.	.	.	.	.	.	.	.	.
M. rickardsi	.	.	.	.	.	.	.	.	.	.	cf.	cf.	.	.	.	.	.	cf.	•	.	cf.
M. proteus	.	.	.	.	.	.	.	.	.	.	cf.	•	.	.	.	.	.	.	.	.	.
Str. whitei	.	.	.	.	.	.	.	.	.	.	.	.	•	.	.	.	.	.	.	.	.
Mcl? galaensis	.	.	.	.	.	.	.	.	.	.	.	•	.	.	.	.	.	.	.	.	.
Pe. tenuis s.l.	.	.	.	.	.	.	.	.	.	.	.	.	cf.	.	.	.	.	.	.	.	.
M. crispus	.	.	.	.	.	.	.	.	.	.	.	.	.	cf.	cf.	•	.	•	•	.	•
M. exiguus	.	.	.	.	.	.	.	.	.	.	.	.	.	cf.	cf.	•	cf.	.	.	.	.
M. aff. *pergracilis*	.	.	.	.	.	.	.	.	.	.	.	.	.	.	.	.	.	.	.	•	.
M. discus	.	.	.	.	.	.	.	.	.	.	.	.	.	.	.	.	.	.	.	.	cf.

Key to subzonal symbols:

turriculatus Biozone (T): ru = *runcinatus*, ren = *renaudi*, u = *utilis*, p = *proteus*, c = *carnicus* subzones.
crispus Biozone (C): g = *galaensis*, c = *crispus* subzones.

Localities: see map on facing page

1 shore exposure [SN 8705 6380]; 2 shore exposure [SN 8704 6390]; 3 shore exposure [SN 8707 6404]; 4 quarry [SN 8669 6350]; 5 shore exposure [SN 8703 6445]; 6 trackside exposure [SN 8687 6468]; 7 trackside exposure [SN 8689 6470]; 8 trackside exposure [SN 8675 6475]; 9 trackside exposure [SN 8614 6502]; 10 trackside exposure [SN 8628 6535]; 11 shore exposure [SN 8570 6526]; 12 shore exposure [SN 8563 6533]; 13 shore exposure [SN 8508 6545]; 14 shore exposure [SN 8450 6548]; 15 shore exposure [SN 8481 6547]; 16 shore exposure [SN 8385 6550]; 17 shore exposure [SN 8355 6576]; 18 shore exposure [SN 8340 6596]; 19 stream section [SN 8220 6502]; 20 stream section [SN 8204 6513]; 21 stream section [SN 8263 6708]. For key to range chart symbols see Appendix 3.

Away from the contacts with the Caban Conglomerate, Ystrad Meurig Grits and Devil's Bridge formations the oxic facies of the Rhayader Mudstones is similar to that of the Derwenlas Formation, predominantly comprising thinly interbedded, variably silt-laminated turbidite mudstones (type E) and burrowed hemipelagites. Type D turbidite mudstones and type Cii turbidite sandstone/mudstone couplets are also present locally. Although the two diagenetic subfacies present in the Derwenlas Formation also occur in the Rhayader Mudstones, the pervasively oxidised, pale grey-green subfacies predominates (Plate 11a). North of the Wye valley in the Rhayader sheet area the latter subfacies is pale grey but otherwise similar. Strongly colour-banded subfacies (Plate 11d) mudstones are largely restricted to levels adjacent to units of anoxic facies. A further distinctive diagenetic subfacies only occurs in the River Wye [SN 9700 6751], near Rhayader. There, up to 7 m of purple, silt-laminated mudstones pass down, with interbanding and spotting, into pale grey-green mudstones.

Throughout much of the formation, the type E turbidite mudstones comprise base-cut-out units (T_{4-7}) with diffuse siltstone laminae and streaks in their lower parts (Plate 11d). In the pale grey-green subfacies it is difficult to distinguish such units from hemipelagites, due to the masking effects of the penecontemporaneous oxidation.

In the Rhayader sheet area a silt-rich subfacies (Plates 8c and 11a), containing abundant strongly silt-laminated, top-cut-out type E turbidite mudstones (T_{0-3}), is locally present in the oxic facies. Developed in the

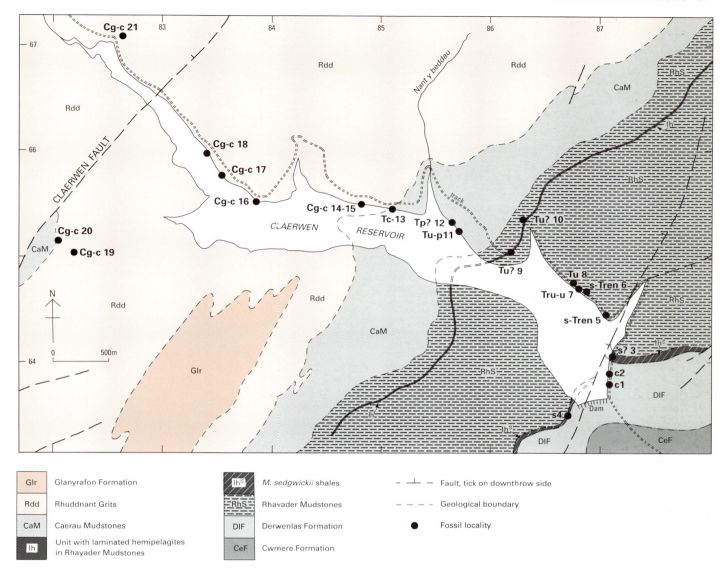

Glr	Glanyrafon Formation
Rdd	Rhuddnant Grits
CaM	Caerau Mudstones
Ih	Unit with laminated hemipelagites in Rhayader Mudstones

Ih^S	*M. sedgwickii* shales
RhS	Rhavader Mudstones
	Derwenlas Formation
DIF	
CeF	Cwmere Formation

– ⊥ – Fault, tick on downthrow side

– – – Geological boundary

● Fossil locality

lowermost part and in the upper half of the formation, it forms part of fining-upwards and coarsening-upwards sequences respectively (Figure 16). The higher silt-rich beds account for much of the formation's craggy out-crop north of the Elan valley, where they commonly pass into a sandstone-rich subfacies containing packets of turbidite sandstone/mudstone couplets (type Cii) (Plates 11e and f). This silt- and sand-rich level may be the lateral equivalent of the Devil's Bridge Formation. Gradational contacts with the Devil's Bridge Formation in the Llanilar sheet area are also characterised by the incoming of the silt-rich subfacies, together with scattered and thin packets of turbidite sandstone/mudstone couplets (type C). Similar type C turbidites also occur in the Rhayader sheet area, in the lateral pass-age into the Gafallt Shales.

Although thinner and less common, units of anoxic facies in the Rhayader Mudstones are similar to those in the Derwenlas Formation, comprising very thinly inter-bedded turbidite mudstones (type D) and laminated hemipelagites (Plates 11b and c). Apart from the *M. sedg-wickii* shales, thick mappable units of anoxic facies are confined to the upper third of the formation (Figure 16). Thinner units, generally less than 2 m thick, occur widely scattered throughout.

BIOSTRATIGRAPHY

The graptolite assemblages of the *M. sedgwickii* shales confirm that the base of the Rhayader Mudstones throughout the district falls within the *sedgwickii* Biozone (Tables 9–13). Assemblages from near the top of the *M. sedgwickii* shales show them to extend up into the *halli* Biozone [e.g. SN 6815 6570]. The rest of the Rhayader Mudstones range from the *halli* to the *turriculatus* Bio-zone but the top of the formation is markedly dia-chronous. In the Rhayader sheet area, graptolites from anoxic-facies units near the top of the formation on the eastern limb of the Tylwch Anticline [e.g. SN 9727 7151] are of the *proteus* Subzone and of the earlier *johnsonae* Subzone on the western limb [e.g. SN 9528 7621]. Of the

Table 12   Distribution of graptolites in the Caban Conglomerate Formation and coeval mudstone formations. See Figures 18 and 21 for locality maps and Figure 19 for stratigraphical position of localities.

Biozone groupings by locality:

Localities	Biozone
1–8	persculptus
9–17	acuminatus
18	upper acuminatus (atavus-cyphus)
19–20	atavus
21	atavus-cyphus
22–24	acinaces-cyphus
25	cyphus
26–28	triangulatus
29–30	convolutus
31	sedgwickii/halli
32–33	turriculatus (gemmatus Subzone)
34	turriculatus

Distribution chart (localities 1–34):

Taxon	1	2	3	4	5	6	7	8	9	10	11	12	13	14	15	16	17	18	19	20	21	22	23	24	25	26	27	28	29	30	31	32	33	34
N? parvulus	cf.	•		?	•	?	•	cf.	•																									
N? persculptus		?																																
N. normalis		?	?	•	•	•	cf.	cf.	•		•		?		•	?	cf.		•	•	•													
N. angustus							cf.				•		?		•	cf.			cf.	cf.														
N. medius								??				cf.																						
Ak. ascensus										•	•	•		•																				
'D.' modestus s.l.											?							?																
Pa. acuminatus												cf.	•	•																				
'Orthograptus' cabanensis															•	•																		
'D.' aff. modestus primus																aff.																		
Rh. extenuatus																		•																
At. atavus																			cf.	??		cf.	•											
La. acinaces																																		
Rh. toernquisti																						•	•	•	•									
Me. hughesi sensu B & R																						cf.	?											
N. rectangularis																							cf.											
Pribylograptus sp.																							?											
At. strachani																									cf.									
Me. undulatus																									•								•	
G. tamariscus s.l.																									?									
M. t. triangulatus																										cf.	cf.	cf.						
M. convolutus																													•	•				
M. lobiferus																													?	•				
N? scalaris																													?					
M. denticulatus s.s																														cf.				
Cl. retroversus																														cf.				
M. sedgwickii																															•	cf.	cf.	
Pseudoplegmatogr. sp.																															?			
M. contortus																															?			
P. variabilis																																	•	
P. renaudi																																	?aff.	•
M. turriculatus s.l.																																		•
M. gemmatus																																		cf.
Pe. kurcki																																		•
M. sp. 3																																		•

Formation / Member / Facies (by locality):

Locality	Formation / Member / Facies
1	Mottled Mudstone Member
2	Cerig Gwynion Grits
3–8	Dyffryn Flags
9–12	Cwmere Formation
13–17	Dyffryn Flags
18–20	Cwmere Formation
21	Caban Conglomerate (second sequence)
22–25	Cwmere Formation
26–28	Dyffryn Flags
29	Gafallt Shales
30	Sedgwickii Grits
31	M. sedgwickii Shales
32–33	Gafallt Shales
34	Rhayader Mudstones

Localities

1 crag [SN 8991 6205]; 2 stream section [SN 9063 6013]; 3 quarry [SN 9076 6031]; 4 crag [SN 8864 6260]; 5 roadside quarry [SN 8855 6268]; 6 crag [SN 8865 6268]; 7 crag [SN 9662 6546]; 8 quarry [SO 0003 6839]; 9 stream section [SN 8919 6072]; 10 shore exposure [SN 9223 6391]; 11 gully [SN 8703 6200]; 12 crag [SN 8554 6120]; 13 trackside exposure [SN 9198 6317]; 14 trackside exposure [SN 9195 6297]; 15 trackside exposure [SN 9194 6296]; 16 quarry [SO 0169 7019]; 17 stream section [SN 924 637]; 18 quarry [SN 9240 6442]; 19 crag [SN 8310 5774]; 20 crag [SN 9000 5640]; 21 stream section [SN 8916 6031]; 22 stream section [SN 8588 6165]; 23 stream section [SN 8917 6023]; 24 roadside quarry [SN 9244 6462]; 25 stream section [SN 8925 6011]; 26 stream section [SN 8926 6009]; 27 crag [SN 9256 6424]; 28 crag [SN 9277 6447]; 29 crag [SN 9151 6600]; 30 crag [SN 940 647]; 31 crag [SN 9404 6466]; 32 crag [SN 9222 6772]; 33 crag [SN 9243 6488]; 34 stream section [SN 9379 6607]. For key to range chart symbols see Appendix 3.

intervening subzones, only *utilis* Subzone faunas are abundant; the *gemmatus* Subzone is only known from one locality [SN 9380 6609] (Table 12) and the *runcinatus* and *renaudi* subzones have not been proved (Tables 11 and 12).

In the Llanilar sheet area, graptolites from anoxic facies units near the top of the Rhayader Mudstones, south of the Ystwyth Fault, are indicative of the *gemmatus* Subzone at Gwenhafdre [SN 6744 6705] and of the *runcinatus* to *renaudi* subzonal interval at Bwlchyddwyallt [SN 7109 6309]. Although no graptolites have been collected from the thin development of the formation in the north-east of the Llanilar sheet area, sequences of similar thickness just to the north, in the Aberystwyth district, are no younger than the *halli* Biozone (Loydell, 1991).

The *M. sedgwickii* shales have yielded moderately diverse acritarch assemblages (Appendix 2bii and iii) including *Ammonidium* sp. A, *A. listeri*, *Cymatiosphaera octoplana*, *Diexallophasis sanpetrensis*, *Domasia limaciformis*, *D. trispinosa*, *Elektoriskos williereae*, *Lophosphaeridium* sp. A, *L. parverarum*, *Multiplicisphaeridium neahgae*, *M. raspum*, *Oppilatala eoplanktonica* and *Tylotopalla*; this assemblage indicates a late Aeronian to Telychian age.

The remainder of the Rhayader Mudstones have yielded diverse acritarch assemblages (Appendix 2bi, ii and iii), except in the lower part where only rare sphaeromorphs are present. Common species include *Ammonidium listeri*, *A. microcladum*, *A.* sp. A, *Cymatiosphaera* sp. A and *Visbysphaera microspinosa*. Other taxa present include *Domasia* spp., *Elektoriskos williereae*, *Multiplicisphaeridium martiniae* and *Salopidium granuliferum*. These floras suggest a latest Aeronian to late Telychian age and are referable to the *A. microcladum/D. estillis* to *G. encantador* biozones. The most abundant and diverse assemblage (MPA 29285) is recorded from the the lower part of the *turriculatus* Biozone, probably the *gemmatus* Subzone, in Nant yr Haidd, where at least 44 acritarch taxa are present at [SN 9379 6607].

A small number of recycled acritarchs were recovered from the Rhayader Mudstones, including *Cristallinium cambriense* from the Cambrian, *Stelliferidium simplex* from the early Ordovician, and *Stellechinatum* cf. *brachyscolum* from the mid- to late Ordovician.

DETAILS

Rhayader sheet area

M. sedgwickii shales   In Rhayader, the *M. sedgwickii* shales are exposed in Rhyd-hir Brook [SN 9746 6781] and have yielded a graptolite fauna indicative of either the *sedgwickii* or *halli* Biozone (Table 9, locality G). To the south-west, anoxic-facies mudstones containing only indeterminate graptolites are overlain by colour-banded, oxic-facies mudstones in a disused railway cutting [SN 9625 6655] north of Glan Elan.

Impersistent *M. sedgwickii* shales are developed in the Caban Conglomerate Formation near Elan village. Immediately west of the Foel Fault they occur within the crop of the Gafallt Shales (Figure 21d, locality 32; Table 12). On the northern slopes of Carn Gafallt they are assumed to occupy a marked slack above the fourth sequence of Caban Conglomerates; exposures in the overlying Sedgwickii Grits yield graptolites from the *sedgwickii–halli* biozonal interval (p.106).

West of the Carreg-ddu reservoir, the *M. sedgwickii* shales are exposed in a series of crags [SN 9063 6571 to SN 9060 6558] south-west of Tynllidiart. A fossiliferous section through the upper part of the *M. sedgwickii* shales and their gradational junction with the overlying oxic facies mudstones is exposed in Nant Methan [SN 8982 6559 to SN 8987 6558]. The presence of *Monograptus sedgwickii*, *M.* cf. *contortus* and *Metaclimacograptus undulatus* indicates either the *sedgwickii* or the *halli* Biozone. Farther south-west, the *M. sedgwickii* shales are well exposed along the eastern shore of the Claerwen reservoir [SN 8707 6403 to SN 8710 6412], where much small-scale folding and faulting account for a wide crop (Table 11). They are also exposed on the west side of the reservoir, in a quarry [SN 8669 6350], from where abundant *Monograptus sedgwickii* indicate the *sedgwickii* Biozone (Table 11, locality 4). Traceable as a prominent slack south-west of the reservoir, the southernmost sections in this unit are crags along Nant Garregfelen [SN 8619 6290].

Strata above the M. sedgwickii shales   In Afon Irfon [SN 8290 5916 to SN 8312 6010] 200 m of Rhayader Mudstones are well exposed overlying the *M.sedgwickii* shales, which crop out just to the south of the district. Silt-rich, colour-banded and burrow-mottled subfacies form the lowest part of the sequence. They pass up into silt-depleted, pale grey-green subfacies mudstones in the uppermost part. Graptolites from thin units of anoxic facies, scattered throughout the sequence, confirm the presence of the *turriculatus* Biozone, but cannot be assigned to individual subzones.

The crags and track section along the northern shore of the Claerwen reservoir provide the most complete section through the Rhayader Mudstones in the district (Figure 16 and Table 11). Above the *M. sedgwickii* shales, silt-laminated, oxic-facies mudstones, colour banded and burrow mottled in the lower part, succeed to the north as far as the first inlet [SN 8746 6482], which follows the line of the Llyn Carw Fault. West of the fault, the section provides an unbroken sequence through the downfaulted middle and upper parts of the formation. The middle of the formation [SN 8710 6440 to SN 8669 6477] comprises silt-impoverished, green, oxic-facies mudstones in which turbidite mudstones (type E) and hemipelagites are commonly indistinguishable. In this sequence there are at least three thin (less than 2 m) rusty-weathering units of anoxic-facies mudstones. Graptolites from the lowest unit (Table 11, locality 6) are no younger than the *renaudi* Subzone, whilst those from the uppermost unit (Table 11, locality 8) are of the *utilis* Subzone. To the north-west [beyond SN 8669 6477], strongly silt-laminated, pale grey-green, oxic-facies mudstones gradually appear. On the western side of the second inlet [SN 8629 6535 to SN 8614 6502], packets up to 3 m thick of turbidite sandstone/mudstone couplets (type Cii), with sandstones to 3 cm thick, record the acme of this coarsening-upwards sequence. They are succeeded by a slack-forming unit of anoxic facies, about 10 m thick, exposed in crags above [SN 8628 6520] and below [SN 8605 6499] the track. The unit has only yielded long-ranging Telychian graptolites, possibly of *utilis* Subzone age (Table 11, localities 9 and 10). The uppermost part of the formation is exposed on the reservoir shore [SN 8598 6496 to SN 8588 6503] and comprises silt-laminated, strongly colour-banded and burrow-mottled oxic-facies mudstones.

The gradational junction with the underlying Gafallt Shales of the Caban Conglomerate Formation is well exposed in the craggy ground north of the Elan valley. There, the base of the Rhayader Mudstones is diachronous, younging westwards. The junction is taken at the appearance of a sequence of turbidite mudstones (type E), in which the turbidite sandstone/mudstone couplets (type Cii) that characterise the Gafallt

Biozones

convolutus spans columns; sub-headings: lower | undivided | upper* | upper**

Column biozone labels (top, italic): *acuminatus*, *acinaces-cyphus*, *triangulatus/magnus*, *magnus*, *leptotheca*, *convolutus* (lower, undivided, upper*, upper**), *sedgwickii/halli*, *turriculatus* Biozone *gemmatus* Subzone

Graptolite localities

1 2 3 4 5 6 7 8 9 10 11 12 13 14 15 16 17 18 19 20 21 22 23 24 25 26 27 28 29 30 31 32 33 34 35 36 37

Species (rows):

- N? parvulus
- Ak. ascensus
- D. modestus
- N. normalis
- N. medius
- Pa. acuminatus
- Rh. toernquisti
- Di. confertus confertus
- La. acinaces
- Me. hughesi sensu Bulman & Rickards
- triangulate monograptid
- M. triangulatus fimbriatus
- M. triangulatus triangulatus
- M. chrysalis
- At. strachani
- Ra. longispinus
- N.? magnus
- Me. undulatus
- Pgl. barriei
- M. argenteus
- M. denticulatus sensu Sudbury, 1958
- Co. gregarius gregarius
- M. involutus
- 'G.' sinulatus sinuatus
- M. lobiferus lobiferus
- M. imago
- Ra. peregrinus
- Me. hughesi s.s.
- "O." insectiformis
- M. millepeda
- P. jaculum
- N? scalaris
- M. convolutus
- M. clingani
- M. limatulus
- M. denticulatus s.s.
- M. urceolinus
- M. lobiferus harpago
- Ra. spina sensu Rickards, 1970
- "O." bellulus
- Cl. retroversus
- M. decipiens
- P. regularis regularis
- Pgl. sp. 1 Rickards
- G. incertus
- G. tamariscus angulatus
- Pe. ovatoelongatus
- M. capillaris
- Ra. hybridus hybridus
- G. tamariscus tamariscus
- M. undulatus
- M. sp.
- Ce. cometa extrema
- Mcl. crenularis
- Pe. minor
- Pr. leptotheca
- M. sedgwickii
- La. tenuis
- P. concinnus
- P. regularis latus
- M. contortus
- M. urceolus
- P. variabilis
- Str. ansulosus
- M. capillaris of Loydell 1993
- P. sp. 2 Loydell 1993
- Pe. kurcki
- P. nudus s.l.
- M. turriculatus s.l.
- M. gemmatus
- M. halli
- Str. pseudoruncinatus
- Str. sp. of Loydell 1993
- P. renaudi
- G. fastigatus
- Ra. fugax
- Ra. maximus
- Pe. altissimus s.l.
- N? nebula
- Str. filiformis
- G. elegans

Formations / Members / Facies

- CHWAREL GOCH FACIES
- CWMERE FORMATION
- YSTRAD MEURIG GRITS FACIES (first sequence)
- DERWENLAS FORMATION
- YSTRAD MEURIG GRITS FACIES (second sequence)
- DERWENLAS FORMATION
- YSTRAD MEURIG GRITS FACIES (third sequence)
- DERWENLAS FORMATION
- M. sedgwickii SHALES
- YSTRAD MEURIG GRITS FACIES (fourth sequence)
- DEVIL'S BRIDGE FMTN.
- RHAYADER MUDSTONES

Table 13 Distribution of graptolite taxa in the Teifi Anticline (see Figures 22 and 47 for maps of localities and Figure 23 for stratigraphical position of localities) (facing page).

Localities

1 quarry [SN 6906 6655]; 2 roadside exposure [SN 7272 7197]; 3 stream section [SN 7612 7275]; 4 stream section [SN 7616 7287]; 5 stream section [SN 7614 7273]; 6 roadside exposure [SN 7324 7203]; 7 Hendre quarry, first/second levels [SN 7216 6952]; 8 Hendre quarry, north face of topmost lift [SN 7200 6987]; 9 stream section [SN 7553 7228]; 10 crag [SN 7174 7152]; 11 road section [SN 7098 7190]; 12 crag [SN 757 723]; 13 crag near river bank [SN 7581 7293]; 14 roadside exposure [SN 6987 7172]; 15 trackside exposure [SN 6960 7118]; 16 crag [SN 7342 7096]; 17 mine working [SN 7355 6999]; 18 track section [SN 7104 6735]; 19 railway cutting [SN 7094 6729]; 20 stream section [SN 7552 7247]; 21 stream section [SN 7557 7267]; 22 trackside exposure [SN 7059 7089]; 23 trackside exposure [SN 6854 6957]; 24 crag [SN 7134 6925]; 25 track section [SN 7175 7111]; 26 trackside exposures [SN 7026 7100]; 27 roadside exposure [SN 6898 7147]; 28 Cwm Rheidol No. 1 borehole [SN 7302 7833] (513.79 m to 526.00 m); 29 river and adjacent track section [SN 7668 7298]; 30 stream section [SN 7557 7286]; 31 trackside exposure [SN 7189 7222]; 32 trackside exposure [SN 7153 7234]; 33 trackside exposure [SN 6851 7086]; 34 quarry [SN 6862 6607]; 35 Cwm Rheidol No. 1 borehole [SN 7302 7833] (498.04 m); 36 quarry [SN 7630 7243]; 37 section in Grogwynion Mine [SN 7131 7252]. * localities assigned an upper *convolutus* biozonal age on the basis of thickness or positional criteria; ** localities assigned an upper *convolutus* biozonal age on the basis of graptolite taxa.

For key to range chart symbols see Appendix 3.

Shales, are largely absent. The best sections include the northern part of Graig Dolfaenog [SN 9190 6724] and crags [SN 9320 6540 to SN 9298 6530] above the Water Works. At the latter locality, silt-rich subfacies of the Rhayader Mudstones overlying the Gafallt Shales gradually pass up into silt-depleted subfacies, as seen in the lower reaches of Nant yr Haidd [SN 9380 6600 to SN 9370 6641], north of the Elan Valley Hotel. There, a 5 m-thick unit of anoxic facies, exposed beneath the aqueduct [SN 9380 6609], contains low *turriculatus* Biozone graptolites, probably belonging to the *gemmatus* Subzone (Table 12, locality 34).

The westward transition from the Gafallt Shales into the Rhayader Mudstones is not exposed, due to the Carreg-ddu Fault. West of the fault, river and roadside sections below the Penygarreg dam expose oxic facies, comprising silt-impover-ished, pale grey-green mudstones. Thereabouts, thin units of anoxic facies are well seen in the river [SN 9120 6738] and along the road [SN 9096 6733 to SN 9107 6732]. A limited graptolite assemblage from the latter unit comprised only long-ranging Telychian forms. To the west, the upper part of the formation is in silt-rich subfacies and forms the crags [SN 9100 6760 to SN 8950 6875] on the northern side of the Penygar-reg reservoir. The thick anoxic-facies unit seen in the upper-most part of the Claerwen reservoir section (Figure 16) is exposed along the shore [SN 8952 6888] of the Craig Goch reservoir. It is overlain by silt-rich, strongly colour-banded and burrow-mottled oxic-facies mudstones exposed in crags [SN 8913 6857], south of the mountain road. Similar mudstones are seen adjacent to the Caerau Mudstones on the reservoir headland [SN 8932 6893], to the north.

The Wye valley between Rhayader and Safn-y-coed [SN 9285 7325], provides a semicontinuous section through the Rhayader Mudstones in the core of the Tylwch Anticline. South of Rhayader the lowest part of the formation is poorly exposed. The waterfalls [SN 9684 6790] at Rhayader expose oxic mudstones

of the silt-depleted, pale grey-green subfacies, which extend up river in reefs [to SN 9682 6830]. A 2 m-thick anoxic-facies unit containing only long-ranging graptolites crops out farther upstream [SN 9680 6836] and marks the entry of the upper silt-rich subfacies, which prevails throughout the remainder of the Wye valley. There, the subfacies commonly contains scattered sandstone-based type E turbidites and is well exposed in the abundant crags along the valley sides as far north as Garreg Lwyd [SN 942 733]. A cutting [SN 9629 6929] in these oxic mudstones on the A470 road contains a 2 m-thick anoxic unit that has yielded a monospecific graptolite assemblage of the long-ranging *Monograptus marri*. Oxic, colour-banded and burrow-mottled subfacies forming the uppermost part of the formation are exposed on Graig Safn-y-coed [SN 933 726] and in road cuttings [SN 9325 7342 to SN 9296 7362] to the north.

North-east of the Wye valley, only the upper silt-rich part of the formation is exposed. Pale grey, silt-laminated oxic-facies mudstones, in contrast to the grey-green hues of equivalent levels farther south, prevail throughout this area and are well exposed in the crags of Cefn Lletyhywel [SN 960 731]. There, thin units of anoxic facies [at SN 9658 7286], have yielded *utilis* Subzone graptolites. Bundles of turbidite sandstone/mudstone couplets (type Cii) are also common in the formation north-east of the Wye valley and are exposed in crags on Mynydd Perthi [SN 954 728], Llethr Llwyn [SN 958 738], Pyllau Clais [SN 967 743], Craig yr Eryr [SN 961 750] and Cefn Pen-lan [SN 9545 7522]. At the last locality, graptolites recovered from a thin unit of anoxic facies indicate the *utilis* Subzone. The stream of Marcheini Fawr [SN 9419 7442 to SN 9400 7454] exposes the oxic, colour-banded subfacies at the top of the formation on the western limb of the Tylwch Anticline. This subfacies occurs above and below a 10 m-thick unit of anoxic facies, exposed in an adjacent track cutting [SN 9429 7420]. This unit appears to correlate with the one seen at a similar level in the Claerwen and Craig Goch reservoir sections. It has yielded graptolites indicative of either the *utilis* or the *johnsonae* Subzone. The upper, strongly colour-banded, silt-laminated oxic-facies mudstones dominate exposures in the northernmost part of the Tylwch Anticline. They are well exposed in the valley of Afon Dulas, where they form the crags of Craig Gellidywyll, Cefn Llech, Garreg y Gwynt and the southern slopes of Wennallt. On the western limb of the Tylwch Anticline, anoxic-facies units within this sequence are exposed in crags [SN 9561 7741] and in a quarry [SN 9507 7702]. At the latter locality, 4 m of oxic-facies mudstones separate a lower 2.5 m-thick anoxic unit from an upper 3 m-thick one, both of which contain *utilis* Subzone graptolites. However, two similar units of anoxic-facies, in a quarry [SN 9528 7621] west of Cwmyrychen, have yielded *johnsonae* Subzone graptolites. On the eastern limb of Tylwch Anticline, an anoxic-facies unit in the colour-banded and burrow-mottled subfacies at the top of the formation is exposed in a disused quarry [SN 9727 7151] north-west of Ty Canol, and has yielded graptolites, no older than the *proteus* Subzone.

North-east of Rhayader, in Rhyd-hir Brook [SN 9859 6875 to SN 9828 6868], colour-banded and burrow-mottled, oxic-facies mudstones overlying the *M. sedgwickii* shales, pass upwards into variably silt-laminated, largely unbanded pale grey subfacies mudstones. Just north of the Cwmysgawen Fault, pale grey, silt-laminated, oxic-facies mudstones with bundles of turbidite sandstone/mudstone couplets (type Cii) succeed silt-depleted, oxic-facies mudstones in stream sections [SN 9806 6866 to SN 9793 6906 and SN 9903 6926 to SN 9887 6967]. East of Beili-Neuadd, the Rhyd-hir Brook [SN 9978 6966 to SN 9967 6975] and its tributary [SN 9984 6963 to SN 9995 6985], constitute the easternmost sections in the Rhayader Mudstones in the district. There, colour-banded and burrow-mottled oxic-facies mudstones pass up into the Caerau Mudstones.

Llanilar sheet area

South of the Ystwyth Fault   On the eastern limb of the Teifi Anticline in Coed Bwlchgwallter, the southern bank of Afon Ystwyth [SN 7668 7298] and an adjacent track section (Table 13, locality 29), expose the *M. sedgwickii* shales, there about 15 m thick, overlying the Hafod Member of the Ystrad Meurig Grits Formation. Further sections in the *M. sedgwickii* shales are exposed in Nant Ffin [SN 7558 7293] and [SN 7557 7286; Table 13, locality 30], where they again succeed the Hafod Member.

South-west of Coed Bwlchgwallter, crags on Banc Melyn [SN 750 706], east of Glogfawr Mine, expose a sequence of weakly silt-laminated, green-grey oxic-facies mudstones, in which scattered, thin turbidite sandstone/mudstone couplets (type Cii) occur in the upper part and form a transition into the adjacent outliers of the succeeding Devil's Bridge Formation. Around Cwm Mawr Mine, Nant Lluest exposes sections [SN 7337 6714, SN 7373 6733 and SN 7380 6739] in the *M. sedgwickii* shales which contain graptolites of the *sedgwickii–halli* biozonal interval. At the westernmost and easternmost localities, an upward passage into the succeeding colour-banded and burrow-mottled, oxic-facies mudstones is exposed. Graptolites of probable *sedgwickii* Biozone age have also been recovered from the eastern tips [SN 7455 6637] of the Abbey Consols Mine, north of Strata Florida Abbey, suggesting that the *M. sedgwickii* shales were encountered in the underground workings.

South of Pontrhydfendigaid crags and small quarries [SN 7333 6554] near Twyn provide a transect through the *M. sedgwickii* shales into the succeeding oxic-facies mudstones. Intermittent sections in Nant Gorffen [SN 7157 6377 to SN 7157 6440] south of Bron-mwyn, and a quarry [SN 7153 6331] expose oxic-facies mudstones, with scattered turbidite sandstone/mudstone couplets (type Cii) that reflect proximity to the Devil's Bridge Formation. Farther south, a quarry [SN 7221 6229] exposes the gradational contact with the Devil's Bridge Formation. There, the uppermost part of the Rhayader Mudstones is in oxic facies. In contrast, in a quarry [SN 7109 6309] to the east of Bwlchyddwyallt, a thick unit of anoxic facies forms the uppermost part of the formation, suggesting that the base of the Devil's Bridge Formation is diachronous in this area. The quarry exposes about 15 m of thinly interbedded, siltstone-based mudstone turbidites (type Di) and laminated hemipelagites. Thicker silts and thin sandstone beds appear in the upper part and mark the gradation into the Devil's Bridge Formation. The anoxic facies unit has yielded graptolites of the *runcinatus–renaudi* subzonal interval, including *Monograptus runcinatus*, *Pristiograptus renaudi* and *Streptograptus* cf. *plumosus*. Lower, predominantly oxic facies of the Rhayader Mudstones are poorly exposed in streams [SN 7075 6292 to SN 7072 6323 and SN 7070 6346 to SN 7055 6352], south of Maes-elwed Farm. A quarry [SN 7095 6390] and crags [SN 7099 6420] near the farm contain thin units of anoxic facies.

On the western limb of the Teifi Anticline, a poorly exposed track section [SN 6815 6570], east of Penlan, exposes anoxic facies which have yielded *Monograptus halli* and *M. involutus* of the *halli* Biozone, suggestive of the uppermost part of the *M. sedgwickii* shales. To the north-east, a quarry [SN 6862 6607] exposes the lowest 10 m of the *M. sedgwickii* shales overlying oxic facies of the Derwenlas Formation. Graptolites of the *sedgwickii–halli* biozonal interval have been recovered from the anoxic division (Table 13, locality 34). Near to the top of the Rhayader Mudstones, a quarry [SN 6744 6705] at Gwenhafdre exposes colour-banded and burrow-mottled, silt-laminated oxic-facies mudstones, succeeded by anoxic-facies mudstones. The latter contain *gemmatus* Subzone graptolites including *Mono-graptus gemmatus*, *M. halli*, *Pristiograptus renaudi*, *Rastrites fugax* and *Glyptograptus fastigatus*. This unit of anoxic facies compares with that recognised beneath the Devil's Bridge Formation on the eastern limb of the Teifi Anticline. In an otherwise poorly exposed tract to the north, Nant y Ffin [SN 6806 6801 to SN 6794 6819], above its confluence with Nant Rhydgaled, provides sections in silt-depleted, colour-banded and burrow-mottled, oxic-facies mudstones. Near the Ystwyth Fault, the disused railway cuttings [SN 6839 7069 to SN 6796 7100] south of Pen-y-bont provide an extensive section through the formation. Variably silt-laminated, diffusely banded, green-grey oxic-facies mudstones forming the lower part of the formation are exposed in the eastern cuttings. A folded sequence of oxic, silt-rich subfacies forming the upper part of the formation occurs in the western cuttings.

North of the Ystwyth Fault   Sections in anticlinal cores around Llanafan include Llwynprenteg Farm [SN 6872 7170], a road cutting [SN 6860 7229], and Nant Pant-yr-haidd [SN 6887 7255 to SN 6901 7265]. In the last named, gradational contacts between silt-rich, colour-banded oxic-facies mudstones and the succeeding Devil's Bridge Formation are seen. To the east, track sections [SN 7189 7222 and SN 7153 7234] at Grog-wynion Mine expose the *M. sedgwickii* shales and their passage into overlying oxic-facies mudstones. Graptolite assemblages obtained from both localities are indicative of the *sedgwickii* Biozone (Table 13, localities 31 and 32). Opencast mineral workings [SN 7131 7252], to the north, expose colour-banded oxic-facies mudstones passing up into about 7 m of anoxic-facies mudstones, which are overlain by the Devil's Bridge Formation. The anoxic interval, in common with sections at this level south of the Ystwyth Fault, has provided a diverse assemblage of the *gemmatus* Subzone (Table 13, locality 37).

In Cwm Rheidol No. 1 borehole [SN 7302 7833], facies of the Ystrad Meurig Grits Formation replace the *M. sedgwickii* shales. They are overlain by about 25 m (true thickness) of Rhayader Mudstones (Appendix 1; Figure 23).

Outcrops of Rhayader Mudstones forming periclinal inliers along the northern margin of the Llanilar sheet area are largely extrapolated from sections described in the Aberystwyth district (Cave and Hains, 1986). In the westernmost inlier, the gorge [SN 7489 7906 to SN 7435 7780] of Afon Rheidol near Parson's Bridge exposes a complete section through the attenuated sequence of the Rhayader Mudstones. The *M. sedgwickii* shales, exposed just north of the district, pass up into pale and darker grey colour-banded and strongly silt-laminated oxic-facies mudstones seen both above and below the bridge. These in turn are abruptly overlain downriver by the basal sandstone-dominated facies of the Devil's Bridge Formation.

Inliers of colour-banded oxic mudstones to the east of Devil's Bridge are exposed in a stream [SN 7520 7634 to SN 7512 7639], west of Bodcol, and in crags [SN 7599 7687] west of Bodcol Mine. In both sections, the thin sandstones of type Cii turbidites enter several metres below the feature-forming packet of thicker sandstones taken, for mapping convenience, as the base of the Devil's Bridge Formation. Sections in the inlier west of Mynach Vale Mine are provided by the Mynach [SN 7702 7762 to SN 7722 7760], and in a nearby quarry [SN 7724 7865] the transition into the Devil's Bridge Formation is exposed.

Depositional model

The latest Hirnantian to early Telychian mudstone-dominated succession of the district (Figure 17) is similar in setting and geometry to the wedges of fine-grained deposits termed slope-aprons (Stow and Piper, 1984; Stow, 1985; 1986; Stow et al., 1985; Pickering et al., 1989), which are accumulating along the steeper gradients that mark the margins of modern oceanic and intracratonic basins. The palaeotectonic setting of this succession compares closely with those accumulating and preserved within modern intracratonic basins or basins within rifted continental margins (Case, 1974; Pickering et al., 1989). In such basins, slope-apron systems can form in water depths significantly shallower than those that prevail in their oceanic counterparts.

Slope-aprons are constructed of one or more turbidite systems, each recording a cycle of sedimentation controlled by contemporary movements in sea level. Such systems compare with the Type III turbidite systems of Mutti (1985) and Mutti and Normark (1987).

SLOPE-APRON GEOMETRY AND SEDIMENT SOURCE

Although much of the proximal portions of these mud-dominated systems are not preserved due to intra-Telychian erosion across the Tywi Lineament, the laterally extensive median to distal parts of the sequence show progressive westward thinning (Figures 13 and 34). Across the district most formations thin westwards by as much as 84 per cent. An exception to this is the Derwenlas Formation in the core of the Teifi Anticline. There an anomalously thick sequence may reflect higher rates of sedimentation adjacent to the depositional sites of the Ystrad Meurig Grits. To the north, however, in the Aberystwyth district, the Derwenlas Formation conforms to the regional trend of westward thinning (Cave and Hains, 1986).

The Tycwtta Mudstones may represent a remnant of the proximal part of the slope-apron. It is about 160 per cent thicker than its correlative, the Cwmere Formation, just 7 km to the west.

The source of the mud-dominated sediments supplied to the basin during the Hirnantian to early Telychian remains largely circumstantial because of the absence of palaeocurrent indicators. However, an easterly source is suggested by the westward thinning of the component formations and the palaeocurrent vectors and geometries of the associated coarse-grained deposits (Caban Conglomerate, Ystrad Meurig Grits and Devil's Bridge formations).

Geochemical analyses of mudstones associated with southerly supplied facies show that there are distinct compositional differences between these and the coeval slope-apron mudstones (Ball et al., 1992), suggesting a different, and by inference easterly, provenance for the latter.

EVIDENCE FOR SLOPE FAILURE AND SEDIMENTARY CREEP

Modern and ancient slope-aprons are characterised by slope failure and the downslope movement of intact masses of sediment. Slides, transported on detachment surfaces parallel to the bedding, and slumps generated by rotational failure surfaces, mainly occur in areas of steep gradients and/or high sedimentation rates (Pickering et al., 1989); they typify the proximal regions of slope-aprons. The abundance of disturbed units in the Tycwtta Mudstones suggests deposition in such a setting.

Large-scale slumping and sliding are less important in the median and distal parts of slope-aprons, but smaller-scale movements and sedimentary creep (cf. Carter, 1975; Stow, 1986; Pickering et al., 1989) result in common synsedimentary faults with displacements ranging from millimetres to several metres, bedding-parallel slide planes, and sediment convolution. Such phenomena are widely observed in the slope-apron sequences west of the Tywi Anticline.

SLOPE-APRON COARSE-GRAINED SYSTEMS

Coarse-grained clastic sediment is transported across slope-aprons in narrow channels which debouch on to the median and distal parts to form sand-dominated lobes (Figure 17). In the proximal part of the slope-apron the Allt-y-clych Conglomerate appears to represent a series of laterally restricted channel deposit. The Caban Conglomerate Formation, situated on the median part of the slope-apron, also records deposition in a series of laterally restricted channels but displays, in addition, sandy overbank and proximal sandy-lobe facies. Expanded slope-apron sequences flanking the channelised portions of the formation in the Rhayader area (Figure 19) represents levee-like accumulations which received sediment from both overbank and normal slope-apron turbidity currents (see p.115). Sandy-lobe deposits emplaced on to distal portions of the slope-apron are represented by the Ystrad Meurig Grits and Devil's Bridge formations.

SLOPE-APRON MUDSTONES

The mud- and silt-dominated facies characteristic of modern slope-aprons have been reviewed by Stow and Piper (1984), and Stow (1985) and compare closely with those described from the district.

The thicker mudstone turbidites (type D) record the input of sediment from low-concentration turbidity currents and represent the most rapidly deposited elements of the slope-apron systems of the district. The very thinly bedded (type D) and variably silt-laminated (type E) mudstone turbidites reflect slower rates of sedimentation from slower-moving, very low-concentration turbidity currents, possibly evolved from storm- or flood-replenished nepheloid layers (Stow and Bowen, 1980; Tyler and Woodcock, 1987). The high proportion of hemipelagite reflects, as in modern analogues, the importance of very slow background deposition from suspension. Moreover, the intergradation of these various deposits underlies the process-continuum characteristic of such mud-dominated slope-aprons (Stow and Piper, 1984).

EVOLUTION OF THE HIRNANTIAN TO TELYCHIAN SLOPE-APRON

Long-term variations in the oxicity of bottom waters recorded in the mudstone-dominated Hirnantian to Telychian succession of the district cannot readily be matched in modern slope-apron deposits (cf. Stow and Piper, 1984), for anoxic bottom conditions are not well represented in the deposits of present-day, well-aerated seas (Jenkyns, 1986; but see Thornton, 1984).

Major changes from anoxic to oxic facies in Hirnantian to Telychian times appear to correlate with the sea-

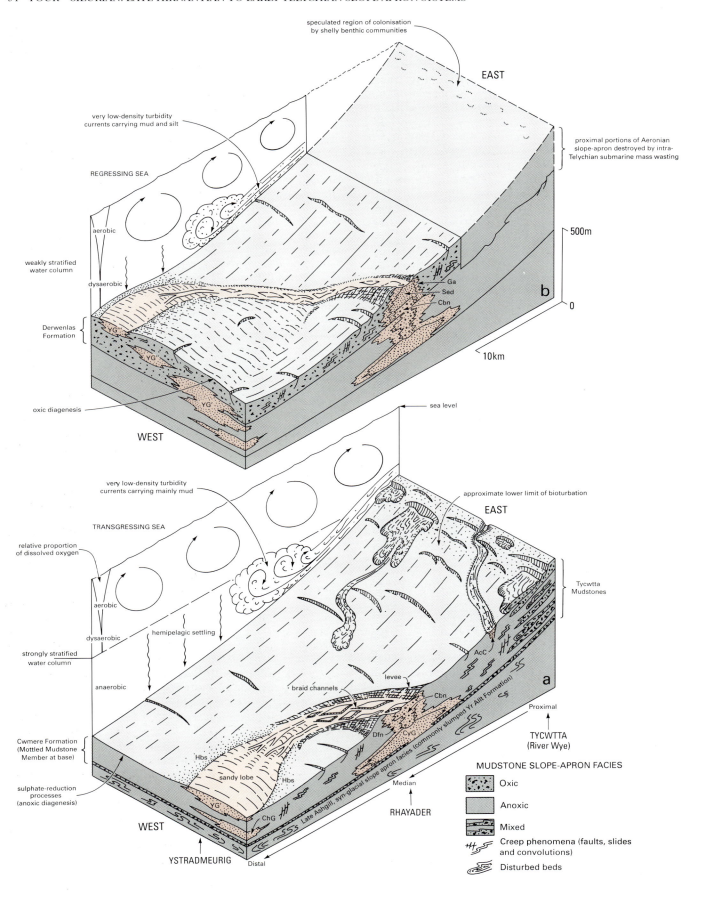

speculated region of colonisation
by shelly benthic communities

EAST

very low-density turbidity
currents carrying mud and silt

proximal portions of Aeronian
slope-apron destroyed by intra-
Telychian submarine mass wasting

REGRESSING SEA

aerobic

500m

weakly stratified
water column

dysaerobic

Ga
Sed
Cbn

b

Derwenlas
Formation

YG'''

0

YG''

10km

oxic diagenesis

YG'

WEST

sea level

very low-density turbidity
currents carrying mainly mud

approximate lower limit of bioturbation

TRANSGRESSING SEA

EAST

relative proportion
of dissolved oxygen

Tycwtta
Mudstones

aerobic

dysaerobic

hemipelagic settling

strongly stratified
water column

anaerobic

AcC

a

braid channels

levee

Cbn

Proximal

Cwmere Formation
(Mottled Mudstone
Member at base)

Dfn

CyG

Late Ashgill, syn-glacial slope apron facies (commonly slumped Yr Allt Formation)

TYCWTTA
(River Wye)

Hbs

Median

sulphate-reduction
processes (anoxic diagenesis)

sandy lobe

Hbs

RHAYADER

YG'

WEST

ChG

YSTRADMEURIG

Distal

MUDSTONE SLOPE-APRON FACIES

Oxic

Anoxic

Mixed

Creep phenomena (faults, slides
and convolutions)

Disturbed beds

level changes widely documented within coeval cratonic shelf successions (Johnson et al., 1985; 1991) (Figure 39). Major sequences of anoxic facies equate with periods of marine transgression and deepening, and thick sequences of oxic facies with periods of falling sea level. Thus, each major anoxic/oxic-facies couplet comprises a turbidite system that correlates with a coeval shelf depositional system bounded by non-sequences. Two such turbidite systems are recognised in the district, a late Hirnantian to mid-Aeronian system (Cwmere, Tycwtta and Derwenlas formations) initiated by post-glacial transgression, and a late Aeronian to early Telychian system (Rhayader Mudstones) which developed in response to the widespread *M. sedgwickii* transgression (Figure 39).

The worldwide rise in sea level, which followed the Hirnantian glaciation, caused a substantial decrease in sedimentation rates on the oversteepened, synglacial slope-apron and resulted in the deposition of the less silty mudstone turbidites and hemipelagites of the Tycwtta Mudstones and Cwmere Formation. Following a short oxic interval, during which the Mottled Mudstone Member was deposited, anoxic bottom conditions prevailed across median to distal parts of the slope-apron until early Aeronian times and account for the remainder of the Cwmere Formation (Figure 17a).

The oceanographic processes which led to this and other Lower Palaeozoic periods of basinal anoxia have been assessed by Leggett (1980), and derive from models by Berry and Wilde (1978) and Jenkyns (1980; 1986). The drowning of marginal shelf areas promoted an increase in the volume of organic matter supplied to basinal waters. Oxidation of the organic matter reduced the level of dissolved oxygen. In addition, the outflow of warm surface waters from expanded shelf areas favoured the formation of a thermally stratified water column and inhibited the transfer of oxygen to the cold bottom waters from the surface layers. The effect of these processes was to create a zone of oxygen depletion which, impinging on the sea floor, established conditions favourable to the reduction of sulphate and the preservation of organic carbon, but inimical to the bottom-dwelling fauna (Leggett, 1980; Curtis 1980; 1987).

Abundant evidence of slumping within the coeval Tycwtta Mudstones suggests that they were deposited in a proximal slope-apron setting, although contemporane-ous movements on the Garth Fault may have triggered some of the disturbed beds. The presence of intervals with burrowed hemipelagites in the Tycwtta Mudstones suggests that this formation was deposited high on the slope-apron, where fluctuations in the upper levels of the zone of oxygen depletion would have allowed oxic bottom conditions to prevail periodically.

The predominantly oxic facies of the Derwenlas Formation equates with the mid-Aeronian marine regression, an event which attained its acme during the *convolutus* Biozone (Johnson et al., 1985; 1991). The withdrawal of the sea from the shelf reduced the effects of thermal stratification in the basin and allowed bottom waters to be replenished with oxygen from surface levels (Figure 17b). The resulting rapid and widespread introduction of oxic, but probably dysaerobic, sea-bed conditions permitted burrowing animals to inhabit the slope-apron surface, and sustained oxidising early diagenetic processes in the sea-bed sediments. Contemporaneous increases in the silt content of turbidites (type E) reflect the rejuvenation of source areas and the basinward migration of shelf facies belts. The prevalence of complete or top-cut-out, strongly silt-laminated, type E turbidite mudstones in the upper part of the Derwenlas Formation in the Rhayader sheet area coincides with the period of lowest sea level. The thin units of anoxic facies in the Derwenlas Formation record periodic reintroduction of anoxic bottom conditions. Although some of these appear to be widespread in the Welsh Basin, it is not known whether they are due to parochial factors (e.g. Thornton, 1984) or to widespread transgressive pulses.

Deposition of the *M. sedgwickii* shales marked the onset of a major eustatic transgression (e.g. Leggett et al., 1981; Johnson et al., 1991), and the return of anoxic bottom conditions throughout the Southern Welsh Basin during the *sedgwickii* and early part of the *halli* biozones. Although widespread, this transgression was of short duration and the succeeding parts of the Rhayader Mudstones record the onset of a substantial period of predominantly oxic-facies deposition. Although the presence of silt-rich subfacies within this sequence accords with an increase in the supply of coarser detritus to the slope-apron during a regression, the detailed distribution of such subfacies within the district may principally reflect the changing positions of transport pathways followed by the coeval coarse-grained turbidites. Thus the fining-upwards, lower part of the sequence (Figure 16) may relate to the waning of the Caban Conglomerate system. Shutdown of this supply route allowed silt-depleted subfacies briefly to accumulate in the Rhayader area. The return of the silt-rich subfacies in the upper parts of the Rhayader Mudstones in the Rhayader sheet area relates to the progressive expansion of the easterly derived, sandy Hafdre Formation (Mackie and Smallwood, 1987) south of the district, and the blanketing of distal portions of the slope-apron (Llanilar sheet area) by the coeval Devil's Bridge Formation. This increased supply of coarser detritus probably reflects the onset of Telychian tectonism, which culminated in the abandonment of slope-apron deposition in the district.

Figure 17 Diagram illustrating geometry and facies relationships of the latest Hirnantian–Aeronian slope-apron systems in the district, and the processes active during sedimentation (facing page):

a in the *cyphus* Biozone, a period of marine transgression and anoxic bottom conditions; **b** in the *convolutus* Biozone, a period of marine regression and oxic bottom conditions. Abbreviations: AcC — Allt-y-clych Conglomerate; Cbn — Caban Conglomerate facies; ChG — Chwarel Goch facies; CyG — Cerig Gwynion Grits; Dfn — Dyffryn Flags; Ga — Gafallt Shales; Hbs — Henblas facies; Sed — Sedgwickii Grits; YG′, YG″, YG‴ — first, second and third sequences of Ystrad Meurig Grits facies.

In the westernmost part of the district, slope-apron deposition ceased during the *gemmatus* Subzone, as it was replaced by the sandstone-lobe system of the Aberystwyth Grits Group generated tectonically and supplied from the south. Early to mid-Telychian times saw the eastward migration of similar lobes (Cwmystwyth Grits Group) and the complementary withdrawal of slope-apron conditions. In the eastern parts of the slope-apron, which survived the longest, the increased incidence of anoxic intervals in the upper parts of the Rhayader Mudstones may provide evidence for a transgressive deepening in the *utilus–johnsonae* subzonal interval. Such an event, given the current uncertainties in the correlation of graptolite and brachiopod biostratigraphies (Johnson et al., 1991), may represent the *turriculatus* transgression reported from both distant and adjacent shelf successions by Cocks et al. (1984) and Johnson et al. (1991). However, the pattern of alternating oxic and anoxic facies now established and maintained throughout this and the succeeding biozone is not consistent with reported eustatic movements of sea level (Figure 39) (see Caerau Mudstones) and suggests that basinal facies changes were now largely controlled by tectonism.

LOCALISED COARSE-GRAINED TURBIDITE FORMATIONS

Between the late Hirnantian and early Telychian, coarse-grained turbidites, sourced from the east, invaded the contemporaneous mud-dominated slope-apron deposits of the district. These coarse-grained deposits form a series of lenticular bodies that intertongue with the slope-apron mudstones. Although conglomerates and sandstones predominate, more muddy facies form an integral part of their architecture. Two largely contemporaneous but geographically separated formations are recognised, the Caban Conglomerate Formation of the Rhayader sheet area and the Ystrad Meurig Grits Formation of the Llanilar sheet area. A further coarse clastic division, the late Rhuddanian, Allt-y-clych Conglomerate, which is restricted to the eastern limb of the Tywi Anticline, has already been described (p.81).

Caban Conglomerate Formation

The Caban Conglomerate Formation comprises thick-bedded, turbidite conglomerates and sandstones, together with thinner-bedded turbidite sandstones, mudstones and hemipelagites (Plates 12 and 13). It crops out on the north-western limb of the Tywi Anticline and around the Rhiwnant Anticline (Figure 18). In a 2.5 km-long belt around Carn Gafallt [SN 941 647], it comprises a 700 m-thick sequence occupying the stratigraphic interval elsewhere represented by the Cwmere Formation, the Derwenlas Formation and the lower part of the Rhayader Mudstones. To the north-east and south-east of Carn Gafallt the formation breaks down into a series of tongues interleaved with these mudstone formations.

The formation takes its name from the area just to the north of the Caban-coch dam [SN 9250 6470] in the Elan valley (Plate 12; Figure 21d), south-west of Rhayader. There, Lapworth (1900) described two prominent conglomerates and intervening shales as the Caban Conglomerates, the lower part of his Caban Group. As well as Lapworth's Caban Conglomerates, this new formation includes his Cerig Gwynion Grits, Dyffryn Flags, *Monograptus–Sedgwickii* Grits and Gafallt Shales together with the Drygarn Conglomerate of Davies (1926). The complex stratigraphical relationships now recognised within the formation (Figure 19) have precluded the continued use of Lapworth's (1900) terms as lithostratigraphical subdivisions (cf. Kelling and Woollands, 1969). It is convenient, however, to retain these names for the five recurrent facies from which the formation is constructed, each consisting of different combinations of the various turbidite and hemipelagite types.

Cerig Gwynion Grits facies

The facies predominantly comprises a suite of medium- to thick-bedded turbidite sandstones (types Bi and Ci) that is confined to, and takes its name from, the Cerig Gwynion Grits of Lapworth (1900). It equates with 'Facies B' of Kelling and Woollands (1969). Subordinate thin-bedded turbidite sandstones and mudstones (types Cii and D) occur scattered throughout the facies as thin packets. Intercalated hemipelagites, mostly laminated, are sparsely developed.

The Cerig Gwynion Grits occupy the lowest part of the formation, cropping out on both the north-western limb of the Tywi Anticline, between Gaufron and the southern boundary of the district, and around the north-eastern part of the Rhiwnant Anticline (Figure 18). Along most of its crop the facies forms a single unit, but on the north-western limb of the Rhiwnant Anticline, west of Craig-y-Bwlch [SN 8940 6260], this splits into two tongues.

The maximum thickness of the unit, about 110 m, occurs on the north-western limb of the Tywi Anticline, south-west of Carn Gafallt, from which point the grits thin gradually to zero, both to the north-east and to the south-west. Kelling and Woollands (1969) maintained that the thickest development of the Cerig Gwynion Grits comprised three thick tongues of conglomerate on the north-western limb of the Tywi Anticline, around Nant Cedni (Cedney) [SN 901 564], 4 km south of the district. Reconnaissance work has shown, however, that these strata are of Caban Conglomerates facies developed at a higher horizon, and that the Cerig Gwynion Grits do not reach so far to the south-west (Figure 18).

On the north-western limb of the Tywi Anticline between Gwastedyn Hill [SN 9860 6680] and the southern margin of the district, Cerig Gwynion Grits sharply overlie the Mottled Mudstone Member of the Cwmere Formation (Figure 20; Plate 12a). Their base is taken at the base of the first thick-bedded turbidite sandstone (type Bi). Elsewhere, sequences of Dyffryn Flags of varying thickness (locally too thin to be depicted on the map) intervene between the base of the facies and the Mottled Mudstone Member, indicating that the base of the facies is diachronous. The junction between the Cerig Gwynion Grits and these units of Dyffryn Flags

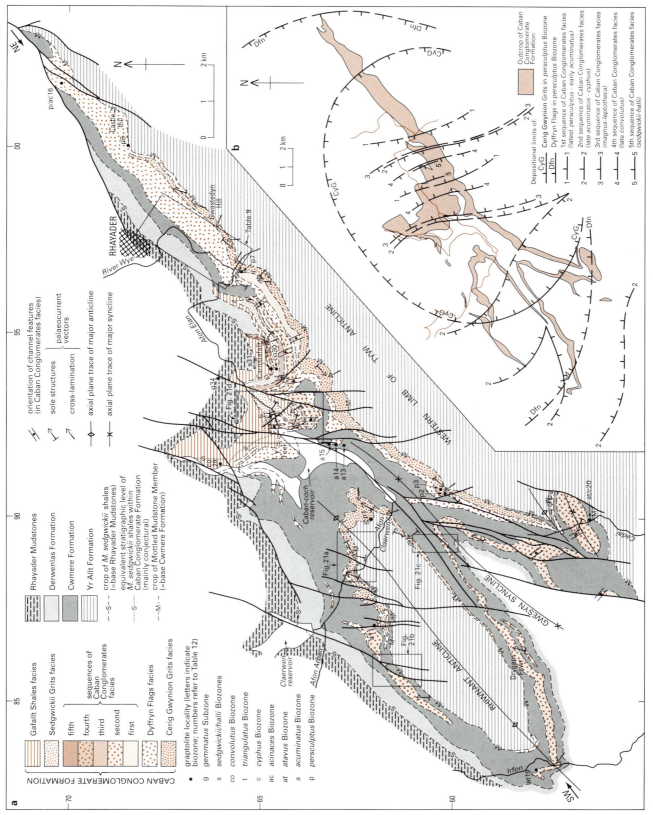

Figure 18   The geology of the Caban Conglomerate Formation.

a   Geological map of the Caban Conglomerate Formation showing palaeocurrent vectors and the position of critical graptolite localities (linework south of grid line 60 is largely based on reconnaissance mapping and aerial photograph interpretation and may be subject to future amendment); arrows SW and NE indicate the line of section in Figure 19.   **b**   Diagram of the suggested depositional limits of the Cerig Gwynion Grits facies and coeval Dyffryn Flags facies, and the five sequences of Caban Conglomerates facies.

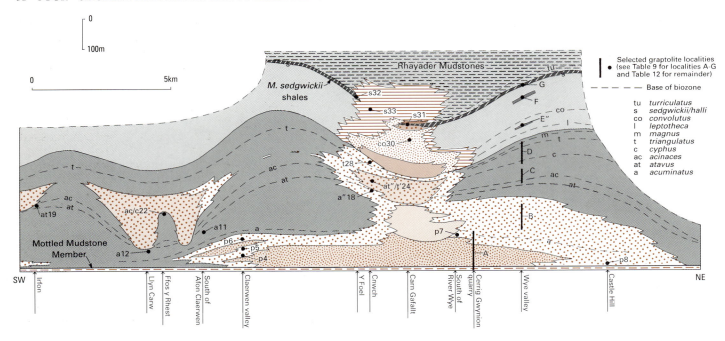

Figure 19   Architecture of the Caban Conglomerate Formation along the SW–NE line of section in Figure 18. For key to formations and facies see Figure 18. Localities E and F are sited in units with laminated hemipelagites in Derwenlas Formation.

is either sharp or gradational over several metres. Where the boundary is gradational, widely spaced, thick-bedded turbidite sandstones (types Bi and Ci) appear in the Dyffryn Flags and increase in abundance rapidly upwards; the base of the Cerig Gwynion Grits is taken where they become predominant.

Except in the area south-west of Carn Gafallt, where they are overlain gradationally by the first sequence of Caban Conglomerates facies, the Cerig Gwynion Grits are overlain either sharply or gradationally by Dyffryn Flags. At their north-eastern and south-western limits the facies passes laterally into Dyffryn Flags.

Individual type Bi turbidite sandstones range from 0.15 to 3.5 m thick (Plate 7d). They commonly occur in stacked sequences, comprising as many as 12 turbidites, and displaying high levels of amalgamation (Figure 20; Plate 12a). Sandstone bases are commonly erosional, although the depth of channelling seldom exceeds 50 cm. Sole marks are restricted to scattered large flutes and more abundant longitudinal grooves. The sandstones comprise medium- to coarse-grained quartz arenites. In the middle and upper parts of the facies, between Gwastedyn Hill and the Dulas valley, some sandstones exhibit conglomeratic bases, and pebbly sandstones are locally present. Scattered granules and mudstone rip-up clasts are more widely distributed. The sandstones are typically massive or normally graded. However, diffuse parallel lamination may be developed locally and a well-defined parallel lamination is present in the uppermost parts of some beds. Dish structures are common. Rare, cleaved, high-matrix sandstones (type Bii) are locally present.

The type Ci turbidite sandstones occur either as top-cut-out or base-cut-out units. Top-cut-out units only contain either T_{ab} or T_b divisions and comprise medium- to thick-bedded, medium- to coarse-grained sandstones up to a metre thick. Such units mainly occur interbedded with type Bi turbidite sandstones. Base-cut-out units comprise sandstone/mudstone couplets lacking the T_a division. The basal sandstones are generally less than 40 cm thick and fine to medium grained.

Packets similar to Dyffryn Flags occur in the Cerig Gwynion Grits and predominantly comprise thin-bedded turbidite sandstone/mudstone couplets (type Cii) with scattered mudstone turbidites (type D) and hemi-pelagites. Individual packets are up to 5 m thick. The basal fine- to coarse-grained sandstones of the type Cii turbidites range up to 8 cm thick, whilst lenticular bedded, commonly coarse-grained sandstones form the bases of the mudstone turbidites (type Di). Scattered hemipelagites, are mainly laminated, and typically very thin (<5 mm). Between 15 and 25 per cent of the Cerig Gwynion Grits facies may be made up of these thin-bedded packets.

In the Cerig Gwynion Grits both individual thick-bedded turbidite sandstones (type Bi) and packets of strata composed of various other types of turbidite are commonly lenticular, wedging out over distances of tens of metres (Plate 12a). This lateral wedging appears to record depositional thickness variations and the amalgamation of type Bi turbidite sandstones.

Within the crop of the Cerig Gwynion Grits the percentage of type Bi turbidite sandstones remains fairly constant between 75 and 85 per cent. The facies dies out

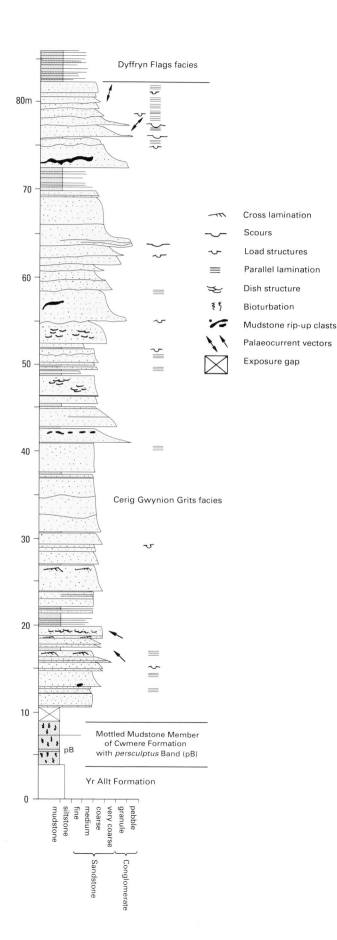

Figure 20 legend:

Cross lamination

Scours

Load structures

Parallel lamination

Dish structure

Bioturbation

Mudstone rip-up clasts

Palaeocurrent vectors

Exposure gap

laterally by the progressive wedging out of individual type Bi turbidite sandstones. Although the average thickness of these sandstones and the occurrence of amalgamation both decrease towards the extremities of the crop, there is no corresponding increase in thickness in the intercalations of thin-bedded turbidites. Although both coarsening/thickening-upwards and thinning/fining-upwards sequences are locally present in the Cerig Gwynion Grits, no overall pattern can be distinguished for the whole crop.

Sole marks indicate northerly directed transport with a dominant trend towards the north-north-east (Figure 18). However, subsidiary transport towards the south-south-west is also indicated. In contrast to directions obtained from sole structures, those from cross-lamination indicate transport towards the west-north-west or north-west.

BIOSTRATIGRAPHY

The sequence of Cerig Gwynion Grits in the district probably lies entirely within the *persculptus* Biozone. Graptolite faunas of this biozone have been recovered from the lower and upper parts of the sequence in the Wye valley (Table 9) whilst those from the Marchnant valley (Figure 18a; Table 12, localities 2 and 3) are probably also of this age. Elsewhere, support for a *persculptus* Biozone age comes from the graptolites collected from the enveloping Dyffryn Flags in the Claerwen valley (Figure 21a; Table 12).

DETAILS

North-western limb of Twyi Anticline

Cerig Gwynion Grits, close to a lateral transition with Dyffryn Flags, are exposed in a roadside quarry [SN 9974 6826] near Gaufron. There, about 5 m of thin-bedded turbidite sandstones and mudstones (types Cii and D) are overlain by about 10 m of thicker-bedded sandstone turbidites (base-cut-out type Ci), in stacked sequences, separated by thin units of thin-bedded turbidites, similar to those that underlie them. The type Ci sandstones are parallel and cross-laminated, medium grained and up to 30 cm thick.

To the south-west, thick-bedded sandstone turbidites (type Bi) first appear in crags [SN 9952 6763] at the north-eastern end of Gwastedyn Hill, where a 1.5 m-thick bed of coarse-grained pebbly sandstone is exposed just above the base of the Cerig Gwynion Grits. Farther south-west, on Gwastedyn Hill, crags [e.g. SN 9857 6693 to SN 9863 6687 and SN 9808 6634 to SN 9817 6626] of medium- to coarse-grained turbidite sandstones, in beds 0.5 to 2 m thick (type Bi), provide several semi-continuous sections through the facies.

Cerig Gwynion quarry [SN 9700 6565], at the south-eastern end of Gwastedyn Hill, exposes a complete section through the type development of the facies (Figure 20; Plate 12a). The section displays a thickening- and coarsening-upwards sequence, overlain in the uppermost part by a thinning- and fining-upwards sequence. The latter includes several top-cut-out type Ci sandstone turbidites, up to 85 cm thick. Pebbly sandstones (type Bi) are restricted to the middle

Figure 20   Graphic log of the Cerig Gwynion Grits facies in Cerig Gwynion quarry [SN 969 656].

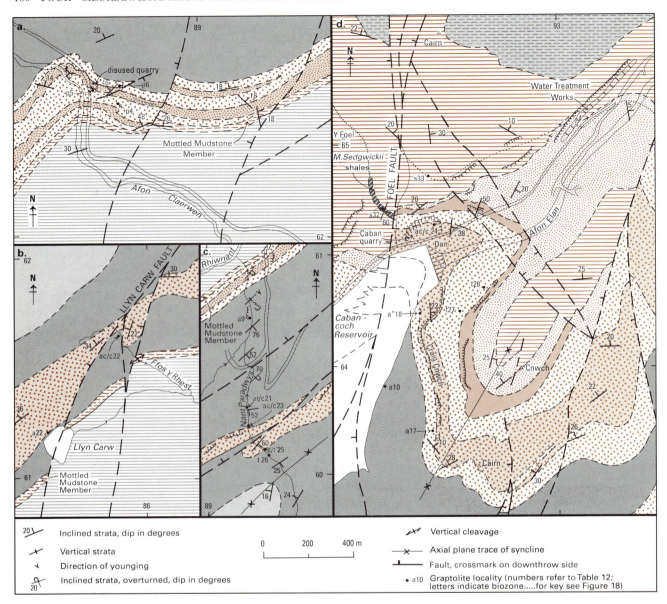

Figure 21   Geological maps showing the position of selected graptolite localities in the Caban Conglomerate and adjacent formations: **a** Claerwen valley;   **b** Llyn Carw;   **c** Nant Paradwys;   **d** Elan valley. For key to formations and facies see Figures 18 and 19.

and upper part of the section. Both stacked sequences and individual thick-bedded turbidite sandstones (type Bi) exhibit lateral wedging, well seen in the lower part of the sequence on the lowest level in the quarry. In the quarry, the Cerig Gwynion Grits thin north-eastwards, from 79 to 71 m, as a 4 m-thick bundle of thick-bedded turbidite sandstones (type Bi), at the top of the sequence, tongues out into the Dyffryn Flags. Graptolite faunas from 5 m above the base [SN 9718 6564], and 10 to 15 m from the top [SN 9729 6575] of the Cerig Gwynion Grits sequence in the quarry, are of the *persculptus* Biozone (Table 9).

On the south-west side of the Wye valley, crags [SN 9677 6534 to SN 9666 6546] expose medium- to thick-bedded turbidite sandstones (type Bi) in a succession similar to that at Cerrig Gwynion quarry. Two well-defined slacks mark intervals composed predominantly of thin-bedded turbidites (types Cii and D).

A semicontinuous section through the Cerig Gwynion Grits, where they are overlain by the first sequence of Caban Conglomerate facies, is exposed in crags [SN 9494 6377 to SN 9489 6395] south-east of Carn Gafallt. Medium- and coarse-grained turbidite sandstones (type Bi), with scattered granules in the upper part of the sequence, occur in beds up to a metre thick. A unit of thin-bedded turbidites (types Cii and D), occupying a marked slack in the middle of the sequence, is exposed in crags [SN 9492 6389].

Between Talwrn [SN 9410 6410] and the southern boundary of the district, the Cerig Gwynion Grits lack pebbly units. A nearly complete section through the facies is exposed in the stream section [SN 9069 6010 to SN 9060 6035] and two adjacent quarries in the Marchnant valley. Medium- to coarse-grained turbidite sandstones (type Bi) in beds up to 1.5 m thick dominate the sequence. Two prominent units of thin-bedded turbidites (types Cii and D) occur near the base and top of the sequence and are 3 m and 2 m thick respectively. The lower unit in the stream [SN 9603 6013] and in the eastern quarry [SN 9076 6031] has yielded probable *persculptus* Biozone graptolites (Figure 18a: Table 12, localities 2 and 3). A sharp contact between Cerig Gwynion Grits and overlying Dyffryn Flags is exposed in Nant Rhyd Goch [SN 9162 6158].

Rhiwnant Anticline

On the south-eastern limb of the anticline, Nant yr Ych [SN 8781 5992 to SN 8782 5988] is the most southerly section in Cerig Gwynion Grits near to their lateral passage into Dyffryn Flags. The facies here is 5 m thick and comprises medium-grained turbidite sandstones up to 1 m thick (types Bi and Ci) separated by subordinate thin-bedded turbidites (types Cii and D). It is underlain and overlain gradationally by Dyffryn Flags. About 250 m to the south-west the Cerig Gwynion Grits pass laterally into Dyffryn Flags.

In Nant Paradwys [SN 8917 6095 to SN 8917 6089] (Figure 21c) the Cerig Gwynion Grits are about 26 m thick and comprise medium- to thick- bedded turbidite sandstones (type Bi), up to 1.5 m thick, commonly in stacked sequences, separated by thin-bedded turbidites (types Cii and D) in packets up to 1.2 m thick. A few of the coarser-grained sandstones (type Bi) contain scattered granules in their basal portions, whilst some exhibit large grooves and crescentic flutes on their soles. The grits are gradationally underlain and overlain by thin units of Dyffryn Flags.

On the northern side of the Claerwen valley, opposite Llanerch Cawr, the Cerig Gwynion Grits are 56 m thick and form prominent crags [SN 9010 6180 to SN 9020 6177]. Gradationally overlying 14 m of Dyffryn Flags, they largely comprise medium- to coarse-grained turbidite sandstones (type Bi) up to 5 m thick. The sandstones occur either singly or in stacked sequences, separated by packets of thin-bedded turbidites (types Cii and D) up to 1 m thick. The facies is sharply overlain by 3 m of Dyffryn Flags.

Farther west, crags [SN 8850 6270 to SN 8930 6265] on the northern slopes of the Claerwen valley, and Afon Claerwen [SN 8847 6252 to SN 8838 6278] (Plates 7d and 9) expose sections through two local tongues of Cerig Gwynion Grits and associated Dyffryn Flags (Figure 21a). The lower tongue, about 14 m thick, coarsens and thickens upwards from a gradational contact with underlying Dyffryn Flags. A 2 to 3 m-thick, coarse-grained turbidite sandstone (type Bi), which caps this sequence, is sharply overlain by Dyffryn Flags. The upper tongue, first appearing as crags [around SN 8930 6270], rapidly thickens westwards to about 12 m. Both lower and upper contacts with Dyffryn Flags are sharp. It also comprises a coarsening- and thickening-upwards sequence and is capped by a 2 m-thick, coarse-grained turbidite sandstone (type Bi). Cleaved, high matrix sandstones (type Bii), up to a metre thick, occur in both tongues.

West of Afon Claerwen, the two tongues of Cerig Gwynion Grits are last well exposed in Llwydnant, which follows the line of a north-north-east-trending fault. In the west bank [SN 8765 6188 to SN 8769 6197] the lower tongue is about 12 m thick, whilst 150 m downstream on the east bank [SN 8770 6213] the upper tongue is of similar thickness. Both tongues mainly comprise turbidite sandstones (type Bi), 0.3 to 1 m thick, with scattered thin packets of thin-bedded turbidites (types Ci and D). Farther west, exposure is poor and the failure of both tongues is inferred from features.

Caban Conglomerates facies

The facies is typified by and named after the Caban Conglomerates of Lapworth (1900). It comprises medium to thick-bedded turbidite couplets of pebble or cobble conglomerate and sandstone (type Ai), occurring as stacked and commonly amalgamated sequences (Plates 7b and 13a, b and d). Rare very thin packets of turbidite sandstone/mudstone couplets (type Cii and Di) occur scattered throughout. The facies also embraces the Drygarn Conglomerate of Davies (1926) and is equivalent to 'Facies A' of Kelling and Woollands (1969).

There are five stratigraphically discrete sequences of Caban Conglomerates facies in the Caban Conglomerate Formation (Figures 18 and 19). The first and second sequences occur at stratigraphic levels equivalent to parts of the Cwmere Formation, whilst the third and fourth sequences occur at levels equivalent to the lower half and uppermost part of the Derwenlas Formation respectively. The fifth sequence overlies the *M. sedgwickii* shales at the base of the Rhayader Mudstones. Most of the sequences have sharp basal contacts that are commonly erosion surfaces.

The crop of the first sequence of Caban Conglomerates facies is restricted to a 3 km-long belt on the north-western limb of the Tywi Anticline, between Talwrn [SN 9410 6410] and the Wye valley. There, it overlies Cerig Gwynion Grits with either gradational or sharp contacts. Up to 110 m thick, it is overlain by, and tongues out westwards and north-eastwards into Dyffryn Flags.

The second sequence of Caban Conglomerates facies crops out in three areas. Between the Wye valley and the Caban-coch reservoir, where it corresponds to the Lower Conglomerate of Lapworth (1900), it is up to 110 m thick, and is completely enveloped by Dyffryn Flags. Along the north-western limb of the Rhiwnant Anticline and on both limbs of the Gwesyn Syncline, where it corresponds to the Drygarn Conglomerate of Davies (1928), it ranges up to 180 m thick and is largely enveloped by Cwmere Formation.

The third sequence, the Upper Caban Conglomerate of Lapworth (1900), is only present between the Wye valley and the Caban-coch reservoir. Up to 50 m thick, it overlies Dyffryn Flags and passes both laterally and upwards into Sedgwickii Grits. Broadly its base equates with the boundary between the Cwmere and Derwenlas formations, although where strongly erosive, as in the Elan valley, it lies in the uppermost part of the Cwmere Formation.

The fourth and fifth sequences crop out only near the summit and on the northern slopes of Carn Gafallt [SN 932 647], where an intervening slack is assumed to be occupied by the *M. sedgwickii* shales (p.89). They comprise the thinnest and most restricted of the sequences of Caban Conglomerates facies. The fourth sequence is up to 15 m thick and largely overlies Sedgwickii Grits. The fifth sequence, up to 5 m thick, is overlain by and passes laterally into Sedgwickii Grits. Both sequences are finer grained than the others, with granule and pebble conglomerates predominating.

Individual conglomerate/sandstone couplets (type Ai) in the Caban Conglomerates facies are up to 6 m thick. They commonly rest on a scoured surface and in many cases occupy irregular, steep-sided channels (Plate 13a). As a result, individual turbidite units are markedly lenticular, and cannot be traced laterally for more than about 100 m. Rare sole marks include large grooves and irregular load casts. The clast-supported conglomerates range from pebble-cobble to granule-rich varieties with maximum clast sizes up to 25 cm. The coarser-grained examples are commonly disorganised, whereas the finer grades commonly exhibit inversely graded lower divisions and normally graded upper divisions (Plates 7b and 13d). Sandstones occupying the upper part of the couplets are coarse and pebbly and, though mainly structureless, locally display parallel lamination and grain-size layering. Mudstone rip-up clasts up to a metre in length and commonly concentrated in swarms are locally present in the couplets. Extrabasinal clasts in the conglomerates consist predominantly of quartz, quartzite and indurated sandstone. Scattered igneous pebbles comprise acid and basic volcanic rocks, largely dominated by rhyolites, together with granites, granophyres and diorites (Davies and Platt, 1933; Evans, 1992). Derived shelly material in the conglomerates includes corals, stromatoporoids, brachiopods and crinoid debris.

Thin turbidite sandstone/mudstone couplets (types Cii and Di) in packets up to 60 cm thick are locally preserved between the thick conglomeratic turbidites. The basal sandstones of the couplets are commonly coarse grained. Hemipelagites have not been found in the Caban Conglomerates facies.

Palaeocurrent vectors obtained from the Caban Conglomerates facies give a persistent trend towards the north-north-west (Figure 18).

BIOSTRATIGRAPHY

The first sequence of Caban Conglomerates facies ranges from the late *persculptus* Biozone into the early *acuminatus* Biozone. On the south-west side of the Wye valley it overlies Dyffryn Flags of probable *persculptus* Biozone age (Figure 18a, locality 7). Constraints on its upper age limit come from more widespread graptolite data in the enveloping facies (Figure 19).

The second sequence of Caban Conglomerates facies ranges from the late *acuminatus* Biozone through to about the *cyphus* Biozone. In the Rhiwnant Anticline the base is markedly diachronous; graptolites from the underlying strata are of the *acuminatus* Biozone at Llyn Carw (Figure 21b) and referable either to the *acinaces* or *cyphus* Biozone in Ffos y Rhestr (Figure 21b) and Nant Paradwys (Figure 21c). South of the district, in the Irfon valley (Figure 18a, locality 19; Table 12) the *atavus* Biozone has been proved immediately beneath the second sequence. On the north-western limb of the Tywi Anticline in the Elan valley, the underlying strata contain probable upper *acuminatus* Biozone graptolites (Figure 21d, locality 18; Table 12). Graptolites have only been

recovered from the middle part of the second sequence in the Elan valley, and probably belong either to the *acinaces* or *cyphus* Biozone (Figure 21a, locality 24; Table 12). Graptolites from 40 m above the top of the sequence in Nant Paradwys (Figure 21c, locality 26; Table 12) and 50 m above in the Elan valley (Figure 21d, locality 27; Table 12) are of *triangulatus* Biozone age, suggesting that the upper limit of the second sequence probably lies within the *cyphus* Biozone.

The third sequence of Caban Conglomerates facies probably ranges from the *magnus* to the *leptotheca* Biozone. As *triangulatus* Biozone graptolites occur in Dyffryn Flags about 10 m below the base of the third sequence in the Elan valley (Figure 21d, locality 26; Table 12), it is likely that the base of the conglomerates lies within the *magnus* Biozone. The base is certainly either at or below a level equivalent to the Cwmere–Derwenlas Formation junction, which lies within the *magnus* Biozone in the Rhayader sheet area. The top of the third sequence is only constrained by a *convolutus* Biozone fauna in the overlying Sedgwickii Grits (Figure 19, locality 30).

The fourth and fifth sequences of Caban Conglomerates facies, cropping out below and above the *M. sedgwickii* shales, are therefore of the late *convolutus* Biozone and *sedgwickii/halli* biozones respectively.

DETAILS

First sequence

Crags [SN 9492 6395 to SN 9468 6410] south-east of Carn Gafallt provide a semicontinuous section through the thickest part of the sequence. Beds of pebble-cobble conglomerate, up to a metre thick, are interbedded with subordinate beds of pebbly sandstone, locally with conglomeratic layers.

On the south-western slopes of the Wye valley, scattered crags [SN 9662 6545 to SN 9639 6540] provide a section through the sequence as it tongues out north-eastwards into Dyffryn Flags. Overlying Dyffryn Flags with a sharp contact, the lower part of the sequence forms a north-north-east-trending ridge that comprises a coarsening-upwards sequence of thick-bedded, locally pebbly, coarse-grained sandstones, and pebble conglomerates, the latter in units up to 2.5 m thick. To the west, slacks formed by tongues of Dyffryn Flags separate two smaller ridges with crags of medium- to thick-bedded, locally pebbly sandstone.

Second sequence

Between the Wye valley and the Caban-coch reservoir, the second sequence is well exposed in the line of crags [SN 9240 6410] on the north-west slope of Cnwch. Caban-coch quarry [SN 9240 9460] at the northern end of the Caban-coch dam exposes the upper part of the second sequence, passing up gradationally into Dyffryn Flags (Plates 7b, 12b, 13a and d). About 40 m of cobble and pebble conglomerates with subordinate coarse-grained, commonly pebbly sandstones, form a series of amalgamated, thick-bedded, turbidite conglomerate/sandstone couplets (type Ai). The bases of individual couplets are scoured and define broad channels, up to 3 m deep. One such channel, at the base of a cobble conglomerate in the lower part of the north-east side of the quarry, can be traced laterally for 20 m, displaying steep and locally undercut sides. Both normal and reverse grading, and a variety of loading phenomena, occur in the conglomerates and sandstones. Large, vertically orientated, mudstone clasts up to 2 m across occur in one bed in the north-east of the quarry. Also present are thin lenticular units, up to 60 cm thick, of thin-bedded turbidite sandstone/mudstone couplets (type Cii and Di). One such unit near the base of the quarry has yielded a graptolite fauna probably referable either to the *acinaces* or *cyphus* Biozone (Figure 21d; Table 12, locality 24). A quarry [SN 9240 6430] on the south side of the dam exposes the sharp basal contact of the second sequence of Caban Conglomerates facies.

Along the north-western limb of the Rhiwnant Anticline, the second sequence is well exposed on Cerrig Llwyd y Rhestr [SN 852 610]. The lower third is dominated by thick-bedded pebbly sandstones, the remainder by pebble conglomerates.

In the Gwesyn Syncline, the second sequence is repeated in Nant Paradwys [SN 8917 6022 to SN 8925 6012] by faulting. The base is exposed below a waterfall [SN 8917 6022], above which 55 m of thick-bedded, coarse pebble conglomerates, with rare units of pebbly sandstone, pass up gradationally into Dyffryn Flags.

Third sequence

The sequence is well exposed as a line of crags on the western slopes of Cnwch [SN 9265 6430], and in crags [SN 9260 6471] on the northern side of the Elan valley, immediately east of Caban-coch quarry (Plate 12b). At the latter locality, a series of deep scours cut into the Dyffryn Flags defines the base of the sequence. Pebble-cobble conglomerates predominate but at the top, thick-bedded sandstones form part of a gradational passage up into the Sedgwickii Grits.

Fourth sequence

This sequence forms a narrow east–west ridge on Carn Gafallt. A crag [SN 9430 6470] displays medium-bedded, pebbly, coarse-grained sandstones overlain by sandy pebble conglomerates.

Fifth sequence

Pebbly sandstones and sandy conglomerates are locally exposed in crags [e.g. SN 9445 6477] along a narrow east–west-trending ridge on Carn Gafallt.

Dyffryn Flags facies

The facies of thin-bedded turbidite sandstone/mudstone couplets (type C) and laminated hemipelagites, which overlies the Cerig Gwynion Grits on Gwastedyn Hill, constitutes the type development of the Dyffryn Flags of Lapworth (1900). Although this stratigraphically restricted sequence represents the thickest development (185 m), the facies name is applied to mappable sequences of comparable lithology throughout the lower part of the Caban Conglomerate Formation. It includes, for example, the 'Intermediate Shales' of Lapworth (1900). The facies is restricted to those parts of the Caban Conglomerate Formation that are coeval with the Cwmere Formation. It typically occupies a transitional position between the coarser-grained Cerig Gwynion Grits and Caban Conglomerates facies and the enveloping slope-apron mudstones of the Cwmere Formation (Figures 18 and 19). Some developments of the facies are locally too thin to be depicted on the map.

Contacts between Dyffryn Flags and the enveloping Cwmere Formation are typically gradational. Dyffryn Flags are recognised where thin turbidite sandstones (type C) exceed 10 per cent of the sequence. Similarly, contacts between Dyffryn Flags and Cerig Gwynion Grits are also commonly gradational. There, Dyffryn Flags are recognised where thick-bedded turbidite sandstones (type Bi) are sparsely developed or absent, in sequences of thin-bedded sandstone/mudstone couplets (type C). Where Dyffryn Flags succeed Caban Conglomerates, the junction is drawn at the disappearance of conglomeratic turbidites within gradational sequences commonly several metres thick.

In the stacked sequences of thin-bedded turbidite sandstone/mudstone couplets (type C) that characterise the Dyffryn Flags, type Cii are by far the most abundant, their fine- to medium-grained, basal sandstones commonly ranging up to 5 cm thick. Scattered type Ci turbidites with sandstones locally up to 45 cm thick, and rare type Bi sandstones up to 65 cm thick, occur most frequently in areas of vertical or lateral transition into Cerig Gwynion Grits or Caban Conglomerates facies. Flute and groove casts are mainly restricted to the soles of these thicker sandstones. Shelly debris is locally present in the sandstones of type C couplets. The percentage of sandstone in the Dyffryn Flags varies between 10 and 75 per cent, with higher proportions characterising vertical and lateral transitions into the coarser-grained facies. Thin turbidite mudstones (mainly type Di and rare Dii) occur scattered throughout the facies but increase in abundance in zones of transition with the Cwmere Formation.

Laminated hemipelagites are present throughout, except where Dyffryn Flags overlie the Mottled Mudstone Member and burrowed hemipelagites locally persist into the basal few metres of the sandy facies. Hemipelagites of either type rarely exceed a centimetre in thickness.

BIOSTRATIGRAPHY

In its type area, in the Wye valley, the Dyffryn Flags ranges from the *persculptus* into the *acuminatus* Biozone (Table 9). However, on Carn Gafallt, sequences adjacent to Caban Conglomerates facies span an interval from the *persculptus* to the *magnus* Biozone (Figure 19). Only the *persculptus*, *acuminatus* and *triangulatus* biozones are proved; the rest are inferred.

DETAILS

Rhiwnant Anticline

On the north-western limb of the anticline, west of the Claerwen valley, the second sequence of Caban Conglomerates passes laterally across a north-trending fault into a tongue of Dyffryn Flags, which gives rise to the ridge of Brach y Wern [SN 8706 6260]. Scattered crags at the southern end of this ridge expose thick-bedded turbidite sandstones (type Bi) and, at the northern end, more typical turbidite sandstone/mudstone couplets (type C). Farther east, the same horizon is recognised on the eastern side of the Claerwen valley in crags [SN 8765 6307 to SN 8815 6325] that expose a 30 m-thick packet of Dyffryn Flags containing 10 to 20 per cent of sandstone (Figure 18a).

Craig Cwm-Clyd [SN 8900 6250], on the northern side of the Claerwen valley, provides a semicontinuous sequence through units of Dyffryn Flags enveloping two local tongues of Cerig Gwynion Grits. The lowest development gradationally overlies the Mottled Mudstone Member and comprises a sequence, 14 m thick, thickening- and coarsening-upwards and gradationally overlain by the lower tongue of Cerig Gwynion Grits. A further 35 m of Dyffryn Flags, forming a marked slack, occur between the two tongues of Cerig Gwynion Grits and exhibit sharp lower and upper contacts. Probable *persculptus* Biozone graptolites occur in a roadside quarry [SN 8855 6268] at the top of this sequence (Figure 21a, locality 5; Table 12). A higher 30 m-thick sequence of Dyffryn Flags overlies the upper tongue of Cerig Gwynion Grits and in turn passes up gradationally into Cwmere Formation mudstones. Crags at the base of this upper sequence contain graptolites referable to the *persculptus* or possibly to the *acuminatus* Biozone (Figure 21a, locality 6; Table 12).

Sequences of Dyffryn Flags, too thin to be portrayed on the map, occur beneath and above the Cerig Gwynion Grits on the north side of the Claerwen valley [SN 9015 6180], opposite Llanerch Cawr. The lower sequence is 14 m thick and gradationally overlies the Mottled Mudstone Member, whilst the upper one is only 4 m thick and passes rapidly upwards into Cwmere Formation mudstones. Similar sequences of the facies persist at these same stratigraphic levels along the south-eastern limb of the anticline to the point [SN 8760 5975], where the Cerig Gwynion Grits pass laterally into Dyffryn Flags (Figure 18). A thin sequence of Dyffryn Flags exposed just south of the district, in Nant yr Ast [SN 8725 5941], confirms the lateral failure of the Cerig Gwynion Grits.

North-western limb of Tywi Anticline

Sections in Dyffryn Flags overlying the Cerig Gwynion Grits, between the southern edge of the district and Talwrn [SN 9410 6410], include the Marchnant [SN 9060 6035 to SN 9058 6056] and Nant Rhyd-goch [SN 9163 6159 to SN 9154 6171]. In the former, Dyffryn Flags overlie Cerig Gwynion Grits with a gradational contact and contain abundant turbidite sandstone/mudstone couplets (type Cii) with scattered sandstones (type Ci) up to 15 cm thick. A packet composed mainly of mudstone turbidites (type D), is present in the middle of the sequence. The junction with the overlying Cwmere Formation mudstones is a sharp contact.

On the western side of the Wye valley, a thin tongue of Dyffryn Flags forms a marked slack between the Cerig Gwynion Grits below and the first sequence of Caban Conglomerates facies above. The upper part of this tongue is exposed in low crags [SN 9662 6546] below Graig Allt-y-bont, where it contains scattered turbidite sandstones (type Ci) up to 10 cm thick and rare type Bi turbidite sandstones up to 60 cm thick. Probable *persculptus* Biozone graptolites have been obtained (Figure 18a; Table 12, locality 7).

North of the Wye valley, the gradational junction of Dyffryn Flags with Cerig Gwynion Grits is best seen in Cerrig Gwynion quarry [SN 9700 6575] (Figure 20). There, *persculptus* Biozone graptolites occur at the base [SN 9722 6577] and *acuminatus* Biozone graptolites about 40 m above the base of the sequence [SN 9723 6597] (Table 9, locality A). A semicontinuous section through the remainder of the Dyffryn Flags sequence is displayed in Prysg stream [SN 9835 6684 to SN 9800 6722], Lapworth's (1900) type locality for his Dyffryn Flags. The fining- and thinning-upward sequence largely comprises turbidite sandstone/mudstone couplets (type C), with type Ci couplets restricted to the lower half. Towards the middle part [around SN 9822 6700], the sequence is punctuated by sandstone-rich bundles, which probably correlate with the first sequence of Caban Conglomerates facies. For a short distance

above this level, shelly sandstones in the facies are commonly decalcified, and represent the 'Rottenstone Beds' of Lapworth (1900). A graptolite fauna referable to the *acuminatus* Biozone is present in the upper part of the sequence [SN 9806 6719] (Table 9, locality B) about 30 m below the top of the Dyffryn Flags.

North-east of Gaufron [SN 999 681], the Dyffryn Flags overlie the Mottled Mudstone Member. There, the lower part of the sequence, which represents the lateral replacement of the Cerig Gwynion Grits (last seen 300 m to the west), is exposed in a quarry [SO 0004 6839]. Scattered turbidite sandstone/mudstone couplets (type Ci), with sandstones up to 30 cm thick, occur in a sequence dominated by type Cii couplets. The proportion of sandstone is about 30 per cent. Graptolites from the quarry probably belong to the *persculptus* Biozone (Figure 18a, locality 8; Table 12). A quarry [SO 0169 7019] north of Vaynor, is typical of the upper part of the Dyffryn Flags sequence in this area. Rare type Ci turbidite sandstone/mudstone couplets are present but the proportion of sandstone is only 20 per cent. Graptolites, possibly of the *acuminatus* Biozone, have been obtained (Figure 18a, locality 16, Table 12). Scattered sections on Lan Goch [SN 023 702], north of Vaynor, expose the feather edge of the Dyffryn Flags sequence as it passes laterally eastwards into Cwmere Formation mudstones. There the sandstone percentage has fallen to about 10 per cent. Turbidite sandstone/mudstone couplets (type Cii) and turbidite mudstones (type D) are abundant, while type Ci couplets are sparsely developed.

Sections in Dyffryn Flags underlying the second sequence of Caban Conglomerates facies, between Talwrn [SN 9410 6410] and the Caban-coch reservoir, include a gully [SN 9240 6370 to SN 9244 6371] on the western side of Craig Cnwch. There, the sequence exhibits an upward increase in the percentage of sandstone towards the conglomerates. Graptolites, although not diagnostic, suggest an *acuminatus* Biozone age (Figure 21d, locality 17; Table 12). A less sandy succession is seen directly beneath the conglomerates, in the quarry [SN 9238 6423] on the south bank of the Caban-coch reservoir. Normalograptid graptolites from the section (Figure 21d, locality 18; Table 12) have a range within the span of the upper *acuminatus* to the *cyphus* Biozone. However, they are possibly not younger than the *acuminatus* Biozone, for in the Wye valley section (Figure 19; Table 9) pre-*acinaces* faunas are dominated by *normalis* group graptolites, with many collections containing no other taxa.

The Dyffryn Flags that lie between the second and third sequences of Caban Conglomerates facies (Intermediate Shales of Lapworth, 1900) are best displayed in crags [SN 9250 6466], just east of Caban-coch quarry (Plate 12b). Only the upper half of the sequence is exposed. It comprises type Cii and Di couplets; scattered type Ci couplets, with basal sandstones up to 30 cm thick, appear in the uppermost part. Intercalated laminated hemipelagites are less than a millimetre thick. A similar sequence on the south side of the Elan valley has yielded *triangulatus* Biozone graptolites from crags 10 to 20 m below the third sequence of Caban Conglomerates facies (Figure 21d, localities 27 and 28; Table 12).

Dyffryn Flags developed beneath Sedgwickii Grits and Gafallt Shales, west of the Foel Fault, are well exposed on Craig y Foel [SN 9160 6400 to SN 9205 6446]. Gradationally overlying Cwmere Formation mudstones, the sequence initially thickens and coarsens upwards. About the middle of the sequence [around SN 9185 6426], type Ci turbidite sandstone/mudstone couplets become common and rare type Bi turbidite sandstones up to 50 cm thick appear. Above this level, both diminish in importance upwards. This burst of thicker sandstones is probably coeval with the second sequence of Caban

Conglomerates facies seen to the east in Caban-coch quarry [SN 9240 9460]. Sharp contacts with overlying Sedgwickii Grits and Gafallt Shales are exposed in the crags.

The western shore [SN 9083 6440 to SN 9099 6462] of the Carreg-ddu reservoir and an adjacent track section [SN 9068 6440 to SN 9099 6462] expose three thin packets of Dyffryn Flags high in the Cwmere Formation (Figure 18). The basal sandstones of the type Cii turbidites are up to 6 cm thick and commonly medium to coarse grained. The packets are laterally equivalent to the second sequence of Caban Conglomerates.

An outlier of Dyffryn Flags, comprising up to 60 per cent of sandstone, is exposed in a track section [SN 9194 6295 to SN 9200 6217] on Graig Fawr. Its lateral equivalence to the second sequence of Caban Conglomerates facies is suggested by abundant type Ci turbidite sandstone/mudstone couplets, with basal sandstones up to 45 cm thick, and by the presence of *acuminatus* Biozone graptolites (Figure 18a, localities 14 and 15; Table 12).

Sedgwickii Grits facies

The facies broadly equates with the '*Monograptus-Sedgwickii* Grits' of Lapworth (1900), from which it takes its name. However, graptolites recorded during the present survey show that most of the facies predates the *sedgwickii* Biozone. Sedgwickii Grits predominantly comprise type Ci turbidite sandstone/mudstone couplets (Plate 13E). Subordinate type Cii couplets and thick-bedded turbidite conglomerates (type Ai), turbidite sandstones (type Bi) and turbidite mudstones (type Di) are locally present. Intercalated hemipelagites are mainly burrowed and thus the facies is largely oxic.

Restricted to the upper half of the Caban Conglomerate Formation (Figure 19), Sedgwickii Grits crop out between the Carreg-ddu reservoir and the Wye valley, where they largely overlie the third sequence of Caban Conglomerates facies (Plate 12B). From their thickest development of about 100 m around Carn Gafallt, they thin along strike both to the north-east and south-west, tonguing out in both directions into Gafallt Shales. Along most of their crop they pass upwards into Gafallt Shales. However, on the northern slopes of Carn Gafallt a marked slack occurs in the uppermost part of the Sedgwickii Grits, which, traced eastwards, separates the fourth and fifth sequences of Caban Conglomerates. The slack is assumed to be occupied by the *M. sedgwickii* shales (p.89).

The sandstones of individual type Ci sandstone/mudstone couplets are fine to coarse grained, 5 to 40 cm thick and locally amalgamated. Shell and crinoid debris is abundant (Plate 13C), either scattered throughout or concentrated in basal lags, and is frequently decalcified to rottenstone lenses. Sandstone soles commonly display flutes, grooves and trace fossil casts. Although the typical Bouma T_{bcde} sequences of type Ci couplets are present, middle-cut-out T_{be} sequences, in which the sandstones are entirely parallel laminated, are widespread. The mudstone components of these couplets are typically subordinate.

Individual type Cii sandstone/mudstone couplets are 2 to 12 cm thick. The basal sandstones, generally less than 5 cm thick, are predominantly fine grained but coarse examples, rich in bioclastic debris, are locally present.

The sandstones mainly lack erosive sole structures but display abundant trace fossil casts.

Scattered, thick, coarse-grained and locally pebbly turbidite sandstones (type Bi), up to a metre thick, occur in the lowest part of the Sedgwickii Grits between Carn Gafallt and immediately west of the Foel Fault. Similar but thinner type Bi sandstones are also present in the middle part of the sequence, in its thickest development around Carn Gafallt, where they comprise a 4 m-thick sequence, coarsening and thickening upwards, of stacked, coarse pebbly and shelly sandstones. A few thin, type Bi sandstones, up to 50 cm thick, commonly ungraded and with markedly channelled bases, are also present in the uppermost part of the Sedgwickii Grits around Elan village.

Two 1.5 m-thick packets of thin conglomeratic turbidites (types Ai and Aii), separated by 14 m of turbidite sandstone/mudstone couplets (type C) and subordinate thin, type Bi sandstones, occur near the top of the Sedgwickii Grits around Elan village. Tonguing out to the south-west at the Water Treatment Works [SN 9310 6510], they probably lie at the same stratigraphic level as the fourth sequence of Caban Conglomerates facies on Carn Gafallt. Individual thin type Ai and Aii turbidites range from 0.3 m to 1 m thick and occur interbedded with thin turbidite mudstones (type Di). The type Ai conglomerate/sandstone couplets have sharp tops and bases, and consist largely of ungraded clast-supported granule and pebble-cobble conglomerate. Both Ai and Aii types of turbidite are markedly lenticular around the Water Treatment Works.

Scattered siltstone/mudstone couplets (type Di) up to 50 cm thick occur scattered throughout the facies but in the upper part, west of Carn Gafallt, such turbidites form packets several metres thick, interleaved with the dominant sandstone-rich parts of the facies.

The interbedded hemipelagites are predominantly of the pale grey burrow-mottled variety. However, in the middle and uppermost part of the Sedgwickii Grits in the Elan valley, very thin (about 1 mm), laminated hemipelagites occur associated with packets of thick-bedded turbidite mudstones (type D). None has yielded graptolites.

Sedgwickii Grits overlie the third sequence of Caban Conglomerates facies with a gradational contact, their base being taken where turbidite sandstone/mudstone couplets (type C) become predominant over thick-bedded sandstones (type Bi). West of the Foel Fault on Craig y Foel, Sedgwickii Grits overlie Dyffryn Flags with a sharp contact; thick sandstones (type Bi) occur in the basal part. There, the base of the Sedgwickii Grits coincides with the entry of burrowed hemipelagites, suggesting an horizon equivalent to the contact between the Cwmere and Derwenlas formations in the enveloping mudstone succession.

The proportion of sandstone in the facies varies from 70 to 30 per cent, with sections typically comprising both thinning- and fining-upwards sequences. However, around Carn Gafallt, where the facies is at its thickest, this gross trend is interrupted by local coarsening-upwards sequences. Towards the lateral limits of the facies, thick sandstones (type Bi) disappear, type Ci couplets become thinner and subordinate to type Cii, and the proportion of sandstone decreases to about 30 per cent.

Palaeocurrent data from sole marks and cross-lamination, collected from the Sedgwickii Grits, indicate transport towards the north-west, a direction consistent with their geometry.

BIOSTRATIGRAPHY

The Sedgwickii Grits sequence east of the Foel Fault probably belongs mainly to the *convolutus* Biozone. Graptolites of this age occur about 65 m above the base of the sequence on Carn Gafallt (Figure 18a, locality 30). However, exposures [SN 9403 6468] on Carn Gafallt above the *M. sedgwickii* shales-slack yield graptolites of the *sedgwickii–halli* biozonal interval (Figure 19; Table 12, locality 31). Sequences of the facies west of the Foel Fault are earlier (*magnus* to *leptotheca* biozones), their bases resting either on or just above a level equivalent to the contact between the Cwmere and Derwenlas formations.

DETAILS

The crest of a ridge [SN 9580 6558] south of Dolifor Farm provides the most north-easterly section in Sedgwickii Grits. There, the basal sandstones of type Ci and subordinate type Cii turbidite sandstone/mudstone couplets are up to 6 cm and 2 cm thick respectively and account for 30 per cent of the sequence.

Crags on the southern slopes of Carn Gafallt provide a discontinuous section through the thickest development of the facies. The junction with the underlying third sequence of Caban Conglomerates facies is not exposed; coarse-grained turbidite sandstones (type Bi), up to 60 cm thick, occur in crags [SN 9374 6455] close to the base. In the middle of the section, crags [SN 9404 6452] expose a sequence at least 4 m thick, coarsening and thickening upwards. There, 3 m of stacked, coarse, shelly, decalcified turbidite sandstones (type Bi), up to 30 cm thick and in places pebbly and cross-bedded, are capped by over 40 cm of sandy, granule conglomerate, with abundant brachiopod and bryozoan debris. Slightly higher and about 60 m above the base of the facies, crags [SN 9404 6457] of thin-bedded sandstones have yielded *convolutus* Biozone graptolites (Figure 18a, locality 30; Table 12). Crags [SN 9404 6460] near the top of the section exhibit stacked type Cii and subordinate type Ci sandstone/mudstone couplets. The sandstones, there comprising 50 per cent of the sequence, are fine grained, shelly and up to 5 cm thick. A slump unit, 2.4 m thick, with slump fold axes orientated 138° and indicating transport to the south-west, occurs in crags [SN 9435 6466] at a similar level to the east. A section [SN 9403 6468], just north of the summit of Carn Gafallt on the northern side of the *M. sedgwickii* shales-slack, comprises oxic-facies silt-laminated turbidite mudtones (type E) with scattered sandstones (type Cii couplets) up to 2 cm thick. The siltstones and sandstones have yielded graptolites of the *sedgwickii–halli* biozonal interval (Figure 18 and 19, locality 31; Table 11).

The road section [SN 9286 6478 to SN 9302 6498] in the Elan valley and an adjacent section [SN 9300 6500 to SN 9310 6515] behind the Water Treatment Works exposes a semicontinuous section through the upper half of the local Sedgwickii Grits sequence (Plate 12b). Cuttings near the base of the road section expose a thick packet of mudstone turbidites (type Di), each ranging up to 50 cm thick, interbedded with both burrowed and laminated hemipelagites. The remainder of the road section exposes the more typical turbidite sandstone/mudstone couplets (type C) with sandstones up to 40 cm thick. The topmost 30 m of the facies is displayed in the Water Treatment Works. At the base, three beds of ungraded pebble-cobble conglomerate (type Ai) separated by turbidite mudstones (type D) form a unit 1.2 m thick, which lenses out rapidly to the west. It is overlain by 14 m of stacked sandstone/mudstone couplets (type C). The type Ci sandstones are mainly coarse grained, ungraded and up to 0.3 m thick and have markedly channelled bases. Evidence for synsedimentary wet-sediment disruption is widespread. The sandstones are overlain by a further conglomeratic unit, 1.5 m thick, comprising a type Ai overlain by a type Aii conglomeratic turbidite. In the latter, the basal pebble-cobble conglomerate is overlain by coarse-grained laminated sandstone, and this in turn by mudstone. Both conglomerates lens out rapidly into mudstone to the west. This conglomeratic unit is overlain by about 8 m of turbidite mudstones with scattered turbidite sandstones up to 2 cm thick (mainly type Di with a few type Cii). The presence of milli-metre-thick hemipelagites suggests that this mudstone packet may correlate with the *M. sedgwickii* shales. It is overlain by about 5 m of sandstone/mudstone couplets (type C), in which the effects of wet-sediment destratification are widespread. A massive, coarse-grained sandstone (type Bi) 50 cm thick caps the sequence, and is overlain by Gafallt Shales.

The steep crags on the north-west [SN 9270 6430] and south-east [SN 9290 6405] sides of Cnwch display sections through Sedgwickii Grits (Plate 13e). To the north, a complete section through the facies is present, on a buttress [SN 9260 6477] to the north of Caban-coch dam (Plate 12b). There, thick, coarse-grained sandstones (type Bi) with scattered granules occur in the lowermost part of the grits, where they succeed the third sequence of Caban Conglomerates facies. Above this level, a sequence of sandstone/mudstone couplets (type C), in which sandstones are up to 40 cm thick (but on average 15 cm) thins and fines upwards into the Gafallt Shales.

Immediately west of the Foel Fault, at the base of a prominent buttress [SN 9204 6442] on the lower slopes of Craig y Foel, 30 m of Sedgwickii Grits are exposed overlying Dyffryn Flags with a sharp contact. A few coarse-grained sandstones (type Bi), up to a metre thick, occur in the basal part but otherwise the facies comprises a fining- and thinning-upwards sequence of thinner-bedded sandstone/mudstone couplets (type C) that passes upwards into Gafallt Shales. Traced north-westwards up the buttress, the sequence of Sedgwickii Grits thins markedly, the lower and upper parts passing laterally into Gafallt Shales. Above the buttress, at the top of Craig y Foel [SN 9195 6449] and there overlying Gafallt Shales, the Sedgwickii Grits are reduced to 10 m of sandstone/mudstone couplets (mainly type Cii), in which the sandstones are up to 10 cm thick.

Sedgwickii Grits fail north of Craig y Foel, reappearing on Creigiau Dolfolau [SN 9180 9200 to SN 9147 6565], on the eastern side of the Carreg-ddu reservoir. There, overlying Gafallt Shales with a sharp contact, they comprise 25 m of turbidite sandstone/mudstone couplets (mainly type Cii), in which the sandstones are up to 20 cm thick. The proportion of sandstone declines from 70 to 50 per cent as the facies passes up into Gafallt Shales.

Gafallt Shales facies

The facies broadly equates with the Gafallt Shales of Lapworth (1900) and largely comprises silt-laminated, burrow mottled, pale and dark grey, colour-banded, oxic-facies mudstones with variably abundant thin sandstones.

In detail it represents thinly intercalated turbidite mud-stones (type E) and burrowed hemipelagites interbedded with locally abundant turbidite sandstone/mudstone couplets (type C) and sparse, thin, high-matrix sand-stones (type Bii). Rare thin units with laminated hemi-pelagites are locally developed.

Cropping out between the Carreg-ddu reservoir and the Wye valley, Gafallt Shales typically intervene between the coarser facies of the Caban Conglomerate Formation (Caban Conglomerates facies and Sedgwickii Grits) and the enveloping, oxic slope-apron mudstones of the Derwenlas Formation and Rhayader Mudstones. The vertical and lateral contacts displayed by the facies (Figure 19) reflect its transitional setting in the for-mation. It represents the oxic equivalent of the earlier anoxic Dyffryn Flags facies. The western limit of Gafallt Shales, although in detail involving complex interdigita-tion, for convenience has been taken at the Carreg-ddu reservoir.

The thickest development of Gafallt Shales, about 250 m, occurs west of the Foel Fault, where they span an interval elsewhere equivalent to the Derwenlas For-mation and the lower part of the Rhayader Mudstones. In the central part of the crop around Carn Gafallt, about 170 m succeed the the Sedgwickii Grits.

Contacts with underlying Sedgwickii Grits are grada-tional, the junction being taken at the incoming of very thin turbidite mudstones (type E) and the disappearance of thick sandstone/mudstone couplets (type Ci). In contrast, west of the Foel Fault, where Gafallt Shales are overlain by tongues of Sedgwickii Grits, the contacts are sharp. The basal contact with both Dyffryn Flags and Cwmere Formation mudstones is sharp. As it coincides with the upward change from laminated to burrowed hemipelagites, it may be correlated with the contact between the Cwmere and Derwenlas formations. Lateral and vertical transitions into the Derwenlas Formation and Rhayader Mudstones are gradational, being taken at the disappearance of type C sandstone/mudstone couplets.

In Gafallt Shales the thinly intercalated silt-laminated turbidite mudstones (type E) and burrowed hemi-pelagites compare with the oxic facies of the Derwenlas Formation and Rhayader Mudstones. Colour banding is prevalent, but more thoroughly oxidised, pale grey sequences are locally developed, notably in the upper parts of the facies adjacent to the Rhayader Mudstones. The type E turbidite mudstones mainly comprise top-cut-out, strongly silt-laminated units (T_{0-3}) with individual siltstone laminae rarely exceeding 5 mm in thickness.

Two impersistent units with laminated hemipelagites occur in the facies; they are laterally equivalent to units of anoxic facies in the coeval slope-apron mudstone for-mations. The lower unit, a metre thick, has been recog-nised only in the lower part of the facies adjacent to the Carreg-ddu reservoir [SN 9150 6600] and is a correla-tive of one of the *convolutus* bands in the Derwenlas Formation. The upper one has been recognised at several localities on the north side of the Elan valley, east of the Foel Fault. At one point [SN 9243 6488] (Figure 21d, locality 33) it has yielded graptolites of the *sedg-wickii–halli* biozonal interval and is clearly laterally equiv-alent to the *M sedgwickii* shales to the west (p.89).

The most abundant turbidite sandstone/mudstone couplets in the Gafallt Shales are of the Cii type. Scattered throughout, they comprise thin, fine- to locally coarse-grained sandstones up to 2 cm thick, overlain by mudstones up to 12 cm thick. Less common type Ci couplets exhibit fine- to coarse-grained sandstones, 5 to 30 cm thick, locally with shelly bases. Some exhibit marked convolute lamination.

The distribution of type C turbidite sandstone/mudstone couplets is laterally variable in the Gafallt Shales. East of the Foel Fault, the Gafallt Shales comprise a thinning- and fining-upwards sequence, in which type Ci couplets become less abundant and thinner upwards, and the overall sandstone percentage decreases from about 30 per cent near the base, to about 15 per cent near the top. West of the Foel Fault, on Y Foel, type Ci couplets are largely absent. Above the tongue of Sedg-wickii Grits, adjacent to the west side of the fault, Gafallt Shales thin and fine upwards into a unit of anoxic facies equivalent to the *M. sedgwickii* shales; immediately above and below this unit, type C couplets are absent.

Packets of type C couplets in the upper parts of the Gafallt Shales sequence, north of Y Foel on Graig Dolfaenog [SN 917 665], reflect a complex and pro-tracted interdigitation with the enveloping Derwenlas Formation and Rhayader Mudstones. The packets locally contain up to 50 per cent sandstone.

Upper levels of the Gafallt Shales on the northern part of Y Foel and Graig Dolfaenog contain scattered high-matrix (type Bii) turbidites. Consisting of massive, cleaved, very muddy, siltstones or fine-grained sand-stones, they range in thickness from 4 to 20 cm. In the same area [SN 9160 6602] a single, 10 cm-thick, muddy debrite with phosphate pebble clasts has been noted in the lower part of the sequence.

Although no palaeocurrent data have been collected from the Gafallt Shales, largely due to the scarcity of sole structures, the geometry of the facies unit suggests that currents flowed towards the north or north-west.

BIOSTRATIGRAPHY

The Gafallt Shales facies ranges in age from the *magnus* Biozone to the early *turriculatus* Biozone but is markedly diachronous (Figure 19). West of the Foel Fault, it overlies Dyffryn Flags at a level equivalent to the contact between the Cwmere and Derwenlas formations, which is known to lie within the *magnus* Biozone in the district. The *convolutus* Biozone has been proved in the lower unit of anoxic facies developed in the lower third of the sequence in this area (Figure 18a, locality 29; Table 12).

The *M. sedgwickii* shales are present within the crop of the middle part of the Gafallt Shales immediately west of the Foel Fault (p.89). Furthermore the impersistent upper unit of anoxic facies, laterally equivalent to the *M. sedgwickii* shales, has yielded graptolites from the *sedgwickii–halli* biozonal interval (Figure 21d, locality 33; Table 12) from east of the Foel Fault. The remaining Gafallt Shales are undated but a probable *gemmatus* Bio-zone fauna in the lowermost part of the overlying

Rhayader Mudstones in the Elan valley (Figure 18a, locality 34; Table 12) provides an upper limit.

DETAILS

On the northern slopes of Carn Gafallt a stream bank [SN 6790 6514] south-west of Fron-dorddu exposes sandstone/mudstone couplets (type Ci) in which the sandstones are up to 7 cm thick.

Crags east of the Foel Fault, on the northern side of the Elan valley, provide extensive exposure in Gafallt Shales (Plate 12b). The lower part, including the junction with the Sedgwickii Grits, is exposed in a cutting in the Water Treatment Works [SN 9310 6516 to SN 9333 6540]. Largely comprising pale and dark grey, colour-banded, oxic-facies mudstones with many type Cii sandstone/mudstone couplets, it also contains scattered type Ci couplets with coarse-grained basal sandstones rarely exceeding 10 cm. The higher parts of the sequence and the gradational junction with the overlying Rhayader Mudstones are well exposed in crags above the Water Treatment Works. Farther south-west, crags around and including Craig Gigfran [SN 9260 6480] display a much muddier sequence, above the Sedgwickii Grits, than that seen to the east. Type Ci couplets are rare and confined to the lower part. Crags [SN 9242 6486] higher in the sequence, comprising turbidite mudstones and thin sandstone/mudstone couplets (type Cii) intercalated with sparse, millimetre-thick laminated hemipelagites, have yielded *sedgwickii–halli* Biozone graptolites (Figure 21d, locality 33; Table 11) suggesting that these strata are the correlatives of the *M. sedgwickii* shales. The overlying sequence is similarly muddy.

Immediately west of the Foel Fault the sequence of Gafallt Shales above the tongue of Sedgwickii Grits is exposed in crags [SN 9208 6458 to SN 9222 6470].

Crags on the western slope of Y Foel, above the Carreg-ddu reservoir provide numerous sections in the basal part of the Gafallt Shales between the Dyffryn Flags and the overlying tongue of Sedgwickii Grits. A near-continuous section [SN 9175 6509] is displayed at the southern end, in which turbidite mudstones (type E) and scattered type Cii couplets, with basal sandstones up to 2 cm thick, are intercalated with burrowed hemipelagites. The sequence above the Sedgwickii Grits is exposed in numerous crags on the flat-topped Y Foel ridge. Type Cii sandstone/mudstone couplets, in which basal sandstones are about 2 cm thick, are variably abundant in weakly colour-banded, oxic-facies mudstones. A few, thin, high-matrix sandstones and muddy siltstones (type Bii) are locally present [e.g. SN 9294 6547].

North of the north-east-trending splay of the Foel Fault, the lower part of the Gafallt Shales is exposed in a road section [SN 9138 6602 to SN 9132 6625], in which strongly colour-banded, oxic-facies mudstones contain abundant sandstone/mudstone couplets (mainly type Cii). The basal sandstones of scattered type Ci couplets are up to 7 cm thick, commonly coarse grained and decalcified. Similar but higher strata are exposed in the adjacent Nant Dolifolau, where [at SN 9150 6660] a metre-thick unit with laminated hemipelagites has yielded *convolutus* Biozone graptolites (Figure 18 and Table 11). Rare, high-matrix sandstones (type Bii), up to 20 cm thick, appear upstream at higher stratigraphic levels. At the confluence [SN 9160 6602] of Nant Dolifolau and Nant Rhinghyll, a 10 cm-thick debris flow with phosphate pebbles is exposed.

The upper two-thirds of the Gafallt Shales and the junction with the Rhayader Mudstones are exposed in the semicontinuous section [SN 915 661 to SN 916 672] on Graig Dolfaenog, along the eastern side of the Carreg-ddu reservoir. Pale and dark grey, colour-banded, oxic-facies mudstones occur throughout but the colour banding becomes less distinct towards the top. Scattered high-matrix sandstones (type Bii) occur in the lower part. Type Cii sandstone/mudstone couplets are common throughout but type Ci couplets are largely confined to the upper part, where sandstone-rich packets containing up to 50 per cent sandstones are developed.

Ystrad Meurig Grits Formation

Jones (1922, p.5) recognised 'a group of massive grits of great thickness' in the Teifi Anticline near Ystradmeurig, in the Llanilar sheet area; he subsequently named them the Strata Florida Grits (Jones, 1938). Following informal usage by James and James (1969), the term Ystrad Meurig Grits was formally applied by Cave (1979). An account of the formation based on the preliminary results of the recent survey has been presented by Cave (1991; 1992). Previous authors followed Jones (1938) in assuming that the Ystrad Meurig Grits Formation occupied a level between the *magnus* and *leptotheca* graptolite bands. To the north, in the Aberystwyth district, Cave and Hains (1986) and Strong (1979) recognised two separate sandstone bodies in the *convolutus* and *sedgwickii* biozones respectively. They suggested that these shared a common source with the Ystrad Meurig Grits Formation and that both were related to the Caban Conglomerate Formation of the Rhayader area, a proposal first made by James and James (1969). The recent survey has shown that, in the Teifi Anticline, the Ystrad Meurig Grits Formation comprises a complex body of sandstone turbidites coeval with parts of the Cwmere, Derwenlas and Rhayader Mudstones formations and collectively spanning an interval from the *acuminatus* to the *sedgwickii* Biozone (Figure 23).

The crop of the Ystrad Meurig Grits Formation is confined to the Teifi Anticline, south of the Ystwyth Fault, where its architecture, combined with folding and faulting, has produced a complex outcrop pattern (Figure 22). North of the Ystwyth Fault, the Ystrad Meurig Grits Formation has been proved in Cwm Rheidol No. 1 borehole [SN 7302 7833] (Appendix 1). Comparable and contemporary strata in the Aberystwyth district, included in the Derwenlas Formation and Cwmsymlog Formation (Rhayader Mudstones of this account) by Cave and Hains (1986), are here regarded as contiguous parts of the formation.

The Ystrad Meurig Grits Formation comprises thick- to medium-bedded turbidite sandstones together with thinner-bedded turbidite sandstones, mudstones and hemipelagites. It has been divided into three separate and partly recurrent facies: Chwarel Goch facies; Ystrad Meurig Grits facies; and Henblas facies. A distinctive unit of thin-bedded, shelly turbidite sandstones, phosphatic conglomerates, turbidite mudstones and laminated hemipelagites comprises the Hafod Member. It forms the top of the formation in the Coed Bwlchgwallter [SN 763 727] area.

Chwarel Goch facies

The Chwarel Goch facies comprises thin-bedded turbidite sandstones and mudstones and laminated hemi-

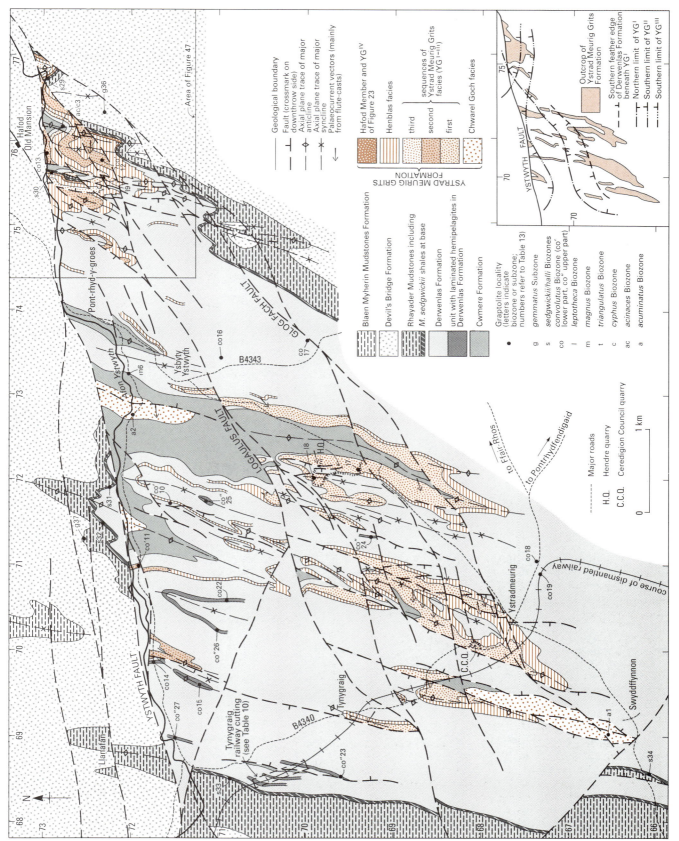

Figure 22 Geological map of the Ystrad Meurig Grits Formation showing palaeocurrent vectors, the position of critical graptolite localities and (inset) the suggested depositional limits of the three outcropping sequences of Ystrad Meurig Grits facies.

pelagites. It interdigitates with the Cwmere Formation and constitutes the oldest and most discrete component of the Ystrad Meurig Grits Formation. It is at no point seen in contact with other parts of the formation. Its crop is restricted to cores of anticlines adjacent to the south side of the Ystwyth Fault and, farther south, to a fault-bounded sequence on Drysgol [SN 694 675], west of Ystradmeurig. Nowhere is its base exposed. Named after a quarry [SN 7276 7204] in Coed Maenarthur, the type section of the Chwarel Goch facies is the gorge of Afon Ystwyth [SN 7301 7198 to SN 7270 7206], where about 70 m of it are overlain with a gradational contact by Cwmere Formation mudstones. On Drysgol [SN 694 675] at least 150 m of Chwarel Goch facies are exposed beneath the Cwmere Formation.

Thinly bedded sandstone/mudstone couplets (mainly type Cii) and siltstone-based mudstone turbidites (type Di) form the dominant resedimented components of the facies. The fine-grained sandstones are commonly pyritic, typically less than 3 cm thick and rarely constitute more than 30 per cent of any section (Plate 14a). Type Di mudstone turbidites normally form a minor element of the facies, but increase in abundance near the gradational contact with the Cwmere Formation. The intercalated laminated hemipelagites are on average less than 2 cm thick. The Chwarel Goch facies compares lithologically with the Dyffryn Flags facies of the Caban Conglomerate Formation.

Biostratigraphy

Two graptolite assemblages from the Chwarel Goch facies, one from just below the contact with the Cwmere Formation, are diagnostic of the *acuminatus* Biozone (Table 13, localities 1 and 2).

Details

The facies is best exposed in the type section, the gorge of Afon Ystwyth [SN 7301 7198 to SN 7270 7206]. To the south, a road section [SN 7272 7197], near to the contact with the Cwmere Formation, contains *acuminatus* Biozone graptolites (Figure 22, locality 2 and Table 13). In crags [SN 7273 7166] farther south, flute casts suggest that the depositing turbidity currents flowed from an easterly sector.

In the core of an anticline, west of the Hendre Anticline, a forestry cutting [SN 7155 7197] exposes 17 m of Chwarel Goch facies in which cross-laminated sandstones up to 7 cm thick display westerly directed flute casts.

In the faulted crop at Drysgol, numerous crags [SN 6932 6702 to SN 6945 6757] along the ridge comprise siltstone/mudstone couplets (type Di) with abundant sandstone/mudstone couplets (mainly type Cii), in which the sandstones are up to 8 cm thick. To the south, several small quarries [SN 6906 6655] about Tynddraenen yield assemblages dominated by *Normalograptus parvulus*, together with *Akidograptus ascensus*, suggesting an early *acuminatus* Biozone age (Table 13, locality 1).

Ystrad Meurig Grits facies

The Ystrad Meurig Grits facies comprises medium- to thick-bedded turbidite sandstones with subordinate turbidite mudstones and mainly burrowed hemipelagites (Plate 14b). Four sequences of the facies occur at stratigraphically discrete and physically separate levels in the Ystrad Meurig Grits Formation. These sequences occur between the upper part of the Cwmere Formation and the lowermost part of the Rhayader Mudstones (Figure 23). Only three of the sequences crop out in the district. Although the fourth and youngest only crops out in the Aberystwyth district (Strong, 1979; Cave and Hains, 1986), it was proved in the Cwm Rheidol No. 1 borehole [SN 7302 7833] north of the Ystwyth Fault.

The first sequence of Ystrad Meurig Grits facies mainly crops out between Ystradmeurig and Ysbyty Ystwyth but also east of Pont-rhyd-y-groes in the Coed Bwlchgwallter area (Figure 22). Around Ystradmeurig and Tancnwch [SN 716 685], at the southern limits of its crop, the first sequence is about 150 m thick. However, it thins rapidly northwards to disappear in less than 3 km, passing laterally into a coeval sequence of Henblas facies. In concert with the northward thinning, the base of the first sequence becomes progressively younger in the same direction. In the south it lies within the uppermost part of the Cwmere Formation, but traced north it transgresses the contact between the Cwmere and Derwenlas formations, to lie, at its northern extremities, about 30 m above this horizon (Figure 23). In the Coed Bwlchgwallter area, the first sequence of Ystrad Meurig Grits facies is 30 m thick and lies an equal distance above the base of the Derwenlas Formation, thus providing an east to west dimension to the thickness and facies changes (Figure 23). In most places the first sequence is overlain by Henblas facies.

The crop of the second sequence of Ystrad Meurig Grits facies is confined to the area between the Logaulas and Ystwyth faults. The sequence displays marked thickness variations. In the Coed Bwlchgwallter area, where it gives rise to the prominent ridges on Bryn [SN 757 723] and at Storehouse [SN 753 725], this sequence is up to 100 m thick and both succeeds and is succeeded by Henblas facies. Farther west, in Coed Craigyrogof [SN 698 714 and SN 710 718], it is less than 40 m thick, overlies Derwenlas Formation mudstones, and passes rapidly southwards into Henblas facies.

The third sequence of Ystrad Meurig Grits facies only occurs in the Coed Bwlchgwallter area, in isolated fault slices. The full sequence is nowhere seen, and the outcrops may in reality embrace more than one development of the facies. Estimated to be about 70 m thick and to be both underlain and overlain by Henblas facies, it appears to span a stratigraphical level equivalent to the upper part of the Derwenlas Formation (Figure 23).

A sandstone-dominated unit in the Cwm Rheidol No. 1 borehole [SN 7302 7833], 24 m thick (true thickness), represents the fourth sequence of the Ystrad Meurig Grits facies. Graptolite assemblages from the borehole confirm that this sequence is younger than those south of the Ystwyth Fault, and that it spans a stratigraphic interval probably equivalent to the uppermost part of the Derwenlas Formation and the overlying *M. sedgwickii* shales. The sequence therefore correlates with the Hafod Member, south of the Ystwyth Fault, and the thinner but comparable sandstone facies described from the Aberystwyth district by Strong (1979; see also Cave and Hains, 1986).

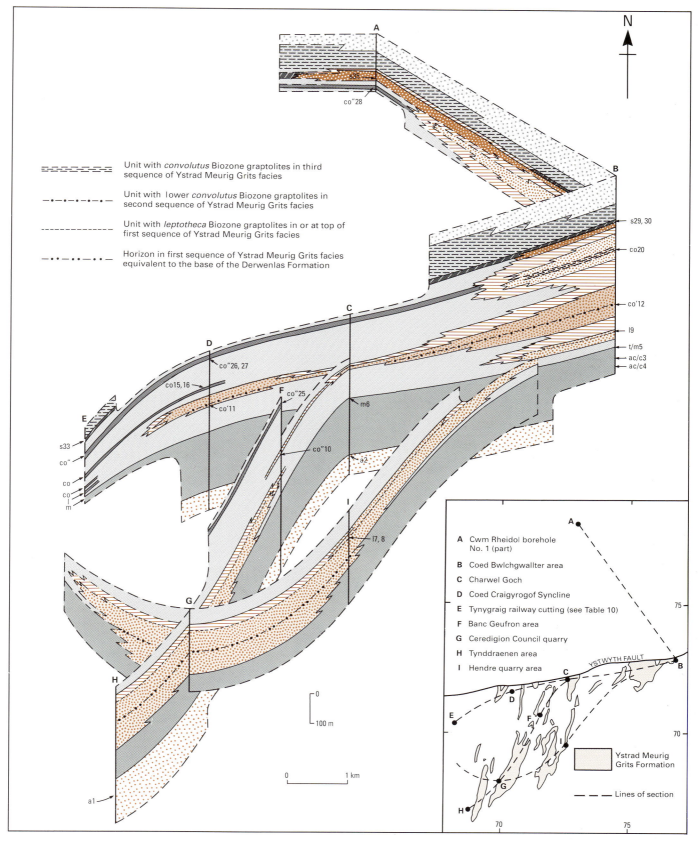

Unit with *convolutus* Biozone graptolites in third
sequence of Ystrad Meurig Grits facies

Unit with lower *convolutus* Biozone graptolites in
second sequence of Ystrad Meurig Grits facies

Unit with *leptotheca* Biozone graptolites in or at top of
first sequence of Ystrad Meurig Grits facies

Horizon in first sequence of Ystrad Meurig Grits facies
equivalent to the base of the Derwenlas Formation

A Cwm Rheidol borehole
No. 1 (part)
B Coed Bwlchgwallter area
C Charwel Goch
D Coed Craigyrogof Syncline
E Tynygraig railway cutting (see Table 10)
F Banc Geufron area
G Ceredigion Council quarry
H Tynddraenen area
I Hendre quarry area

YSTWYTH FAULT

Ystrad Meurig
Grits Formation

Lines of section

Figure 23   Architecture of the Ystrad Meurig Grits Formation. For key to formations and facies
see Figure 22.

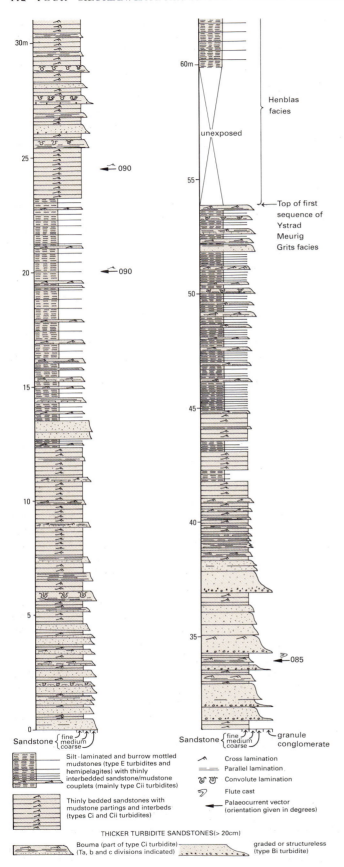

Sandstone { fine medium coarse

Sandstone { fine medium coarse } granule conglomerate

Silt-laminated and burrow mottled mudstones (type E turbidites and hemipelagites) with thinly interbedded sandstone/mudstone couplets (mainly type Cii turbidites)

Thinly bedded sandstones with mudstone partings and interbeds (types Ci and Cii turbidites)

⌐ Cross lamination
= Parallel lamination
Convolute lamination
Flute cast
← Palaeocurrent vector (orientation given in degrees)

THICKER TURBIDITE SANDSTONES (> 20cm)

Bouma (part of type Ci turbidite) (Ta, b and c divisions indicated)

graded or structureless (type Bi turbidite)

The Ystrad Meurig Grits facies comprises a varied suite of turbidites, dominated by sandstone/mudstone couplets (type C), in which mudstone is typically subordinate to medium-bedded, fine- to medium-grained sandstone (Figure 24). The couplets exhibit T_{a-e}, T_{b-e} and T_{c-e} Bouma sequences, as well as top-cut-out varieties, the product of local amalgamation (Plate 14b and c). Sandstones displaying pervasive convolute lamination are common. Thick and locally very thick, structureless, medium- to coarse-grained sandstones and rare pebbly sandstones (type Bi) also occur (Plate 14f). More muddy intervals comparable to Henblas facies commonly occur between individual or stacked sequences of the thicker turbidites. They are present throughout but increase in thickness and frequency upwards. Such units are up to several metres thick, and comprise thinly interbedded turbidite sandstone/mudstone couplets (type Cii) and silt-laminated, pale grey and green, colour-banded and burrow-mottled mudstones (top-cut-out type E turbidite mudstones and burrowed hemipelagites). Up to three discrete levels of rusty-weathering anoxic facies are locally present in the Ystrad Meurig Grits facies. Graptolites occur both in the laminated hemipelagites and in the interbedded, commonly pyritic, turbidite sandstones.

At several localities, notably in the first sequence of the facies north of Ystradmeurig and in the second sequence on Bryn and at Storehouse, the various turbidite components are arranged in crudely thinning- and fining-upwards sequences. The thickest sandstones (type Bi) are concentrated in the lower parts of such sequences, where levels of amalgamation of the predominant type C sandstones are also high.

BIOSTRATIGRAPHY

Graptolites from three localities in the first sequence of Ystrad Meurig Grits facies probably come from a single horizon, and belong to the *leptotheca* Biozone (Table 13, localities 7 to 9). Assemblages from the second sequence are of *convolutus* Biozone age, and moreover suggest the lower part of the biozone (Table 13, localities 10 to 12). The third sequence also belongs to the *convolutus* Biozone (Table 13, locality 20). The base of the fourth sequence, in the Cwm Rheidol No. 1 borehole (Appendix 1), lies about 7 m above a unit of anoxic-facies mudstones in the Derwenlas Formation, which contains graptolites indicative of the upper part of the *convolutus* Biozone (Figure 23; Table 13, locality 28). As a fragment of *M.* cf. *sedgwickii* has been obtained from the sandstone sequence (Table 13, locality 35) and the *M. sedgwickii* shales are absent from the borehole, it is likely that the fourth sequence of Ystrad Meurig Grits facies spans the equivalent stratigraphic interval here.

Figure 24 Composite graphic log of the upper part of the first sequence of Ystrad Meurig Grits facies on Craig Ystradmeurig [SN 705 687].

DETAILS

First sequence

In the Coed Bwlchgwallter area, Nant Bwlchgwallter [SN 7615 7268] exposes much of the 30 m-thick sequence. At the base, thick beds of massive, in places conglomeratic and amalgamated sandstone (type Bi) overlie mudstones of the Derwenlas Formation. Thinner-bedded turbidite sandstone/mudstone couplets (type C) and intervals of silt-laminated, colour-banded, oxic-facies mudstones prevail in the upper parts of the section forming a transition into the overlying Henblas facies. To the west, such thinner-bedded lithologies are seen again in Nant Ffin, where rusty-weathering, pyritic sandstones and dark grey, silt-laminated mudstones, exposed south of a mineralised fault [SN 7553 7228], have yielded graptolites indicative of the *leptotheca* Biozone (Figure 22, locality 9 and Table 13).

To the south-west, the first sequence is exposed in crags [SN 7369 7183] north-east of Ysbyty Ystwyth, and to the north of Ysguboriau [SN 7284 7046]. At the latter locality a thinning-upwards sequence, about 45 m thick, displays flute-casts directed towards the south-west.

The first sequence is extensively exposed in crags and quarries to the north of Ystradmeurig. The basal part is seen in an adit [SN 7225 6900] and in a road cutting [SN 7213 6907]. In both sections, massive conglomeratic turbidite sandstone beds (type Bi) succeed Cwmere Formation mudstones. Hendre quarry [SN 7215 6965] exposes the most extensive section through the sequence. Tectonic disturbance in the upper levels of the quarry are considerable (Plates 22a and 25), but the abandoned southern face exposes the upper 50 m of the first sequence in an undisturbed section. Massive, pebbly and amalgamated sandstones (type Bi) (Plate 14f) are confined to the lower 25 m of a thinning- and fining-upwards sequence. A graptolitic level near the top of the sequence has yielded *leptotheca* Biozone assemblages from two localities in the quarry (Figure 22 and Table 13, localities 7 and 8). Farther west, sections in the upper part of the sequence and its passage into succeeding Henblas facies, are provided by Craig Ystradmeurig [SN 7040 6842 to SN 7060 6879] (Figure 24) and the disused Ceredigion Council quarry [SN 7008 6800] (Plates 14b and c). A graphic log of the latter locality is figured in Smith et al. (1991); the graptolitic level of Hendre quarry is exposed there. Smaller quarries [SN 6973 6859 and SN 6921 6929] expose 4 m- and 20 m-thick sections respectively, and an upper contact with Henblas facies is seen in the disused railway tunnel [SN 6925 6944] at Tynygraig.

Exposures close to the northern limits of the first sequence of Ystrad Meurig Grits facies are seen on the southern flank of Mynydd Bach, in a track section [SN 7177 7032] and in a quarry [SN 7128 7026]. Despite its peripheral position, the sequence at both localities includes thick beds of granule-rich sandstone (type Bi) and amalgamated, planar-laminated sandstones (top-cut-out type Ci).

Second sequence

In the Coed Bwlchgwallter area, crags along the north-eastern flank of Bryn [SN 757 723] and north of Storehouse [SN 753 725] expose a 100 m-thick, thinning- and fining-upwards sequence. On Bryn, crags [SN 7573 7235] about 40 m above the base of the sequence expose pyritic sandstones, with dark grey mudstone laminae, and interbeds which have yielded *convolutus* Biozone graptolites (Figure 22, locality 12 and Table 13); the presence of abundant *Coronograptus gregarius gregarius* hints at the lower part of the biozone. The second sequence is exposed again, in crags [SN 7486 7352 and SN 7457 7203] east of Pont-rhyd-y-groes.

A road cutting [SN 7099 7191] north of Coed Craigyrogof, exposes 30 m of amalgamated turbidite sandstones in units up to 36 cm thick in a sequence of thinly interbedded sandstone/mudstone couplets (type C) and laminated siltstones. In the lower half, dark grey intercalated hemipelagites yield graptolites belonging to the lower *convolutus* Biozone (Figure 22, locality 11 and Table 13). A kilometre to the west, roadside quarries [SN 6995 7175] expose the upper 7 m of the second sequence and its passage into the overlying Derwenlas Formation.

Third sequence

Nant Ffin [SN 7553 7246] exposes a 30 m-thick faulted sequence of interbedded, coarse- to fine-grained, pyritic turbidite sandstones, up to 20 cm thick, and dark grey graptolitic hemipelagites. This anoxic sequence is far thicker than any identified in either of the first two sequences of the Ystrad Meurig Grits facies. It also differs palaeontologically in containing only long-ranging taxa of the *convolutus* Biozone (Figure 22, locality 20 and Table 13). Its thickness suggests that it may correlate with the thick, upper unit of anoxic facies in the local Derwenlas Formation, as seen in the Tynygraig railway cutting (see Figure 23). Strata thought to belong to the top of the third sequence of Ystrad Meurig Grits facies occur about 25 m below the Hafod Member in a forestry track [SN 7678 7294] adjacent to Afon Ystwyth. There, 8 m of sandstone/mudstone couplets (type C) with parallel-laminated basal sandstones up to 7 cm thick, are intercalated with pale, colour-banded and silt-laminated, oxic mudstones.

Fourth sequence

This has been recorded in the district only from Cwm Rheidol No. 1 borehole (Appendix 1).

Henblas facies

The Henblas facies comprises pale grey to green, colour-banded and burrow-mottled, oxic-facies mudstones with abundant siltstone laminae and cross-laminated ripple lenses together with scattered thin sandstones (Plate 14d). It forms a lateral and vertical transition zone between the sandstone-dominated sequences of Ystrad Meurig Grits facies and the enveloping mudstones of the Derwenlas Formation. The distribution of the Henblas facies in the Teifi Anticline is shown in Figures 22 and 23. The facies is named after Henblas Farm [SN 7040 6762], to the north of which it is well exposed on Craig Ystradmeurig [SN 705 685], overlying the first sequence of Ystrad Meurig Grits facies. Contacts with the other facies of the Ystrad Meurig Grits Formation and the Derwenlas Formation are highly diachronous, and this is reflected in the marked thickness variations exhibited by the facies (Figure 23). Individual units of the Henblas facies are up to 60 m thick, the thickest occurring in the Coed Bwlchgwallter area [SN 755 725].

In detail the Henblas facies comprises strongly silt-laminated, turbidite mudstones (top-cut-out type E), thinly interbedded with burrowed hemipelagites that are rarely separately distinguishable. Thin-bedded sandstone/mudstone couplets (type Cii) with basal sandstones up to 3 cm thick are common. They decline in importance towards contacts with the Derwenlas Formation, the volume of top-present and base-cut-out, type E turbidite mudstones increasing concomitantly. Thicker, sandstone/mudstone couplets (type Ci) are rare.

No graptolite faunas have been obtained from the Henblas facies.

DETAILS

About 20 m of Henblas facies overlying the first sequence of Ystrad Meurig Grits facies is exposed around Craig Ystrad Meurig [SN 705 685]. To the west, the facies is seen in crags [SN 6940 6795] at Llwynmalus. To the east, less than 5 m of the facies are present at Hendre quarry, but thicker sequences are inferred from exposures [e.g. SN 7365 7173] north-east of Ysbyty Ystwyth. Henblas facies in forestry sections [about SN 7190 7060] are regarded as laterally equivalent to the first sequence of Ystrad Meurig Grits facies.

Forestry sections [SN 7084 7117, SN 7123 7077 and SN 7211 7152] on Mynydd Bach are interpreted as coeval to the second sequence of Ystrad Meurig Grits. A quarry [SN 7175 7153] in these strata exhibits thinly bedded, sandstone/mudstone couplets (type Cii) and silt-laminated mudstones, overlying dark grey mudstones with laminated hemipelagites. The latter contain graptolites indicative of the lower part of the *convolutus* Biozone (Figure 22, locality 10 and Table 13). In the Coed Bwlchgwallter area, Henblas facies overlying the second sequence of Ystrad Meurig Grits facies is well seen in crags [SN 7587 7237] on Bryn. Strata between the third sequence of Ystrad Meurig Grits facies and the Hafod Member, exposed in a cutting [SN 7673 7294] south of Afon Ystwyth, are included in Henblas facies even though intervals of silt-depleted, pale grey and colour-banded mudstones, akin to the Derwenlas Formation, are present.

Hafod Member

This comprises a distinctive unit of shelly sandstones, phosphatic conglomerates and mudstones, only present at the top of the Ystrad Meurig Grits Formation in the Coed Bwlchgwallter area [SN 755 725]. The member forms a tongue in the lower part of the *M. sedgwickii* shales, and the type section is on the south bank of Afon Ystwyth [SN 7669 7297], 800 m east-south-east of the former Hafod Mansion.

The type section exposes the whole of the 10 m-thick member, which overlies with a sharp contact colour-banded, burrow-mottled mudstones and thin sandstones of the Henblas facies. The basal 3 m of the member comprise rusty-weathering, calcareous and pyritic, parallel- and cross-laminated shelly sandstones, up to 20 cm thick, with basal conglomeratic lenses (type Aii turbidites), interbedded with dark grey turbidite mudstones. The bioclastic debris in the sandstones locally forms pods of recrystallised limestone, which weather to rottenstone. The conglomeratic portions of these type Aii turbidites are also rich in shell debris, together with quartz and abundant lithic clasts. The latter dominantly consist of black, well-rounded, discoidal pebbles and granules of phosphatised mudstone, but tabular cobbles of laminated sandstone and rounded pebbles of acid volcanic rocks are also present. The remainder of the member comprises turbidite sandstone/mudstone couplets (type C) with predominantly non-calcareous basal sandstones, normally less than 4 cm thick, interbedded with mudstone turbidites (type Dii) and laminated hemi-

pelagites. The proportion of sandstone decreases upwards to the gradational contact with the overlying *M. sedgwickii* shales.

Although no graptolites were recovered from the Hafod Member, its stratigraphical position confirms it is of *sedgwickii* Biozone age.

DETAILS

In addition to the type section [SN 7669 7297], upper levels of the member and its passage into the *M. sedgwickii* shales are exposed in Nant Ffin [SN 7558 7293], close to its confluence with Afon Ystwyth.

Depositional model

Similarities between the coeval Caban Conglomerate and Ystrad Meurig Grits formations suggest that they are representatives of a single easterly sourced, turbidite sub-system, which punctuated the Hirnantian to early Telychian mud-dominated slope-apron sequence of the district. Their outcrops define a depositional belt in which coarse-grained sediments were effectively confined.

Despite intervals of pronounced lateral expansion, the Caban Conglomerate Formation, south of Rhayader, can be seen as a nested channel sequence, composed predominantly of proximal resedimented facies (Figure 19). It marks the site of a stable and long-lived submarine channel which, viewed regionally, enabled large volumes of sandy detritus to bypass the median portions of the coeval slope-apron and to supply the distal turbidite lobes of the Ystrad Meurig Grits Formation. The latter, in marked contrast, reveals a pattern of abandonment and switching which, in its northwards progression, compares with the modern Navy Fan offshore of California (Normark et al., 1979). This is the product of distributary avulsion on a turbidite fan, the position of successive lobes being determined by the sea-floor relief generated by earlier lobes.

FACIES INTERPRETATION

An environmental assessment of the facies of the Caban Conglomerate Formation has been presented by Kelling and Woollands (1969) and, although turbidite nomenclature has changed, their interpretation is broadly confirmed here. The conglomeratic portions of the formation have been described in detail by Holroyd (1978) and Kelling and Holroyd (1978).

Kelling and Woollands (1969) showed that the Caban Conglomerate Formation accumulated contemporaneously with the enveloping mudstone sequence. The coarsest facies, the Caban Conglomerates, were recognised as channel fills. The high levels of amalgamation, complex grading, irregular geometry and commonly steep erosive margins displayed by the type Ai coarse-grained turbidites, suggest deposition from surging and switching gravel-laden flows confined to tracts up to 4 km wide. The facies can be compared with other ancient braided, base-of-slope channel systems described by Pickering et al. (1989) and Hein and Walker (1982). The channels are envisaged to have debouched on to lobe-like constructional features, characterised in their proximal regions by the thick-bedded turbidite sand-stones (type Bi) of the Cerig Gwynion Grits, and distally by the predominantly Bouma turbidites of the Sedgwickii Grits and the Ystrad Meurig Grits facies. Finer-grained facies, Dyffryn Flags, Gafallt Shales or their western equivalents, were deposited at the fringes of these lobes and also occur as flanking deposits of the proximal, channellised parts of the subsystem. In the latter setting, they represent the inner portions of confining levees, predominantly composed at thickened sequences of slope-apron mudstones (Figure 19; see p.93).

SUBSYSTEM EVOLUTION

The results of the recent survey, and especially the recognition of the Ystrad Meurig Grits Formation as the western correlative of the Caban Conglomerate Formation, has enabled the evolutionary history proposed by Kelling and Woollands (1969) for the Caban-coch region to be revised (Figures 17, 18 and 19).

The subsystem was initiated during the *persculptus* Biozone with the introduction and aggradation of the Cerig Gwynion Grits in the Rhayader area (Figures 18 and 19). The crop of the Cerig Gywynion Grits indicates that the facies has a lobe-like geometry, with a depositional focus on line with the later channelised portions of the formation (Figures 17a, 18b and 19). However, the wide spread of palaeocurrent directions suggests a more complex depositional pattern of coalescing and interdigitating sand bodies, collectively sourced from the south-east.

In latest *persculptus* Biozone times, the conglomeratic feeder channel advanced into the basin causing an abrupt decline in the supply of coarse clastic sediment to the exposed, eastern portions of the Cerig Gwynion Grits lobe. This event is marked along the lobe flanks by a rapid upward passage into fine-grained facies (Dyffryn Flags) (Figure 19). The narrow tract of Caban Conglomerates developed in the Carn Gafallt area, above the Cerig Gwynion Grits lobe, confirms that the channel remained active throughout the early part of the *acuminatus* Biozone. Overbank deposition by Bouma turbidites contributed to the flanking belts (levee deposits) of Dyffryn Flags, which accumulated on either side of the channel throughout this period. The distal Chwarel Goch facies, of *acuminatus* Biozone age, forms the lowest exposed part of the Ystrad Meurig Grits Formation.

The late *acuminatus* to *cyphus* biozonal interval saw a major change in the volume of detritus supplied. The pre-existing, conglomeratic feeder channel of the Carn Gafallt area expanded rapidly to form a broad, braided system, which now constitutes the second sequence of Caban Conglomerates facies (Figure 19). This interval also saw the onset of Caban Conglomerates facies deposition in the Rhiwnant Anticline. There, pronounced local thickening of the conglomerate sequence defines a bifurcating pattern of early channels, active probably in the early *acuminatus* Biozone (Figure 18b). Entering abruptly above mudstones of the Cwmere Formation, these conglomerate-filled channels record a major avulsion of the pre-existing coarse clastic supply route. During the *atavus*

and *cyphus* biozones, the early channels in the Rhiwnant area also expanded into a braided system, accounting for the laterally extensive portions of the second sequence of Caban Conglomerates facies. The coeval lobe deposits of this episode, though not separately distinguishable, are represented by the lowest parts of the first sequence of Ystrad Meurig Grits facies, which are coeval with Cwmere Formation mudstones. The width of the contemporary gravel supply paths to the east suggests that a large portion of this terminal lobe (and probably others) lies at depth to the south of Ystradmeurig.

Following the cessation of conglomeratic deposition in the Caban and Rhiwnant areas in the *triangulatus* Biozone, a single, broad, conglomeratic pathway was re-established above the long-lived north-eastern sector of the subsystem in *magnus–leptotheca* Biozone times (Figures 18b and 19); this is the third sequence of Caban Conglomerates facies. Bouma turbidites contributed to the flanking sequences of Sedgwickii Grits and Gafallt Shales by overbank deposition. Those parts of the first sequence of Ystrad Meurig Grits facies that invaded the Derwenlas Formation define the limits of contemporary lobe deposition, and the associated Henblas facies defines its distal fringe (Figure 23).

The upward passage of the third sequence of Caban Conglomerates facies into the lobe-like Sedgwickii Grits marked the onset of a long-term decline in the grade and volume of coarse clastic detritus supplied to the subsystem. Although the *convolutus* Biozone saw a steady contraction in the depositional tract of the coarser facies, the thickening-upward sequences, with type B and locally thin conglomeratic turbidites, which developed in the axial part of the Sedgwickii Grits, testify to periodic resurgences in supply. Such sequences may correlate down-current with the discrete lobe facies of the second and third sequences of Ystrad Meurig Grits facies. The latter are individually much thicker than their Caban counterparts. The Ystrad Meurig Grits and Derwenlas formations form an upper Aeronian sequence which is over twice as thick as that in the Caban-coch area. This westwards expansion confirms the constructional character of the Ystrad Meurig lobes, but implies low net deposition along the supply route through the Caban-coch area. There, levees built of Gafallt Shales may have helped to confine the flow and to aid the efficient transfer of sand- and silt-grade detritus to the western lobes.

The thin fourth and fifth sequences of Caban Conglomerates facies associated with the *M. sedgwickii* shales record the brief and final re-establishment of a narrow, coarse, clastic pathway within the Caban-coch region. The event was manifested in the west by the deposition of the Hafod Member and the fourth sequence of Ystrad Meurig Grits.

The early part of the *turriculatus* Biozone saw the demise of coarse clastic deposition within the Caban/Ystrad Meurig tract. Its final phases were marked, in the Caban-coch area, by a contracting belt of Gafallt Shales. During this time, sand-carrying turbidity currents routed via relict channels may have continued to contribute sediment to western portions of the contemporary slope-apron, the Devil's Bridge Formation.

SUBSYSTEM SOURCE AND EXTERNAL CONTROLS

Both the geometry and palaeocurrent indicators of the Caban Conglomerate Formation suggest that its long-lived supply route, the submarine canyon envisaged by Kelling and Woollands (1969), lay to the south-east of Carn Gafallt, with a southern branch briefly supplying the Rhiwnant area. The sediment sources probably lay in slope-apron, shelf and terrestrial environments to the east. The Caban Conglomerates facies contain Precambrian igneous and metamorphic clasts (Davies and Platt, 1933; Evans, 1992), and abundant shells. However, the supply route across the basin margin is unknown. The continuous crop of the Tycwtta Mudstones, on the south-eastern limb of the Tywi Anticline, provides a major constraint. Although the Allt-y-clych Conglomerate has a similar clast assemblage and therefore possibly the same source as the Caban Conglomerates facies, its extensive linear crop (e.g. Lockley, 1978; 1983; James, 1983), precludes the presence of even a mudstone-filled feeder channel in this region. The coeval and largely complete Rhuddanian shelf sequences in the Garth (Andrew, 1925) and Llandovery (Cocks et al., 1984) areas also provide no evidence for possible supply routes. Strike-slip displacement along the Tywi Lineament may provide a possible solution; the Caban Conglomerate Formation may have been offset by many kilometres from its, as yet unrecognised, supply route.

The subsystem was deposited during a period of eustatic changes in sea level (Figure 39). Deposition of the lower parts of the Caban Conglomerate Formation was maintained throughout the late Hirnantian to Rhuddanian postglacial transgression (Figure 17a). Although transgressions are normally regarded as periods when coarse-grained deep-sea systems contract and are abandoned (Stow et al., 1985), the period of maximum rudite supply to the Caban Conglomerate Formation coincided with the transgressive maximum in the *cyphus* Biozone (Johnson et al., 1991). The subsequent phase of conglomerate deposition coincided with the beginning of the Aeronian regression. Furthermore, the acme of this regression, in the *convolutus* Biozone, saw a contraction in the depositional tract in the Cabancoch area (Figure 17b). Deposition in the proximal part of the subsystem thus appears to have been out of step with the contemporary movements of sea level, and the fluctuations in sediment supply may therefore reflect independent tectonic uplift of source areas and the operation of efficient shelf and slope-apron bypass systems.

Although variations in the volume of coarse clastic supply to the subsystem fail to reflect sea-level movements, changes in the composition of the sediment may be related to such movements. A basinward migration of shallow-water benthic communities took place during the Aeronian regression, and their availability for resedimention is reflected in the abundance of shell debris in the Sedgwickii Grits. Furthermore, the phosphatised clast assemblages of the Hafod Member are compatible with an interval of phosphate diagenesis sustained briefly along the basin margin at the onset of the *M. sedgwickii* transgression (cf. Jenkyns, 1986, p.393) (see also Cwm Barn Formation).

EARLY TELYCHIAN SANDY TURBIDITE FACIES

Devil's Bridge Formation

The Devil's Bridge Formation appears to occupy a transitional setting in the basinal succession between the easterly sourced slope-apron systems and the southerly derived sandstone-lobe systems (see Chapter 5). Lithologically it resembles parts of the southerly derived systems, being deposited contemporaneously with their initiation and early growth. However, palaeocurrent indicators and heavy mineral assemblages obtained from the formation suggest that it was derived, at least in part, from an easterly quadrant.

The Devil's Bridge Formation as defined by Cave and Hains (1986) is expanded here to include the Dolwen Mudstones of Jones (1909). The formation is characterised by rhythmically interbedded thin turbidite sandstones and thicker turbidite mudstones. It crops out mainly in the Llanilar sheet area, where it occurs along both limbs of the Teifi Anticline to the south of the Ystwyth Fault and occupies a broad tract to the north (Figure 25). In the district the formation overlies the Rhayader Mudstones although, as in the Aberystwyth district (Cave and Hains, 1986; Loydell, 1991), the base is locally very diachronous. Although the Devil's Bridge Formation may possibly pass laterally into the partly coeval, southerly derived sandstone-lobe system of the Aberystwyth Grits Group, it is separated everywhere at outcrop by the fringing mudstone facies of the lobe, the Borth Mudstones Formation.

In the north-eastern part of the Llanilar sheet area, where the underlying Rhayader Mudstones are thinnest (Figure 25), the Devil's Bridge Formation attains its maximum thickness of about 600 m. Over much of the remaining crop it is about 470 m thick. Along the western limb of the Teifi Anticline it passes laterally into the Borth Mudstones and, in the district, is probably absent west of the Glandyfi Lineament.

The Devil's Bridge Formation is largely composed of thin turbidite sandstone/mudstone couplets (mainly type Cii) (Plate 15a). Interbedded hemipelagites are mainly burrowed and associated phosphatic nodules are common. Units with laminated hemipelagites are rare, and less than 3 m thick.

The basal sandstones of the type Cii couplets are fine grained, cross- or convolute-laminated (T_c divisions) and rarely exceed 5 cm in thickness (Plate 15a). Their soles commonly exhibit trace fossils (hypichnia). Scattered thicker (up to 20 cm), fine- to medium-grained basal sandstones of type Ci couplets, with fluted soles and parallel-laminated T_b divisions, are also present. The sandstones of type C couplets commonly contain a high proportion of finely comminuted shell debris, and weathered surfaces locally display the leached moulds of crinoid ossicles and ribbed brachiopod valves. The grey turbidite mudstones of the couplets are typically thicker

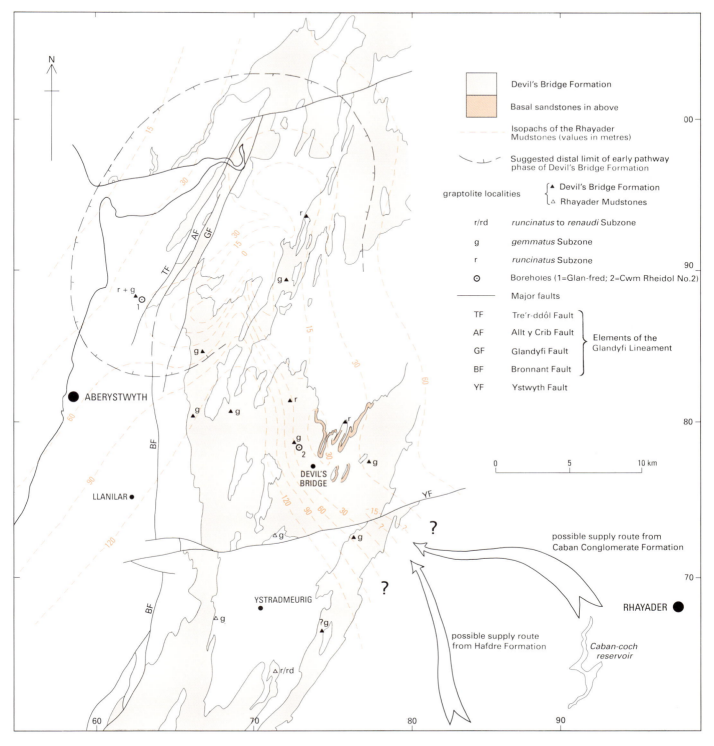

Figure 25   Diagram illustrating the form of the early pathway phase of the Devil's Bridge Formation. Isopachs for the Rhayader Mudstones are based on thickness data from this survey, Cave and Hains (1986), and Pratt et al. (1995); *runcinatus* and *gemmatus* subzone graptolite localities include those of Loydell (1991). No account is taken of Acadian shortening, which may, in part, account for the marked geniculation in isopach trend observed north-west of Devil's Bridge. Rhayader Mudstones are thinnest or absent where the depositional tract of the early pathway coincides with that of the preceding fourth sequence of Ystrad Meurig Grits facies.

Bronwenllwyd [SN 6641 6631], Caebalcog [SN 6693 6682], and north of Pantyddafad [at SN 6750 6908]. In these sections, thin-bedded sandstone/mudstone couplets (type Cii) are up to 10 cm thick; sandstones, rarely greater than 3 cm thick, comprise between 20 and 40 per cent of the succession. Just south of the Ystwyth Fault, near Wenallt, disused railway cuttings [SN 6781 7105 to SN 6694 7141] contain up to 20 per cent sandstone. Mudstone-dominated units, with individual turbidite mudstones (type D) up to 40 cm thick and with a sandstone content of less than 10 per cent, are locally present.

Sections in the westernmost part of the Teifi Anticline commonly contain less than 15 per cent sandstone, and the dominant thin- to thick-bedded mudstones include abundant type D turbidites. In the south, a mudstone-rich sequence, predominantly in oxic facies, is exposed in a stream [SN 6130 6064 to SN 6112 6061] to the east of Meini-gwynion, and a stream [SN 6135 6107 to SN 6109 6128] east of Cilcert. Thin, laminated hemipelagites in the latter section [at SN 6112 6125] (Figure 27, locality 1), have yielded *Pristiograptus* cf. *renaudi*, suggestive of the lower *turriculatus* Biozone. Facies transitional with the Borth Mudstones are exposed for 1.5 km in Afon Aeron [SN 6354 6250 to SN 6296 6364]. Farther north, comparable facies are seen in quarries near Blaen-yr-esgair [at SN 6536 6568], east of Penycastell [at SN 6681 6754] and south-west of Brynarth [at SN 6653 6959]. A quarry [SN 6568 7138] east of Cwm-byr, also in these transitional facies, has yielded the graptolite *M*. cf. *planus*, suggestive of either the *renaudi* or *utilis* subzones.

Depositional model

The distribution and dating of the Devil's Bridge Formation demonstrate that it was deposited in the west of the district contemporaneously with the Rhayader Mudstones to the east. In detail the formation records two separate phases of deposition; an initial, laterally restricted early pathway phase followed by a widespread blanket phase.

EARLY PATHWAY PHASE

Isopachs of the Rhayader Mudstones demonstrate that it is thinnest beneath the earliest sandstone-rich part of the Devil's Bridge Formation (Figure 25). This earliest-formed part of the Devil's Bridge Formation was deposited in a laterally restricted north-west-trending tract that provided an early pathway for sand-carrying turbidity currents across the north-eastern corner of the Llanilar sheet area and into the Aberystwyth district (Figure 34).

Also deposited in this tract were an overlying 100 m of Devil's Bridge Formation containing pale, green-grey, silt-laminated turbidite mudstones (type E), the 'lower Devil's Bridge Formation' of Cave and Hains (1986). These strata demonstrate that in this pathway, Bouma turbidites of the Devil's Bridge Formation were deposited contemporaneously with fine-grained turbidites typical of the adjacent oxic slope-apron facies (Rhayader Mudstones). Graptolite dating confirms that these lower levels of the formation do not record the subsequent infilling of large-scale erosional features, but reflect an expanding depositional belt on the slope-apron, which was receiving Bouma turbidites in the period between the *runcinatus* Subzone and the early *gemmatus* Subzone.

Turbidites supplied via this pathway appear to account for the whole of the Devil's Bridge Formation west of the Glandyfi Fault in the Aberystwyth district (Figure 25). An upward passage into the Borth Mudstones as early as the *gemmatus* Subzone demonstrates that the cessation of sand supply to this area broadly coincided with the onset of the blanket phase of deposition in the east (Figure 34).

The position and orientation of this tract of early Devil's Bridge Formation could suggest a continued northward migration of the sandy lobes of the Ystrad Meurig Grits Formation, supplied via the Caban/Ystrad Meurig feeder system (Cave and Hains, 1986). However, the latter was in decline during the early *turriculatus* Biozone (p.115) and the flow of sediment through the system was seemingly insufficient to supply a depositional tract that extended across to the western reaches of the basin. Sediment routed via the Caban/Ystrad Meurig tract may have contributed to the earliest levels of the early pathway sequence, but was ultimately subordinate to the feeder system which evolved to supply the blanket phase.

BLANKET PHASE

During the *gemmatus* Subzone, turbidity currents supplying the Devil's Bridge Formation spread out over a much wider area from north of the Aberystwyth district to south of the Llanilar sheet area (Figure 34). In contrast to the earlier phase, their depositional limit was restricted to the east of the Glandyfi Lineament. Palaeocurrent and heavy-mineral data from the Devil's Bridge Formation suggest that sediment continued to be derived predominantly from the east (Morton et al., 1992). Outcrop and sedimentological data suggest three possiblities for the location of the feeder system for the Devil's Bridge Formation: that it lay to the north of the district; that it traversed the Rhayader Mudstones slope-apron, but is now infilled by mudstone; or that it lay to the south of the district. Evidence to support the first two possibilities is lacking. However, south of the district and east of the Central Wales Syncline, Mackie and Smallwood (1987) have described a sequence of sandy Bouma turbidites (Hafdre Formation), in the Llyn Brianne area of the Builth Wells district. The lower part of this formation is coeval with the Devil's Bridge Formation and was derived from the east and south-south-east. It may therefore represent a new sediment supply route for the Devil's Bridge Formation, replacing the earlier and now abandoned Caban feeder system to the north.

The lack of a Caban-type coarse clastic corridor supplying the (lower) Hafdre and Devil's Bridge formations suggests that the turbidity currents were sourced from a wide sector of the contemporary shelf. To the east, complete sequences confirm that a broad shelf was available both to store sediment and to act as a source for turbidity currents (Ziegler et al., 1968; Bridges, 1975; Cocks et al., 1984). Further support for the idea of a linear shelf source is provided by the abundant shelly debris in the Devil's Bridge Formation turbidites (Cave and Hains, 1986).

Although the source of the formation is therefore the same as the contiguous slope-apron facies (Rhayader Mudstones), the volume of turbiditic deposits introduced via this new route during the blanket phase suggests that the sediment source was more active, and/or of far greater extent, than that which had previously supplied easterly derived detritus to the basin. As the blanket phase of the Devil's Bridge Formation (and coeval parts of the Hafdre Formation) broadly coincides with the period during which large volumes of tectonically generated detritus were introduced in the west of the district (Aberystwyth Grits Group; see Chapter 5), it may reflect the same episode of source rejuventation (Soper and Woodcock, 1990). The growing influence of tectonism during this period is also reflected in the confinement of the blanket phase to east of the Glandyfi Lineament, suggesting the existence of a structurally generated barrier to turbidite sand deposition, not present during the preceding pathway phase. Wilson et al. (1992) have suggested that this feature records footwall uplift on the Bronnant Fault (and its splays) during the *gemmatus* Subzone (Figure 34).

FIVE

Silurian: Telychian sandstone-lobe systems

The Telychian Stage encompassed major changes in basinal sedimentation patterns in the district. The west to north-westerly prograding mud-dominated slope-apron facies were progressively replaced from west to east by sand-rich turbidite systems fed from the south. Although sandstone-rich facies (Aberystwyth and Cwmystwyth Grits groups) predominate, associated turbiditic mudstone facies (Borth Mudstones Formation, Blaen Myherin Mudstones Formation, Caerau Mudstones and Adail Mudstones) are important in defining the architecture of these systems (Figures 12 and 13).

MUDSTONE FORMATIONS

At outcrop in the district, the oldest southerly supplied Telychian deposits are always represented by distinctive mudstone facies, which become progressively younger eastwards. The lateral continuity of these facies across the Teifi Anticline and beneath the Central Wales Syncline, though suspected, cannot be demonstrated and therefore their separate outcrops have been given different formational names. From west to east they comprise the Borth Mudstones Formation, the Blaen Myherin Mudstones Formation and the Caerau Mudstones. Similar mudstone facies, the Adail Mudstones, locally overlie the eastern part of the Aberystwyth Grits Group.

Borth Mudstones Formation

The Borth Mudstones, first defined in the Aberystwyth district (Cave, 1975) and later described in detail by Cave and Hains (1986), consist of thin- to medium-bedded turbidite mudstones (type Dii with subordinate Di) and less common turbidite sandstone/mudstone couplets (type Cii). Intercalated hemipelagites are predominantly burrowed, but laminated examples are also present.

The Borth Mudstones crop out principally in a structurally complex north–south tract between the Bronnant Fault and the Devil's Bridge Formation (Figure 26). They are in part the lateral equivalent of the Devil's Bridge Formation and in part younger than it (Figure 12). At outcrop, the contact of the two formations is commonly gradational or interdigitating and, at most localities, complicated by folding.

The relationship of the Borth Mudstones to the succeeding Aberystwyth Grits Group, shown to be a simple superposition in the coast section south of Borth in the Aberystwyth district by Cave and Hains (1986), can rarely be demonstrated in the present district, as the Bronnant Fault and its splays separate the two in most places. Only in the south of the district, near Ffos-yr-odyn [SN 625 635] and in Gwenffrwd (Cwm Bwlch) [SN 597 607], is the Aberystwyth Grits Group observed overlying Borth Mudstones (Figure 27).

Cave and Hains (1986) suggested a maximum thickness of at least 350 m for the Borth Mudstones of the Aberystwyth district. In the Llanilar sheet area, the sequence is complete only in the Gwenffrwd fault slice [SN 615 610] where an estimated 600 m of strata are present.

The turbidite mudstones (type Dii) which comprise the bulk of the formation range up to 30 cm thick (Plate 8b). Basal laminated siltstones of type Di turbidites rarely exceed a centimetre in thickness. Scattered thin turbidite sandstone/mudstone couplets (type Cii) are common, but the combined thickness of their basal fine-grained sandstones seldom exceeds 10 per cent of any section. Such sandstones are typically less than 2 cm thick, but rare examples up to 11 cm thick have been recorded. Although they occupy varying proportions of complete couplets, their thickness appears unrelated to that of the succeeding mudstone.

Scattered high-matrix sandstones (type Bii), ranging in thickness from 15 to 75 cm, occur in the western and presumably youngest parts of the main Borth Mudstones crop, from Lledrod [SN 645 704] northwards. They are either structureless and cleaved throughout or exhibit a well-segregated, massive, or more rarely parallel-laminated, basal part. Similar beds, 'Harp Rock Type' turbidites, occur in the upper part of the formation to the north of the district, near Borth [SN 595 875] (Cave and Hains, 1986, p.95).

BIOSTRATIGRAPHY

In their type area in the Aberystwyth district, the Borth Mudstones have yielded graptolite assemblages indicative of the *gemmatus* to *renaudi* subzones of the *turriculatus* Biozone (Loydell, 1989; 1991). In the Llanilar sheet area the *renaudi* Subzone has been proved near the base of the Borth Mudstones south of the Ystwyth Fault [SN 6584 7110] (Figure 27, locality 3), whilst the *utilis* Subzone is suggested by graptolites from the highest part of the formation beneath the Aberystwyth Grits in the Bronnant Fault zone, in the Aeron Fechan [SN 6284 6364] (Figure 27, locality 4). The latter date confirms that parts of the formation were deposited contemporaneously with both the Devil's Bridge Formation and the Aberystwyth Grits Group.

DETAILS

The lower part of the Borth Mudstones and their passage into the underlying Devil's Bridge Formation are well exposed in sections in Nant Cadwn [SN 6095 6120] and its tributaries [SN 6105 6143] east of Cilcert, in the stream [SN 6638 7031] east of Hafod-weunog, in Afon Ystwyth [SN 6668 7300] below

Trawsgoed Bridge, and in a section [SN 6624 7830] to the east of Pant-y-crûg. Intercalated laminated hemipelagites in the lower part of the formation, exposed in a forestry cutting [SN 6584 7110](Figure 27, locality 3) east-south-east of Cwm-byr, have yielded *Pristiograptus renaudi*, *Monograptus cavei*, *M. halli* and *Streptograptus plumosus*, indicative of the *renaudi* Subzone; the presence of *M. marri* may suggest a high level in this subzone.

In the south of the district, a section [SN 5965 5995 to SN 5968 6075] in Gwenffrwd (Cwm Bwlch) exposes the upper part of the Borth Mudstones and its contact with the Aberystwyth Grits. A graptolite fauna from the thin sandstones in this sequence, about 200 m south-east of the upper contact [SN 5971 6067 to SN 5968 6068](Figure 27, locality 2), includes *Monograptus planus* and *M.* cf. *halli*, indicative of an early *turriculatus* Biozone age. The presence of graptolites of the *sedgwickii /halli* group, together with the absence of *utilis* Subzone indicators, may possibly suggest that the *renaudi* Subzone is present. Graptolites of the *utilis* Subzone, including *Monograptus* cf. *utilis*, *M. halli*, *M. planus*? and *Normalograptus*? cf. *scalaris*, were obtained close to the top of the formation in the Aeron Fechan [SN 6284 6364], east of Ffos-yr-odyn (Figure 27, locality 4).

Exposures with scattered high-matrix sandstones (type Bii), typical of the upper parts of the Borth Mudstones in the northern part of their main crop, occur in road cuttings [SN 6443 7042 and SN 6427 7055] at Lledrod, stream sections [SN 6465 7076 to SN 6475 7073] east of Rhyd-lŵyd farm, a quarry [SN 6465 7514] east of Glanystwyth, and a quarry [SN 6444 7882] east-south-east of Capel Seion.

Blaen Myherin Mudstones Formation

The type section of the Blaen Myherin Mudstones is the spectacular gorge [SN 8005 7853 to SN 8027 7841] of Nant Rhuddnant in the north-west corner of the Rhayader sheet area, 1.1 km south of Blaen Myherin farm (Table 16; Figure 45). The formation, first recognised by Jones (1909), is lithologically similar to the Borth Mudstones and consists mainly of thin- to thick-bedded turbidite mudstones (type Dii) and siltstone/ mudstone couplets (type Di) interbedded with laminated and burrowed hemipelagites (Plate 15b). Scattered turbidite sandstone/mudstone couplets (type Cii) occur but never exceed more than 10 per cent of any section.

The principal outcrop of the Blaen Myherin Mudstones lies along the eastern limb of the Teifi Anticline where it occupies a single, continuous, north-north-east-trending belt that extends from the type section to the southern margin of the district, east of Tregaron. Farther west, to the north and east of Llanafan [SN 690 728] and at Trewaun [SN 646 603], outliers of similar mudstone have been assigned to the formation. Rare high-matrix sandstone turbidites (type Bii) occur in the mudstones of these outliers.

The base of the formation is everywhere gradational with the underlying Devil's Bridge Formation; in its type section it has been taken at the base of a 15 m-thick packet of turbidite mudstones with laminated hemipelagites exposed 20 m upstream from the confluence with Nant y Ffin [SN 8005 7853] (Figure 45).

Along its principal crop the formation maintains a thickness of about 100 m; it thickens northwards, beyond the northern margin of the district, and westwards, where more than 500 m are estimated to be present in the Trewaun and Llanafan outliers.

BIOSTRATIGRAPHY

Graptolite assemblages from the basal 15 m and from the middle of the Blaen Myherin Mudstones in Nant Rhuddnant (Table 16) belong to the *johnsonae* Subzone. However, about 3 km farther north, beyond the northern boundary of the district, the base of the formation is earlier, for the lower part has yielded *utilis* Subzone graptolites from a forestry section [SN 8149 8143] (Figure 29, locality 1) on Peraidd Fynydd. This reflects the north-eastwards failure of the Devil's Bridge Formation in the Llanidloes district. Farther east, sections in the upper part of the formation along the A44 road [SN 8578 8198 to SN 8641 8177] (Figure 29, locality 2) have yielded assemblages spanning the *turriculatus–crispus* biozonal boundary.

No graptolites have been recovered from the western outliers of Blaen Myherin Mudstones, but the underlying Devil's Bridge Formation has yielded *utilis* Subzone assemblages.

The Blaen Myherin Mudstones have yielded moderately diverse latest Aeronian/earliest Telychian to late Telychian acritarch assemblages referable to the *A. microcladum* to *G. encantador* biozonal interval (Appendix 2bi). They include *Ammonidium listeri*, *A.* sp. A, *Cymatiosphaera* sp. A, *Dictyotidium biscutulatum*, *D.* sp. A, *Elektoriskos williereae*, *Lophosphaeridium* sp. A and *L. parverarum*.

DETAILS

Llanilar sheet area

The formation is well exposed in Glasffrwd stream [SN 7530 6491 to SN 7535 6467], a roadside quarry [SN 7676 6806] north of Tynddol, and a stream [SN 7709 6866] on Llethr Brith, where the contact with the overlying Rhuddnant Grits is observed. A 30 m-thick section [SN 7807 7114], along a forestry cutting north of Trawsallt, shows mudstone turbidites up to 10 cm thick intercalated with burrow-mottled hemipelagites. Laminated hemipelagites are exposed in stream sections [SN 793 731] along Nant Milwyn, near Esgair Milwyn, and [SN 789 736] west of Tynewydd.

Exposures in the western outliers of Blaen Myherin Mudstones include a stream section [SN 6914 7764 to SN 6967 7757] south of Cennant, outcrops [e.g. SN 687 755] along Nant Magwr and its tributary west of Llanerch Pentir, and Nant y Fawnog [SN 7100 7736 to SN 7148 7730] south of Bwlchcrwys. Sections in the southernmost outlier occur along a track [SN 651 615] east of Tŷnant and in a quarry [SN 6448 6018] south-south-west of Trewaun.

Rhayader sheet area

South of the Ystwyth Fault the Blaen Myherin Mudstones crop out in crags [SN 8035 7426] on the northern slopes of Graig Goch. North of the Ystwyth Fault, a complete but partly inaccessible sequence is displayed in the type section along the gorge of Nant Rhuddnant [SN 8005 7853 to SN 8027 7841] (Figure 45). A 15 m-thick unit with laminated hemipelagites at the base and a similar thinner unit towards the middle of the section have yielded graptolites indicative of the *johnsonae* Subzone (Table 16, localities 9 to 11).

Caerau Mudstones

The mudstones that crop out beneath the Cwmystwyth Grits Group east of the Central Wales Lineament (Figure 12) are here assigned to the Caerau Mudstones of W D V Jones (1945). The formation contains the same suite of turbidite mudstones and sandstones (mainly type D with scattered Cii) with intercalated burrowed and laminated hemipelagites, as the Borth Mudstones and Blaen Myherin Mudstones. However, in addition, the Caerau Mudstones contain units of very thinly interbedded turbidite mudstones (type E) and burrowed hemipelagites, similar to the earlier Telychian, oxic-facies, slope-apron mudstones.

In the district the Caerau Mudstones are confined to the Rhayader sheet area where they occur principally in two crops flanking the Rhayader Mudstones of the core of the Tylwch Anticline. The western crop, extending from the northern edge of the district near Bronheulwen [SN 950 787] to Drum Nantyrhelyg [SN 818 597] in the south, locally intertongues with the Rhuddnant Grits. Isolated inliers of the formation occur along Nant Brwynog [SN 818 650], the Claerwen reservoir [SN 842 656], at several points along Nant Cletwr and its tributaries [e.g. SN 880 686] and over a more extensive area south of Llangurig [SN 905 785]. An isolated crop of strata assigned to the Caerau Mudstones, overthrust by Rhayader Mudstones, occurs north-east of the Penygarreg reservoir close to the axis of the Tylwch Anticline [SN 915 680]. The eastern crop of the formation terminates on the eastern flank of the Waun Marteg Syncline, against the Cwmysgawen Fault on Cwm-hir Bank [SO 027 709].

The base of the formation, along both flanks of the Tylwch Anticline, is taken at the incoming of abundant packets of interbedded thin- to thick- bedded grey turbidite mudstones (type D) and laminated hemipelagites, above the silt-laminated, colour-banded, oxic-facies mudstones typical of the Rhayader Mudstones. Oxic-facies mudstones persist as interbeds into the lower part of the Caerau Mudstones, however. Farther east, between the Brynscolfa and Cwmysgawen faults [SO 007 701], the Caerau Mudstones overlie the turbidite sandstone/mudstone couplets which form the upper part of the Cwmbarn Formation. The contact is gradational.

On the western limb of the Tylwch Anticline the Caerau Mudstones are overlain either by the Glanyrafon Formation or the Rhuddnant Grits. However, south of Graig Gellidywyll [SN 944 767], a sequence of Rhuddnant Grits divides the Caerau Mudstones into lower and upper tongues. The upper tongue thins and fails south of Nant Cletwr [SN 877 691]. On the eastern limb of the Tylwch Anticline the Caerau Mudstones are mainly overlain by the Glanyrafon Formation but farther east, on the eastern limb of the Waun Marteg Syncline, upper parts of the formation pass laterally into the Dolgau Mudstones (Figure 12).

In the western crop of the Caerau Mudstones the intercalation of Rhuddnant Grits gives rise to marked thickness variations. In the south, the formation maintains a thickness of about 100 m. Around Craig Goch reservoir the formation, including the intercalation of Rhuddnant Grits, is about 300 m thick, with the lower tongue, below the grits, about 120 m and the upper tongue just less than 100 m thick. North of the River Wye, around Dolhelfa [SN 930 740], the lower tongue increases to 150 m and the upper to 200 m, in a sequence about 400 m thick. The eastern crop of the formation thins from about 350 m through Pant-y-dwr [SN 983 748], to just under 300 m on the eastern limb of the Waun Marteg Syncline [SN 995 698].

The Caerau Mudstones consist largely of thin- to thick-bedded turbidite mudstones (types Di and Dii) with scattered turbidite sandstone/mudstone couplets (type Cii) and interbedded hemipelagites of both varieties. By contrast with the Borth Mudstones and Blaen Myherin Mudstones, laminated hemipelagites are more common. In the lower tongue of the formation in the west, and in the lower half of the formation in its eastern crop, colour-banded, very thin turbidite mudstones (type E) and burrowed hemipelagites occur as abundant packets up to many metres thick (Plates 15c and d). The type E turbidite mudstones are typically strongly silt-laminated and include top-cut-out (T_{0-3}) varieties. Colour banding resulting from penecontemporaneous oxidation of both hemipelagitic and turbiditic lithologies is well developed. This lithology is closely similar to that of the underlying Claerwen Group.

Scattered high-matrix turbidite sandstones (type Bii), up to 30 cm thick are present throughout the upper tongue of the formation west of the Tylwch Anticline, but none has been recorded in the sequence east of this structure.

BIOSTRATIGRAPHY

The Caerau Mudstones range in age from the late *turriculatus* to the mid-*crispus* Biozone. However, the base and top of the formation are diachronous, both becoming younger in a northward and eastward direction, west of the lateral passage into the Dolgau Mudstones.

On the western limb of the Tylwch Anticline, south of the Wye valley, the formation ranges from the *proteus* Subzone, or possibly slightly earlier, to the *carnicus* Subzone, as demonstrated in the Claerwen reservoir section (Table 11). To the north, the lowest part of the lower tongue of the formation has yielded probable *carnicus* Subzone faunas in the Wye valley [SN 9277 7300] and just north of the district [at SN 9540 7932], whilst the upper part contains *galaensis* Subzone faunas in the Wye valley [at SN 9340 7415] (Figure 29, locality 3). The upper tongue is everywhere within the *crispus* Biozone, with possible *galaensis* Subzone faunas occurring east of Cwmgwary [at SN 9418 7624] and undoubted *crispus* Subzone assemblages in the Wye valley [SN 9331 7456 to SN 9264 7421].

On the eastern limb of the Tylwch Anticline the lower part of the formation is in the *carnicus* Subzone, as demonstrated by the Afon Dulas section (Table 14). The middle part of the formation has yielded *galaensis* Subzone faunas at the latter locality (Table 14) and elsewhere, but upper levels may range into the *sartorius* Subzone.

Acritarch assemblages from the Caerau Mudstones are sparse and of limited diversity (Appendix 2bii, iii and iv). A mid to late Telychian age is indicated by *G. encantador* Biozone floras comprising *Ammonidium listeri, A. microcladum, A.* sp. A, *Cymatiosphaera* sp. A, *Dictyotidium biscutulatum, Domasia limaciformis, Elektoriskos williereae, Eupoikilofusa striatifera, Helosphaeridium echiniforme, Moyeria cabottii, Multiplicisphaeridium fisheri, Oppilatala eoplanktonica* and *Salopidium granuliferum.* Recycled taxa present include *Stelliferidium* sp. from the early to mid-Ordovician, and *Baltisphaeridium* sp., *Peteinosphaeridium* sp., and *Stellechinatum brachyscolum* from the mid to late Ordovician.

DETAILS

Western crop, lower tongue   Sections through the southern part of the lower tongue of the Caerau Mudstones are exposed in cuttings [SN 8260 6005 to SN 8186 6021] in Tywi Forest, in shoreline outcrops and adjacent crags [SN 8588 6502 to SN 8510 6546] along the Claerwen reservoir, and in Nant y Gadair [SN 8632 6559 to SN 8723 6689]. These exposures display packages of thin- to thick-bedded turbidite mudstones (types Di and Dii), which commonly contain laminated hemipelagites, interbedded with packages dominated by thinner-bedded and paler turbidite mudstones (type Ei), and burrow-mottled hemipelagites. The lower part [SN 8563 6533] of the Claerwen reservoir section has yielded probable *proteus* Subzone graptolites (Table 11, localities 11 and 12). The uppermost part of the formation, both in the Tywi Forest section [at SN 8203 6014] and on the shore [SN 8508 6545] of the Claerwen reservoir (Table 11, locality 13), has yielded graptolites of the *carnicus* Subzone.

The lower tongue is again exposed along Nant Gris [SN 8842 6800 to SN 8810 6791], south-south-east of Craig Goch dam, and its base is seen on the headland [SN 893 690] north of the dam, where probable *proteus* Subzone graptolites have been recovered [SN 8945 6915] (Figure 32, locality 1). The presence of the *carnicus* Subzone is suggested by *M. carnicus* together with a lack of *crispus* Biozone indicators in a graptolite assemblage recovered from a stream section [SN 9277 7300] on the south side of the Wye valley, near Tymawr. North of the River Wye, a trackside quarry [SN 9340 7415] (Figure 29, locality 3) near Dolhelfa-ganol, about 40 m below the top of the lower tongue, has yielded *Monoclimacis? galaensis, Streptograptus* cf. *exiguus, S. mustadi?, Monograptus discus?,* and *M. rickardsi,* indicative of the *galaensis* Subzone of the *crispus* Biozone. Farther north-east, a section [SN 9394 7460 to SN 9388 7482] through the lower tongue is well exposed along Marcheini Fawr. Crags [near SN 947 766] on the west side of Craig Gellidywyll provide some of the northernmost exposures of the lower tongue. A possible *carnicus* Subzone fauna, including *M. carnicus?* and *M. proteus?,* has been obtained near the base of the formation from a small quarry [SN 9450 7932] at Ole-ddu, just north of the district. A quarry [SN 9448 7649] containing both burrow-mottled and laminated hemipelagites near the top of the lower tongue has yielded *carnicus* Subzone graptolites, which include *Monograptus carnicus, M. rickardsi,* cf. *Monoclimacis? galaensis, Streptograptus* sp. nov. and *Petalolithus* cf. *tenuis.*

Inliers of the lower tongue, west of its main crop, are exposed along Nant Brwynog [SN 8183 6492 to SN 8201 6512] and on the shore [SN 8424 6600] of Claerwen reservoir. Inliers crop out along Nant Cletwr above a thrust contact with the Rhuddnant Grits [at SN 8843 6883] (Figure 32, locality 5), and in a quarry [SN 8802 6893] (Figure 32, locality 6), where

graptolites referable either to the lower or middle part of the *crispus* Biozone have been recovered.

An outlier, tentatively assigned to the lower tongue of Caerau Mudstones, crops out in track cuttings [SN 9153 6764 to SN 9125 6764] north-east of Penygarreg dam, and in exposures [SN 9110 6801 to SN 9122 6818] along Nant Ffos-rhodd. Laminated hemipelagites are abundant and contain graptolites which include *Monograptus utilis, M. cavei* and *Streptograptus plumosus,* of *utilis* Subzone age. However, the presence of *Streptograptus johnsonae* in the highest part of the section suggests that the *johnsonae* Subzone is also present. This dating suggests that these sections equate with anoxic facies included elsewhere (p.89) at the top of the Rhayader Mudstones.

Western crop, upper tongue   Sections in the upper tongue occur in a road cutting [SN 8861 6919] and in quarries, crags and stream exposures [e.g. SN 8884 6992] in the vicinity of Hirnant farm. These sections display burrow mottled hemipelagites interbedded with thin- to thick-bedded turbidite mudstones (types Di and Dii). Farther north, the Caerau Mudstones are exposed in a gorge and quarry section [SN 8900 7045 to SN 8889 7050], along Nant Torclawdd. There, the formation displays tectonic contacts with the Glanyrafon Formation both to the east and west. North-east of Craig Goch reservoir, close to the Rhayader Mountain Road, a disused quarry [SN 9115 7134] exposes turbidite mudstones (type Di) up to 20 cm thick and scattered basal sandstones of type Cii turbidites. Both burrow-mottled and laminated hemipelagites are present, and the latter have yielded graptolites of the *crispus* Biozone.

North of the River Wye, on Banc Dolhelfa, a track section [SN 9331 7456 to SN 9264 7421] provides an almost continuous section through the upper half of the upper tongue; scattered high-matrix sandstones (type Bii) up to 30 cm thick are interbedded with turbidite mudstones and hemipelagites. The section has yielded graptolites of the *crispus* Subzone of the *crispus* Biozone, which include *Monograptus crispus, M. discus* and *Streptograptus loydelli.* A graptolite assemblage, including *M. discus,* from a stream section [SN 9418 7624] east of Cwmgwary indicates the *crispus* Biozone or younger. The presence of a densely thecate form of *M. crispus* hints that a *galaensis* Subzone age is most likely.

Eastern crop   The Afon Dulas [SN 9758 7764 to SN 9825 7774], east of Pont Rhyd Griffith, displays packets of very thin-bedded, silt-based turbidite mudstones (type E) and burrow-mottled hemipelagites interbedded with packets of thin- to thick-bedded turbidite mudstones (type D) and laminated hemipelagites. Graptolite assemblages from the section (Table 14) range from the *carnicus* Subzone of the *turriculatus* Biozone to the *galaensis* Subzone of the *crispus* Biozone. Graptolites of the latter subzone have also been recovered from a quarry [SN 9852 7515] in Pant-y-dwr, and the *galaensis* or *crispus* Subzone is represented in a track cutting [SN 9798 7079] east of Tynshimley.

Rhyd-hir Brook [SN 9962 6981 to SN 9952 7001], south of the Brynscolfa Fault, exposes the upper part of the formation and its contact with the succeeding tongue of Glanyrafon Formation. North of the fault, on the eastern limb of the syncline, a complete sequence through the formation is exposed in the stream [SO 0069 7008 to SO 0043 7035] west of Brynscolfa farm. There the formation gradationally overlies the Cwmbarn Formation. Significant exposures between the Bwlch-y-sarnau and Cwmysgawen faults include a quarry [SO 0172 7092] and crags [SO 0238 7077]. The latter has yielded long-ranging Telychian graptolites; the former, graptolites of late *turriculatus* Biozone (probably *carnicus* Subzone) age. An assemblage including *Monograptus proteus, M. carnicus?, Streptograptus storchi* and cf. *Monoclimacis galaensis,* recovered from a forestry

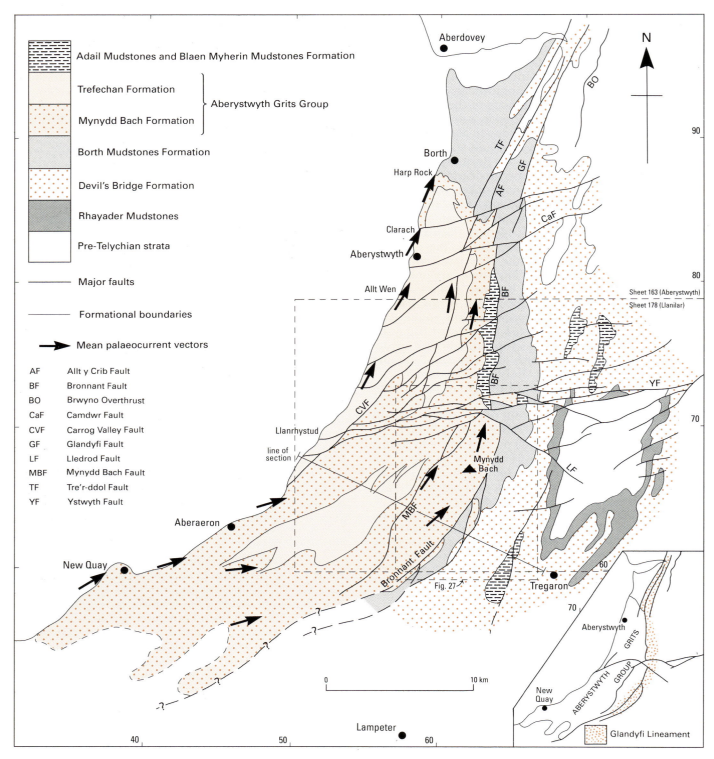

Figure 26   Geological map of the Aberystwyth Grits Group showing mean palaeocurrent vectors and (inset) the position of the Glandyfi Lineament. Palaeocurrent vectors include data from Wood and Smith (1959), Ankatell and Lovell (1976) and Ankatell and Smith (1992); the southern boundary of the group is based on Craig (1987). Line of section refers to C–C′ in Figure 44, and profiles in Figure 33a.

Acritarch assemblages from the Caerau Mudstones are sparse and of limited diversity (Appendix 2bii, iii and iv). A mid to late Telychian age is indicated by *G. encantador* Biozone floras comprising *Ammonidium listeri*, *A. microcladum*, *A.* sp. A, *Cymatiosphaera* sp. A, *Dictyotidium biscutulatum*, *Domasia limaciformis*, *Elektoriskos williereae*, *Eupoikilofusa striatifera*, *Helosphaeridium echiniforme*, *Moyeria cabottii*, *Multiplicisphaeridium fisheri*, *Oppilatala eoplanktonica* and *Salopidium granuliferum*. Recycled taxa present include *Stelliferidium* sp. from the early to mid-Ordovician, and *Baltisphaeridium* sp., *Peteinosphaeridium* sp., and *Stellechinatum brachyscolum* from the mid to late Ordovician.

DETAILS

Western crop, lower tongue   Sections through the southern part of the lower tongue of the Caerau Mudstones are exposed in cuttings [SN 8260 6005 to SN 8186 6021] in Tywi Forest, in shoreline outcrops and adjacent crags [SN 8588 6502 to SN 8510 6546] along the Claerwen reservoir, and in Nant y Gadair [SN 8632 6559 to SN 8723 6689]. These exposures display packages of thin- to thick-bedded turbidite mudstones (types Di and Dii), which commonly contain laminated hemipelagites, interbedded with packages dominated by thinner-bedded and paler turbidite mudstones (type Ei), and burrow-mottled hemipelagites. The lower part [SN 8563 6533] of the Claerwen reservoir section has yielded probable *proteus* Subzone graptolites (Table 11, localities 11 and 12). The uppermost part of the formation, both in the Tywi Forest section [at SN 8203 6014] and on the shore [SN 8508 6545] of the Claerwen reservoir (Table 11, locality 13), has yielded graptolites of the *carnicus* Subzone.

The lower tongue is again exposed along Nant Gris [SN 8842 6800 to SN 8810 6791], south-south-east of Craig Goch dam, and its base is seen on the headland [SN 893 690] north of the dam, where probable *proteus* Subzone graptolites have been recovered [SN 8945 6915] (Figure 32, locality 1). The presence of the *carnicus* Subzone is suggested by *M. carnicus* together with a lack of *crispus* Biozone indicators in a graptolite assemblage recovered from a stream section [SN 9277 7300] on the south side of the Wye valley, near Tymawr. North of the River Wye, a trackside quarry [SN 9340 7415] (Figure 29, locality 3) near Dolhelfa-ganol, about 40 m below the top of the lower tongue, has yielded *Monoclimacis? galaensis*, *Streptograptus* cf. *exiguus*, *S. mustadi?*, *Monograptus discus?*, and *M. rickardsi*, indicative of the *galaensis* Subzone of the *crispus* Biozone. Farther north-east, a section [SN 9394 7460 to SN 9388 7482] through the lower tongue is well exposed along Marcheini Fawr. Crags [near SN 947 766] on the west side of Craig Gellidywyll provide some of the northernmost exposures of the lower tongue. A possible *carnicus* Subzone fauna, including *M. carnicus?* and *M. proteus?*, has been obtained near the base of the formation from a small quarry [SN 9450 7932] at Ole-ddu, just north of the district. A quarry [SN 9448 7649] containing both burrow-mottled and laminated hemipelagites near the top of the lower tongue has yielded *carnicus* Subzone graptolites, which include *Monograptus carnicus*, *M. rickardsi*, cf. *Monoclimacis? galaensis*, *Streptograptus* sp. nov. and *Petalolithus* cf. *tenuis*.

Inliers of the lower tongue, west of its main crop, are exposed along Nant Brwynog [SN 8183 6492 to SN 8201 6512] and on the shore [SN 8424 6600] of Claerwen reservoir. Inliers crop out along Nant Cletwr above a thrust contact with the Rhuddnant Grits [at SN 8843 6883] (Figure 32, locality 5), and in a quarry [SN 8802 6893] (Figure 32, locality 6), where

graptolites referable either to the lower or middle part of the *crispus* Biozone have been recovered.

An outlier, tentatively assigned to the lower tongue of Caerau Mudstones, crops out in track cuttings [SN 9153 6764 to SN 9125 6764] north-east of Penygarreg dam, and in exposures [SN 9110 6801 to SN 9122 6818] along Nant Ffos-rhodd. Laminated hemipelagites are abundant and contain graptolites which include *Monograptus utilis*, *M. cavei* and *Streptograptus plumosus*, of *utilis* Subzone age. However, the presence of *Streptograptus johnsonae* in the highest part of the section suggests that the *johnsonae* Subzone is also present. This dating suggests that these sections equate with anoxic facies included elsewhere (p.89) at the top of the Rhayader Mudstones.

Western crop, upper tongue   Sections in the upper tongue occur in a road cutting [SN 8861 6919] and in quarries, crags and stream exposures [e.g. SN 8884 6992] in the vicinity of Hirnant farm. These sections display burrow mottled hemipelagites interbedded with thin- to thick-bedded turbidite mudstones (types Di and Dii). Farther north, the Caerau Mudstones are exposed in a gorge and quarry section [SN 8900 7045 to SN 8889 7050], along Nant Torclawdd. There, the formation displays tectonic contacts with the Glanyrafon Formation both to the east and west. North-east of Craig Goch reservoir, close to the Rhayader Mountain Road, a disused quarry [SN 9115 7134] exposes turbidite mudstones (type Di) up to 20 cm thick and scattered basal sandstones of type Cii turbidites. Both burrow-mottled and laminated hemipelagites are present, and the latter have yielded graptolites of the *crispus* Biozone.

North of the River Wye, on Banc Dolhelfa, a track section [SN 9331 7456 to SN 9264 7421] provides an almost continuous section through the upper half of the upper tongue; scattered high-matrix sandstones (type Bii) up to 30 cm thick are interbedded with turbidite mudstones and hemipelagites. The section has yielded graptolites of the *crispus* Subzone of the *crispus* Biozone, which include *Monograptus crispus*, *M. discus* and *Streptograptus loydelli*. A graptolite assemblage, including *M. discus*, from a stream section [SN 9418 7624] east of Cwmgwary indicates the *crispus* Biozone or younger. The presence of a densely thecate form of *M. crispus* hints that a *galaensis* Subzone age is most likely.

Eastern crop   The Afon Dulas [SN 9758 7764 to SN 9825 7774], east of Pont Rhyd Griffith, displays packets of very thin-bedded, silt-based turbidite mudstones (type E) and burrow-mottled hemipelagites interbedded with packets of thin- to thick-bedded turbidite mudstones (type D) and laminated hemipelagites. Graptolite assemblages from the section (Table 14) range from the *carnicus* Subzone of the *turriculatus* Biozone to the *galaensis* Subzone of the *crispus* Biozone. Graptolites of the latter subzone have also been recovered from a quarry [SN 9852 7515] in Pant-y-dwr, and the *galaensis* or *crispus* Subzone is represented in a track cutting [SN 9798 7079] east of Tynshimley.

Rhyd-hir Brook [SN 9962 6981 to SN 9952 7001], south of the Brynscolfa Fault, exposes the upper part of the formation and its contact with the succeeding tongue of Glanyrafon Formation. North of the fault, on the eastern limb of the syncline, a complete sequence through the formation is exposed in the stream [SO 0069 7008 to SO 0043 7035] west of Brynscolfa farm. There the formation gradationally overlies the Cwmbarn Formation. Significant exposures between the Bwlch-y-sarnau and Cwmysgawen faults include a quarry [SO 0172 7092] and crags [SO 0238 7077]. The latter has yielded long-ranging Telychian graptolites; the former, graptolites of late *turriculatus* Biozone (probably *carnicus* Subzone) age. An assemblage including *Monograptus proteus*, *M. carnicus?*, *Streptograptus storchi* and cf. *Monoclimacis galaensis*, recovered from a forestry

Table 14   Distribution of graptolites in the Caerau Mudstones in Afon Dulas near Cenarth Mill.

	Graptolite localities						
	1 2	3	4 5	6		7
Mcl? galaensis	●					
S. storchi	●	●				
M. carnicus	●	●		●		
M marri	?			cf.		
Pristiograptus sp.	● ●	●		●		●
M. rickardsi	●	●				
Pe. tenuis s.l.		●	● ●			●
M. discus				●		
S. exiguus				●		
M. clintonensis						●
	carnicus Subzone			*galaensis* Subzone	*sartorius* Subzone	
	turriculatus Biozone			*crispus* Biozone		

Localities (see map below)

1 stream section [SN 9771 7762]; 2 trackside exposure [SN 9771 7765]; 3 trackside exposure [SN 9781 7764]; 4 south bank of river [SN 9792 7762]; 5 south bank of river [SN 9794 7762]; 6 valley side exposure [SN 9799 7762]; 7 railway cutting [SN 9858 7796] (Glanyrafon Formation). For key to range chart symbols see Appendix 3.

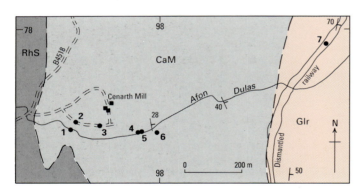

Glr	Glanyrafon Formation
CaM	Caerau Mudstones
RhS	Rhayader Mudstones
50 ⟋	Inclined strata, dip in degrees

● Fossil locality

section [SO 0197 7136] on Cwm-hir Bank, also suggests the *carnicus* Subzone. South of the Cwmysgawen Fault, in a roadside quarry [SO 0222 7063] at Lan Goch, strata assigned to the Caerau Mudstones have provided microflora with ranges from mid-Aeronian to late Telychian, including the spore *Ambitisporites*.

Adail Mudstones

This newly recognised formation is composed predominantly of turbidite mudstones and overlies the Mynydd Bach Formation of the Aberystwyth Grits Group. Its type section is a tributary [SN 6294 7348 to SN 6339 7324] of Nant Adail, where its base is taken at the top of a packet of thick-bedded sandstones included in the Mynydd Bach Formation (p.131). The top of the formation is not present in the district.

The outcrop of the Adail Mudstones occupies a north–south belt, predominantly to the north of the Ystwyth Fault, trending through New Cross [SN 634 770] to the northern edge of the district (Figure 26). The western limit of the outcrop is the conformable junction with the Mynydd Bach Formation; the eastern limit is the Bronnant Fault. The formation attains a thickness in excess of 200 m in this region.

The formation predominantly comprises thick-bedded, grey, turbidite mudstones (type D) with subsidiary thin-bedded turbidite sandstone/mudstone couplets (type Cii). Rare thicker sandstones (types Ci and Bii) occur scattered in the lower parts of the formation. Intercalated hemipelagites are of the burrowed type. The formation includes strata that in the south of the Aberystwyth district were previously included in the Borth Mudstones. As the two formations are lithologically identical, they are separated using structural criteria (p.177) where they are juxtaposed across the Bronnant Fault. No fossils have been recorded from the Adail Mudstones.

DETAILS

The lower part of the type section of the Adail Mudstones, including the base, is exposed in Cwm Hendre-haidd [SN 6294 7348 to SN 6301 7346]. Upstream [SN 6319 7339 to SN 6339 7324], scattered thin-bedded turbidite sandstone/mudstone couplets (type Cii) and rare, massive high-matrix sandstones (type Bii) up to 35 cm thick, are present. To the south the formation is exposed in a stream [SN 6296 7130] west of Abernac. North of Afon Ystwyth, the formation is exposed at New Cross garage [SN 6336 7703], and in nearby track cuttings [SN 6304 7698]. Further sections are provided at Pen-y-wern [SN 6360 7660] and near Glasdir farm [at SN 6369 7615]. Farther north, a cutting [SN 6331 7840] exposes a section close to the base of the formation. Additional exposures in this vicinity occur near Nant-yrhydd [SN 6383 7863], and near Capel Seion [SN 6339 7910].

SAND-RICH TURBIDITE GROUPS

The Aberystwyth Grits and Cwmystwyth Grits groups define a suite of southerly derived, sandstone-rich, turbidite systems which advanced progressively across the Southern Welsh Basin during the Telychian Stage.

ABERYSTWYTH GRITS GROUP

The Aberystwyth Grits are the oldest and most westerly of these sandstone-rich turbidite systems. They mainly comprise medium- to thin-bedded, turbidite sandstone/ mudstone couplets (types Ci and Cii) with both laminated and burrowed hemipelagites. However, over wide areas of the district, these Bouma turbidites are augmented by locally predominant thick-bedded sandstone turbidites (type B). Sequences in which the latter are abundant have been mapped as the Mynydd Bach Formation, and those comprising almost exclusively Bouma turbidites, as the Trefechan Formation.

Although the Aberystwyth Grits were first described by Keeping (1881), their base was not formally defined until much later (Cave, 1976). A summary of the earlier work on the group (at that time classified as a formation), and detailed descriptions of its outcrops in the Aberystwyth district, have been provided by Cave and Hains (1986).

Although the group overlies the Borth Mudstones in coastal sections in the Aberystwyth district (Cave and Hains, 1986), the line traditionally taken to mark the outcrop of the base inland (e.g. Wood and Smith, 1959; Lovell, 1970; Loydell, 1991) is now known to be the Bronnant Fault, along much of its length (Figures 26 and 27) (Wilson et al., 1992). In the district, the group is largely restricted to the west of the Bronnant Fault zone. Normal contacts with the Borth Mudstones are only seen, south of the Ystwyth Fault, between splays of the Bronnant Fault. Moreover, across much of the eastern part of their outcrop the Aberystwyth Grits young eastwards so that what were previously considered to be the oldest parts are now seen to occupy a high level in the group. In this eastern area, north of the Ystwyth Fault, the group is overlain by the Adail Mudstones (Figure 26).

This structural re-evaluation of the group has important implications for estimates of its thickness. The steep, easterly dipping sequence adjacent to the Bronnant Fault has a minimum thickness of about 2.5 km. Although the much thinner estimate of 300 m by Cave and Hains (1986) for the Aberystwyth district may be an underestimate due to unrecognised faults (cf. Wilson et al., 1992), it appears to reflect a genuine westward and northward thinning of the group (Figures 12 and 44).

The diverse trace fossil assemblages of the group, typically in the form of semirelief casts (hypichnia) on the bases of sandstones, have been documented by Crimes and Crossley (1991).

Mynydd Bach Formation

The formation is named after the ridge of Mynydd Bach [SN 610 660] where, between the Mynydd Bach and Bronnant faults, it is at least 2.5 km thick (Figure 27). The base of the Mynydd Bach Formation is not exposed west of the Bronnant Fault, but in the Bronnant Fault zone, sections in Gwenffrwd (Cwm Bwlch) [SN 5968 6075] and Aeron Fechan [SN 6275 6365] expose the contact with the underlying Borth Mudstones. The outcrop of Mynydd Bach Formation north of the Ystwyth Fault, which extends through Llanilar to the northern edge of the district, occupies the upper part of the Aberystwyth Grits and is about 600 m thick. In this region, the formation passes up into the Adail Mudstones.

The thick sequence of Mynydd Bach Formation on Mynydd Bach is coeval with the Trefechan Formation on the coast in the Aberystwyth district. This lateral westwards passage is achieved by complex intertonguing of the two formations. Sequences of Mynydd Bach Formation which occupy a broadly anticlinal outcrop east of the Carrog

Valley Fault appear to equate with the lower part of the thick eastern sequence of the formation (Figure 44). This extensive lower tongue of Mynydd Bach Formation correlates with the lower parts of the Aberystwyth Grits in the Aberystwyth district, in which 'Harp Rock Type' turbidites (type Bii) are locally abundant (Cave and Hains, 1986).

Both varieties of type B turbidite sandstones occur in the Mynydd Bach Formation. Medium- to thick-bedded, high-matrix, turbidite sandstones (type Bii) (Plates 16b and f) are locally abundant. They include the beds described by Wood and Smith (1959) as 'fragmented', 'slurried' or displaying 'prolapsed bedding' (Plate 7f). They do not form stacked sequences (stacking of more than two is rare), but occur concentrated in packets of strata, from less than 1 m up to tens of metres thick, that contain two or more such sandstones separated by Bouma turbidites (mainly type C but locally type D). These packets in turn are separated by thicker or thinner intervals in which type B turbidite sandstones are rare.

Medium- to very thick-bedded, mud-depleted sandstones and pebbly sandstones (type Bi) are the least common in the formation. They typically form stacked sequences, up to 20 m thick, in which individual beds are either amalgamated or separated by thin, impersistent mudstone beds (Figure 28a, Plate 16a). Such sequences are rare, randomly distributed and appear to be laterally discontinuous.

Along the eastern limits of the Mynydd Bach Formation, an upward decrease in the proportion of sandstone marks a protracted transition into the succeeding Adail Mudstones.

Palaeocurrent vectors obtained from the formation, east of the Mynydd Bach Fault, parallel the arcuate trend of the Bronnant Fault and swing from a regional southwesterly trend in the south to a south-south-westerly trend in the north (Figure 26).

BIOSTRATIGRAPHY

Graptolites recovered from the underlying Borth Mudstones in the Bronnant Fault zone in Aeron Fechan [SN 6284 6364] (*utilis* Subzone) and in Gwennffrwd [SN 5968 6068] (probably *renaudi* Subzone), suggest that the base of the Mynydd Bach Formation becomes older westwards (Figure 27). This trend, if continued west of the Bronnant Fault, would suggest that the base of the thick sequence on Mynydd Bach may be as early as the *gemmatus* Subzone; graptolites of this subzone have been reported from the lowest part of the Aberystwyth Grits in the Newquay district (Loydell, 1991). However, assemblages recovered from Mynydd Bach show that most of the 2.5 km-thick succession, east of the Mynydd Bach Fault, falls within the *utilis* Subzone, with the *johnsonae* Subzone probably present [SN 6483 7169] 300 m below the faulted top of this sequence (Figure 27). In the district the Mynydd Bach Formation west of the Mynydd Bach Fault remains undated.

DETAILS

Mynydd Bach, north of the Bontnewydd road   This is the type area for the formation. Packets of thick-bedded structureless

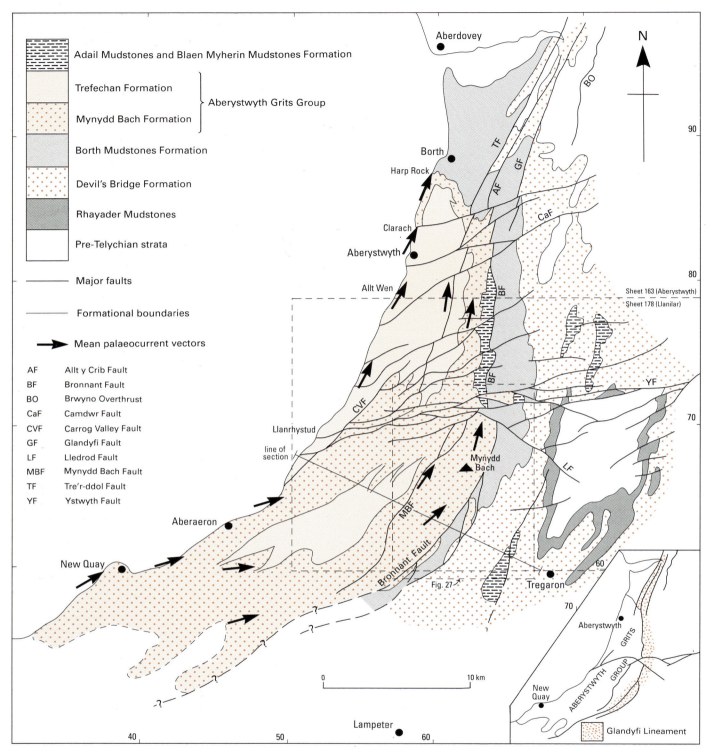

Figure 26   Geological map of the Aberystwyth Grits Group showing mean palaeocurrent vectors and (inset) the position of the Glandyfi Lineament. Palaeocurrent vectors include data from Wood and Smith (1959), Ankatell and Lovell (1976) and Ankatell and Smith (1992); the southern boundary of the group is based on Craig (1987). Line of section refers to C–C' in Figure 44, and profiles in Figure 33a.

Figure 27   Geological map of the Mynydd Bach area showing critical graptolite localities. Line of section is part of that shown on Figure 26.

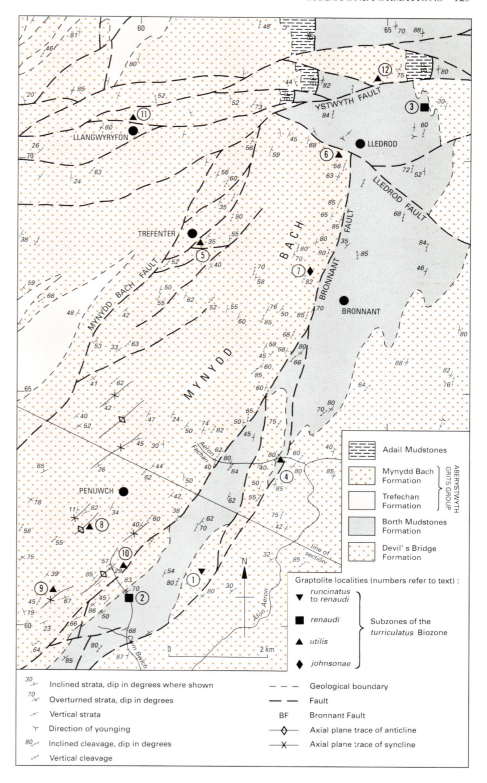

sandstones (type Bii) interbedded with Bouma turbidites (type C) are exposed in numerous crag and quarry sections across the north-western flank of Mynydd Bach, between Tanyglogau [SN 6157 6998] and Pwll-clai [SN 6289 7079], and to the south on Pant-y-gwair, between Cofadail School [SN 6048 6780] and Hafod Ithel [SN 6090 6880]. There, the sandstones range from massive, mud-depleted varieties (type Bi) to cleaved, locally granule-rich, feldspathic, high-matrix varieties (type Bii) (Plate 16b). High-matrix sandstones exposed in a disused quarry [SN 6180 6918] display contorted rip-up clasts, and have yielded the long-ranging Telychian graptolite *M. marri*. East of Trefenter, extensive sections are seen in track cuttings [between SN 6116 6821 and SN 6180 6836]. Graptolites obtained from the westernmost section [SN 6116 6821] include *Monograptus* cf. *utilis* and *Streptograptus plumosus*; these suggest a *utilis* Subzone age (Figure 27, locality 5). Disused quarries [SN 5970 6673, SN 5964 6610 and SN 5944 6606] provide further sections in the formation.

A disused quarry [SN 6400 7015] on the eastern flank of Mynydd Bach, close to the Lledrod Fault, exposes cleaved high-matrix sandstones (type Bii) up to 1.1 m thick. Graptolites obtained from a 1 cm-thick, planar-laminated sandstone, located 4 m above the base of the quarry, include *M. turriculatus*, *M. marri*, *Rastrites* cf. *linnaei* and *Streptograptus barrandei*, indicative of the *utilis* Subzone (Figure 27, locality 6). To the south, a stream [SN 6386 6885 to SN 6401 6880] north of Navy Hall provides a continuous section through the eastern part of the Mynydd Bach succession. Packets of turbidite sandstones (type B), with individual beds ranging up to 50 cm thick, occur in a sequence dominated by turbidite mudstones (type D). The latter are exposed in the nearby roadside quarry [SN 6404 6880]. Graptolites of probable *johnsonae* Subzone age, including *Streptograptus johnsonae* and *M. turriculatus*, were recovered from a quarry [SN 6339 6746] south-west of Navy Hall (Figure 27, locality 7). A road section [SN 6348 6686] close to the Bronnant Fault exposes parallel-laminated sandstones (type Ci) up to 25 cm thick. A packet of coarse-grained, locally amalgamated turbidite sandstones (type Bi), in the eastern (upper) part of the Mynydd Bach Formation, is exposed in Bontnewydd

School quarry [SN 6224 6527]. There, massive to weakly laminated sandstones are up to 1.2 m thick.

Mynydd Bach, south of the Bontnewydd road   Several large quarries are situated in this poorly exposed region. A quarry [SN 6085 6320] north of Brynamlwg, displays a range of turbidite

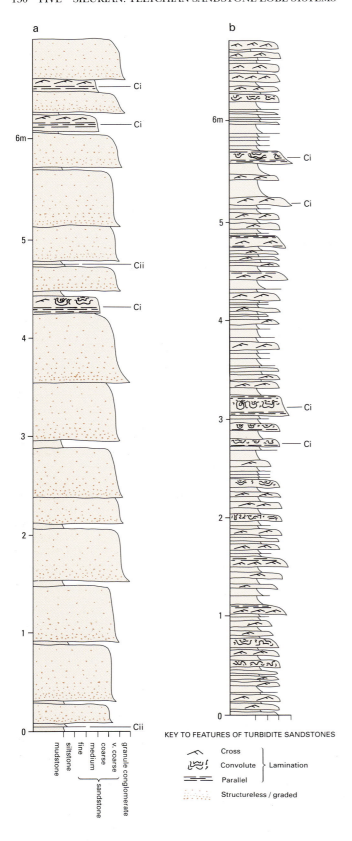

a

6m

5

4

3

2

1

0

Ci
Ci
Cii
Ci
Cii

mudstone
siltstone
fine
medium
coarse
v. coarse
granule conglomerate

sandstone

b

6m

5

4

3

2

1

0

Ci
Ci
Ci
Ci

KEY TO FEATURES OF TURBIDITE SANDSTONES

Cross ⌢
Convolute   } Lamination
Parallel ═
Structureless / graded ⦂

sandstones from mud-depleted, amalgamated units up to 3 m thick (type Bi), some with pebble and granule-rich bases, to more muddy varieties which contain contorted rip-up clasts and pseudonodules. Thin-bedded turbidite sandstone/mudstone couplets (type C) predominate in a quarry [SN 5722 6278] on the west side of Moelfryn, close to the Mynydd Bach Fault, but scattered pseudonodular high matrix sandstones (type Bii), up to 1 m thick, are also present. Craig Fawr quarry [SN 5895 6212] (Figure 28a, Plate 16a), west of Penuwch, exposes a sequence of commonly amalgamated, mud-depleted sandstones (type Bi) and thin mudstone interbeds. Coarse-grained sandstones with granule conglomerate bases form massive beds up to 2 m thick. Several sandstone beds display multiple grading and parallel- and cross-laminated tops. Laminated hemipelagites exposed in the eastern face of the quarry have yielded *utilis* Subzone graptolites including *M. utilis, M. marri, Streptograptus barrandei, S. plumosus?* and *S. johnsonae* (Figure 27, locality 8). A quarry [SN 5821 6075] south of Crynfryn has yielded the graptolites *M. involutus, M. halli, M. marri, M.* cf. *planus, M. utilis?, Streptograptus plumosus* and *Pristiograptus bjerringus*, again indicative of the *utilis* Subzone (Figure 27, locality 9). To the east, the lower part of the Mynydd Bach sequence is seen in the upper reaches of Gwenffrwd stream [SN 5956 6171 to SN 5939 6119] and in the nearby quarry [SN 5971 6125] at Henbant, where graptolites of the *utilis* Subzone are present (Figure 27, locality 10).

East of the Bronnant Fault   Gwenffrwd (Cwm Bwlch) [SN 5945 6093 to SN 5968 6075] (Figure 27) provides a section through the base of the formation. There a sequence of steeply dipping, westward-younging, thick-bedded sandstones (type B) overlies the Borth Mudstones. Farther north-east, the formation is exposed between eastern splays of the Bronnant Fault in Nant Ffynnon Geitho [SN 6188 6184 to SN 6171 6201] south-west of Cwrt Mawr, and Aeron Fechan [SN 6280 6366 to SN 6224 6348] east and south of Ffos-yr-odyn, where a contact with Borth Mudstones is observed (p.123).

West of the Mynydd Bach Fault   This area is extensively drift covered and exposure is restricted to river sections or high ground. At Trichrug [SN 542 599], coarse-grained, feldspathic, high-matrix sandstones (type Bii), commonly over 50 cm thick, graded and rich in mudstone clasts, are exposed in small disused quarries [SN 5395 6006; SN 5409 6049; SN 6008 5529] and at Plas-y-Bryniau [SN 5632 6153]. The bed and valley sides of Afon Cledan provide an almost continuous section from Llanon [SN 5160 6681] to Rhydydorth [SN 5268 6611]. There, packets of thick- to very thick-bedded sandstones (type Bii) are common throughout a folded, but predominantly westward-younging, sequence. Similar strata are exposed in crags [SN 5178 6667] near Llanon, where gutter-casted bases and amalgamated beds are abundant. To the east, a disused quarry [SN 5143 6664] near Llanon exposes an amalgamated sequence of graded sandstones (type Bi) in which individual sandstones are up to 1 m thick. Over 10 m of sandstones (type Bi) up to

Figure 28   Representative graphic logs of the Aberystwyth Grits Group:

a   Mynydd Bach Formation, Craig Fawr quarry [SN 5895 6212], Penuwch; turbidite sandstones are of type Bi unless indicated otherwise.   **b**   Trefechan Formation, Allt Wen [SN 5766 7955]; turbidite sandstone/mudstone couplets are of type Cii unless indicated otherwise.

1.75 m thick are displayed in a quarry [SN 5154 6458] near Garnfoel. Similar lithologies are seen in Afon Peris, notably between Allt-goch [SN 5359 6715] and Glan Peris [SN 5573 6726], in its tributary Nant Cennin [north of SN 5544 6673], and again in Afon Peris, east of Pantyrhogfaen [SN 5665 6660]. Throughout these sections the Mynydd Bach Formation intertongues with the Trefechan Formation.

North of the Ystwyth Fault Intermittent sections in the formation, adjacent to the Carrog Valley Fault, are exposed in Afon Carrog [SN 5615 7197], south of Llanddeiniol. To the east, it crops out in crags [SN 562 714] near Gilfachau, around Pencwm [e.g. SN 573 706], and in a quarry [SN 5755 7125] east of Bryngwyn. Exposures [SN 5809 7079] at Grip include amalgamated thick-bedded, mud-depleted sandstones (type Bi) in which individual beds exceed 1.0 m. In Llangwyryfon quarry [SN 5980 7082] (Figure 27, locality 11), parallel-laminated sandstones, parts of type Ci couplets, predominate. The section has yielded graptolites of the *utilis* Subzone (Loydell, 1989). Sections farther north, where the formation intertongues with Trefechan Formation, are provided by a track floor [SN 5674 7392 to SN 5663 7372] at Carrog Farm and by disused quarries [SN 5774 7339, SN 5839 7313 and SN 5871 7379] east of Cefn-graigwen farm.

East of Llangwyryfon, between the splays of the Ystwyth Fault, the Mynydd Bach Formation is exposed in a quarry [SN 6021 7089] west of Rhandir-isaf, at Brynchwith [SN 6096 7138], at Gorslas [SN 6184 7125], and at point nearby [SN 6237 7133]. The uppermost parts of the formation in these fault slices are seen in sections [SN 6272 7146 to SN 6273 7122] along a tributary of Afon Wyre, west of Abernac, and in a quarry [SN 6317 7161] to the north. Thick-bedded sandstones (type B) are present in both these exposures, interbedded with turbidite mudstones (type D). At a roadside quarry [SN 6483 7169] below Gaer Fawr, turbidite mudstones (type D) are locally interbedded with thin-bedded sandstone/mudstone couplets (type C) and high-matrix sandstones (type Bii)(Plate 16f), the latter up to 65 cm thick. Laminated hemipelagites have yielded *utilis* Subzone graptolites including *M. turriculatus, M. marri, M. planus* and *Normalograptus? scalaris* (Figure 27, locality 12).

South of Llanilar, in Nant Adail [SN 6235 7487 to SN 6244 7386], mudstones (type D) augment the standard suite of sandstone-bearing turbidites and are commonly capped by pale burrow-mottled oxidised layers with associated phosphatic lenses. At one place [SN 6267 7406] graptolites have been collected from an interval with laminated hemipelagites; they include *M. marri, M.* cf. *planus Normalograptus? scalaris* and *Petalolithus altissimus* sensu lato, indicative of the *utilis* Subzone. Mudstone-dominated facies in the central part of the section form part of the transition into the overlying Adail Mudstones. In Cwm Hendre-haidd [SN 6286 7356 and SN 6292 7351], a packet of thick-bedded sandstones (type B) marks the top of the Mynydd Bach Formation. North of Afon Ystwyth thick-bedded sandstones (type B) are exposed in quarries [SN 6258 7576, SN 6181 7655 and SN 6280 7895] to the east of Glenydd, south-west of Pencraig and south of Gors respectively, and in a track section [SN 6280 7895 to SN 6304 7885] at Cefn-llech. However, in many sections in this northern area thinner-bedded Bouma turbidites are dominant, for example in track cuttings [SN 6239 7599] at Pantygwyfol, stream exposures [SN 6189 7587 to SN 6195 7617] farther west, and a quarry [SN 6291 7615] to the east of New Cross.

Trefechan Formation

The Trefechan Formation crops out in two areas. The northern area is centred on Llanfarian, and includes the coastal strip and ground west of the Carrog Valley Fault. The poorly exposed southern area lies west of the Mynydd Bach Fault, and is centred on Rhos Cilcennin (Figure 26). The formation is named after the coastal cliffs and quarries around Trefechan [SN 583 811], in the Aberystwyth district.

The Trefechan Formation predominantly comprises Bouma turbidite sandstone/mudstone couplets (type C, with type Cii predominant), subordinate turbidite mudstones (type Di), and intercalated laminated and burrowed hemipelagites (Figure 28b)(Plates 8b and 16c, d, e and g). Rare, thick-bedded turbidite sandstones (types Bi and Bii) are locally present. Although localised thinning-upwards sequences are reported from the Trefechan Formation on the coast between Llanryhstud and Clarach (Wood and Smith, 1959; Crimes and Crossley, 1980), in general there appears little systematic variation in either turbidite sandstone or sandstone/mudstone couplet thickness. The thicker, basal sandstones of type Ci couplets range up to 25 cm thick whereas those of type Cii couplets are typically less than 10 cm thick. However, both varieties appear to be randomly distributed (Wood and Smith, 1959; Lovell, 1970; Crimes and Crossley, 1991). The total proportion of sandstone, in the coastal exposures of the district, varies between 30 and 60 per cent (Lovell, 1970).

Palaeocurrent data obtained principally from the sandstones confirm the trends established by previous workers (e.g. Wood and Smith, 1959). Bottom structures (flutes and grooves) and cross-lamination indicate a south-south-westerly derivation (Figure 26). Westerly to west-north-westerly derivation vectors have been reported from near Monk's Cave [SN 550 733], from a thin sequence of massive, thick-bedded turbidite sandstones (type B) (Crimes and Crossley, 1980; Wood and Smith, 1959).

Trace-fossil orientations in the Aberystwyth Grits, in particular of the net-like burrow systems of *Paleodictyon* and *Squamodictyon* preserved in semirelief on the bases of the turbidite sandstones, suggest that semi-permanent bottom currents also flowed from the south-south-west parallel to the bulk turbidite flowpath (Crimes and Crossley, 1980). The trace-fossil orientations maintain this trend even where the associated turbidite sandstones yield deviant flow vectors, for example at Monk's Cave [SN 548 733].

BIOSTRATIGRAPHY

In the district, the Trefechan Formation of coastal sections has only yielded graptolite assemblages of the *proteus* Subzone, but *utilis* and *johnsonae* Subzone graptolites occur immediately to the north (Loydell, 1989; 1991). The presence of the *utilis* and *johnsonae* subzones confirms the lateral passage of those parts of the Trefechan Formation into the thick sequence of Mynydd Bach Formation between the Mynydd Bach and Bronnant faults. The *proteus* Subzone faunas have only been obtained west of the Carrog Valley Fault; these strata may correlate with uppermost, undated levels of the Mynydd Bach Formation farther east. A possible *proteus* Subzone fauna occurs in the cliffs [SN 5335 7040] south of Carreg Ti-pw. Loydell's (1991) record of grapto-lites possibly referable to the *crispus* Biozone from this vicinity [SN 5342 7053] has not been confirmed.

Southern area   In the west, sections are restricted to Nant Camel [SN 5152 6024 to SN 5177 6032] west of Cilcennin, Nant Erthig [west of SN 5101 6142] and downriver [from SN 5113 6306] in Afon Arth west of Pennant. To the north-west, the formation is exposed at Pencerig Peris [SN 5681 6628] in Nant Peris, where it overlies the Mynydd Bach Formation.

Northern and coastal areas   The cliffs between Llanrhystud [SN 534 705] and Pantyrallad [SN 556 748], and between Morfa Bychan [SN 5660 7770] and Allt Wen [SN 5750 7930] (Figure 28b, Plate 16d and e), provide the principal sections in the district. Turbidite sandstone/mudstone couplets (types Ci and Cii) predominate. However, to the north and south of Monk's Cave [SN 5553 7445], a packet of massive thick-bedded sandstones (type B) is repeated by folding and faulting. Grapto-lites of *proteus* Subzone age have been reported from the cliffs at Monk's Cave and sections north of Morfa Bychan (Loydell, 1989; 1991). South of Carreg Ti-pw [at SN 5335 7040], lami-nated hemipelagites have yielded graptolites dominated by the long-ranging *M. marri*. Loydell (1991) reported rare specimens of *Streptograptus mustadi* from this vicinity.

Quarries [SN 5745 7483, SN 5773 7430, SN 5802 7690] east of Glan-rhos, at Cefnmelgoed and at Chancery respectively, expose typical Bouma turbidites (types Ci and Cii). Farther east, near Tan-y-graig [at SN 5903 7582], north-east of Bryngwyn [at SN 6047 7468] and along Cwm Moch [SN 6123 7468 to SN 6123 7441], widely scattered thick-bedded sand-stones (type B) are also present and suggest a region of transi-tion with the Mynydd Bach Formation. North of Afon Ystwyth, thick-bedded sandstones are again present in a quarry [SN 5930 7822] south of Bach-y-rhiw, at a point [SN 6138 7732] west of Cwmhwylog, and along the track [SN 6114 7619] to Lanlwyd, but are absent from extensive exposures [SN 6119 7546] below Coed y Castell, and from the inverted sections [SN 6032 7626, SN 6067 7676, SN 6089 7854] at Craig-y-bwlch, Penlanlas and north-east of Tynywern respectively.

CWMYSTWYTH GRITS GROUP

The term Cwmystwyth Formation, later amended to Cwmystwyth Grits (Jones, 1938), was first used by Jones (1922) to describe the thick sequence of thin- and thick-bedded turbidite sandstones and mudstones in the core and limbs of the Central Wales Syncline, as typified by the exposures in the eastern part of Cwm Ystwyth [SN 809 747 to SN 854 758] (Figure 31). Later, W D V Jones (1945) used the name Moelfre Group to describe this sequence and equivalent strata on the eastern limb of the Tylwch Anticline. Local names have been used for parts of the Cwmystwyth Grits that contain abundant thick sandstones. Thus, to the north of Cwm Ystwyth, the lower part was termed the Rhuddnant Group (including the Rhuddnant Grits) by Jones (1909), and north-west of Rhandirmwyn, in the Builth district, Davies (1933) named the upper part the Pysgotwr Grits.

The group is separated from the predominantly earlier Aberystwyth Grits Group by the Teifi Anticline; it crops out in the district principally in the core of the Central Wales Syncline and along the eastern limb of the Tylwch Anticline (Figure 29). It mainly comprises thin-bedded, Bouma turbidite sandstone/mudstone couplets (type Cii) with subordinate hemipelagites, but abundant medium- to thick-bedded, high-matrix turbidite sand-stones (type Bii) characterise discrete levels in the group and thereby form the basis for its subdivision into three formations (Figure 12). The Rhuddnant Grits and the higher Pysgotwr Grits are both characterised by the presence of commonly feldspathic high-matrix sand-stones and confined to the Central Wales Syncline. The remainder of the group, virtually free of such type Bii sandstones, comprises the Glanyrafon Formation. It inter-tongues with the Rhuddnant Grits and Pysgotwr Grits. In the core of the Central Wales Syncline the Glanyrafon Formation lies between the Rhuddnant and Pysgotwr grits and also overlies the latter. In the Waun Marteg Syn-cline the Glanyrafon Formation forms the sole represen-tative of the group (Figure 12).

The thickness of the group and its constituent forma-tions vary considerably; some of the changes take place across major faults. The maximum thickness of the group, estimated at about 2.4 km, occurs on the western limb of the Central Wales Syncline, west of the Claerwen Fault, where all three component formations are pres-ent. To the east, in the Waun Marteg Syncline, the group is 900 m thick and thins eastwards to a feather edge in the region of the Tywi Lineament.

Rhuddnant Grits

The formation takes its name from sections in Nant Rhuddnant [SN 8028 7842 to SN 8050 7842] (Figure 45; Table 16) north of Cwmystwyth, in the Rhayader sheet area, where Jones (1909) proposed the terms Rhuddnant Shales and Rhuddnant Grits for strata containing packets of thick high-matrix sandstone beds (his speckly grits) that succeed the Blaen Myherin Mudstones. The base of the formation, as defined here, is taken at the base of the first packet of such sandstones (type Bii turbidites).

The main outcrop of the formation occupies a broad tract along the Central Wales Syncline (Figure 29). Along the western limb, the Rhuddnant Grits succeed the Blaen Myherin Mudstones and are over 1.5 km in thickness; they form the oldest, thickest and westernmost com-ponent of the Cwmystwyth Grits Group. On the eastern limb, the oldest parts of the Rhuddnant Grits overlie the Caerau Mudstones. Less than 800 m are present south of the Claerwen reservoir. However, to the north and north-east the formation exhibits complex intertonguing, the lower parts with the Caerau Mudstones and the upper parts with the Glanyrafon Formation (Figures 12 and 29). At Nant Cletwr [SN 886 688], at the southern end of the Craig Goch reservoir, the formation is confined to a 200 m-thick sequence interleaved with Caerau Mudstones (Figure 29). This tongue thins north-eastwards to a feather edge near Cwmgary [SN 943 767], north of the Wye valley. A small outlier of Rhuddnant Grits occurs on the western limb of the Teifi Anticline at Trewaun [SN 645 600], at the southern edge of the Llanilar sheet area, but is not exposed in the district.

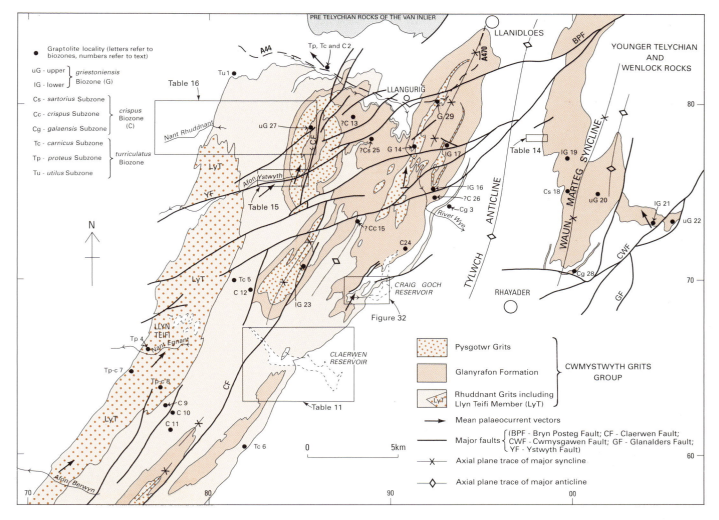

Figure 29 Geological map of the Cwmystwyth Grits Group showing mean palaeocurrent vectors (including data from Clayton, 1992 and Smith, 1987a) and the position of graptolite localities described in the text.

The Rhuddnant Grits predominantly comprise Bouma, thin-bedded, turbidite sandstone/mudstone couplets (type Cii) with subordinate, but locally abundant, thin- to thick-bedded, high-matrix, turbidite sandstones (type Bii) (Figure 31, Plates 7e and 17a, b and c). Thin, burrowed and laminated hemipelagites occur between the turbidites. The high-matrix sandstones are confined to packets of strata up to 50 m thick. Each packet displays between 3 and 30 such beds interbedded with thin-bedded, Bouma turbidites. Individual high-matrix sandstones range from 0.1 to 1.5 m in thickness, and uncommon amalgamated units up to 3 m. The thickness and distribution of high-matrix sandstone beds, both within packets and throughout the formation, are random, and there is no systematic variation in the thickness of the packets (Clayton, 1992). Intervals lacking high-matrix sandstones are typically less than 30 m thick in the western part of the outcrop, but are thicker and more prevalent in the eastern part, reflecting the lateral passage into the Glanyrafon Formation.

Turbidite mudstones (type D) occur scattered throughout the Rhuddnant Grits but are most abundant in regions of vertical and lateral transition with adjacent mudstone formations. They form, in addition, thick mappable sequences in which turbidite sandstone/mudstone couplets (type Cii) and high-matrix sandstones are sparsely developed. Such sequences are confined to the western limb of the Central Wales Syncline and there appears to be little pattern to their distribution.

LLYN TEIFI MEMBER

This member forms the lower part of the formation on the western limb of the Central Wales Syncline between the southern edge of the district and just south of Nant Rhuddnant [SN 7963 7743]. It takes its name from the area around Llyn Teifi [SN 785 675] where it is well exposed in extensive crag sections. The member is characterised by a greater abundance of high-matrix sandstones than the rest of the formation (Figure 30, Plate 17a). Stacking and amalgamation of high-matrix sandstone beds is more common and intervals without such beds are less abundant. The member is about a kilometre thick in the south, thinning northwards to about 400 m in Cwm Ystwyth, beyond which it fails to the north-east.

Bottom structures on high-matrix sandstones (type Bii), both in and adjacent to the district (Clayton, 1992), suggest that the high-concentration turbidity currents consistently flowed from the south-south-west (Figure 29). Current vectors derived from cross-lamination in the Bouma turbidites (type Cii) are more variable. South of the district, current vectors obtained from type Bii and Cii turbidites are parallel, but in northern sections, vectors obtained from type Cii turbidites suggest that depositing currents flowed from between the south and south-east (Clayton, 1992).

BIOSTRATIGRAPHY

On the western limb of the Central Wales Syncline the basal kilometre of the Rhuddnant Grits spans the late *turriculatus* Biozone (*proteus* and *carnicus* subzones), while the top 450 m encompass much of the *crispus* Biozone (*galaensis* to early *sartorius* subzones). As *proteus* Subzone faunas have only been found about 200 m above the base of the formation in the Afon Mwyro valley [SN 7658 6578] (Figure 29, locality 4), east of Pontrhydfendigaid, and in Nant Rhuddnant (Table 16, locality 12) at an unknown level close to the base of the Rhuddnant Grits (Rigby, 1980), it is possible that the base of the formation may range down into the *johnsonae* Subzone. Just north of the district on the western limb of the Central Wales Syncline, however, the base of the *crispus* Biozone occurs in the upper part of the Blaen Myherin Mudstones in cuttings [SN 8578 8198 to SN 8641 8177] along the A44 road (Figure 29, locality 4). This demonstrates that the base of the Rhuddnant Grits youngs markedly northwards in this area. The *carnicus* Subzone has been proved in the middle part of the formation in the Nant Rhuddnant valley and in Afon Diliw (Table 16, localities 14 and 15), and in Cwm Ystwyth (Table 15, localities 1 to 3). The *crispus* Biozone is about 450 m thick in Cwm Ystwyth, where all three subzones have been proved (Table 15).

The Llyn Teifi Member probably ranges from *johnsonae* Subzone (see above) well into the *carnicus* Subzone, for faunas of the *proteus* Subzone occur 200 m above the base [SN 7568 6578] (Figure 29, locality 4), and those of the *carnicus* Subzone about 50 m above the top, in Nant-y-Fign [SN 8150 6956] (Figure 29, locality 5).

In contrast to the western limb, most of the formation on the eastern limb of the Central Wales Syncline lies within the *crispus* Biozone, though between the southern edge of the district and the River Wye, the base falls within the latest *turriculatus* Biozone (*carnicus* Subzone). Graptolites of this subzone enter at the base of the formation, south of Moel Prysgau [SN 8197 6035] (Figure 29, locality 6), whilst to the north, in the Claerwen reservoir section, the formation overlies Caerau Mudstones of *carnicus* Subzone age (Table 11). The lowest *crispus* Biozone fauna occurs 65 m above the base of the formation in Nant Cletwr (Figure 32, locality

4), adjacent to the Craig Goch reservoir, and at a similar level in the Claerwen reservoir section (Table 11). The upper parts of the formation have yielded faunas referable either to the *galaensis* or *crispus* subzones (Table 11). North of the Wye valley, the surviving tongue of Rhuddnant Grits lies within the lower part of the *crispus* Biozone, as demonstrated by graptolite faunas from the enveloping Caerau Mudstones (see p.124).

Acritarch floras from the Rhuddnant Grits are moderately diverse (Appendix 2bi and ii), and include common *Ammonidium* spp., *Domasia* spp., *Elektoriskos williereae*, *Lophosphaeridium* sp. A, *Micrhystridium* spp., *Multiplicisphaeridium martiniae*, *Multiplicisphaeridium* spp., and *Oppilatala eoplanktonica*. Also present are *Cymatiosphaera* sp. A, *Dictyotidium stellatum*, *Lophosphaeridium* sp. B, *Oppilatala grahni*, and *Salopidium granuliferum*. *Gracilisphaeridium encantador* occurs from the *proteus* Subzone onwards. These floras are typical of the Telychian *D. estillis* and *G. encantador* biozones, and the presence of *D. stellatum*, *E. williereae*, *L.* sp. B and *M. martiniae* in the upper part of the formation in Cwmystwyth indicates the middle part of the *G. encantador* Biozone.

Recycled acritarchs recovered from the Rhuddnant Grits include *Acanthodiacrodium* spp., *Cymatiogalea* and *Vulcanisphaera* sp. from the early Ordovician, and *Cleithronetrum* cf. *cancellatum*, *Frankea sartbernadensis*, *Orthosphaeridium bispinosum*, *Peteinosphaeridium nudum* and *Peteinosphaeridium trifurcatum breviradiatum* from the mid to late Ordovician.

DETAILS

Llanilar sheet area   Extensive sections through the Llyn Teifi Member are exposed [SN 761 657 to SN 795 661] in the valleys of Afon Mwyro and its tributary Nant Egnant (Figure 30, Plate 17a), east of Strata Florida Abbey. These sections are characterised by abundant, thick, commonly amalgamated and graded, high-matrix sandstones (type Bii). On the high craggy ground to the south of Nant Egnant, the member is well seen on Banc Mawr [SN 728 602], Garn Gron [SN 740 611], Carn Fflur [SN 746 624], Bryn-y-crofftau [SN 747 635] and Pen-y-maen Berthgoed [SN 765 647], and in numerous track cuttings and quarries in Tywi Forest. Graptolites have been collected from sections [SN 7548 6466] (Figure 29, locality 7) near the western entrance of Tywi Forest, and to the east [SN 7740 6384] (Figure 29, locality 8). Both localities are of probable late *turriculatus* Biozone (*proteus*–*carnicus* subzones) age. Crags [SN 7658 6578] along Afon Mwyro, some 200 m above the base of the member, have yielded graptolites which include *Monograptus turriculatus*, *M. proteus* and *Petalolithus tenuis*, indicative of the *proteus* Subzone (Figure 29, locality 4).

North of Nant Egnant the member is well exposed in its type area near Llyn Teifi [SN 785 675]. Sections [SN 774 681 to SN 784 683] (Plate 17b) in this folded sequence are seen along the mountain road east of Ffair Rhos. Farther north, the base of the member is present in a stream [SN 7709 6866] draining Llethr Brith. Near the bottom of a waterfall section [SN 7837 7078], to the north of Trawsallt, a 35 cm-thick high-matrix sandstone displays a channelised top. North of the Ystwyth Fault the Llyn Teifi Member is exposed on Glog [SN 792 745] and across the southern flanks [SN 795 751] of Pen Lan-fawr.

Sections in the remaining parts of the Rhuddnant Grits, in the south-east of the Llanilar sheet area, are provided by quarries and track cuttings in Tywi Forest. Thin-bedded

Figure 30

Graphic log of the
Llyn Teifi Member
of the Rhuddnant
Grits in the
Egnant gorge
[SN 777 662].

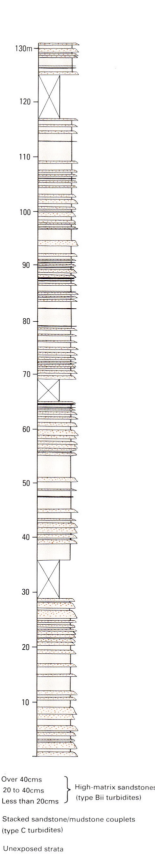

Over 40cms
20 to 40cms } High-matrix sandstones
Less than 20cms } (type Bii turbidites)

Stacked sandstone/mudstone couplets
(type C turbidites)

Unexposed strata

sandstone/mudstone couplets (type Cii) predominate in this part of the formation; high-matrix sandstones (type Bii) are more widely spaced than in the Llyn Teifi Member. Laminated hemipelagites are common in this region and have yielded graptolites at several locations. For example, north of Esgair Llyn-du [at SN 7763 6289] (Figure 29, locality 9) and [at SN 7789 6274] (Figure 29, locality 10), graptolites include *Streptograptus exiguus, Pseudoplegmatograptus* cf.*obesus* and *Petalolithus* cf. *wilsoni,* indicative of the *crispus* Biozone. Farther south, a track cutting [SN 7790 6124] (Figure 29, locality 11) contains *Monograptus discus* and *M.* cf. *crispus,* taxa which confirm that upper levels of the formation range into the *crispus* Biozone. Beyond the forest, the Rhuddnant Grits are well exposed in crags [near SN 792 631], to the north and west of Llyn Gorast.

Rhayader sheet area, west of the Claerwen Fault   The Llyn Teifi Member is well exposed on Creigiau Canol [SN 803 688], Domen Milwyn [SN 808 720] and Banc Hir [SN 805 728]. To the north, in the Ystwyth valley, the basal contact with the Blaen Myherin Mudstones is seen in the steep craggy slopes [SN 8015 7410 to SN 8090 7440] of Graig Goch. The cliff section [SN 812 739] of the glacial cirque of Graig Ddu (Figure 48) exposes a mudstone-dominated unit up to 50 m thick, which consists predominantly of thick-bedded turbidite mudstones (type D) intercalated at the base with sandstone-based couplets (type C). Thicker sandstones (type Bii) (Plate 17c) occur close to the gradational upper and lower contacts of the unit. A comparable mudstone sequence occurs at Cwmystwyth Mine, on the north side of the Ystwyth valley, and crops out in Nant Watcyn [SN 8055 7494] and the crags to the west. The crags [SN 802 748] which overlook the mine expose a folded sequence in the sandstone-dominated facies of the member. Farther north, the frequency of high-matrix sandstones diminishes, as the member grades into the undivided parts of the formation, as seen in sections [SN 8025 7799] near Rhestr Cerrig.

Succeeding the Llyn Teifi Member in the south, the upper parts of the Rhuddnant Grits form crags at Crug Gynon [SN 802 638], and near Banc-y-llyn [SN 807 653] and Cefn Brwynog [SN 817 660]. There, high-matrix sandstones, locally up to 1 m thick, are subordinate to turbidite sandstone/mudstone couplets (type C). To the north, the mountain road from Ffair Rhos [SN 740 680] to the Claerwen reservoir [SN 820 672] continues (see p.134) the traverse through the western limb of the Central Wales Syncline. Near the ford [SN 803 678] over Afon Claerddu, a fold-repeated mudstone unit is exposed. To the east, a quarry [SN 8112 6736] displays 10 m of thin-bedded sandstone/mudstone couplets (type Cii) comprising over 40 per cent sandstone. High-matrix sandstones up to 1.1 m thick form a packet exposed in a roadside quarry [SN 8219 6717] near Claerwen Farm, and in crags to the north. Graptolites collected from nearby exposures [SN 8263 6708] (Table 11, locality 21) in the Ddwynant are indicative of either the *galaensis* or the *crispus* Subzone.

The mudstone unit observed along the mountain road is seen again in Nant-y-Fign [SN 8150 6956] (Figure 29, locality 5). This section has yielded the graptolites cf. *Monoclimacis? galaensis, Streptograptus storchi, S. tenuis* and *Monograptus* cf. *carnicus,* indicative of the *carnicus* Subzone. However, 150 m downstream (east), laminated hemipelagites in the higher parts of the formation contain *Monograptus crispus, M. discus* and cf. *Monoclimacis? galaensis,* of *galaensis* Subzone age. Farther downstream [SN 8217 6929] (Figure 29, locality 12), graptolites of the *crispus* Biozone were collected, and an assemblage of longer-ranging Telychian forms was found [at SN 8230 6871].

The eastern end of Cwm Ystwyth provides a transect through the whole of the Rhuddnant Grits on the western limb of the Central Wales Syncline (Figure 29). Sections [SN 820 747 and SN 830 747] above the Llyn Teifi Member are exposed on the south side of the valley in streams draining Byrlymau Elan and in the cirque of Craig Cwmtinwen. Along the valley floor [SN 8268 7545 to SN 8410 7554], outcrops provided by the river and the parallel leat (water channel) for Cwmystwyth Mine afford almost complete exposure through the upper 800 m of the formation (Clayton, 1992) (Figure 31; Table 15). Farther upstream [SN 8432 7561] at the site of a broken dam the upper few metres are again exposed in anticlinal cores. The high-matrix sandstones (type Bii) are up to 2 m thick, and concentrated in packets, up to 50 m thick, which contain between 2 and 30 such beds. The thickness and spacing of the high-matrix sandstones appear random. The high-matrix sandstones are predominantly separated by thin-bedded sandstone/mudstone couplets (type Cii) but, in the lower reaches of the section, mudstone turbidites (type D) commonly occur. Table 15 provides details of the graptolites collected from this section. The lower part of the section is of *carnicus* Subzone age. Some 450 m below the top of the formation (Table 15, locality 4), graptolites of the *galaensis* Subzone appear, whilst *crispus* Subzone taxa appear at higher levels (Table 15, localities 12 to 14). The uppermost parts of the formation have yielded *sartorius* subzone graptolites (Table 15, localities 16 and 20).

In the north-west corner of the Rhayader sheet area, the Rhuddnant Grits succeed the Blaen Myherin Mudstones in Nant Rhuddnant [SN 8028 7842 to SN 8050 7842] (Figure 45). Medium- to thick-bedded high-matrix sandstones (type Bii) occur scattered throughout the sequence. The basal part of the formation is dominated by mudstone turbidites (type D) (the 'Rhuddnant Shales' of Jones, 1909), but higher in the section, sandstone-based turbidites (type C) increase in abundance (the 'Rhuddnant Grits' of Jones, 1909). The latter are well seen in waterworn surfaces at the eastern end [SN 8045 7840] of the section. To the east, a major forestry road between Cripiau Nantmelyn [SN 8079 7846] and Afon Dulas [SN 8426 7737], and the parallel Afon Fechan [east of SN 8294 7877], provide numerous sections in the Rhuddnant Grits. Graptolites collected by Rigby (1980) along this section have been reassessed and augmented during the current survey (Table 16). Taxa probably from Nant Rhuddnant and from a locality about a kilometre to the east (Table 16, localities 12 and 13) are possibly of *proteus* Subzone age. However, a probable *carnicus* Subzone assemblage has been collected from a nearby cutting [SN 8165 7880] (Table 16, locality 14) in laminated hemipelagites and mudstone turbidites (type D). Graptolites of the *crispus* Biozone occur in cuttings [SN 8277 7822] (Table 16, locality 16) and throughout the remainder of the section. A trackside quarry [SN 8413 7789] (Table 16, locality 21) has yielded an assemblage indicative of the *crispus* Subzone. Nearby waterfalls [SN 8420 7766] in Afon Diliw are formed by a steeply dipping packet of high-matrix sandstones (type Bii); one of these displays the lifting of large rip-up clasts from the underlying strata. Groove-casts on the bases of the thin-bedded sandstones (type Cii) indicate sediment transport from the south.

Rhayader sheet area, east of the Claerwen Fault   On the eastern limb of the Central Wales Syncline, the Rhuddnant Grits are exposed along forestry cuttings [SN 8008 6128 and SN 8077 6240], north of Esgair Saeson. In this area, thin-bedded sandstone/mudstone turbidite couplets (type Cii) predominate, with laminated hemipelagites locally abundant; high-matrix sandstones (type Bii) are thin and widely spaced, though seldom absent from any extensive section. East of the

syncline in the Glanyrafon Formation near Moel Prysgani [SN 806 611], exposures [SN 8197 6035] have yielded *carnicus* Subzone graptolites, which include cf. *Monoclimacis? galaensis*, *Streptograptus storchi*, *S.* sp. nov., *S. exiguus* and *M. proteus*, (Figure 29, locality 6).

Farther north, the formation is intermittently exposed [between SN 830 665 and SN 850 655] along the northern shore of the Claerwen reservoir. There, graptolites collected from near the base of the formation include rare *M.* cf. *crispus*, suggesting a *galaensis* Subzone age (Table 11, localities 14 and 15). Localities higher in the formation are referable either to the *galaensis* or the *crispus* subzones (Table 11). An extensive section [SN 8541 6594 to SN 8560 6683] in Nant y Beddau exposes the lower part of the formation and the transition into the underlying Caerau Mudstones (Table 11). Thin-bedded sandstone/mudstone couplets (type Cii) enter some 9 m below the lowest high-matrix sandstone (type Bii).

Nant Cletwr and its tributaries [SN 864 694 to SN 886 688], draining Clawdd Du Mawr and Clawdd-du-bach, contain abundant outcrops of the Rhuddnant Grits. Sections [SN 8874 6890 to SN 8845 6883] through the lower part of the formation are exposed in the eastern reaches of Nant Cletwr and on the shore of the Craig Goch reservoir (Figure 32). This westward-dipping sequence includes very muddy, well-cleaved high-matrix sandstones (type Bii) within a predominantly mudstone (type D) background. Thin-bedded sandstone/mudstone couplets (type Cii) are more common upstream between more massive high-matrix sandstones. Packets of strata with either laminated or burrowed hemipelagites are present in the sequence. The base of the *crispus* Biozone is taken at the appearance of *M. crispus* and *M. discus* [SN 8852 6886] (Figure 32, locality 4), 70 m west (upstream) of the road bridge, above strata containing only long-ranging taxa typical of the uppermost *turriculatus* Biozone (Figure 32, localities 2 and 3). To the west, thrusts (Figure 32, locality 2) and folds repeat the sequence on the north side of the Cletwr valley. Farther north, in cuttings [SN 8735 7365] along the Rhayader Mountain Road, a packet of high-matrix sandstones (type Bii), repeated by folding, marks the local top to the Rhuddnant Grits. To the north and east the Rhuddnant Grits are limited to discrete packets of high-matrix sandstones in the Glanyrafon Formation, for example along Nant Hirin [SN 8821 7185], in Afon Elan [SN 8840 7267 and SN 8803 7302], and in the gorge [SN 8800 7890] (Figure 29, locality 13) east of Glangwy. Hemipelagites at the latter locality have yielded graptolites which include *Monograptus crispus?* and *Streptograptus exiguus*, suggesting the *crispus* Biozone.

On the western limb of the Tylwch Anticline the Rhuddnant Grits intertongue with the Caerau Mudstones. Crags of high-matrix sandstone occur on the slopes east of Craig Goch reservoir [around SN 903 697] and on Geufron [SN 915 710] to the north-east. North of the Wye valley high-matrix sandstones are exposed in a stream [SN 9284 7393] and in crags [SN 9340 7428] near Dolhelfa. Farther north [SN 9340 7428], high-matrix sandstone debris occurs to the west of Craig Gellidywyll.

High-matrix sandstones are exposed in the cores of three periclines in the River Wye [near SN 917 751], to the west of Neuadd-ddu. These belong to a tongue of Rhuddnant Grits which intervenes between the Caerau Mudstones and Glanyr-

Figure 31   Composite graphic log of the Cwmystwyth Grits Group in the eastern end of Cwm Ystwyth [SN 8269 7546 to SN 8528 7573] derived, in part, from logs by Smith (1988) and Clayton (1992) (facing page).

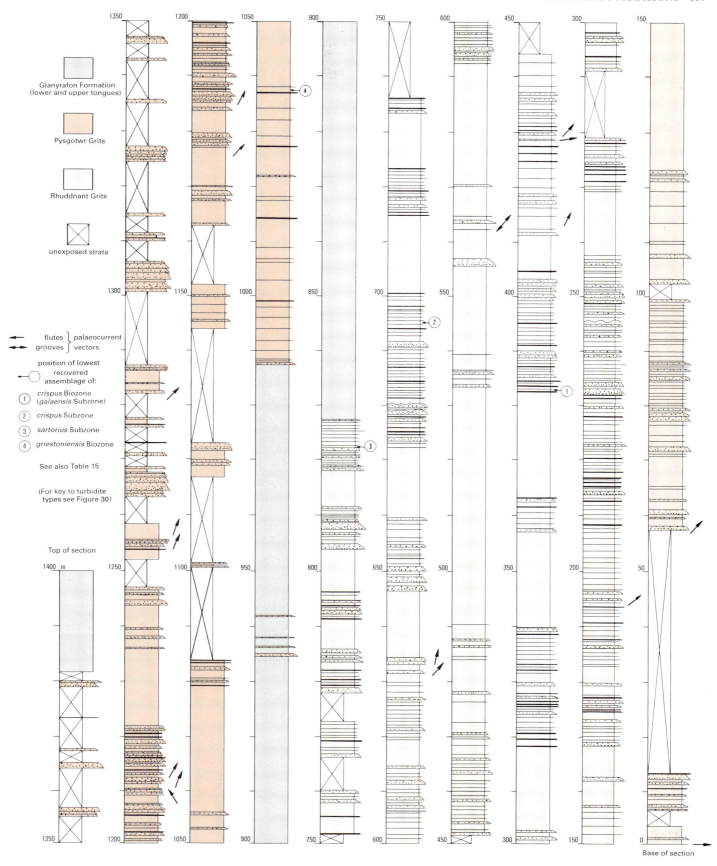

Glanyrafon Formation
(lower and upper tongues)

Pysgotwr Grits

Rhuddnant Grits

unexposed strata

flutes ⎫ palaeocurrent
grooves ⎭ vectors

position of lowest
recovered
assemblage of:

① *crispus* Biozone
  (*galaensis* Subzone)

② *crispus* Subzone

③ *sartorius* Subzone

④ *griestoniensis* Biozone

See also Table 15

(For key to turbidite
types see Figure 30)

Top of section

Base of section

Table 15   Distribution of graptolites in the Cwmystwyth Grits Group in the eastern end of Cwm Ystwyth.

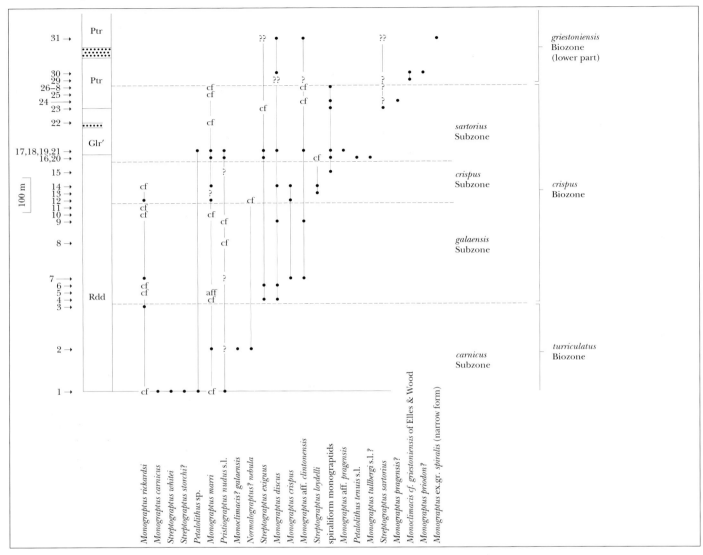

Localities: see map on facing page

1 south bank of river [SN 8274 7547]; 2 north bank of river [SN 8301 7541]; 3 river section [SN 8326 7541]; 4 river section [SN 8330 7544]; 5 leat section [SN 8330 7547]; 6 north bank of river [SN 8334 7545]; 7 leat section [SN 8334 7547]; 8 scree [SN 8360 7547]; 9 leat section [SN 8377 7543]; 10 leat section [SN 8379 7543]; 11 leat section [SN 8385 7545]; 12 leat section [SN 8388 7547]; 13 leat section [SN 8391 7548]; 14 leat section [SN 8393 7548]; 15 river section [SN 8407 7549]; 16 river section [SN 8409 7552]; 17 river section [SN 8422 7559]; 18 river section [SN 8423 7559]; 19 north bank of river [SN 8431 7561]; 20 north bank of river [SN 8433 7562]; 21 north bank of river [SN 8435 7563]; 22 south bank of river [SN 8454 7569]; 23 stream section [SN 8458 7571]; 24 north bank of river [SN 8462 7568]; 25 north bank of river [SN 8464 7566]; 26 east bank of stream [SN 8463 7575]; 27 west bank of stream [SN 8465 7579]; 28 west bank of stream [SN 8465 7581]; 29 south bank of river [SN 8472 7567]; 30 north bank of river [SN 8478 7565]; 31 roadside exposure [SN 852 757]. For key to range chart symbols see Appendix 3.

afon Formation, and may be traced to the northern margin of the district. A section [SN 9068 7797] in Nant-y-clochfaen exposes high-matrix sandstones (type Bii) scattered in a facies of sandstone/mudstone couplets (type Cii). Basal flute casts indicate that the turbidity currents flowed from the south-west. Intercalated hemipelagites are largely of the laminated variety. Disused quarries [SN 9011 7873 and SN 9023 7889] afford additional sections in these strata.

Pysgotwr Grits

The Pysgotwr Grits take their name from sections north-west of Rhandirmwyn in the adjacent Builth district, described by Davies (1933) and re-examined by Mackie and Smallwood (1987) and Smith (1987a; 1988). The formation appears to equate directly with the geographi-cally separate Talerddig (Grits) Group of Wood (1906)

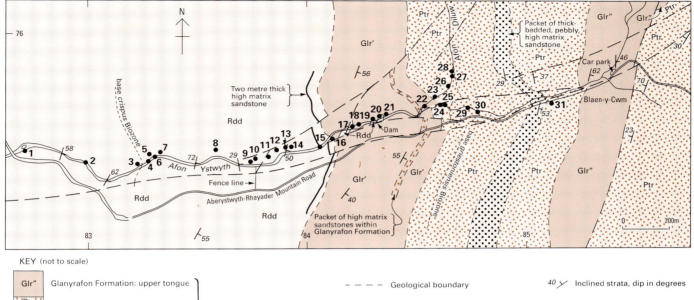

KEY (not to scale)

Glr″	Glanyrafon Formation: upper tongue
Ptr / Ptr	Pysgotwr Grits Formation
Glr′ / Glr′	Glanyrafon Formation: lower tongue
Rdd	Rhuddnant Grits Formation

CWYMSTWYTH GRITS GROUP

- - - - Geological boundary

-------- Base of biozone

———— Fault

3
● Fossil locality

40 ⟋ Inclined strata, dip in degrees

and Bassett (1955), of the Trannon area in the adjacent Llanidloes district, and with the Pen-y-Gelli Grit of the Bala district.

The formation is very similar lithologically to the Rhuddnant Grits, comprising thin-bedded, Bouma turbidite sandstone/mudstone couplets (type Cii) with subordinate but locally abundant thin- to thick-bedded, high-matrix, turbidite sandstones (type Bii), arranged in packets (Figure 31, Plate 17e). In the district the Pysgotwr Grits overlie the Glanyrafon Formation, their base being taken at the entry of the lowest packet of high-matrix sandstones.

The Pysgotwr Grits crop out in three principal areas in the district. Two occur in separate synclinal outliers along the core of the Central Wales Syncline, centred on Llyn Cerrigllwydion and the upper reaches of Afon Ystwyth near Glan Fedwen, and the third occupies a further series of synclinal outliers within the eastern limb of the major syncline, centred on the River Wye, near Tyncoed.

In the western outcrops, the complete formation is only exposed in the syncline centred on Glan Fedwen, where it is about 350 m thick. The sequence thins eastwards to the Wye valley, where about 180 m are present west of Pant-y-drain [SN 910 764], and to less than 30 m farther east around Neuadd-ddu [SN 921 753]. North-north-east of the Wye valley, around the northern boundary of the district, the formation is less than 10 m thick.

The high-matrix sandstones (type Bii) (Plate 18a) vary from fine to very coarse grained and locally display pebbly and granule conglomerate in their basal portions. The beds are up to 1.5 m thick but amalgamated units range up to 6 m in thickness. The thickness and spacing

of the high-matrix sandstone beds and the packets in which they occur appear random, though sandstone amalgamation is more common in the western outcrops (Smith, 1987a). A feature-forming, 35 m-thick packet containing very thick, and commonly amalgamated, high-matrix sandstones, including pebbly and granule-rich examples, occurs in the middle of the formation in the western outcrops.

The mudstones of the turbidite sandstone/mudstone couplets (type Cii) are grey in the lower part of the formation, but become progressively paler and greener upwards. These paler mudstones are commonly difficult to distinguish from intercalated burrowed hemipelagites, their colouration reflecting the effects of extensive penecontemporaneous oxidation of the underlying turbidite mudstone. Laminated hemipelagites are rare but have been noted at the base of the formation in the eastern crop at Dolfach [SN 9120 7720].

Palaeocurrent vectors from the Pysgotwr Grits (Figure 29) in the core of the Central Wales Syncline confirm the south-south-westerly derivation established by Smith (1987a). Vectors derived from sequences farther east, in the Wye valley, suggest a greater influence by currents that flowed from south to north.

BIOSTRATIGRAPHY

Graptolite assemblages from the Pysgotwr Grits in Cwm Ystwyth indicate that the basal 50 m belong to the *sartorius* Subzone of the *crispus* Biozone, whilst the remainder belong to the *griestoniensis* Biozone (Table 15). Farther east in the Wye valley, at Dolfach [SN 9120 7720] (Figure 29, locality 14), the basal beds have yielded

Table 16   Distribution of Telychian graptolites in the area around Nant Rhuddnant and Afon Dilw.

Localities (• on map)	Tg 1	Tr-u 2	Tu 3	Tu 4	Tu 5	Tu 6	Tu 7	Tu 8	Tjo 9	Tjo 10	Tjo 11	Tp? 12	Tp? 13	Tc 14	Tc 15	C* 16	C 17	C 18	C* 19	C* 20	C 21	?C 22	?C 23	?C 24	G 25
Monograptus turriculatus s.l.	cf.	·	cf.	•	•	•	•	·	•	·	•	·	·	•	•	·	·	·	·	·	·	·	·	·	·
Pristiograptus nudus s.l.	cf.	·	·	cf.	•	·	·	·	·	·	·	·	·	cf.	·	·	cf.	·	·	·	·	·	·	·	cf.
Streptograptus pseudoruncinatus	•	cf.	·	·	·	·	·	·	·	·	·	·	·	·	·	·	·	·	·	·	·	·	·	·	·
Streptograptus filiformis	?	·	·	·	·	·	·	·	·	·	·	·	•	·	·	·	·	·	·	·	·	·	·	·	·
Glyptograptus elegans	?·	·	·	·	·	·	·	·	·	·	·	·	·	·	·	·	·	·	·	·	·	·	·	·	·
Glytograptus auritus	cf.·	·	·	·	·	·	·	·	·	·	·	·	·	·	·	·	·	·	·	·	·	·	·	·	·
Monograptus utilis	·	·	•	•	•	•	cf.	•	·	·	·	·	·	·	·	·	·	·	·	·	·	·	·	·	·
Monograptus planus	·	·	?	·	cf.	cf.	·	?	·	·	·	·	·	·	·	·	·	·	·	·	·	·	·	·	·
Streptograptus plumosus	·	·	•	•	•	•	cf.	cf.	·	·	·	·	·	·	·	·	·	·	·	·	·	·	·	·	·
Pseudoplegmatograptus obesus	·	·	?	?	·	·	·	·	·	·	·	·	·	·	·	·	·	·	·	·	·	·	·	·	·
Monograptus cavei	·	·	·	•	·	·	·	·	·	·	·	·	·	·	·	·	·	·	·	·	·	·	·	·	·
Monograptus bjerreskovae	·	·	·	cf.	·	cf.	·	?	·	·	cf.	·	·	·	·	·	·	·	·	·	·	·	·	·	·
Monograptus halli	·	·	·	cf.	·	·	·	·	·	·	·	·	·	·	·	·	·	·	·	·	·	·	·	·	·
Monograptus marri	·	·	·	?	·	cf.	·	•	•	cf.	cf.	cf.	·	·	·	cf.	•	•	·	·	·	·	·	·	·
Glyptograptus tamariscus	·	·	·	·	?	·	·	·	·	·	·	·	·	·	·	·	·	·	·	·	·	·	·	·	·
Petalolithus tenuis s.l.	·	·	·	·	·	·	cf.	·	·	·	·	·	·	·	·	·	·	·	·	·	·	·	·	·	·
Streptograptus johnsonae	·	·	·	·	·	·	·	·	•	•	·	·	·	·	·	·	·	·	·	·	·	·	·	·	·
Monograptus acus	·	·	·	·	·	·	·	·	·	·	·	cf.	·	·	·	·	·	·	·	·	·	·	·	·	·
Monograptus becki?	·	·	·	·	·	·	·	·	·	·	·	•	·	·	·	·	·	·	·	·	·	·	·	·	·
Monoclimacis? galaensis	·	·	·	·	·	·	·	·	·	·	·	·	·	cf.	•	·	·	·	·	·	·	·	·	·	·
Streptograptus storchi	·	·	·	·	·	·	·	·	·	·	·	·	·	•	•	·	·	·	·	·	·	·	·	·	·
Rastrites distans	·	·	·	·	·	·	·	·	·	·	·	·	·	·	cf.	·	·	·	·	·	·	·	·	·	·
Monograptus discus	·	·	·	·	·	·	·	·	·	·	·	·	·	·	·	•	·	·	•	•	cf.	·	•	•	·
Monograptus crispus	·	·	·	·	·	·	·	·	·	·	·	·	·	·	·	•	•	•	•	•	·	·	·	·	·
Monograptus rickardsi	·	·	·	·	·	·	·	·	·	·	·	·	·	·	·	·	·	·	·	•	·	·	·	·	·
Streptograptus loydelli	ms	·	·	·	·	·	·	·	·	·	·	·	·	·	·	·	·	·	·	·	•	·	·	·	·
Spiraliform monograptids	·	·	·	·	·	·	·	·	·	·	·	·	·	·	·	·	·	·	·	·	·	·	•	•	•
Monograptus clintonensis	·	·	·	·	·	·	·	·	·	·	·	·	·	·	·	·	·	·	·	·	·	·	•	•	•
Monoclimacis griestoniensis	·	·	·	·	·	·	·	·	·	·	·	·	·	·	·	·	·	·	·	·	·	·	·	·	•
Monograptus aff. *progensis?*	·	·	·	·	·	·	·	·	·	·	·	·	·	·	·	·	·	·	·	·	·	·	·	·	•

Subzone / Biozone groupings (top of table):
- *runcinatus-utilis gemmatus* (cols 1–2)
- subzones of the *turriculatus* Biozone: *utilis* (*johnsonae*), *?proteus*, *carnicus* (cols 3–15)
- *crispus* Subzone: *crispus* Biozone *galaensis-crispus* Subzones (cols 16–20); *?crispus* Biozone, *sartorius* Subzone (cols 21–24)
- *griestoniensis* Biozone (upper) (col 25)

* assemblages recorded by J Rigby (1980), specimens missing.

Localities: see map on facing page

1 forestry track [SN 7735 7750]; 2 river section [SN 773 775 to SN 792 778]; 3 track section [SN 7925 7780]; 4 river section [SN 7980 7836]; 5 to 8 river sections [SN 798 784 to SN 8007 7852]; 9 river section [SN 8007 7852]; 10 and 11 river section [between SN 8007 7852 and SN 802 784]; 12 river section [between SN 802 784 and SN 809 785]; 13 track section [SN 8139 7844]; 14 track section [SN 8165 7880]; 15 track section [SN 8384 7941]; 16 track section [SN 8277 7822]; 17 trackside quarry [SN 8278 7821]; 18 track section [SN 8453 7969]; 19 and 20 track sections [between SN 833 781 and SN 843 779]; 21 trackside quarry [SN 8413 7789]; 22 and 23 stream sections [SN 8428 7719]; 24 stream section [SN 8429 7707]; 25 trackside exposure [SN 8545 7831].

For key to range chart symbols see Appendix 3.

griestoniensis Biozone graptolites, indicating that the base of the formation youngs eastwards.

The formation has yielded moderately diverse acritarch assemblages, whose range lies between the *A. microcladum* and *G. encantador* biozones (Appendix 2bi). *Ammonidium listeri, Ammonidium microcladum, Micrhystridium* spp., *Multiplicisphaeridium fisheri* and *Multiplicisphaeridium* spp. are common; also present are *Domasia* spp., *Elektoriskos williereae, Oppilatala eoplanktonica* and *Salopidium granuliferum*. Recycled acritarchs recovered from this formation include *Cymatiogalea* sp. and *Vulcanisphaera britannica* from the early Ordovician, and *Peteinosphaeridium trifurcatum intermedium* from the mid to late Ordovician.

DETAILS

Llyn Cerrigllwydion Syncline   Lower parts of the Pysgotwr Grits are exposed in a periclinal syncline near Llyn Cerrigllwydion Isaf [SN 844 700] and Llyn Cerrigllwydion Uchaf [SN 840 693] (Plates 17e and 23b) and along the upper reaches of Nant Hirin [SN 8496 7072 to SN 8457 7081]. Additional sections [SN 8588 7258 to SN 8586 7232] are seen in the stream draining Llethr Gwngu. In both areas, turbidite sandstone/mudstone couplets (type Cii) predominate but high-matrix sandstones are common. The base of the formation is exposed in Afon Gwngu [SN 8575 7301] and in Nant Dderwen [SN 8627 7356] west of Abergwngu. In the syncline, a feature-forming packet with pebbly high-matrix sandstones is present in the middle of the

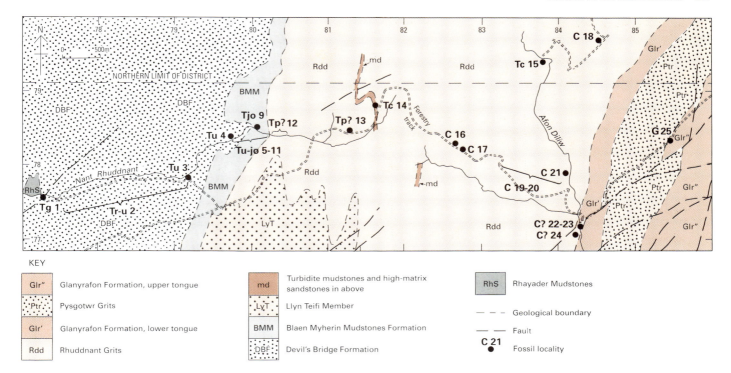

KEY

Glr"	Glanyrafon Formation, upper tongue
Ptr	Pysgotwr Grits
Glr'	Glanyrafon Formation, lower tongue
Rdd	Rhuddnant Grits

md	Turbidite mudstones and high-matrix sandstones in above
LyT	Llyn Teifi Member
BMM	Blaen Myherin Mudstones Formation
DBF	Devil's Bridge Formation

RhS	Rhayader Mudstones
— — —	Geological boundary
— —	Fault
C 21 ●	Fossil locality

formation. It crops out in the crags of Cnapyn Blaendrawsffos [SN 8355 6910], Blaen Rhestr [SN 842 694] and Lloches Lewsyn [SN 845 699]. Thick- to very thick-bedded, coarse-grained, feldspathic, high-matrix sandstones dominate these exposures, and amalgamated units are up to 6 m thick. Basal granule to pebble conglomerates, with rare cobbles up to 15 cm across, are exposed [SN 8398 6910, SN 8358 6887 and SN 8506 7032]. The rounded to subangular clasts consist of quartz and acid volcanic rocks. In the uppermost part of the sequence in the syncline, high-matrix sandstones form crags [SN 8415 6990] to the south-west of Llyn Cerrigllwydion Isaf. Elsewhere [e.g. SN 8477 6999] sequences of thin-bedded sandstones and pale green mudstones (type Cii turbidites) predominate.

Glan Fedwen Syncline   The Pysgotwr Grits are preserved in a periclinal syncline at the headwaters of Afon Ystwyth [SN 855 755], where a semicontinuous section [SN 846 757 to SN 853 757] is present through the western limb (Figure 31). Grapto-lites recovered from this section range from the *sartorius* Sub-zone to the *greistoniensis* Biozone (Table 15). The base of the formation [SN 8459 7569] is marked by the appearance of high-matrix sandstones (type Bii). Sections [SN 8463 7575 to SN 8447 7638] through the lowest parts of the formation are seen in Afon Diliw for some 100 m upstream of the basal contact. Afon Ystwyth [upriver from SN 8474 7566] provides almost unbroken exposure through a sequence of sandstone/mudstone turbidite couplets (type Cii) interbedded with packets of high-matrix sandstone. Mudstones in this sequence become paler and greener upwards. Locally, pebbly sandstones are exposed in the prominent crags [SN 8492 7581] on Craig y Lluest, in the river [SN 8506 7572], and in a track cutting [SN 8500 7563 to SN 8507 7567] south of the mountain road. These sections consist of thick- to very thick-bedded, coarse-grained, feldspathic, high-matrix sandstones (type Bii); the beds are commonly amalgamated and contain granule and pebble-sized clasts. In contrast with equivalent levels farther south (Llyn Cerrigllwydion), conglomeratic lithologies are less widespread, and confined to scattered lenses at the bases of some sandstone

beds. Intercalated thinner-bedded sandstones (type C) include parallel-laminated sandstones up to 20 cm thick in a sequence comprising up to 80 per cent sandstone. Packets of commonly amalgamated, graded and granule-rich high matrix sandstones occur in the upper parts of the formation in the crags [SN 852 758] along the upper reaches of Afon Ystwyth, near Pen Bwlchy-cloddiau [SN 855 778] and in forestry cuttings [SN 8555 7863 to SN 8549 7843].

On the eastern limb of the syncline, south of the mountain road, crags [SN 8548 7552] display a 3 m-thick granule-rich, high-matrix sandstone with a channelled base. To the north, numerous crags [e.g. SN 8574 7598] of high matrix sand-stone are exposed on the western flanks of Glan Fedwen. A packet of thin high-matrix sandstones (type Bii), seen in a road cutting [SN 8595 7533], marks the base of the formation on the eastern limb of the syncline. In an eastern tributary [SN 8587 7637] of Afon Ystwyth the Pysgotwr Grits are overlain by pale green mudstones and thin-bedded sandstones of the Glanyrafon Formation.

Wye valley area   Folded and faulted sequences containing high-matrix sandstones form crags [SN 8907 7492, SN 9012 7599, SN 9054 7669] on Pen-lan-fawr, Trafel-gwyn, and Esgair y Graig. To the east, the base of the formation is seen in Trawsnant [SN 8990 7372 to SN 9029 7319], and in crags [SN 9030 7270] on Cefn Bach which are situated in the core of the eastern syncline. Crags [SN 9098 7649 to SN 9120 7635] to the west of Pant-y-Drain provide a complete section through the formation. There, feature-forming packets of high-matrix sandstones (type Bii) are interbedded with pale green-grey, thin-bedded mudstone/sandstone couplets (type C). Farther east, the River Wye [SN 9167 7574 to SN 9143 7653] exposes a folded 50 m-thick sequence of Pysgotwr Grits and adjacent tongues of the Glanyrafon Formation. The former contains high-matrix sand-stones (type Bii) up to 75 cm thick, which are confined to two packets. North-east of the River Wye, a road cutting and quarry [SN 9120 7740 to SN 9120 7720] (Figure 29, locality 14) at Dolfach provide a section through basal parts of the formation.

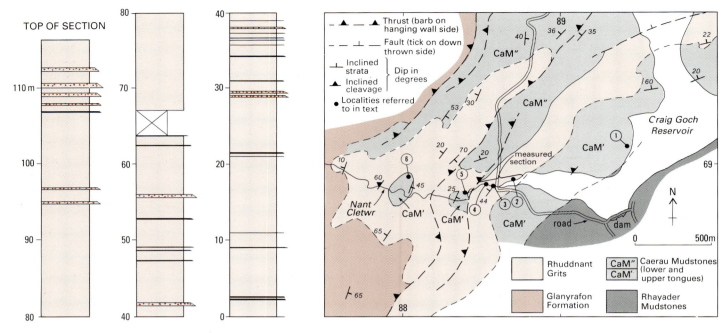

Figure 32 Geological map of Nant Cletwr area, showing localities referred to in text and measured section in Rhuddnant Grits (key as in Figure 30).

Graptolites of the *griestoniensis* Biozone, which include *Monoclimacis griestoniensis*, *Monograptus tullbergi spiraloides*, *M. discus*, *M. priodon* and *Pristiograptus nudus*, have been recovered from the laminated hemipelagites in the quarry. To the north, the Pysgotwr Grits crop out near Craignant [SN 9156 7814] and in quarries [SN 9149 7719 and SN 9142 7737]. The last locality exposes thin-bedded sandstone/mudstone couplets (type C). Farther east, road cuttings [about SN 9165 7663] and crags [SN 920 764 and SN 928 763] near Tyncoed, and west of Cwmgwary, provide additional sections in the formation. A disused quarry [SN 9260 7996] on the A44 road just north of the district yielded a *griestoniensis* Biozone fauna (Figure 29, locality 29). The attenuated north-eastern crop of the formation is exposed in a track cutting [SN 9303 7762], north of Cwm y Saeson, and in crags [SN 9391 7871] on Foel Goch.

Glanyrafon Formation

This formation takes its name from railway cuttings [SN 9858 7794 to SN 9864 7806] near Glanyrafon Halt, on the western limb of the Waun Marteg Syncline. It comprises Bouma turbidite sandstone/mudstone couplets (predominantly type Cii) with very subordinate burrowed and laminated hemipelagites (Plate 17d). High-matrix turbidite sandstones (type Bii) are very rare.

The Glanyrafon Formation crops out in two main areas, one in the core and eastern limb of the Central Wales Syncline, and the other in the Waun Marteg Syncline (Figure 29). In the Central Wales Syncline it comprises two tongues, the lower of which separates the Rhuddnant Grits and Pysgotwr Grits. In the west, in the Llyn Cerrigllywdion and Glan Fedwen synclines, the lower tongue of Glanyrafon Formation is about 150 m thick. The base of this tongue is taken at the incoming of a sequence of Bouma turbidites lacking high-matrix sandstones above the Rhuddnant Grits. In the northern part of the eastern limb of the Central Wales Syncline, the lower tongue thickens rapidly eastwards and northwards due to the thinning of the underlying Rhuddnant Grits and the rising base of the Pysgotwr Grits. To the north of Nant Cletwr [SN 886 688] the lower tongue rests on the upper tongue of Caerau Mudstones. There, the base is taken at the incoming of abundant, thin-bedded, turbidite sandstone/mudstone couplets (type Cii) above the Caerau Mudstones. Across the northern part of the eastern limb of the Central Wales Syncline, the lower tongue of Glanyrafon Formation is about 600 m thick but this includes local thin tongues of Rhuddnant Grits. In the Wye valley area the lower tongue decreases in thickness to less than 500 m.

An upper tongue of Glanyrafon Formation overlies the Pysgotwr Grits in the Glan Fedwen Syncline and in the Wye valley area. Due to the green colour of the interbedded mudstones, Jones (1922) and W D V Jones (1945) assigned these strata to the 'Pale Shales', equating them with the Dolgau Mudstones (Wood, 1906) of the Trannon area in the Llanidloes district. However, on the basis of their sandstone content (commonly more than 15 per cent) they are here included in the Glanyrafon Formation. The upper tongue is 300 m thick in the Glan Fedwen Syncline and 150 m in the Wye valley area, but in both areas the top is not seen.

East of the Tylwch Anticline, the Glanyrafon Formation occupies the core and limbs of the Waun Marteg Syncline, where it is about 750 m thick. West of the St Harmon Fault the formation succeeds the Caerau Mudstones. However, to the east of this fault the lower parts of the Glanyrafon Formation pass laterally into the Dolgau Mudstones (Figure 12). Although a thin basal tongue of Glanyrafon Formation survives at the top of the Caerau Mudstones, most of the formation in this area is underlain by Dolgau Mudstones. This basal tongue fails eastwards, just to the west of the Bwlch-y-sarnau Fault. Between the Bwlch-y-sarnau and Cwmys-gawen faults, the Glanyrafon Formation thins rapidly from about 500 m south of Bwlch-y-sarnau, to less than 100 m on Cwmcynydd Bank [SO 055 730], as both lower and upper levels pass laterally into the Dolgau Mudstones (Figure 12).

The turbidite sandstone/mudstone couplets (type Cii) which dominate the formation are up to 30 cm thick. Rare type Ci couplets with basal sandstones up to 25 cm thick are locally present. The proportion of turbidite sandstone in the formation varies between 12 and 50 per cent, but near to the contacts with contiguous mudstone formations some packets of strata contain less than 5 per cent.

In the Central Wales Syncline, the turbidite mudstones of the lower tongue are predominantly dark to medium grey, whilst in the upper tongue they are pale green. In the lower tongue the intercalated hemipelagites are either laminated or burrow mottled, each predominating in alternating packages of strata. In contrast, laminated hemipelagites are rare in the upper tongue. Comparable variations in turbidite mudstone colour, and hemipelagite distribution, occur in the lower and upper parts of the formation, respectively, in the Waun Marteg Syncline.

Palaeocurrent vectors from both bottom structures and cross-lamination in the turbidite sandstones mainly indicate southerly derivation. However, along the eastern flank of the Central Wales Syncline, some vectors are from south-east; this trend predominates east of the Tylwch Anticline (Smith, 1987a).

BIOSTRATIGRAPHY

In the core and across the eastern limb of the Central Wales Syncline, the lower tongue of Glanyrafon Formation has yielded graptolites indicative of the *crispus* Biozone, in the west, and of the *crispus* and *griestoniensis* biozones in the east. In the core, in the Glan Fedwen Syncline, the lower tongue is entirely within the *sartorius* Subzone (Table 15). However, on the eastern limb of the Central Wales Syncline, in Afon Elan [SN 881 729] (Figure 29, locality 15), near Banc y Defaid, the lower tongue contains assemblages suggesting the older *crispus* Subzone, appearing to confirm the lateral eastward failure of the upper levels of the Rhuddnant Grits (Figure 12). The top of the lower tongue youngs eastwards. On the eastern limb of the Central Wales Syncline, in the Wye valley, lower *griestoniensis* Biozone graptolites occur in the upper part of the lower tongue. East of the Wye valley, on Banc Dolhelfa [SN 924 749] (Figure 29, locality 16) these faunas occur 70 m below the top of the tongue, though the true thickness assignable to this latter biozone is probably greater, since only 2.5 km to the north, in Cwm y Saeson [SN 933 774] (Figure 29, locality 17), they occur about 150 m below the top of the tongue.

The only graptolite assemblage from the upper tongue of the formation was found near the base, in the Glan Fedwen Syncline, at Pen Bwlch-y-cloddiau [SN 8545 7831] (Table 16, locality 25), and probably belongs to the upper part of the *griestoniensis* Biozone.

East of the Tylwch Anticline, graptolite faunas from the Glanyrafon Formation range in age from the early *crispus* Biozone (*galaensis* Subzone) to the latest *griestoniensis* Biozone. Although the uppermost parts of the formation are undated it is likely that they lie within the *crenulata* Biozone.

In the southern part of the Waun Marteg Syncline, near Beili Neuadd [SO 0018 7012] (Figure 29, locality 28), the basal part of the lower tongue of the formation has yielded graptolites of the *galaensis* Subzone (Figure 29, locality 28). The base of the main part of the formation in the Waun Marteg Syncline lies within the mid to late *crispus* Biozone in the west and in the lower *griestoniensis* Biozone in the east. In the west, graptolites from the lower part of the formation east of Pant-y-dwr [SN 9975 7476] (Figure 29, locality 18) belong to the *sartorius* Subzone. Those from the type section [SN 9858 7794 to SN 9854 7806] (Table 14, locality 7) near Glanyrafon Halt, just above the base of the formation, are also probably from this subzone. Only the early *galaensis* Subzone has been proved in the underlying Caerau Mudstones in this area. Farther east [SO 0217 7347] (Figure 35, locality 6) the base of the formation overlies Dolgau Mudstones containing lower *griestoniensis* Biozone graptolites (p.157). The rest of the lower half of the formation in the Waun Marteg Syncline has yielded graptolites of the lower part [SN 9978 7672] (Figure 29, locality 19) and the upper part [SO 0152 7457] (Figure 29, locality 20) of the *griestoniensis* Biozone.

East of the Bwlch-y-sarnau Fault near Fishpool farm [SO 0431 7295] (Figure 29, locality 21), the base of the formation is still within the lower part of the *griestoniensis* Biozone, but 1 km to the east, on Cwmcynydd Bank [SO 0561 7292] (Figure 29, locality 22), it lies within the upper part of that biozone. Graptolites of the *crenulata* Biozone occur at the base of the overlying Dolgau Mudstones at Bwlch-y-sarnau [SO 0296 7456] (Figure 35, locality 2) and, since the Glanyrafon Formation thickens westwards at the expense of the Dolgau Mudstones, it seems probable that the upper part of the formation in the Waun Marteg Syncline area ranges up into that biozone.

Samples from the Glanyrafon Formation in Cwm Ystwyth (Appendix 2bi) proved barren of acritarchs other than sphaeromorphs. However, other samples (Appendix 2bii and iii) have yielded common *Eupoikilofusa striatifera*, *Micrhystridium* spp., *Multiplicisphaeridium* spp. and *Oppilatala eoplanktonica*. Also present are *Ammonidium listeri*, *A. microcladum*, *Dictyotidium biscutulatum*, *Domasia elongata*, *Elektoriskos* cf. *aurora*, *Gracilisphaeridium encantador*, *Lophosphaeridium* sp. A, *Multiplicisphaeridium martiniae*, *Salopidium graniliferum* and *Salopidium* sp. C of Priewalder (1987), which indicate the mid to late Telychian *G. encantador* Biozone. Recycled acritarchs include *Ordovicidium elegantulum*, *Peteinosphaeridium* spp. and *Stellechinatum brachyscolum*, from the mid to late Ordovician.

DETAILS

Lower tongue, west of the Claerwen Fault   Afon Ystwyth between the broken dam [SN 8433 7561] and the confluence [SN 8458 7570] with Afon Diliw provides a complete section through the lower tongue of the Glanyrafon Formation (Figure 31). Sandstone/mudstone couplets (type Cii) predominate throughout this section; the sandstones are rarely greater than 3 cm thick, but form up to 50 per cent of the formation (Plate 17d). A packet of four, high-matrix sandstones in the upper half of the section represents an early tongue of the Pysgotwr Grits. Laminated hemipelagites, dominant in the lower part of the sequence, but subordinate to pale burrowed varieties in the upper part, have yielded graptolites of the *sartorius* Subzone (Table 15).

Nant Dderwen [SN 8598 7371 to SN 8627 7356] provides an additional complete section through the lower tongue of the formation along the western limb of the Llyn Cerrigllwydion Syncline. To the south, extensive sections are exposed in the upper reaches of Nant Hirin [SN 8443 7077] and in the eastern limb of the syncline [SN 8542 7088 to SN 8498 7071]. In a tributary [SN 8502 7050] (Figure 29, locality 23) to the south, graptolites of the lower *griestoniensis* Biozone, which include *Monoclimacis* cf. *griestoniensis* sensu Elles and Wood, *Monograptus discus*, *M. clintonensis*, *M. pergracilis?* and *M.* sp. 3, were collected from strata underlying the basal high-matrix sandstones of the Pysgotwr Grits.

Lower tongue, east of the Claerwen Fault   In the south-west of the Rhayader sheet area the Glanyrafon Formation is preserved in a syncline through Moel Prysgau [SN 815 615] to Esgair Garthen [SN 835 643]. There, it is well exposed in Nant Gwina [SN 8107 6038 to SN 8083 6082] and in a forestry cutting [SN 8137 6084]. Thin-bedded sandstone/mudstone couplets (type Cii) predominate, but the sandstone content is locally less than 20 per cent. Both burrowed and laminated hemipelagites are present. Farther north along Afon Elan the Glanyrafon Formation intertongues with the Rhuddnant Grits. Outcrops near the bridge [between SN 8788 7324 and SN 8884 7235] reveal a broadly anticlinal sequence which includes a fold-repeated packet of high-matrix sandstones. Strata just beneath this packet [at SN 8805 7301], included in the Glanyrafon Formation, have yielded graptolites that suggest the *crispus* Subzone, including *Monograptus crispus*, *M. discus*, *M.* cf. *marri*, *M.* cf. *rickardsi* and *Streptograptus loydelli* (Figure 29, locality 15). In the gorge [SN 8884 7235] at the eastern end of the section, the lower levels of the sequence are dominated by turbidite mudstones and hemipelagites. Down-river, at Pont ar Elan [SN 9031 7155], a section through the sandstone-dominated facies of the formation is exposed, with interbedded hemipelagites well seen on water-worn surfaces. A roadside quarry [SN 9084 7144] to the east has yielded graptolites indicative of the *crispus* Biozone (Figure 29, locality 24). To the north, additional sections in the lower tongue are seen in crags [SN 919 727, SN 912 731] on the eastern flanks of Lan Wen and Cerrig Llwydion, and in Nant yr Ych [SN 9024 7255 to SN 9075 7273].

A section [SN 9169 7518 to SN 9193 7479] through several periclines in the River Wye below Neuadd-ddu exposes Rhuddnant Grits overlain by the lower tongue of the Glanyrafon Formation. Dark grey turbidite mudstones and laminated hemipelagites characterise the lower part of the section, grey-green mudstones and burrowed hemipelagites the upper. West of the river, a forestry track [SN 876 774 to SN 898 783] south of Pant-gwyn Hill provides additional sections in the formation. At one place [SN 8945 7820], graptolites are common but poorly preserved in thin-bedded sandstones (type Cii), and at another [SN 8898 7781] (Figure 29, locality 25), graptolites suggesting the *sartorius* Subzone, including *Monograptus discus*, *M. rickardsi* and *Streptograptus sartorius*, have been collected. East of the river, a quarry [SN 9258 7426] (Figure 29, locality 26) on Banc Dolhelfa has yielded graptolites suggestive of the

crispus Biozone. A nearby track cutting [SN 9270 7425] to the east exposes the transition from the underlying Caerau Mudstones. To the north the upper levels of the lower tongue, exposed in the back scar of a landslip [SN 924 749] (Figure 29, locality 16), have yielded graptolites, including *Monoclimacis* cf. *griestoniensis* sensu Elles and Wood, *Mcl. griestoniensis* sensu stricto?, *Monograptus pragensis pragensis*, *Streptograptus* aff. *loydelli* and *S. sartorius?*, indicative of the lower *griestoniensis* Biozone. A fauna of similar age was obtained from a quarry [SN 9320 7743] (Figure 29, locality 17) north of Cwm y Saeson.

Upper tongue   The upper tongue of the Glanyrafon Formation is exposed in the core of the Glan Fedwen Syncline [SN 8742 7587 to SN 8549 7621]. Thin-bedded sandstone/mudstone couplets (type Cii) predominate in a sequence which consists of 20 to 50 per cent sandstone. The mudstones are distinctively green. Forestry cuttings [SN 8555 7863 to SN 8549 7843] to the north expose the basal contact with the Pysgotwr Grits, and outcrops [SN 8545 7831] about 20 m above the base have yielded graptolites of the upper *griestoniensis* Biozone (Table 16, locality 25; Figure 29, locality 27). Thin-bedded sandstone/mudstone couplets (type Cii) are well seen in a roadside quarry [SN 8587 7883] to the east.

In the River Wye [SN 9162 7643] the upper tongue overlies the Pysgotwr Grits, east of Pant-y-drain. Additional sections to the north are provided by a road cutting [SN 9151 7675], stream sections [SN 9218 7680 and SN 9295 7873], and quarries [SN 9315 7795].

Waun Marteg Syncline   The disused railway cuttings [SN 9858 7794 to SN 9864 7806] south of Glanyrafon Halt comprise the type section of the Glanyrafon Formation. Possible *sartorius* Subzone graptolites have been collected from this section (Table 14, locality 7). There, and in a ravine [SN 9943 7776] to the east, thin-bedded sandstones (type C) comprise over 40 per cent of the sequence; northward-directed flute casts are present at both localities. To the east, Rhydyclwydau Brook [SN 0000 7828 to SN 0070 7858] exposes a folded sequence in the formation. In the upper reaches of the section [SO 0012 7835], mudstones are dominant but track cuttings [SO 0033 7844], north of the brook expose levels with a higher sandstone content. To the south, in a quarry [SN 9975 7476] (Figure 29, locality 18) on the west flank of Moelfre, laminated hemipelagites have yielded *sartorius* Subzone graptolites including *Monograptus clintonensis*, *M. pragensis pragensis*, *Streptograptus sartorius* and *Pristiograptus* cf. *nudus*. Graptolites referable to the lower part of the *griestoniensis* Biozone, including cf. *Monoclimacis* cf. *griestoniensis* sensu Elles and Wood, have been recovered to the north of Cnych Mawr [SN 9978 7672] (Figure 29, locality 19), and a probable upper *griestoniensis* Biozone fauna was found in a trackside section [SO 0153 7458] (Figure 29, locality 20) to the east. Additional sections in the Waun Marteg Syncline are provided by a stream [SN 9958 7274 to SN 9988 7268] east of St Harmon, a quarry [SN 9923 7127] in the west flank of Moel Hywel, a stream [SO 0057 7180] draining Banc Gwyn, and a gully [SO 0132 7354] west of Castell-y-garn. The sandstone content in these sections ranges from 10 to 70 per cent. Cuttings [SN 9948 7084] north of Cwmithig, close to the core of the Waun Marteg Syncline, expose a gradational base to the upper tongue of the formation in this area. There, sandstone-based turbidites (type Cii) enter a sequence of predominantly mudstone turbidites (type D) with interbedded laminated hemipelagites. To the south, a lower tongue of the Glanyrafon Formation is exposed in Rhyd-hir Brook [SN 9949 7006] north of Belli-Neuadd, and in a track cutting [SO 0018 7012] (Figure 29, locality 28) to the east. From the latter locality, *galaensis* Subzone graptolites have been recovered, including *Monograptus crispus*, *M. discus* and *M. turriculatus*.

East of the Bwlch-y-sarnau Fault   The base of the Glanyrafon Formation is exposed opposite the former dam [SO 0431 7295] (Figure 29, locality 21), north of Fishpool farm. There, the sequence consists mainly of thin-bedded sandstone/mudstone couplets (type C), with sandstones up to 12 cm thick. Laminated hemipelagites have yielded graptolites of the lower *griestoniensis* Biozone. Upper levels of the formation are seen in a track section [SO 0288 7450] at Bwlch-y-sarnau, and intermediate levels in a disused quarry [SO 0321 7381] north of Cwm-y-dea. Graptolites of the upper *griestoniensis* Biozone have been collected from a forestry cutting [SO 0561 7292] (Figure 29, locality 22) close to the base of the formation on Cwmcynydd Bank. Farther north, a disused quarry [SO 0562 7351] contains sandstones up to 14 cm thick, which display westerly directed flute casts and cross-lamination.

Depositional model

The introduction of large-scale, sand-rich, southerly sourced turbidite systems during the Telychian, marked an abrupted increase in the grade and volume of detritus being supplied to the Welsh Basin. These systems equate with an angular unconformity, on the southern basin margin in south Pembrokeshire, which formed in respose to tectonic uplift at a time of widespread eustatic transgression of the eastern basin margin (Figures 34 and 39) (Cave, 1979; Smith, 1987a; Clayton, 1992). Soper and Woodcock (1990) have suggested that this uplift was a response to the initial phases of Avalonian and Laurentian plate collision.

FACIES INTERPRETATION

The facies of ancient sandy turbidite systems are best interpreted in terms of the morphological elements present in modern systems (channels, levees, channel-to-lobe transitions, lobes and lobe fringes) rather than the abstract divisions of many submarine fan models (such as upper, mid and lower fan) (Mutti, 1985; Mutti and Normark, 1987). Furthermore, fan models are not considered appropriate in areas where contemporaneous faulting has prevented the development of a complete fan system (Mutti and Normark, 1987; Macdonald, 1986).

Three facies are present in the sandstone-dominated Telychian systems of the district: sandstone-lobe, sandy lobe-fringe and muddy lobe-fringe facies (Figure 13). The terminology used is adapted from Mutti and Normark (1987) and Smith (1987a). An example of the relationship between these three facies is shown in Figure 33b. Although a channel facies has not been reported from these Telychian systems in the Welsh Basin, Smith (1987a) has described a channel-lobe transition facies from proximal parts of the Pysgotwr Grits in the Lampeter district to the south.

Sandstone-lobe facies

The facies is characterised by packets of interbedded high-matrix sandstones (type Bii) and thin-bedded, Bouma turbidites (type Cii). The packets are separated by sequences of thin-bedded Bouma turbidites free of high-matrix sandstones. The sandstone-lobe facies is represented by the Mynydd Bach Formation, the Rhuddnant Grits and the Pysgotwr Grits (Figure 33b).

Sandstone-lobes were originally defined as non-channelised bodies, 3 to 90 m thick, composed of thick-bedded turbidite sandstones alternating with thinner-bedded and finer-grained interlobe deposits (Mutti and Ghibaudo, 1972). The sandstone-lobe facies of the district exhibit many of the characteristics of ancient sandstone-lobe deposits summarised by Mutti and Normark (1977). These include: the occurrence, upcurrent, of a channel-lobe transition facies (exposed south of the district) in the Pysgotwr Grits system; an absence of channels and mud-draped, concave-up scours, although rare tabular scours are present at the junctions of amalgamated high-matrix sandstones; and an abundance of tabular-bedded, coarse-grained turbidites arranged in packets less than 50 m thick, within sequences of fine-grained turbidites hundreds to thousands of metres thick.

The thick sandstone turbidites of such lobes are normally regarded as 'proximal' Bouma turbidites (cf. Mutti and Normark, 1987; Smith, 1987a). Most of the thick sandstones in the Pysgotwr Grits have been interpreted by Smith (1987a) as proximal, middle-cut-out Bouma turbidites in which the T_a division predominates. However, the bimodal thickness distribution of the high-matrix (type Bii) and Bouma (type Cii) sandstones throughout the facies, and the lack of tractional reworking in the upper parts of high-matrix sandstones in mid to distal situations, suggest that the two types of sandstone were deposited by different types of flow (Long, 1966; Clayton, 1992). The high-matrix sandstones are here regarded not as part of the proximal-to-distal continuum of Bouma turbidites, but as the deposits of flows that varied between high-concentration, muddy, coarse-grained turbidity currents and dilute, slurry-like debris flows (see p.46). Thin, but fast and competent flows of this type could have been capable of carrying sand-grade detritus to the distal reaches of the Telychian systems.

Interpretation of the packeted distribution of the high-matrix sandstone rests largely on the geometry of individual beds and packets. Two models have been suggested on the basis of differing evidence from the Pysgotwr and Rhuddnant grits. In the Pysgotwr Grits, Smith (1987a; 1988) reported that the thick sandstones in the packets exhibit a random distribution of thicknesses and cannot be traced across-system with confidence for more than 5 km. However, he was able to correlate some packets in a downcurrent direction for the length of the system. He concluded that each packet represents a sandstone-lobe that developed as a result of tectonic uplift in the source area, which increased the clastic supply to the southern basin margin. In contrast, the intervals of thin-bedded Bouma turbidites between the packets were seen to reflect periods of relative tectonic quiescence. The random thickness patterns within the packets, together with a lack of any overall coarsening-upwards sequence led him to suggest that the sandstone packets grew in an aggradational manner and involved little lateral migration of depositional sites.

By contrast, in the Rhuddnant Grits, Clayton (1992) reported that individual high-matrix sandstones could

a)

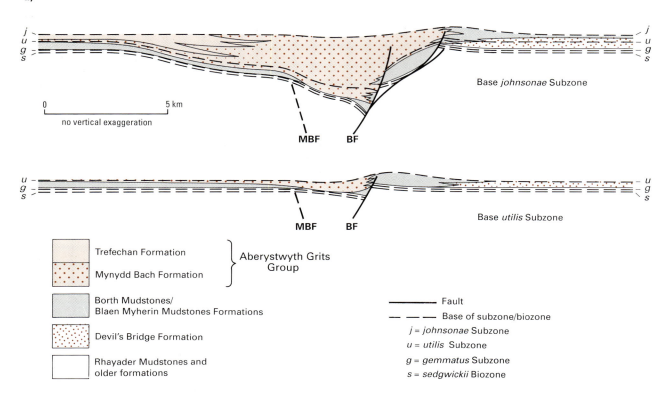

Base *johnsonae* Subzone

0   5 km

no vertical exaggeration

MBF   BF

Base *utilis* Subzone

MBF   BF

Trefechan Formation
Mynydd Bach Formation
} Aberystwyth Grits Group

Borth Mudstones/
Blaen Myherin Mudstones Formations

Devil's Bridge Formation

Rhayader Mudstones and
older formations

———————— Fault

— — — — — Base of subzone/biozone

j = *johnsonae* Subzone
u = *utilis* Subzone
g = *gemmatus* Subzone
s = *sedgwickii* Biozone

b)

A   Aberystwyth
B   Borth
MB   Mynydd Bach
N   New Quay
BF   Bronnant Fault
MBF   Mynydd Bach Fault

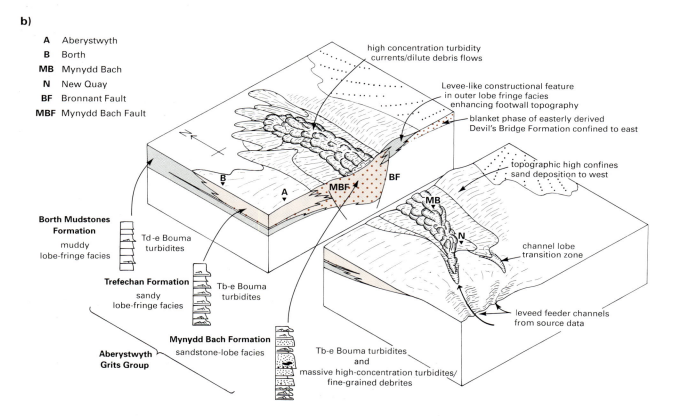

high concentration turbidity
currents/dilute debris flows

Levee-like constructional feature
in outer lobe fringe facies
enhancing footwall topography

blanket phase of easterly derived
Devil's Bridge Formation confined to east

topographic high confines
sand deposition to west

channel lobe
transition zone

leveed feeder channels
from source data

**Borth Mudstones
Formation**
muddy
lobe-fringe facies

Td-e Bouma
turbidites

Trefechan Formation
sandy
lobe-fringe facies

Tb-e Bouma
turbidites

**Aberystwyth
Grits Group**

Mynydd Bach Formation
sandstone-lobe facies

Tb-e Bouma turbidites
and
massive high-concentration turbidites/
fine-grained debrites

not be traced for more than 500 m, and that packets of these beds, each of which he assumed to represent a sandstone-lobe, could not be correlated between sections. He concluded that such lobes were of limited extent and, when abandoned, were covered by interlobe deposits (thin-bedded Bouma turbidites). On this basis he suggested that the packeting was autocyclic in origin, the result of lobe switching following channel avulsion farther south. He concurred, however, that the random thickness distribution of high-matrix sandstones in the packets reflected allocyclic effects, possibly seismic events, in the southern source regions.

Mutti and Normark (1987) have distinguished proximal, median and distal sandstone-lobe facies. Abundant thick to very thick beds and high levels of amalgamation, typical of their proximal settings, are present in the Llyn Teifi Member and parts of the Mynydd Bach Formation. The remainder of the sandstone-lobe facies in the district compares with the deposits of Mutti and Normark's median lobe setting. In these Telychian systems, however, the confining effects of intrabasinal structures promoted marked lateral in addition to downcurrent facies changes. Thus motifs comparable with the proximal and median lobe facies of Mutti and Normark (1987) here correspond to the strongly confined and less confined parts of these systems (Figure 13). Unconfined portions of the Telychian systems compare with the distal lobe facies of Mutti and Normark and are here termed lobe-fringe facies (after Smith, 1987a).

Sandy lobe-fringe facies

This facies comprises thin-bedded Bouma turbidites (type C) and is represented by the Trefechan and Glanyrafon formations. It should be noted that the facies is lithologically identical with strata between individual beds and between packets of high-matrix sandstone in the sandstone-lobe facies. These thin, Bouma turbidites were deposited from low-concentration turbidity currents, triggered by events in the southern source area that were far more frequent (Smith, 1988, estimates about 17 times more frequent) than those responsible for the high-matrix sandstones.

In the district, the facies fringes the western side of the Mynydd Bach Formation sandstone-lobe (Figure 33b) and the eastern side of the Rhuddnant and Pysgotwr grits sandstone-lobes (Figures 13 and 34). The higher proportion of thicker sandstones (type Ci) in the Trefechan Formation suggests that it accumulated in a setting nearer to source than the Glanyrafon Formation (as seen in the district).

Figure 33 Depositional history of the Aberystwyth Grits Group.
a Schematic restored cross-sections for base *utilis* Subzone and base *johnsonae* Subzone times along the line indicated in Figure 26 (see also C–C′ in Figure 44). **b** Three-dimensional schematic model of the Aberystwyth Grits Group turbidite system, illustrating the relationship between turbidite facies and structure.

Muddy lobe-fringe facies

This facies mainly comprises mudstone turbidites (type D) and is represented by the Borth, Blaen Myherin, Caerau and Adail mudstones (Figure 33b). Across much of the district, these mudstone divisions envelope the sandy parts of the systems and record mud accumulation at the sides and in front of the sandstone-lobes or their sandy fringes, as well as earlier deposition of mud. The Bouma turbidites (type D) are genetically related to the Bouma sandstone/mudstone couplets, and in many cases, the same low-concentration flow responsible for a type D turbidite on the muddy fringe of a lobe may have deposited a type C couplet across the associated lobe area and its sandy fringe. Moreover, as the lateral juxtaposition of facies demonstrates, this continuum operated, not only in a proximal to distal sense, but also laterally (Figure 34).

Locally in the upper part of the Borth Mudstones, in the Blaen Myherin Mudstones and in the upper tongue of the Caerau Mudstones, thin argillaceous fine-grained sandstones or muddy siltstones (type Bii) are present. They represent the distal portions of lobe sandstones and are not genetically related to the interbedded Bouma mudstone turbidites.

SYSTEM SOURCES

The textural and compositional immaturity of many of the sandstones in these Telychian systems, in strong contrast to those in the earlier easterly supplied systems, suggests an abundance of first-cycle detritus and relatively short transport paths prior to resedimentation.

The high mud content of these Telychian systems and the large areal extent of their sandstone-lobe facies supports comparison with elongate, high-efficiency turbidite systems (Mutti, 1979; 1985), in which sand from a mud-rich source, typically a prograding delta, is efficiently carried far into the basin via leveed feeder channels. The levees, maintained by overbank mud deposition, largely confine the high-concentration turbidity currents which supply sand to the depositional lobe. The Telychian systems were supplied from a source, possibly a prograding delta, situated on the southern margin of the basin.

It has been suggested that the thick turbidite sandstones (type B) of the sandstone-lobe facies record large but infrequent seismic events which destabilised large portions of the source area, possibly a delta front (Wood and Smith, 1959; Long, 1966; Smith, 1988; Clayton 1992). Mackie and Smallwood (1987) have proposed that the Bouma turbidites (type C) record gravity flows initiated by major storms and were therefore sourced from wide sectors of the shelf. However, Clayton (1992) has suggested that they record frequent slope-failures on the putative delta, in response to normal sedimentation and oversteepening processes. In either case the resultant low-concentration, but thick, slow-moving flows, unconfined by the leveed feeder channels, were able to spread well beyond the lateral depositional limits of the lobe facies (Figure 34).

TECTONIC CONTROLS ON SYSTEM GEOMETRY

The eastward migration of the eastern margin of the southerly derived turbidite systems during the Telychian has long been recognised (Jones, 1956; Bailey, 1969; Zeigler, 1970; Cave, 1979). Although early attempts to explain this trend invoked passive progradational or side-lapping models, James and James (1969) and Smith (1987a) have suggested that the thickness changes and facies distributions were mainly controlled by contemporaneous faulting, and this view has been confirmed and elaborated by the present work (Figure 34).

The preservation of the thick sequences of sandstone-lobe facies (Rhuddnant and Pysgotwr grits and Mynydd Bach Formation) on the downthrow side (west) of the synsedimentary Bronnant and Claerwen faults (Figure 13) suggests that these structures probably gave rise to topographic depressions, along which the high-concentration, lobe-building flows were preferentially directed. There is no evidence (such as slumps and debrites) to suggest the presence of fault scarps along the eastern margins of the depressions, and therefore sea floor relief was probably in the form of linear bulges above the propagating fault tip-lines (Figure 34). The thick, slower moving, low-concentration flows that deposited the lobe-fringe facies were less influenced by these topographical features and this may largely account for their wider geographical distribution.

Proximal-to-distal trends in the Aberystwyth Grits Group, previously identified as occurring in a predominantly downcurrent direction (Walker, 1967; Lovell, 1970; Crimes and Crossley, 1980) are in part oversimplified and partly incorrect. Firstly, new biostratigraphical dating (Loydell, 1991) has demonstrated that many of the trends previously recognised in coastal exposures take place upsequence rather than in a downcurrent direction. Secondly, coastal sections traditionally taken to display proximal characteristics have now been included in sandstone-lobe facies (Mynydd Bach Formation) and inland mapping shows that these extended down-system as far as their contemporary sandy fringe (Trefechan Formation) (Figure 34). This distribution confirms that flow paths of high-concentration turbidity currents and dilute debris flows which supplied the lobe, were controlled by contemporaneous movements on the Bronnant Fault. Sequences showing distal features all occur in the Trefechan Formation and record the greater lateral spread of low-concentration turbidity currents. The Borth Mudstones are even more widely distributed in both downcurrent and lateral senses, the latter reflecting the ability of these low-concentration flows to spread across the topographic feature formed by the sediment-draped fault tip that restricted the deposition of sandy facies. It is therefore evident that in the Aberystwyth Grits Group, lateral (across-flow) facies variations are as important as those that take place downcurrent (Figure 33b).

Topographic features which may have defined the northern limits of deposition of the Telychian systems can be discerned. Deviant palaeocurrent vectors from the northern part of the Aberystwyth Grits Group (Crimes and Crossley, 1980), and the Borth Mudstones (McCann and Pickering, 1989), have been used to suggest a northward-confining gradient to the system, possibly generated by contemporary movements on the Bala

Fault (Smith, 1990). Similarly, the marked south-to-north facies changes in the Rhuddnant Grits between the Ystwyth valley and the A44 road, and strongly deflected palaeocurrent trends in the latter section, may be related to early uplift of the Plynlimon or Van domes (Figure 34). The progressive northerly shift in the limits of the upper parts of the Rhuddnant Grits, and of the Pysgotwr Grits (Smith, 1987a) may reflect either the gradual onlap or bypassing of these features.

The absence of systematic, fining- or coarsening-upwards, and thinning- or thickening-upwards sequences suggests that these Telychian systems did not evolve by the progradation of sandy fans. Instead they can be compared to the aggradational systems of Macdonald (1986), in which sandy (proximal) facies are anchored by bounding faults and are unable to prograde over earlier, less sandy (median to distal) facies.

SYSTEM EVOLUTION

The stratigraphic position of the three Telychian sandstone-lobe systems demonstrates that the synsedimentary faults that influenced the deposition of each system were initiated sequentially from west to east (Figure 34). The movements on each fault were intermittent, however. Periods of activity are represented by the confinement of the sandstone-lobe facies to the west of the fault and periods of quiescence by eastward overstep of the facies across it.

The deposition of the Mynydd Bach Formation (sandstone-lobe facies) was largely confined by the Bronnant Fault during *renaudi* to *johnsonae* Subzone times (Figure 33a). However, there was limited overstep to the east during the ?late *utilis* Subzone. Although lobe-fringe facies (Trefechan Formation and Adail Mudstones) continued to accumulate west of the Bronnant Fault as late as *proteus* Subzone times, the focus of sandstone-lobe deposition moved eastwards during late *johnsonae* Subzone times, and the Llyn Teifi Member of the Rhuddnant Grits was deposited in a depression confined to the east by the precursor Claerwen Fault throughout much of the late *turriculatus* Biozone. The cessation of Llyn Teifi Member deposition in the west, in the late *carnicus* Subzone, coincides with the introduction of sandstone-lobe facies to the east of the Claerwen Fault on the eastern limb of the Central Wales Syncline. Sandstone-lobe facies achieved their greatest lateral extent in the district in the early *crispus* Biozone. Southward and westward contraction of the sandstone-lobe facies during mid-*crispus* Biozone times was followed by the widespread abandonment of sandstone-lobe facies deposition during the late *crispus* Biozone (*sartorius* Sub-

Figure 34   Diagram illustrating the evolution of the Telychian sandstone-lobe systems in sequence from 1 to 4 (facing page)

Cross-sections are derived from Figure 13. Abbreviations:
A — Aberaeron, Ab — Aberystwyth, B — Builth Wells,
L — Llanilar, La — Lampeter, Ll — Llanidloes,
Lla — Llandovery, N — Newquay, Ne — Newtown,
R — Rhayader; BL — Bala Lineament.

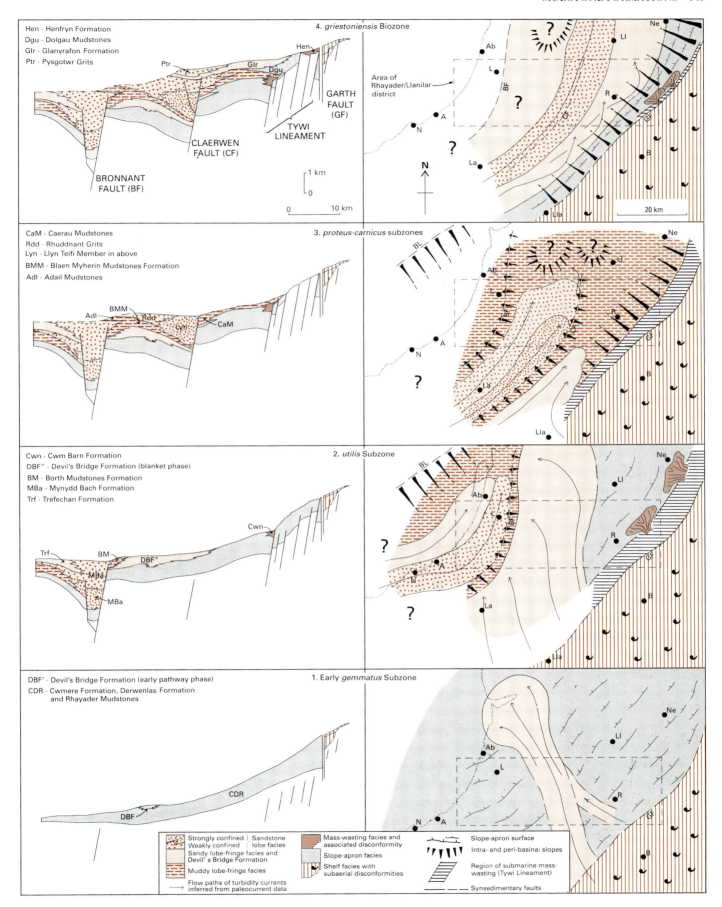

Hen - Henfryn Formation
Dgu - Dolgau Mudstones
Glr - Glanyrafon Formation
Ptr - Pysgotwr Grits

4. *griestoniensis* Biozone

GARTH
FAULT (GF)

TYWI
LINEAMENT

CLAERWEN
FAULT (CF)

BRONNANT
FAULT (BF)

Area of
Rhayader/Llanilar
district

N

1 km

0

0   10 km

20 km

CaM - Caerau Mudstones
Rdd - Rhuddnant Grits
Lyn - Llyn Teifi Member in above
BMM - Blaen Myherin Mudstones Formation
Adl - Adail Mudstones

3. *proteus-carnicus* subzones

Cwn - Cwm Barn Formation
DBF″ - Devil's Bridge Formation (blanket phase)
BM - Borth Mudstones Formation
MBa - Mynydd Bach Formation
Trf - Trefechan Formation

2. *utilis* Subzone

DBF′ - Devil's Bridge Formation (early pathway phase)
CDR - Cwmere Formation, Derwenlas Formation
       and Rhayader Mudstones

1. Early *gemmatus* Subzone

Strongly confined ⎱ Sandstone
Weakly confined ⎰ lobe facies
Sandy lobe-fringe facies and
Devil's Bridge Formation
Muddy lobe-fringe facies

Flow paths of turbidity currents
inferred from paleocurrent data

Mass-wasting facies and
associated disconformity
Slope-apron facies
Shelf facies with
subaerial disconformities

Slope-apron surface
Intra- and peri-basinal slopes
Region of submarine mass-
wasting (Tywi Lineament)
Synsedimentary faults

zone). Sandy lobe-fringe sedimentation continued unabated during this interval (lower tongue of Glanyrafon Formation), perhaps reflecting a period of tectonic quiescence in the southern source area and along the Claerwen Fault.

Sandstone-lobe development resumed in latest *crispus* to *griestoniensis* Biozone times. The position of the thickest and oldest parts of the Pysgotwr Grits and of its coarsest sandstones suggests that renewed activity on the Claerwen Fault may have focussed deposition. However, the eastern limits, initially of the Rhuddnant Grits and subsequently of the Pysgotwr Grits, may reflect passive sidelap against a confining slope, along the western limb of the Tylwch Anticline, inherited from the late Hirnantian to early Telychian slope-apron.

The eastward lateral failure of lobe-fringe facies (Caerau Mudstones and Glanyrafon Formation) during the *crispus* and *griestoniensis* biozones also appears to reflect the confining effect of contemporary west-facing slopes of greater degree and elevation than those envisaged to the west. The location and period of influence of such slopes suggest that, though perhaps inherited from earlier slope-apron gradients, they were principally the product of tectonic oversteepening and contemporaneous faulting along the Tywi Lineament (see Chapter 6). Flow deflection by such gradients may account for the anticlockwise swing in palaeocurrent vectors displayed by the Glanyrafon Formation in this region (Smith, 1987a) (Figure 34). Westward offlap of upper parts of the formation in the early *crenulata* Biozone may also record a tectonic steepening of gradients, but appears principally to reflect a decline in sediment supply from the south. The ensuing failure of lobe-fringe facies deposition in the district signalled the demise of the southerly derived, Telychian sand-rich turbidite systems thoughout the southern Welsh Basin.

SIX

Silurian: Telychian mass wasting and slope-apron systems of the Tywi Anticline

The Telychian succession that crops out around the closure of the Tywi Anticline, north-east of Rhayader (Figure 35), largely comprises a mud-dominated sequence (Dolgau Mudstones) similar to that of the earlier slope-apron facies. West of the Cwmysgawen Fault, it intertongues with the fringing facies of coeval sandstone-lobe systems (Caerau Mudstones and Glanyrafon Formation). However, to the east of the fault, across the Tywi Lineament, the Telychian sequence rests with marked disconformity on earlier Llandovery and Ordovician strata (Figures 12, 35 and 39). Proximal resedimented deposits comprising disturbed beds, debrites and coarse-grained turbidites (Cwm Barn and Henfryn formations) locally overlie this disconformity. The nature of the disconformity and associated resedimented deposits suggests that they relate to an episode of intra-Telychian faulting affecting the Tywi Lineament, a period of fault activity coeval and related to that which controlled the development of the southerly supplied sandstone-lobe systems farther to the west.

Cwm Barn Formation

The Cwm Barn Formation was described, but never formerly named, by Lapworth (1900). It comprises a thinning- and fining-upwards sequence of debrites and thinly interbedded, shelly, phosphatic turbidite conglomerates, sandstones and mudstones (types Aii, C and D). The intergradational nature, and similar clast assemblages, of this suite of turbidites demonstrates that they had a common origin and provenance. The type locality is Cwm Barn quarry [SO 0082 7081] (Figure 35, locality 1) and an adjacent stream section [SO 0080 7000 to SO 0070 7009] at Brynscolfa.

The principal crop of the formation occurs 3 km east-north-east of Rhayader, as a 1.5 km-long, north-east-trending tract, limited to the south-west by the Brynscolfa Fault, to the south-east by the Cwmysgawen Fault and to the east by the Bwlch-y-sarnau Fault (Figure 35). It includes the type section. A conjectural subdrift outcrop lies to the east, close to the junction of the Bwlch-y-sarnau and Cwmysgawen faults. A third, poorly exposed and faulted crop [SO 021 704] is present in the Cwmysgawen Fault zone near Lan Goch.

The base of the formation is everywhere faulted out, except in the Cwmysgawen Fault zone, where it appears to overlie Rhuddanian facies of the Caban Conglomerate Formation (Dyffryn Flags). The formation is at least 125 m thick in the principal crop, but less than 20 m have been recorded in the Cwmysgawen Fault zone, suggesting a marked eastward attenuation.

The lowest exposed beds in the principal crop are seen in Cwm Barn quarry (Figure 36). At the base of the quarry sequence, about 6 m of debrites, in units 0.1 to 5 m thick, comprise clasts up to boulder size dispersed in a silty mudstone matrix. The clasts include abundant well-rounded, discoidal and oblate calcareous nodules, less common siltstones and burrow-mottled mudstones, and rare laminated hemipelagites. The debrite units are separated by beds of parallel-laminated, fine-grained calcareous sandstones up to 20 cm thick.

The debrites are overlain in the quarry by about 8 m of medium grey, silt-laminated mudstones with scattered thin debrite units and rusty-weathering, decalcified sandstones. Disrupted bedding and small slump folds occur in the lower part of the sequence.

Above this, the remainder of the quarry sequence consists of thinly interbedded turbidite conglomerate/ sandstone/mudstone triplets (type Aii) (Plates 7a and 19d), turbidite sandstone/mudstone couplets (type C) and turbidite mudstones (type D). The conglomerates of the type Aii turbidites are poorly graded, clast-supported and up to 1 m thick (Plates 19a and c). They are composed predominantly of moderately well-sorted, locally imbricated, discoidal pebble and granule-grade clasts set in a coarse- to fine-grained sandstone matrix. Scattered cobbles and rare boulders up to 70 cm are present. The bases of the conglomerates rest with sharp and commonly erosional contacts on the underlying beds, showing gutter casts, grooves and small-scale channelling. Amalgamation of conglomerate beds is common. The clasts include abundant reworked and commonly disarticulated shelly fossils, black phosphatised mudstone pebbles, calcareous nodules, vein quartz pebbles and a variety of lithic fragments. The derived fauna comprises brachiopods, trilobites, bivalves, gastropods, ostracods, solitary and colonial corals, bryozoa and crinoid debris. This benthic assemblage contains elements from both shallow- and deep-water Llandovery communities (Ziegler et al., 1968).

The phosphatic clasts are composed of fluorapatite, a mineral phase which may have evolved from a syndepositional precursor carbonate-fluorapatite during low-grade metamorphism. The phosphatic material commonly occurs as a replacive rim surrounding a core of unaltered mudstone, which suggests that these clasts suffered diagenetic replacement prior to transportation in turbidity currents.

Lithic clasts are mainly well-rounded, discoidal, structureless, medium to dark grey mudstone pebbles. Also present are angular blocks of pale grey, burrow-mottled mudstone, fragments of dark grey laminated hemipelagite, rounded pebbles, cobbles and boulders of green feldspathic sandstone containing bioclastic debris, altered coarse-textured igneous lithologies and rare acid volcanic rocks (Plate 19b).

The sandstones of the type Aii and C turbidites are fine to coarse grained and up to 17 cm thick. They are mainly quartzitic, but abundant finely comminuted bioclastic debris and scattered phosphatised mudstone grains also occur. Some sandstones exhibit grading but most are parallel-laminated throughout, with undulatory, convolute and ripple cross-lamination occasionally developed in the upper parts. The mudstones of the type Aii and C turbidites are commonly less than 10 cm thick. Mudstone turbidites (type D) occur either interbedded with the coarser turbidites or in stacked sequences up to 1 m thick. Scattered, weakly burrow-mottled hemipelagites and rare laminated hemipelagites are present.

The sequence above the quarry is exposed in the stream to the west. It comprises a 50 m-thick, thinning- and fining-upwards sequence, dominated by turbidite sandstone/mudstone couplets (mainly type Ci). Conglomeratic turbidites (type Aii) are largely confined to the lowermost part. The percentage of sandstone decreases upwards from 50 per cent at the base to less than 5 per cent at the top, with mudstone turbidites (type D) increasing commensurately and forming a gradational passage into the Caerau Mudstones.

Palaeocurrent vectors at Cwm Barn quarry suggest a wide spread in transport directions. Bottom structures indicate flow paths towards the north-east; clast imbrication fabrics indicate current flow towards the north-west.

BIOSTRATIGRAPHY

Graptolites from the middle part of the Cwm Barn quarry section include *M. utilis* and *M. planus*, indicative of the *utilis* Subzone (Figure 35, locality 1). The proximity of the upper part of the formation to an early *crispus* Biozone locality [SO 0018 7012] (Figure 29, locality 28) in the suceeding strata (p.144), suggests that the Cwm Barn Formation may range into the *johnsonae* or *proteus* subzones. The pre-Telychian graptolite assemblage reported by Lapworth (1900, p.90) from an apparently lower horizon in the

quarry sequence, may have been obtained from a debrite clast of Cwmere Formation.

The conglomerates and sandstones in Cwm Barn quarry have yielded the brachiopods *Atrypa* sp., cf. *Clorinda* sp., *Dolerothis* sp., *Eoplectondonta* cf. *penkillensis*, cf. *Glassina*, *?Hyattidina* sp., *Leangella* cf. *scissa*, *Leptaena* cf. *haverfordensis*, *?Rhynchotrema* sp., cf. *Visbeyella* sp., as well as indeterminate pentamerids, spiriferids and ?resserellids. Rare trilobites comprise *Encrinurus* sp. and a proetid free cheek. The ostracod *Craspedobolbina (Mitrobeyrichia)* sp. (identified by Dr D Siveter, Leicester University) is abundant in some of the sandstones. The shelly fauna cannot be dated with certainty but *E.* cf. *penkillensis* suggests the late Llandovery.

Figure 35 Geological map of the Abbeycwmhir area illustrating the overstep of Ordovician formations within the Tywi Lineament by Llandovery (Telychian) strata. Selected graptolite localities referred to in the text are also shown.

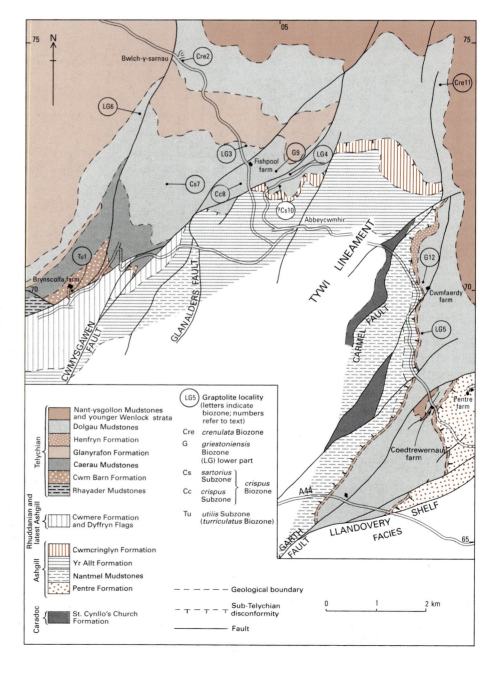

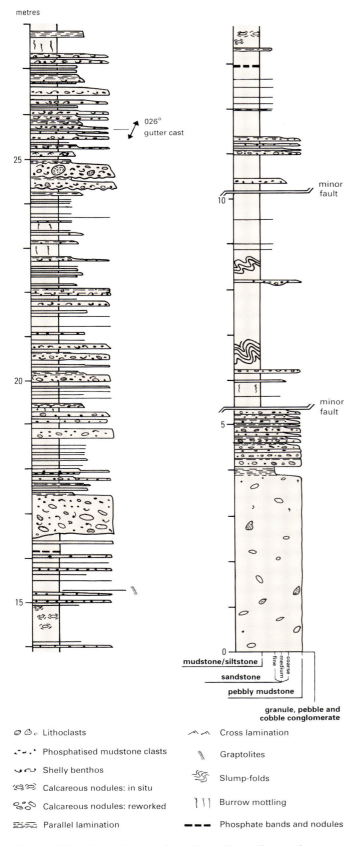

Figure 36 Graphic log of the Cwm Barn Formation at Cwm Barn quarry [SO 0082 7081].

Mudstone clasts from the conglomerates in Cwm Barn quarry (Appendix 2biv) have yielded the Ordovician acritarchs, *Acanthodiacrodium* spp., *Cymatiogalea cristata*, *Elektoriskos williereae* and *Stelliferidium* cf. *simplex*, suggesting a probable Ashgill source, but with recycled early Ordovician (Tremadoc) elements. A sample from a mudstone parting at Lan Goch [SO 0218 7043] has yielded a sparse acritarch assemblage of Aeronian to Telychian age (*A.* sp. B to *G. encantador* Biozone) including *Domasia limaciformis*, *D. trispinosa*, *Lophosphaeridium* cf. sp. A, *Oppilatala eoplanktonica* and *Tylotopalla* spp.

DETAILS

Cwm Barn quarry [SO 0082 7081] and the adjacent stream section [SO 0080 7000 to SO 0070 7009] that together comprise the type locality are described above. Brynscolfa quarry [SO 0089 7010], 100 m to the north-east, exposes a 3 m-thick section with abundant fossiliferous, decalcified turbidite sandstones (type C) and two shelly turbidite conglomerates (type Aii), rich in phosphatised mudstone clasts. One conglomerate contains an 80 cm-diameter boulder of coarse-grained sandstone. A quarry [SO 0131 7043], 650 m north-east of Cwm Barn, exposes 6 m of turbidite sandstone/mudstone couplets (type Ci), locally with slump folds.

A small quarry [SO 0218 7043] at Lan Goch, north of Vaynor, exposes 2 m of rusty-weathering, coarse-grained sandstone and a 20 cm-thick conglomerate. Rare mudstone partings have yielded Aeronian to Telychian acritarchs.

Henfryn Formation

The Henfryn Formation is a new name to describe the late Llandovery heterolithic assemblage of disturbed beds, debrites, conglomerates and sandstones that rest disconformably on Ashgill strata on the eastern limb of the Tywi Anticline. Some of the lithologies assigned to the formation, and the basal disconformity, were first recognised by Roberts (1929, p. 665), although the formation as here defined includes strata which Roberts regarded as Ashgill in age. The formation takes its name from Henfryn quarry [SO 0779 6903] (Figure 35, locality 5; Figure 37), where a complete sequence is exposed.

The principal crop of the formation extends from a faulted northern limit, near Fronrhydnewydd [SO 083 718], to the Garth Fault [SO 058 653] in the south (Figure 35). To the east, a further crop lies between the splays of the Garth Fault between Pentre [SO 085 682] and Gwestre [SO 072 655]. In the drift-covered ground east of Bryn [SO 052 658], a small fault bounded crop is inferred to lie between exposures of Dolgau Mudstones and Ashgill strata.

The degree of disconformity at the base of the formation varies along the crop. In the principal crop, along much of the Clywedog valley, it rests on the Yr Allt Formation, but to the north and south it rests on Nantmel Mudstones. In the eastern crop, between splays of the Garth Fault, it overlies the Pentre Formation. The Henfryn Formation is everywhere succeeded by the Dolgau Mudstones.

The formation exhibits rapid and pronounced thickness variations. Close to the northern limit of the princi-

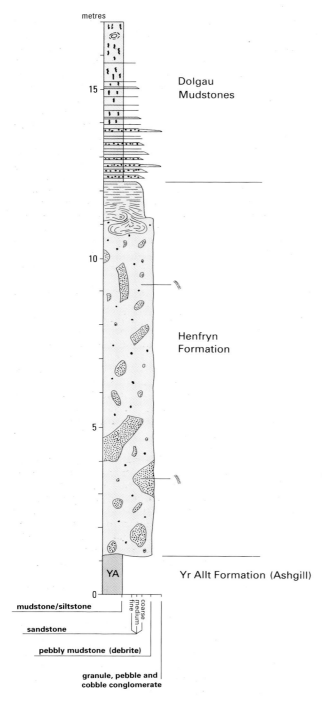

metres

Dolgau
Mudstones

Henfryn
Formation

YA

Yr Allt Formation (Ashgill)

mudstone/siltstone

sandstone

pebbly mudstone (debrite)

granule, pebble and
cobble conglomerate

coarse / medium / fine

Figure 37 Graphic log of the Henfryn Formation
at Henfryn quarry [SO 7779 6903]. For key see
Figure 36.

The formation comprises two groups of deposit:
disturbed beds and debrites, and turbidite conglomerates
and sandstones (types Ai and Bi). Both groups interdigi-
tate and either may locally predominate to the exclusion
of the other; where both groups are developed, conglomerates and sandstones typically occur uppermost.

Disturbed beds comprise slumped, destratified or disaggregated, lithologically uniform masses of the subjacent Ashgill mudstone. In colour and texture, these
beds are not everywhere readily distinguished from, and
have previously been included with, the in-situ Ashgill
formations (Roberts, 1929). However, inclusion of these
cryptic units in the Henfryn Formation is justified by the
occurrence of interbedded debrites containing clasts of
several different rock types. Some of the disturbed beds
may simply be very large clasts within very thick debrites.
Debrites range from silty mudstones with widely
dispersed clasts, to beds in which there is local clast-
support. Recogniseable clasts range up to boulder size,
but outcrop-scale rafts of strata, many metres across, may
be present. The clasts are mainly sedimentary and
include tabular rafts of silty mudstone (Yr Allt Formation) up to 1.5 m across, rounded cobbles of coarse-
grained sandstone, angular blocks of burrow-mottled
mudstone (possibly Cerig Formation) and laminated
hemipelagite. Carbonate and siliceous cone-in-cone
nodules, vein quartz pebbles and less-common conglomeratic and volcanic lithologies have also been recorded.
Fossiliferous clasts have yielded shells and graptolites.
Debrite units up to 10 m thick are recognised and, as at
Henfryn quarry [SO 0779 6903] (Figure 37), may
comprise the bulk of the formation.

The turbidite conglomerates and sandstones in the
Henfryn Formation appear less extensive and laterally
less persistent than the disturbed beds and debrites.
However, some parts of the formation are composed
entirely of such turbidites, notably west of Henfryn [SO
0768 6880]. The conglomerates and pebbly sandstones
(type Ai) typically form massive beds up to 6 m thick.
The conglomerates are clast-supported, poorly sorted,
with commonly rounded and locally imbricated clasts
rarely greater than 12 cm in diameter. The clasts mainly
comprise vein quartz, metaquartzites and igneous lithologies. Contacts between conglomerate and the overlying pebbly sandstone are commonly diffuse, suggesting
both vertical and lateral gradations within single thick
turbidites. Where conglomeratic units overlie pebbly
sandstones, they exhibit either an erosional or a loaded
contact. Medium- to thick-bedded, medium-grained
turbidite sandstones (type Bi) are commonly developed
at the top of the formation (Plate 19E). Although most
units are massive and quartz-rich, there are also highly
micaceous, parallel-laminated units and rare, trough
cross-stratified units. The ubiquitous presence of
abundant sand-grade mudstone clasts imparts a distinctive dark grey colour to the sandstones.

Contacts between the conglomerates and sandstones
on the one hand, and disturbed beds and debrites on the
other, are typically sharp and erosive; this feature, and
their lateral distribution, suggests that the conglomerates
represent channel fills. Contact relationships are locally

pal crop, north-west of Fronrhydnewydd, it attains its
maximum recorded thickness of about 100 m. It thins
markedly both to the north and south. In Henfryn
quarry it is only 14 m thick, and in Camlo Brook under
10 m thick. The eastern succession, between the splays of
the Garth Fault, ranges from 5 to 50 m thick.

complicated by wet-sediment deformation. Large-scale load structures occur, in which huge pillow-like masses of conglomerate and sandstone, up to 30 m across, are enclosed within underlying disturbed mudstones.

Where the Henfryn Formation is dominated by conglomerates and sandstones in the principal crop, the highest part comprises a 3 m-thick thinning- and fining-upwards sequence of turbidite sandstone/mudstone couplets (mainly type Cii). The sandstones are up to 10 cm thick, coarse to fine grained and locally granule-rich. They are composed of quartz and resistant lithic clasts, and in some beds, of abundant dark grey mudstone grains.

Pebble imbrication in the conglomerates in a quarry [SO 0772 6843] south of Henfryn, suggests current flow from a southerly or south-easterly direction.

BIOSTRATIGRAPHY

Graptolites from the matrix of the basal debrite in Henfryn quarry [SO 0779 6903] (Figure 37) are indicative of the early part of the *griestoniensis* Biozone. Longer-ranging Telychian graptolites have been found in a clast in the same deposit.

Roberts (1929, p.660) determined a shelly fauna from debrites north-west of Fronrhydnewydd [SN 0767 7093] as Ashgill in age and placed the deposits in his Camlo Hill Group. A re-examination of his fauna revealed an external mould of the Caradoc brachiopod cf. *Bicuspina spiriferoides*, demonstrating that the assemblage is reworked and was possibly obtained from debrite clasts.

The disturbed beds have yielded late Ordovician acritarchs, as have the clasts and matrix of the debrites. However, the debrites also contain mid to late Telychian acritarchs of the *G. encantador* Biozone (Appendix 2biv).

DETAILS

In the district, the most northerly section in the Henfryn Formation is in a tributary of Bachell Brook [SO 0810 7141], north of Fronrhydnewydd, where about 10 m of pebbly mudstone overlie Nantmel Mudstones. To the south-west, a quarry [SO 0767 7093] exposes 4 m of pebbly mudstones, from which Roberts (1929, p.660) reported a shelly fauna (see above).

A small roadside quarry [SO 0774 7030], south-east of Fronrhydnewydd, exposes about 3 m of black, coarse-grained, conglomeratic sandstones (types Ai and Bi) in beds up to 30 cm thick. The sandstones contain pebbles and granules of vein quartz, metaquartzite, acid volcanic rocks, siltstone and black (?phosphatised) mudstone (Plate 19e). In a stream [SO 0773 7018] by Keeper's Lodge, an eastward-younging section comprises pale, colour-banded mudstones with intercalated laminated hemipelagites of Llandovery aspect, overlain by laminated, silty dark grey mudstones comparable with the Yr Allt Formation and exhibiting irregular and disrupted bedding. However, just upstream at the waterfall, a pebbly mudstone unit is interbedded with mudstones of Yr Allt Formation aspect. Roberts (1929) and Smith (1988) included this section in Ashgill strata but the observed relationships suggest that the Llandovery and Ashgill mudstones are present as slumped masses or large rafts in a debrite sequence of Llandovery age. The Henfryn Formation is succeeded upstream by the Dolgau Mudstones.

Quarries and crags to the south of Cwmfaerdy [around SO 077 694] expose a sequence of intergradational turbidite conglomerates and pebbly sandstones (type Ai) succeeded by a 6 m-thick massive sandstone (type Bi). This sequence thins southwards. The older, northern part (Figure 38, locality 3) of Henfryn quarry [SO 0779 6903] exposes conglomeratic lithologies downfaulted against an underlying pebbly mudstone unit. The mudstone matrix of this unit has yielded *Monoclimacis* cf. *griestoniensis* sensu Elles and Wood and *Monograptus* cf. *priodon*, indicative of the *griestoniensis* Biozone. The newer, southern part of the quarry and adjacent stream section exposes the complete Henfryn Formation sequence in this area (Figure 38, locality 2; Figure 37). Clasts in the basal debrite include large rafts of the underlying Yr Allt Formation silty mudstones, up to 1.5 m across, pale mudstones and laminated hemipelagites (probably Llandovery), cobbles of coarse-grained sandstone, and calcareous and siliceous nodules. A large argillaceous sandstone clast has yielded poorly preserved *Monograptus* cf. *priodon* and *Monoclimacis*? sp., indicative of a Telychian or younger age. The overlying 1 m-thick, coarse-grained, parallel-laminated, micaceous sandstone contains abundant dark grey mudstone grains, and locally exhibits large-scale loading features at its base. Above, a 3 m-thick sequence with turbidite sandstone/mudstone couplets (types Ci and Cii) forms a transition into the Dolgau Mudstones.

A disused quarry [SO 0769 6881] (Figure 38, locality 4) 200 m west of Henfryn exposes about 10 m of mainly massive, dark grey, quartz and mudstone-grain sandstones, commonly with conglomeratic bases (type Bi), overlying the Yr Allt Formation. Conglomerate-filled scours and local trough cross-stratification are present. A quarry [SO 0772 6843] (Figure 38, locality 5) 350 m to the south, exposes about 5 m of massive, poorly sorted conglomerate. Clasts predominantly of vein quartz, metaquartzite and acid volcanic rocks range up to cobble size, and display local imbrication suggestive of current flow from the south or south-east.

Farther south, the formation is exposed only in a track section [SO 0610 6602] on the south side of Camlo Brook, south-east of Carmel, where it overlies the Nantmel Mudstones and comprises a 10 m-thick, unbedded sequence of pebbly mudstone. The larger clasts, ranging up to 50 cm across, comprise sedimentary lithologies including mudstone, coarse-grained (locally shelly) sandstone, and conglomerate. Rounded pebble, granule and coarse sand-grade clasts are predominantly of vein quartz and acid volcanic rocks. They occur in irregular, grain-supported clots or dispersed through the silty mudstone matrix. Acritarchs obtained from a mudstone clast at this locality are indicative of a Telychian age. These strata are abruptly overlain by slump-folded Dolgau Mudstones.

The westward-younging sequence of the Henfryn Formation, cropping out to the south-west of Pentre Farm, is observed at several points along Clywedog Brook. In the brook [SO 0811 6856] (Figure 38, locality 8) at Coedtrewernau, the top few metres of the formation comprise medium to thick beds of structureless, dark grey, medium-grained turbidite sandstone (type Bi), rich in mudstone grains. A similar sequence, exposed in the brook [SO 0818 6705] (Figure 38, locality 7) 200 m south-west of Brynllygoed, is underlain by irregularly bedded and locally destratified dark grey silty mudstones of Yr Allt Formation aspect. Enclosed within these mudstones, near the footbridge [SO 0825 6710] (Figure 38, locality 11), is a mass of the upper sandstone, up to 30 m across and subcircular in plan.

A quarry [SO 0741 6569] east of Gwestre exposes about 20 m of poorly sorted, pebbly mudstone containing disturbed masses of silty mudstone, tabular rafts of coarse-grained sandstone, cobbles of conglomerate and pebbles of acid volcanic rocks. The upper and lower contacts of the deposit are not exposed.

Dolgau Mudstones

The distinctive, predominantly pale green, oxic facies mudstones, which separate Telychian sandy turbidite facies (such as the Cwmystwyth Grits Group) from Wenlock mudstones (Nantysgollon Mudstones) in mid and North Wales, were known to early workers as 'pale slates' or 'Trannon (Pale) Shales'. They were renamed Dolgau Mudstones by Wood (1906) in the Trannon area of the adjacent Llanidloes district.

The Dolgau Mudstones as used here includes all largely sandstone-free sequences of oxic-facies mudstones, occurring between the Caerau Mudstones and the Nant-ysgollon Mudstones. Colour does not form a strict part of this definition. Thus, although the formation includes the pale green mudstones which comprise Roberts' (1929) Trannon Stage, and the Pale Shales of W D V Jones (1945), it also includes the underlying, but in practice inseparable grey and colour-banded oxic facies that overlies and passes laterally into the Caerau Mudstones. The formation dominantly comprises thinly interbedded, variably silt-laminated turbidite mudstones (mainly type E) and burrowed hemipelagites. Scattered thin units of anoxic-facies mudstones commonly occur in the lower part but are rare in the upper part. The formation mainly crops out in the Waun Marteg Syncline and around the nose of the Tywi Anticline.

West of the Cwmysgawen Fault, the Dolgau Mudstones form part of a conformable Llandovery succession inter-tonguing with the Glanyrafon Formation (Figure 12). In this area, the formation occurs as two tongues. In the Waun Marteg Syncline, the lower tongue is interleaved with the lower part of the Glanyrafon Formation, but in the east, the thin underlying division of Glanyrafon Formation fails and the lower tongue of Dolgau Mudstones overlies and, farther east, replaces the upper parts of the Caerau Mudstones (Figure 12). The upper tongue of the formation overlies the Glanyrafon Formation and broadly equates with the 'Pale Shales' of Roberts (1929) and Jones (1945).

East of the Cwmysgawen Fault the two tongues of Dolgau Mudstones merge, due to the eastward failure of the Glanyrafon Formation (Figure 35). In the same area, the formation onlaps older Llandovery strata to rest disconformably on the late Hirnantian (Ashgill) Cwmcringlyn Formation. South of the Cefn-pawl Fault and east of the Carmel Fault, the formation occupies a north–south belt, terminated in the south by the Garth Fault. In this region the formation conformably overlies the Henfryn Formation. Dolgau Mudstones also crop out to the west at Bryn [SO 050 655].

In the south of the Rhayader sheet area, on the eastern limb of the Tywi Anticline, Dolgau Mudstones disconformably overlie the Yr Allt Formation to the west of the Nant-y-fedw Fault [SN 980 595].

The conformable basal contacts of the Dolgau Mudstones are largely gradational. Where the formation overlies the Glanyrafon Formation, the contact is taken at the entrance of oxic-facies mudstones, in which turbidite sandstone/mudstone couplets (type Cii) are rare and the sandstone content is less than 10 per cent. Where it rests

on Caerau Mudstones, the boundary is taken at the entrance of very thin-bedded, predominantly oxic-facies mudstones. Contacts with the Henfryn Formation may be gradational, where the Dolgau Mudstones overlie turbidite sandstone/mudstone couplets (the boundary being taken where the sandstone content decreases to less than 10 per cent), or sharp, where the Dolgau Mudstones overlie debrites or disturbed beds.

In the Waun Marteg Syncline, the lower and upper tongues of Dolgau Mudstones have minimum thicknesses of about 150 m and 130 m, respectively. Both thicken eastwards as the Glanyrafon Formation thins. East of the Bwlch-y-sarnau Fault, the lower tongue is up to 400 m thick, an increase which appears to reflect lateral replacement, not only of the Glanyrafon Formation, but of upper parts of the Caerau Mudstones. In that area, the upper tongue expands to over 500 m thick. East of the Cwmysgawen Fault, the formation is up to 1.2 km thick. By contrast, east of the Carmel Fault only about 400 m rest on the Henfryn Formation. Further attenuation occurs south of the northern splay of the Garth Fault at Coedtrewernau [SO 0805 6760] (Figures 35 and 38), where the formation is fully exposed and 55 m thick. A rapid southwards expansion to about 150 m, east of Lower Coedglasson farm [SO 077 669], is probably due partly to thickness variations of the underlying Henfryn Formation, but principally to the presence of a basal slumped sequence, absent from Coedtrewernau. A minimum thickness of 100 m of Dolgau Mudstones is preserved west of the Nant-y-fedw Fault.

The very thinly interbedded, variably silt-laminated turbidite mudstones (type E) and burrowed hemipelagites (oxic facies) are closely comparable to those in the Claerwen Group. Diagenetic subfacies reflecting differing degrees of contemporaneous oxidation of the mudstones are present (see p.83). They are pale and medium grey banded, and pale grey-green in colour. It is commonly difficult to distinguish turbiditic from hemipelagic components. Thin units of anoxic facies mainly comprise very thin- to thick-bedded turbidite mudstones (type D) and laminated hemipelagites. Scattered turbidite sandstone/mudstone couplets (type Cii) are locally present but their basal sandstones never exceed 10 per cent in any section.

Lithological variations in the formation largely reflect proximity to coeval formations. Dark and pale grey colour-banded oxic facies, together with common thin units of anoxic facies, characterise much of the lower tongue and reflect the lateral and vertical passage from the Caerau Mudstones. Top-cut-out (T_{0-3}) type E turbidite mudstones with abundant, thick (to 5 mm), closely spaced (1 to 3 cm), siltstone laminae, typify parts of the formation in the west subjacent to the Glanyrafon Formation. Scattered sandstone/mudstone couplets (type C), with basal sandstones rarely over 5 cm thick, also occur in this region.

East of the Cwmysgawen Fault, pale and dark grey, colour-banded oxic facies, with scattered thin to very thin units of anoxic facies, again characterise the lower part of the sequence. However, east of the Carmel Fault units of anoxic facies are restricted to the basal few metres, a reflection of continued onlap. The lowermost part of the formation in this area commonly exhibits varying levels of bedding disturbance and slumping, though such effects are absent from the attenuated succession at Coedtrewernau. Scattered sandstone/mudstone couplets (type Cii), with basal sandstones up to 10 cm thick and rich in lithic granules, occur everywhere throughout the lower half of the formation.

The upper tongue in the west and the upper part of the formation, east of the Cwmysgawen Fault, mainly comprise weakly silt-laminated, unbanded, pale green mudstones. East of the Bwlch-y-sarnau Fault, Roberts (1929) separated these mudstones from the lower part of the formation as his 'Pale Shales'. However, the upward change in colour from grey to green is not everywhere maintained, so it cannot be used as a basis for sub-division.

Thin units of anoxic facies occur near the top of the Dolgau Mudstones, in a rapid vertical transition to the succeeding Nant-ysgollon Mudstones (Table 17).

Biostratigraphy

In the Waun Marteg Syncline [SO 0213 7347] (Figure 35, locality 6) early *griestoniensis* Biozone graptolites occur at the top of the lower tongue of Dolgau Mudstones. However, the presence of early *crispus* Biozone graptolites in the underlying Glanyrafon Formation (p.144) suggests that the lower tongue is partly of *crispus* Biozone age. The Glanyrafon Formation, which intervenes between the two tongues of Dolgau Mudstones, contains late *griestoniensis* Biozone graptolites and therefore provides a maximum age for the upper tongue. In the Waun Marteg Syncline the upper tongue has yielded only probable *spiralis* Biozone graptolites (Table 17), from 10 m below its top in Waun Marteg quarry [SO 0093 7706] (Figure 40, locality 1).

East of the Bwlch-y-sarnau Fault, on Cwm-hir Bank [SO 0197 7136], Caerau Mudstones just below the lower tongue of Dolgau Mudstones have yielded a latest *turriculatus* Biozone assemblage. This confirms that the basal Dolgau Mudstones are here laterally equivalent to upper parts of the Caerau Mudstones farther to the west. Graptolites from the middle and upper part of the lower tongue are of the middle to late *crispus* Biozone and early *griestoniensis* Biozone respectively. An assemblage from near the base of the upper tongue at Bwlch-y-sarnau [SO 0296 7455] (Figure 35, locality 2) is of the *crenulata* Biozone. Rapid westward thinning of this tongue suggests that the base may also young in that direction (Figure 12).

Faunas from the disconformable successions, east of the Cwmysgawen Fault, confirm the occurrence of onlap coincidental with the marked overstep of earlier Llandovery and Ashgill strata. Between the Cwmysgawen and Carmel faults, possible late *crispus* and early *griestoniensis* Biozone graptolites occur in the lowest 50 m of the Dolgau Mudstones; there is no evidence for the *turriculatus* Biozone (Figure 35). East of the Carmel Fault, the lower part of the Dolgau Mudstones contains early *griestoniensis* Biozone graptolites, as does the underlying Henfryn Formation (Figures 35 and 38). There is no

evidence for the *crispus* Biozone in this area. Sparse graptolite assemblages from the Dolgau Mudstones, south of the northern splay of the Garth Fault, comprise only long-ranging taxa.

The Dolgau Mudstones lying west of the Nant-y-fedw Fault, just beyond the southern edge of the district, have yielded graptolites indicative of the early *crispus* Biozone and possibly the late *turriculatus* Biozone.

Acritarch assemblages recovered from the Dolgau Mudstones include some of the most abundant and diverse recovered from the district (Appendix 2biii and iv) and indicate the mid to late Telychian *G. encantador* Biozone. Common taxa include *Ammonidium microcladum, A. listeri, Domasia elongata, D. trispinosa, Gracilisphaeridium encantador, Helosphaeridium echiniforme, Micrhystridium inflatum, Multiplicisphaeridium fisheri, Oppilatala eoplanktonica, O. ramusculosa, Salopidium granuliferum* and *Salopidium* sp. C of Priewalder (1987). Other species present include *Ammonidium* sp. A, *A.* sp. B, *Carminella maplewoodensis, Comasphaeridium* cf. *sequestratum, Cymatiosphaera octoplana, C.* sp. A, *Deunffia monospinosa, Dictyotidium stellatum, Domasia limaciformis, Elektoriskos williereae, Lophosphaeridium* sp. B, *Multiplicisphaeridium cladum, M. martiniae, M.* sp. A, *M.* sp. B, *Quadraditum fantasticum, Salopidium* sp. A and *Tylotopalla astrifera*.

Recycled Ordovician taxa are abundant and include *Coryphidium* cf. *tadlum, Cymatiogalea* sp., *Dicrodiacrodium* sp. and *Stelliferidium* sp. from the early Ordovician, and *Arkonia tenuata?, A. virgata, Excultibrachium* aff. *oligocladatum, Frankea* cf. *hamata, Goniosphaeridium splendens, Multiplicisphaeridium continuatum, M. irregulare, Orthosphaeridium bispinosum, Peteinosphaeridium nanofurcatum, P. trifurcatum breviradiatum, Pirea* sp., *Stellechinatum brachyscolum, S. helosum, Veryhachium hamii* and *V. longispinosum* from the mid to late Ordovician.

Details

West of the Cwmysgawen Fault

Lower tongue   Sections in the western limb and core of the Waun Marteg Syncline are exposed in a track cutting [SN 9946 7383] north-east of Baileyhaulwen, in a quarry [SN 9859 7041], and in the access track [SN 9900 7038] to Cwmithig. At these localities thick-bedded turbidite mudstones (type D) up to 15 cm thick, with laminated hemipelagite intervals, predominate. However, intercalated sequences of thinner-bedded, silt-laminated, pale grey and locally colour-banded mudstones (type E) are present in a stream [SN 9932 7065] north of Cwmithig, and in streams to the south [SN 9930 7020 to SN 9920 7029] and west [SN 9964 7014 to SN 9975 7023]. In the eastern limb of the Waun Marteg Syncline, a quarry [SO 0213 7347] (Figure 35, locality 6) in the forest exposes a sequence cut by several thrusts. This locality has yielded the graptolites *Monoclimacis* cf. *griestoniensis* sensu Elles and Wood, *Monograptus discus, M. rickardsi* and *M.* cf. *tullbergi spiraloides*, indicative of the lower *griestoniensis* Biozone.

East of the Bwlch-y-sarnau Fault the lower tongue of Dolgau Mudstones is extensively exposed in forestry track cuttings and quarries on Cwmysgawen Common [SO 035 718] and Cefn-crin [SO 025 727]. Silt-laminated, colour-banded and burrow-mottled, green-grey mudstones are exposed [at SO 0320 7160]. To the north, a track section [SO 0278 7208] (Figure 35, locality 7) with scattered turbidite sandstones, one 35 cm thick, has yielded the graptolites *Monograptus* cf. *marri*, *M. discus*, *M. pragensis pragensis* and *Streptograptus sartorius*, indicative of the *sartorius* Subzone. Some 700 m to the north-west, a stream [SO 0226 7255] exposes laminated hemipelagites from which *griestoniensis* Biozone graptolites have been recovered. Laminated hemipelagites within a locally slumped sequence are exposed in the quarry [SO 0418 7211] (Figure 35, locality 8) at Cwmysgawen farm. The section has yielded the graptolites *Monograptus crispus*, *M. discus*, *M. rickardsi* and *Streptograptus loydelli*, indicative of the *crispus* Subzone. Across the valley, a quarry [SO 0430 7290] (Figure 35, locality 3) north of Fishpool farm exposes silt-laminated turbidite mudstone (type E) and hemipelagites close to the contact with the overlying Glanyrafon Formation. The graptolites *Monoclimacis* cf. *griestoniensis* sensu Elles and Wood, *Monograptus discus*, *M. clintonensis*, *M. rickardsi*, *M. pragensis pragensis*, *M. tullbergi spiraloides?* and *Streptograptus* cf. *exiguus* have been collected and are indicative of the lower part of the *griestoniensis* Biozone. Comparable strata in a quarry [SO 0494 7236] (Figure 35, locality 9) on Cwmcynydd Bank have also yielded *griestoniensis* Biozone graptolites. To the north-east, a quarry [SO 0568 7292] close to the top of the lower tongue includes scattered sandstone/mudstone couplets (type C). The synsedimentary faults which cut this section have been illustrated by Smith (1987a, fig. 12c).

Upper tongue   Sections in the western limb of the Waun Marteg Syncline occur in forestry cuttings [SO 0080 7744 and SO 0087 7726] on Pistyll. To the south, Waun Marteg quarry [SO 0093 7706] (Figure 40, locality 1) exposes the junction of the formation with the Nant-ysgollon Mudstones. Graptolites recovered from the laminated hemipelagites of this section suggest the *spiralis* Biozone (Table 17). Disused quarries [SO 0160 7695] near Waun and 70 m to the north-east, display sections in pale, olive-green, thinly bedded, locally burrow-mottled, turbidite mudstones with common siltstone and fine-grained sandstone laminae (type E). Comparable lithologies are seen east of the Bwlch-y-sarnau Fault in quarries [about SO 0336 7617] at Brondre Fach and, at the base of the division, in

Bwlch-y-sarnau [SN 0296 7455] (Figure 35, locality 2). Thin laminated hemipelagites at the latter locality have yielded *Monoclimacis* cf. *vomerina vomerina*, *Monograptus discus*, *M.* cf. *tullbergi tullbergi*, *M. priodon* and *Pristiograptus nudus*, indicative of the *crenulata* Biozone. From this vicinity Roberts (1929, p.664) reported 'numerous bands, 1 or 2 inches thick, of maroon-coloured mudstones'. He further cited a graptolite fauna (p.664) from a now degraded section [about SO 0352 7408] along the road to Abbeycwmhir, suggestive of the *crenulata* Biozone. To the south, the contact with the underlying Glanyrafon Formation is revealed in a disused quarry [SO 0314 7408]. Farther east, the characteristic green turbidite mudstones with abundant closely spaced siltstone laminae (type E) are well seen in quarries [SO 0525 7381] at Esgair-fawr Farm. In contrast, a nearby quarry [SO 0508 7440] exposes a green turbidite mudstone sequence with few siltstone laminae or laminated hemipelagites. To the east, comparable lithologies are exposed in two quarries [SO 0586 7452, SO 0618 7447]. The former displays nodular phosphate layers and evidence of synsedimentary disturbance.

East of the Cwmysgawen Fault

Olive-green, weakly silt-laminated mudstones, close to the base of the Dolgau Mudstones and its disconformable contact with the Cwmcringlyn Formation, are exposed in a forestry track [SO 0478 7198] on Y Glog. To the north-east, laminated hemipelagites from the same track [SO 0488 7205, SO 0493 7212, SO 0498 7215] (Figure 35, locality 10), and a track [SO 0482 7214] to the south, have yielded long-ranging Telychian graptolites, which may possibly represent the upper levels of the *crispus* Biozone. An additional forestry cutting [SO 0541 7236 to SO 0560 7248] (Figure 35, locality 4) north of Little Park, lies 50 m above the base of the formation. There, pale to medium grey, colour-banded turbidite mudstones, with abundant siltstone laminae (type E) and *Chondrites* burrow mottling, predominate. Thicker structureless mudstone turbidites (type D) and rare sandstone/mudstone couplets (type

Table 17   Distribution of graptolites in the Dolgau Mudstones and Nant-ysgollon Mudstones in Waun Marteg quarry [SN 0093 7707].

For key to range chart symbols see Appendix 3.

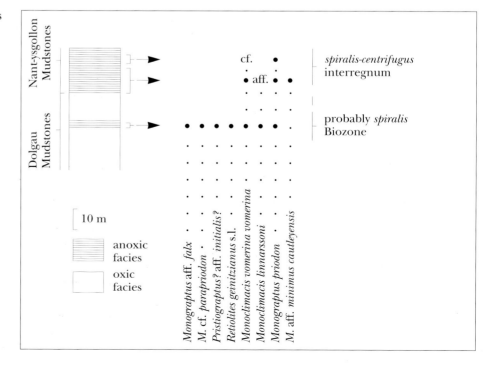

C) are also present; some of the latter are rich in dark grey mudstone granules. Scattered laminated hemipelagite intervals, up to 1 cm thick, have yielded graptolites indicative of the lower *griestoniensis* Biozone. To the east, a forestry section [SO 0668 7302] on Bank Stile exposes massive, grey, argillaceous sandstones that form a local basal division to the formation and are comparable with Henfryn Formation lithologies. Pale grey-green, banded and variably silt-laminated mudstones of the upper levels of the Dolgau Mudstones are exposed in quarries at Porth [SO 0690 7394], Bryn-rhyg [SO 0646 7428] and nearby [at SO 0663 7449]. A cephalon of the trilobite *Ananaspis* or *Acernaspis* was collected from the last locality. At Crychell, pale green mudstones with scattered sandstone/mudstone couplets (type C) and a 10 cm-thick sandstone with abundant black mudstone granules are exposed in a cutting [SO 0778 7440]. Across the valley, laminated hemipelagites exposed in a track cutting [SO 0818 7411] (Figure 35, locality 11) have yielded the graptolites *Monograptus* cf. *crenulata* sensu Elles and Wood, *Monograptus discus, M. tullbergi tullbergi* and *Retiolites geinitzianus*, indicative of the *crenulata* Biozone. The crags just to the east expose the upper part of the formation and the succeeding Nant-ysgollon Mudstones. Farther south, track cuttings [SO 0836 7215 to SO 0855 7204] provide a comparable transect. Grey, burrow-mottled mudstones containing thin coarse-grained pyritic sandstones rich in quartz and dark mudstone granules are exposed in Bachell Brook [SO 0832 7227].

Between the Carmel and Garth faults

Stream sections through the lower part of the Dolgau Mudstones in this area, and the contact with the underlying Henfryn Formation, are exposed in a tributary [SO 0814 7140 to SO 0835 7124] of Bachell Brook and in a tributary [SO 0775 7020 to SO 0800 7037] of Clywedog Brook near Keeper's Lodge. Pale grey and green, colour-banded, locally silt-laminated and burrow-mottled, turbidite mudstones dominate these sections. Widely scattered quartz and lithic granule-rich sandstones are also present in both sections. Laminated hemipelagites in the lower reaches of the Keeper's Lodge section (Figure 35, locality 12) have yielded graptolites of the *griestoniensis* Biozone (Smith, 1988). Nearby road cuttings [SO 0792 7061] at Fronrhydnewydd, and a quarry [SO 0835 7078] at Troed-rhiw-felin, also expose pale green and grey mudstones. A 50 cm-thick unit of anoxic facies at the latter locality has yielded the graptolites *Monograptus* aff. *priodon, M.* ex. gr. *tullbergi* and *Retiolites geinitzianus*, of upper Telychian age.

To the south, the lower part of the Dolgau Mudstones is well exposed in the stream and cuttings about Cwmtelmau [SO 0758 6790]. A well-preserved and diverse assemblage of graptolites collected from Pandy Brook [SO 0843 6847] (Figure 38, locality 6) includes *Monoclimacis* cf. *griestoniensis* sensu Elles and Wood, *Monograptus clintonensis, M. pragensis pragensis* and *Streptograptus sartorius*, indicative of the lower part of the *griestoniensis* Biozone. Sections within the western limb of the faulted syncline through Gwestre [SO 066 657] are provided by Camlo Brook, [SO 0612 6603 to SO 0630 6596]. A slumped sequence in which laminated hemipelagites are common succeeds the Henfryn Formation.

Sections in Clywedog Brook (Figure 38), between splays of the Garth Fault, form part of the eastern limb of the syncline through Gwestre. At Coedtrewernau [SO 0805 6760] (Figure 38, locality 9) a complete westward-dipping sequence of Dolgau Mudstones, overlying the Henfryn Formation, comprises about 55 m of burrow-mottled and banded, pale greenish grey and medium grey turbidite mudstones and hemipelagites with abundant cone-in-cone nodules. They are succeeded by

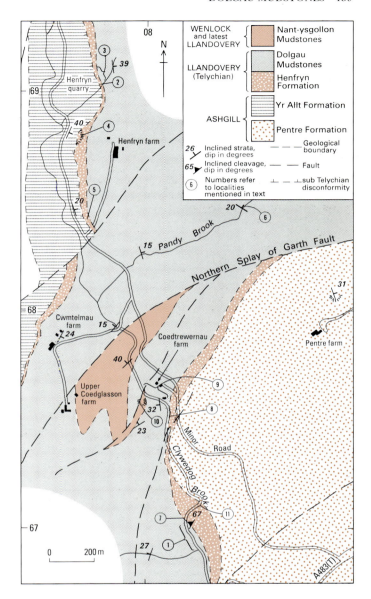

Figure 38   Geological map of the Clywedog brook area showing selected localities mentioned in the text.

laminated hemipelagites of the Nant-ysgollon Mudstones. Upstream [SO 0798 6765], a thrust repeats the sequence and the upper contact is seen again [SO 0795 6873]. The lower part of the formation is exposed in Clywedog Brook [SO 0818 6704 to SO 0822 6694] (Figure 38, locality 1), south of Brynllygoed. The section comprises a slumped sequence of medium grey turbidite mudstones (types D and E) interbedded with common laminated hemipelagites. The latter have yielded the long-ranging graptolite *M. priodon*. The formation is also exposed along the tributary [SO 0820 6689 to SO 0592 6692] of Camlo Brook.

The only section [SO 0504 6558] of Dolgau Mudstones on the most western crop of this area is exposed in a stream west of Bryn. There, westward-dipping pale grey turbidite mudstones with scattered calcareous nodules are interbedded with thin laminated hemipelagites.

Erosional and depositional model

FORM AND ORIGIN OF THE SUB-TELYCHIAN DISCONFORMITY

The presence of a disconformity beneath the Llandovery strata along the Tywi Lineament in the Abbeycwmhir area was first recognised by Roberts (1929). The present survey has shown that it also occurs in the south of the district, west of the Nant-y-fedw Fault.

A reconstruction of the sub-Telychian floor of the Tywi Lineament is shown in Figure 39 inset. Across much of the lineament, Telychian strata overstep in an easterly direction, resting first on earlier Llandovery strata and then progressively on older parts of the local Ashgill succession. The degree of disconformity changes abruptly across the major faults of the lineament. However, along the eastern flank of the lineament, between splays of the Garth Fault, the Telychian oversteps on to Ashgill strata (Pentre Formation) unlike any exposed to the west.

The nature of the overstep confirms an erosional origin for the disconformity (Roberts, 1929; George, 1963; Waters et al., 1992) and does not support the passive, non-depositional slope model of James (1983; 1991). However, in contrast to earlier erosional models, which invoked truncation of an uplifted fold crest (Roberts, 1929; George, 1963), the pattern of overstep clearly illustrates the controlling influence of contemporaneous faulting, with upthrow to the east (Figure 39).

PROCESS AND PRODUCTS OF EROSION

Early erosional models for the disconformity (Roberts, 1929; George, 1963) implied subaerial exposure of the region, but recent workers have argued that the disconformity is of submarine origin, invoking non-deposition (Ziegler, 1970; James, 1983, 1991) or seismically triggered slumping as possible causative processes (Smith and Long, 1969). The principal evidence for a submarine origin for the disconformity is that it is overlain by deep-water resedimented facies of both proximal (Henfryn Formation) and distal (Dolgau Mudstones) aspect, and that it was formed when the shelf to the east was submerged (Chapter 8) (Figures 34 and 39).

The disturbed beds and debrites of the Henfryn Formation, with their slumped masses, large rafts and boulders of Ashgill and pre-disconformity Llandovery lithologies, are the product of submarine mass wasting of steep gradients, formed in response to, and destabilised by, movements on the major faults of the lineament (cf. Smith and Long, 1969). The pattern of overstep demonstrates that the faults intersected the sea floor, where they probably formed a succession of west-facing fault scarps (Figure 34). However, the effects of scour by thermohaline bottom currents, particularly in the oversteepening of gradients, cannot be overlooked as a possible additional factor (Von Rad et al., 1982 and references therein). The formation of the disconformity coincides with basinwide changes in depositional and, by inference, circulatory patterns resulting from the initial Llandovery docking of the Avalonian plate with Laurentia (Soper and Woodcock 1990; Waters et al., 1992).

The exotic clasts in the Henfryn Formation are probably the result of the reworking of earlier conglomerates (such as the Camlo Hill, Dol-y-fan and Allt-y-clych conglomerates), exposed by the mass wasting of fault scarps. Thus the vertical distribution of lithologies and clast types in the Henfryn Formation may reflect, in reverse, the eroded stratigraphy. Slope failures in such coarse-grade and partially sorted deposits would evolve into high-velocity, non-cohesive, high-concentration turbidity currents, capable of substantial channelling and tractional reworking.

The Henfryn Formation is now only preserved between the Carmel and Garth faults. Furthermore, its deposition during the early *griestoniensis* Biozone postdates the onset, in the west, of Dolgau Mudstone accumulation above the disconformity. It is likely, therefore, that these proximal resedimented deposits were confined within the eastern (presumed upper) parts of the faulted Telychian slope complex, against an active fault scarp or within an intra-slope graben (cf. James, 1991). Local thickness variations may reflect the primary depositional relief of coalescing lobes and cones of resedimented material adjacent to fault scarps. However, in such an unstable setting, remobilisation of previously deposited parts of the formation may also account for local attenuation.

THE LOST STRATIGRAPHY OF THE TYWI LINEAMENT

The Henfryn Formation represents the products of the final phase of Telychian slope degradation. By contrast, the earlier, more westerly Cwm Barn Formation, west of the Cwmysgawen Fault, is a conformable element in the complete basinal succession and was formed during the main erosional episode to affect the Tywi Lineament. It therefore provides information on the eroded sequence or 'lost stratigraphy', of the region. The proximal resedimented deposits and abundance of locally derived lithic clasts suggest that they are the products of mass wasting, now preserved in the downfaulted parts of the basin.

Figure 39 The geology of the Tywi Lineament (facing page).

Stratigraphical relationships of Ordovician and Llandovery rocks across the Tywi Lineament; vertical scale is time. Sea-level curve is derived from Cocks and Fortey (1982) and Johnson et al. (1985; 1991). Inset map shows the sub-Telychian floor of the Tywi Lineament. Abbreviations of lithostratigraphic units: Ab/Ac/B, Ca and Cb — Divisions of Andrew (1925); AcC — Allt-y-clych Conglomerate Member; AL — *Acidaspis* Limestone; BMd — Builth Mudstones; BVF — Builth Volcanic Formation; CaM — Caerau Mudstones; Cbn — Caban Conglomerate Formation; CeF — Cwmere Formation; Cer — Cerig Formation; CgF — Cwmcringlyn Formation; CC — Cwm Clyd Formation; Cte — Camnant Mudstones; Cwn — Cwm Barn Formation; DCM — Dol-y-fan Conglomerate; Dgu — Dolgau Mudstones; Hen — Henfryn Formation; LrM — Llanfawr Mudstones; MMM — Mottled Mudstone Member; Msh — *M. sedgwicki* shales; NtM — Nantmel Mudstones; NyG — Nant-ysgollon Mudstones; PdG — Penstrowed Grits; PtrF — Pentre Formation; Rhs — Rhayader Mudstones; TB — Trecoed Beds; Tyc — Tycwtta Mudstones; WF — Wenallt Formation; YA — Yr Allt Formation. * — Ashgill facies of shelf aspect within Tywi Lineament.

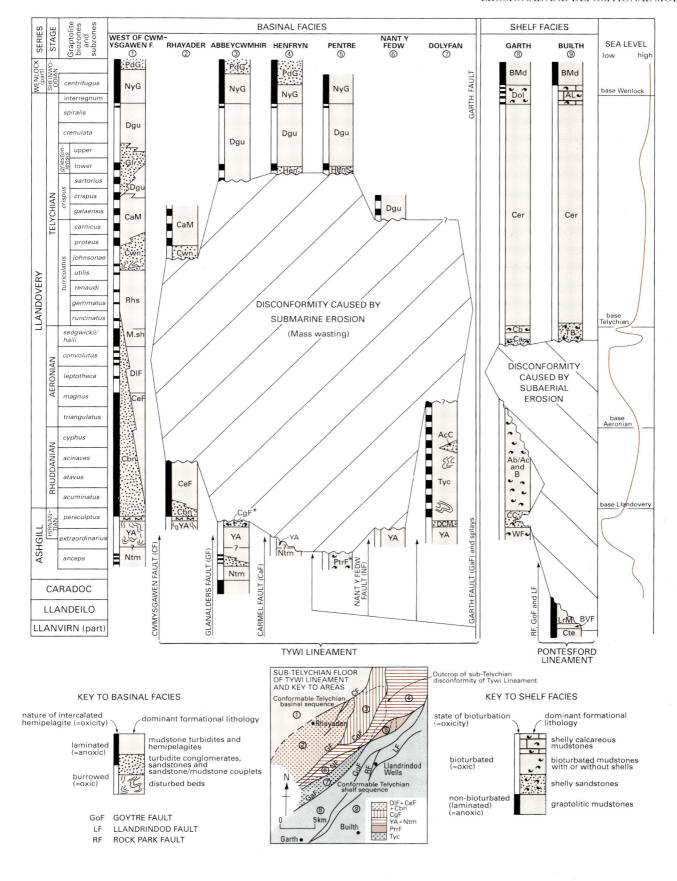

Their position against the Cwmysgawen Fault may reflect an original depositional setting in a hanging-wall trap which facilitated preservation and containment (see Figure 12).

The relationships at Lan Goch, where the Cwm Barn Formation rests disconformably on Dyffryn Flags, and the presence of clasts of graptolitic Cwmere Formation and burrow-mottled Derwenlas Formation in debrites at Cwm Barn quarry, confirm the erosion of proximal parts of Rhuddanian, Aeronian and, by implication, early Telychian slope-apron deposits formerly present to the east of the Cwmysgawen Fault (Figure 34). Less readily explained are the reworked shelly benthos, phosphatised mudstone pebbles and exotic lithic clasts in the conglomeratic turbidites. As the Cwm Barn Formation was deposited during *utilis* to possibly as late as *proteus* Subzone times, it postdates the basal Telychian transgression of the eastern shelf, where poorly fossiliferous Cerig Formation mudstones were accumulating (p.171). It is therefore likely that the shelly fossils and other clasts were supplied not from contemporaneous shelf environments, but from a reworking of pre-existing shelf deposits situated over the Tywi Lineament.

Such deposits may have accumulated during the Aeronian (*convolutus* Biozone) regression, when the contemporary shoreline withdrew to the vicinity of the Garth Fault (Ziegler et al., 1968). Shelf facies, their distribution limited by steep gradients, may then have covered part of the Tywi Lineament. Transgressive reworking of these deposits during the late Aeronian (*sedgwickii* Biozone), to form high-energy shoreface facies comparable to those preserved in eastern shelf sequences, provides a likely origin for the mixed shallow and deep-water shelly assemblages. Moreover, this transgression is likely to have established conditions favourable for phosphatisation (Jenkins, 1986). This speculation is supported by the presence of similar shelly and phosphatic clasts in the contemporaneous Hafod Member of the Ystrad Meurig Grits Formation (p.114). Subsequently buried by upper slope-apron mudstones in the late Aeronian and early Telychian, this envisaged transgressive facies was then exposed and reworked during the ensuing mass-wasting episode to supply the Cwm Barn Formation. The latter is thought to be one of a series of debrite and coarse-grained turbidite cones, formerly developed along the western flank of the Tywi Lineament, but now eroded away (Figure 34).

THE TELYCHIAN EVOLUTION OF THE TYWI LINEAMENT

The presence of Rhuddanian strata beneath the disconformity at Lan Goch confirms that it records an intra-Llandovery rather than an Ashgill event (Roberts, 1929). However, the earliest evidence for the onset of erosion, and hence of fault-controlled uplift across the lineament, is in the mid-Telychian (*utilis* Subzone). Initiation of the erosional episode was therefore broadly coincident with the replacement in conformable basinal sequences of the earlier slope-apron system (Rhayader Mudstones) by southerly supplied mudstone turbidites (Caerau Mudstones). By implication, the synerosional faults of the lineament became active at the same time as fractures such as the precursor Claerwen and Bronnant faults, which controlled the geometry of the southerly supplied sandstone-lobe systems to the west (Figure 34).

Overstep and onlap by the Caerau and Dolgau mudstones record the diachronous eastward cessation of erosion and, by implication, of large-scale movement on the influential faults of the region (Figure 34). The proximal resedimented facies of the Henfryn Formation, and slumped levels in the overlying Dolgau Mudstones, observed along the eastern flank of the lineament, reflect continued instability during the subsequent depositional phase, but fault activity appears largely to have ceased during the early *griestoniensis* Biozone.

The Dolgau Mudstones which blanket the disconformity (or intervening resedimented deposits) are closely comparable to the earlier Claerwen Group, and similarly interpreted as a mud-dominated slope-apron facies. During the Telychian, southerly supplied turbidites (Caerau Mudstones and Glanyrafon Formation) interdigitated with the developing slope-apron system. This reflects the continuing influence of a west-facing Tywi Lineament slope on southerly fed turbidity flows. The contraction of these southerly derived systems and concomitant westward progradation of the Dolgau Mudstones, during the *crenulata* Biozone, signalled the final abandonment in the Llandovery of the lineament as a tectonic and topographic feature.

The Tywi Lineament, therefore, despite its profound effects on the sedimentary sequence and its subsequent Acadian tectonic expression, was active only briefly during the Llandovery. This active episode was broadly coincident with the introduction, growth and migration of the southerly derived, Telychian turbidite systems to the west and was, by inference, a response to the same extrabasinal tectonic events.

The erosional episode records the abandonment and destruction of proximal portions of the preceding slope-apron system, and the transformation of the lineament into a structurally induced slope separating Telychian basinal facies, to the west, from shelf facies to the south-east. The preserved deposits of the region form part of a vertical succession that records the subsidence of this confining slope, the coincident eastward migration of the basin margin, and the resumption of slope-apron deposition.

SEVEN

Silurian: Latest Telychian to Wenlock sandstone-lobe and slope-apron systems

Profound changes in the depositional setting of the basin margin occurred in the latest Telychian to Wenlock. Deposition of slope-apron facies continued above the late Telychian Dolgau Mudstones into the earliest Wenlock (Nant-ysgollon Mudstones) but was rapidly abandoned as a sandstone-lobe system (Penstrowed Grits) invaded from the south. The subsequent demise of this sandstone-lobe in the late Wenlock allowed slope-apron facies (Nantglyn Mudstones) to be re-established. Progradation of this system, coupled with a decline (or shift) in contemporaneous fault activity, brought the eastern part of the basin within the influence of shelf depositional processes during an interval of falling sea level (Mottled Mudstone Member). The formations that record these events are confined to the north-eastern part of the district (Figure 40a).

Nant-ysgollon Mudstones

The formation comprises variable proportions and types of turbidite mudstones and hemipelagites. It derives its name from the 'Nant-ysgollon Shales' of the Trannon area in the Llanidloes district (Wood, 1906) and equates with the lower part of the 'Wenlock Series' as used by Roberts (1929) and the informal 'Pegwn Mudstones' of Jones (1947). It mainly crops out in the north-eastern corner of the Rhayader sheet area (Figure 40a), where it overlies the Dolgau Mudstones with a sharp contact and is succeeded by the Penstrowed Grits. The formation thins eastwards, from about 125 m in the vicinity of Waun Marteg [SO 012 772], to less than 30 m along the eastern edge of the sheet. This is largely a result of an easterly facies change into the Penstrowed Grits (Figure 12). South of the main crop, over 80 m of Nant-ysgollon Mudstones crop out in a small, faulted outlier at Coed-glasson [SO 077 676], between splays of the Garth Fault. Although the top is not seen there, the formation is considerably thicker than the nearest sequence north of the fault.

The dominant lithology of the formation comprises dark grey, brown-weathering, laminated hemipelagites, thinly interbedded with subordinate (less than 30 per cent) mudstone turbidites (type Dii), generally less than 1 cm thick. Although the formation rests with a sharp contact on the Dolgau Mudstones, subordinate paler, burrow-mottled, colour-banded and silt-laminated mudstone turbidites (type E), lithologically similar to those in the underlying Dolgau Mudstones, occur at intervals throughout, as feature-forming units up to 2 m thick. Abundant thick mudstone turbidites (type D) appear in the upper half of the thick western sequence of the formation. They occur interbedded with scattered, medium to thick beds of high-matrix siltstone and sandstone (type Bii turbidites), which increase in abundance both upwards and eastwards, thus providing a vertical and lateral transition into the Penstrowed Grits. Thin slump units are locally present within the formation.

BIOSTRATIGRAPHY

In the west, the lower part of the formation is characterised by the low-diversity graptolite assemblages of the *spiralis/centrifugus* interregnum, as in Waun Marteg quarry [SO 0093 7706] (Table 17 and Figure 40a, locality 1) and a track section [SO 0420 7491 to SO 0424 7489] east of Bwlch-y-sarnau (Figure 40a, locality 4). Faunas from the basal Wenlock *centrifugus* Biozone were recovered about 40 m above the base of the formation in the latter track section. Although in the west the upper part of the formation has yielded possible upper *riccartonensis* Biozone faunas from a forestry quarry [SO 0188 7760] (Figure 40a, locality 15), in the east this biozone appears to be confined to the Penstrowed Grits. The *centrifugus* Biozone has not been recorded from the eastern sequence; it may be present in the Penstrowed Grits.

DETAILS

The base of the formation is exposed in Waun Marteg quarry [SO 0093 7607] (Figure 40a, locality 1). There, 20 m of dark grey, laminated hemipelagites, thinly interbedded with turbidite mudstones less than 1 cm thick, sharply overlie colour-banded and burrowed Dolgau Mudstones. They have yielded graptolites of the *spiralis/centrifugus* interregnum (Table 17).

A slightly higher level is exposed 50 m to the east, in a quarry [SO 0097 7703] where 5 m of similar mudstones are overlain by 1 m of medium grey, burrow-mottled, blocky mudstone.

A forestry quarry [SO 0134 7712] (Figure 40a, locality 2) north of Waun Marteg, in the upper part of the thick western sequence, exposes 10 m of thin- to thick-bedded mudstone turbidites (type D) and several type Bii turbidites interbedded with a few thin laminated hemipelagites. The type Bii turbidites are either thick (up to 65 cm) massive argillaceous siltstones, some with thin basal sandstones, or thin (up to 25 cm) coarse-grained, high-matrix sandstones. The quarry has yielded *Monoclimacis* cf. *flumendosae* sensu lato and *Pristiograptus* cf. *meneghinii meneghinii*, which suggest a level within the upper *riccartonensis* to *flexilis* biozones. Another forestry quarry [SO 0188 7760] (Figure 40a, locality 15) at a similar level, 700 m to the north-east, has yielded poorly preserved specimens of *Monograptus firmus sedberghensis?*, possibly indicative of the upper *riccartonensis* Biozone.

A track section [SO 0420 7491 to SO 0424 7489] (Figure 40a, locality 4), east of Bwlch-y-sarnau, partly exposes the lower 60 m of the formation, which here is about 108 m thick. It comprises thinly interbedded laminated hemipelagites and subordinate very thin mudstone turbidites (type Dii). A few beds of massive, medium grey, burrow-mottled mudstones (type E turbidites and burrowed hemipelagites) up to 2 m thick, locally

Figure 40 Geology of the Wenlock succession.

a Geological map of the basinal Wenlock succession in the district and adjacent areas showing mean palaeocurrent vectors within the Penstrowed Grits including data from Dimberline (1987). **b** Evolution of the Wenlock basinal and shelf systems in the district. Cross-sections are derived from Figure 13. BF — Bwlch-y-sarnau Fault; for key to other abbreviations see Figure 39.

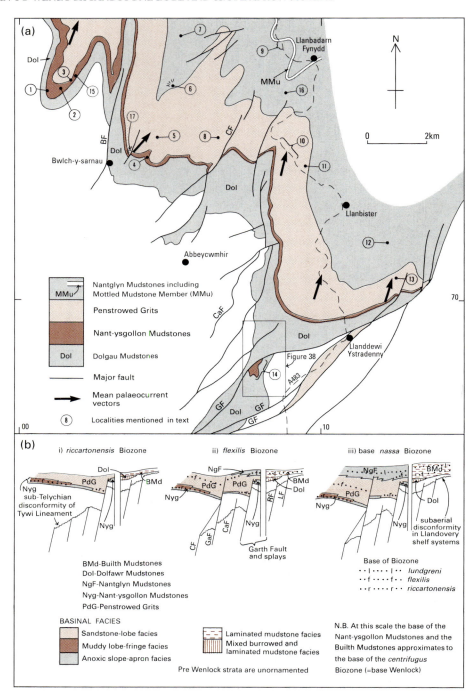

form prominent features. The long-ranging taxa *Monoclimacis vomerina vomerina*, *Monograptus* cf. *priodon*, *M.* aff. *minimus cautleyensis* and *Retiolites geinitzianus* cf. *angustidens*, from 20 m above the base of the formation, suggest an age within the *spiralis* to *centrifugus* biozonal interval. The *centrifugus* Biozone is indicated about 40 m above the base by the presence of *Cyrtograptus centrifugus?*, which occurs together with *Monoclimacis vomerina vomerina?* and cf. *Retiolites geinitzianus* sensu lato. A similar fauna, but including *Monograptus parapriodon?*, was found 43 m above the base. A small quarry [SO 0425 7485] on a continuation of the track exposes the top 14 m of the formation. Massive sandy mudstones (type Bii turbidites), up to 80 cm thick, occur interbedded with mudstone turbidites (type D) and laminated hemipelagites. Two thin high-matrix sandstones (type Bii) occur in the top 2 m.

A quarry [SO 0864 7147] at Llanerch Fraith in the middle part of the thin eastern sequence exposes 10 m of thinly interbedded laminated hemipelagites and very thin turbidite mudstones (type Dii). Several very thin slump units are present. The mudstones have yielded *Monograptus* aff. *priodon*, (perhaps an early *M. priodon* to *M. flemingii* transient), *Monoclimacis vomerina vomerina* and cf. *Retiolites angustidens angustidens*, suggesting an age within the *crenulata* to *centrifugus* biozonal range. The base of the formation is exposed in a track section [SO 0854 7118] 300 m to the south.

The Coedglasson outlier (Figure 38), situated to the south of the northern splay of the Garth Fault, is well exposed in Clywedog Brook [SO 0793 6775 to SO 0789 6790] and its western tributary [SO 0778 6765 to SO 0794 6773]. It comprises laminated hemipelagites and subordinate, very thin turbidite mudstones (type Dii). The contact with pale and medium grey, colour-banded and burrow-mottled mudstones of the Dolgau Mudstones is seen at the southern end of the section and also 120 m farther downstream (south) [SO 0799 6763], where the basal part of the formation is repeated by a

thrust. At the downstream locality, the contact is gradational over a metre, but to the north it is disturbed by a bedding plane slide. A graptolite fauna from the bank [SO 0797 6784], about 10 m above the base of the northern section, comprising *Monograptus* cf. *priodon*, *M.* aff. *cultellus?* and *Monoclimacis vomerina gracilis?*, lies within the *crenulata* to *centrifugus* biozonal range. A quarry [SO 0798 6759] (Figure 38, locality 10) at a similar stratigraphic level, adjacent to the brook and below the thrust, has yielded a fauna of the same age range but includes also *Monoclimacis vomerina vomerina* and *Retiolites geinitzianus angustidens* from 2 to 4 m above the base of the quarry. The presence of *Mcl.* cf. *vomerina basilica*, from a level 6 to 8 m higher in the quarry, may indicate a younger age within the *centrifugus* to *riccartonensis* biozonal interval.

Penstrowed Grits

This sandstone-dominated formation takes its name from Penstrowed quarry, [SO 068 910], near Newtown in the adjacent Montgomery district (Institute of Geological Sciences, 1972). Its regional synonyms include the Fynyddog Grits of Wood (1906), the 'Grit Member' of the Denbighshire Grit and Shale Group of Bassett (1955) and the Castle Vale Formation of Dimberline and Woodcock (1987). Its outcrop within the district is restricted to the north-eastern corner of the Rhayader sheet area (Figure 40a), where it thins from a maximum of over 1100 m west of the Cwmysgawen Fault, to less than 750 m along the eastern margin of the sheet. It disappears 5 km east of the district against a zone of faulting (Geological Survey of Great Britain, 1850), contiguous to the Garth Fault. A detailed sedimentological account of the formation is provided by Dimberline (1987).

The formation comprises thin, sandstone- or siltstone-based Bouma turbidites (types C and Di) and laminated hemipelagites interbedded with subordinate, commonly structureless, medium- to thick-bedded high-matrix turbidite sandstones (type Bii) (Figure 41, Plates 18b and c). Slumped levels and thin bentonitic clays constitute minor components of the formation. The high-matrix sandstones (type Bii), which are mainly grouped in feature-forming packets, range up to 1.5 m thick and embrace a spectrum of mud-rich to mud-depleted varieties. They are typically graded with coarse sand or, rarely, granule-grade bases and fine-grained tops. Mainly feldspathic, they differ from the otherwise comparable high-matrix sandstones of the Aberystwyth Grits and Cwmystwyth Grits groups in containing abundant large mica flakes (up to 4 mm across), leached bioclastic debris and degraded carbonaceous remains. Dimberline and Woodcock (1987) reported fragments of the early terrestrial plants *Prototaxites* and *Pachytheca*. Sandstone-based turbidites (type C), ranging up to 75 cm thick, exhibit Bouma T_{a-e}, T_{b-e} or T_{c-e} sequences in which the capping dark grey mudstone may be either thicker or thinner than the underlying medium- to fine-grained sandstone. The basal siltstones of the mudstone turbidites (type Di), mainly less than 5 cm thick and displaying fading-ripple structures and parallel-lamination, are generally thinner than the overlying turbidite mudstone component which can be up to 50 cm thick. Siltstone-free mudstone turbidites (type Dii) range up to 50 cm in thickness. Although laminated hemipelagites up to 8 cm thick separate most turbidites (Plate 18c), amalgamation of adjacent turbidite beds (mainly Bouma-type) is locally seen.

Slump units, notably in the Castle Vale road cutting [SO 0839 7576 to SO 0926 7466], display westward-facing folds (Dimberline, 1987). Rare, thin (less than 10 cm), pale grey and white mudstones, composed of illite-smectite mixed-layer clays, are interpreted as bentonites representing diagenetically altered volcanic ash layers (Dimberline, 1987).

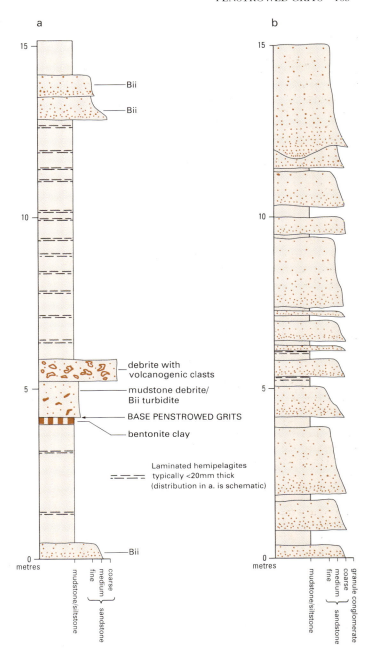

Figure 41   Graphic logs of the Penstrowed Grits.
a   Quarry [SO 0166 7741] exposing the distinctive debrite used to define the local base of the Penstrowed Grits (Figure 40, locality 3); type D turbidites unless otherwise indicated;
b   Quarry [SO 0448 7553] (Figure 40, locality 5); sandstones are all type Bii sandstones.

Palaeocurrent vectors derived from flutes and groove-casts, in accord with Dimberline (1987), are principally towards the north-north-east (Figure 40a).

BIOSTRATIGRAPHY

Within the district, the base of the formation is diachronous, younging westwards. By inference from evidence in the Nant-ysgollen Mudstones, the base of the Penstrowed Grits is assumed to be of *centrifugus* Biozone age in the east (see p.164), whilst in the west the lower parts of the formation have yielded upper *riccartonensis* Biozone faunas [SO 0364 7503 and SO 0166 7741] (Figure 40a, localities 17 and 3). The suggested pattern of westward onlap is in the opposite direction to that described by Dimberline (1987) and Dimberline and Woodcock (1987).

Graptolites of the *flexilis* Biozone have been recovered from the middle [SO 0657 7554] (Figure 40a, locality 8) of the thick western sequence, whilst the top in this area, constrained only by *lundgreni* Biozone assemblages in the succeeding Nantglyn Mudstones, may range into the intervening *ellesae* Biozone. The top of the formation is also diachronous, becoming younger westwards. Dimberline and Woodcock (1987) reported faunas referable either to the late *riccartonensis* or to the *rigidus* Biozone from a locality close to the top of the formation at Llananno [SO 095 745] (Figure 40a, locality 11), immediately east of the district and recorded early *riccartonensis* Biozone graptolites from the overlying Nantglyn Mudstones 4 km farther east. Farther east still, they reported *riccartonensis* Biozone graptolites from near the top of the Penstrowed Grits [SO 125 708] (Figure 40a, locality 13).

DETAILS

East of the Cwmysgawen Fault   The A483 road cutting at Castle Vale [SO 0839 7576 to SO 0926 7466] (Figure 40a, locality 10) exposes the complete spectrum of lithologies present within the formation, including slumped levels and rare bentonites. A graphic log was provided by Dimberline (1987), who reported graptolites referable to the upper *riccartonensis* or *rigidus* biozones.

West of the Cwmysgawen Fault   Lower parts of the thick western sequence of Penstrowed Grits, close to the Cwmysgawen Fault, are extensively exposed in crags and forestry cuttings on the southern flanks [SO 060 750] of Ddyle. To the north, a quarry [SO 0657 7554] (Figure 40a, locality 8) situated between packets of thick sandstones has yielded *Cyrtograptus rigidus cautleyensis*, *C.* cf. *linnarssoni*, *Monograptus flexilis flexilis* and *Monoclimacis flumendosae flumendosae*, an assemblage of *flexilis* Biozone age. The upper part of the formation is well seen about New Well, notably along the stream [SO 0674 7704 to SO 0685 7677] south of Castle Bank, and in trackside exposures [SO 0645 7814 to SO 0604 7801] to the north-west.

Farther west, the base and lower 50 m of the formation are seen in the roadside quarry [SO 0364 7503] north-east of Bwlch-y-sarnau (the Llwybry Gath quarries of Roberts, 1929) (Figure 40a, locality 17, Plate 18b and c). The high-matrix sandstones (type Bii) are rich in carbonaceous and bioclastic debris, large mica flakes and silver-grey phyllite clasts. Upper *riccartonensis* Biozone graptolites have been obtained some 33 m above the basal Bii sandstone turbidite (Dimberline and Woodcock, 1987). About a kilometre to the north-west, a quarry [SO 0448 7553] (Figure 40a, locality 5) exposes a section in the middle part of the formation (Figure 41b).

West of Brondre Fawr, the base of the formation has been taken, for mapping convenience, at the base of a distinctive debrite bed, 1.5 m thick, which contains abundant, dispersed, irregular clasts of chloritised (? altered volcanogenic) material. It is exposed in a small quarry [SO 0279 7894] and crags [SO 0299 7810], west and south of Cwmderw respectively, and a forestry cutting [SO 0166 7741] (Figure 40a, locality 3; Figure 41a) in the Waun Marteg Syncline. The latter section has yielded upper *riccartonensis* Biozone graptolites, including *Monograptus riccartonensis*, *M. radotinensis inclinatus*, *M. ?flexilis belophorus*, *Monoclimacis flumendosae* sensu lato, and *Pristiograptus* sp. A white-weathered bentonitic clay, about 20 cm thick, occurs less than 2 m below the base of the debrite at all these localities. High-matrix sandstones, commonly gradational into sandy mudstones, which persist below the debrite, are arbitrarily included in the Nant-ysgollon Mudstones.

Nantglyn Mudstones

First termed the Nantglyn Flags by McKenny Hughes (1879) in the Vale of Clwyd, North Wales, these mudstones were given group status by Warren et al. (1984) in the Denbigh district. In more recent work in the Corwen district (British Geological Survey, 1993) and the Montgomery district (British Geological Survey, 1994), they have been reclassified as a formation, and are here renamed the Nantglyn Mudstones. In the present district they are synonymous with the Llanbadarn Formation of Dimberline and Woodcock (1987).

Only the lower third of the formation crops out in the district, where it mainly comprises turbidite mudstones with very subordinate very thin turbidite siltstones and sandstones (type D), thinly interbedded with laminated hemipelagites (Plate 18d). Thicker turbidite sandstones (type C) occur in the lower part of the sequence, whilst scattered slump units occur throughout. A distinctive unit of burrowed mudstones, the Mottled Mudstone Member (British Geological Survey, 1994), is the highest stratum in the district. In the western part of the adjacent Montgomery district this member comprises two tongues, and it is the lower of these that crops out in the present district. It correlates with the Lower Mottled Mudstone (Warren et al., 1984) of the Denbigh district.

The formation only crops out in the north-east corner of the Rhayader sheet area, where it is about 540 m thick and poorly exposed. It overlies the Penstrowed Grits with a sharp lithological contact, which is taken at the top of the uppermost thick high-matrix sandstone (type Bii).

Thinly interbedded mudstone turbidites (type D) and laminated hemipelagites, which form the main lithology beneath the Mottled Mudstone Member, have been referred to in North Wales as 'ribbon-banded mudstone' (Boswell, 1926; Warren et al., 1984). Individual turbidite mudstones are up to 8 cm thick, but on average are 2 to 3 cm thick. They generally comprise more than half the lithology, with type Dii turbidite mudstones predominating. The thin basal siltstones or fine-grained sandstones of the type Di mudstone turbidites range up to 1 cm thick, but are mainly less than 5 mm thick. The laminated hemipelagites are on average up to 2 to 3 cm thick but can range up to 6 cm. They are very silty and finely micaceous and, as such, are coarser grained than the turbidite mudstones. It is commonly difficult to distinguish them from the laminated basal siltstones of type Di mudstone turbidites.

The lower third of the Nantglyn Mudstones exposed in the district includes packets of turbidite sandstone/

mudstone couplets (mainly type Cii with subordinate Ci) interbedded with mudstone turbidites (type D) and laminated hemipelagites, the latter increasing in abundance upwards. The basal sandstones of type Ci and Cii couplets are up to 25 and 5 cm thick respectively, whilst the overlying turbidite mudstones are up to 20 cm thick.

Underlying the Mottled Mudstone Member is a 40 m-thick sequence of thinly interbedded mudstone turbidites (type D) and hemipelagites, with abundant fine-grained sandstones up to 8 cm thick. These sandstones exhibit alternating parallel-laminated and cross-laminated, starved-ripple divisions, as well as cut-outs, both at the base and within individual beds. Some contain basal lags of bioclastic debris. The sedimentary structures in these are not consistent with Bouma turbidite deposition but rather with a pulsating or intermittent flow, possibly storm generated (Dimberline, 1987; Tyler and Woodcock, 1987).

The Mottled Mudstone Member rests with a sharp contact on this sequence and comprises thin- to medium-bedded, pale and medium grey-green, colour-banded and burrow-mottled mudstones. Thin ochreous tubules up to several millimetres in diameter are common, but their origin is uncertain. Streaky siltstone laminae and more persistent, very thin parallel-laminated sandstones, up to 1 cm thick, form about 10 per cent of the member. Brachiopods, trilobites and bivalves occur in both the sandstones and mudstones. Scattered carbonate concretions are locally present. The presence of burrows and an indigenous shelly benthos suggests that bottom conditions were oxygenated. The Member can be compared with the shelf facies in the district (Cerig Formation).

Slumped strata occur throughout the formation as scattered units from 30 cm up to tens of metres thick (Plate 18d). A 60 m-thick unit, largely composed of slumped strata, has been mapped at the base of the formation near David's Well [SO 050 773] (Plate 7a).

Palaeocurrent vectors from rare sole structures (Dimberline, 1987) consistently demonstrate transport towards the north.

BIOSTRATIGRAPHY

In the district, the Nantglyn Mudstones beneath the Mottled Mudstone Member range in age from the *flexilis* Biozone, or possibly younger, through to the *lundgreni* Biozone. Faunas of the *flexilis* Biozone occur in the middle of the underlying Penstrowed Grits (see p.166), and probable *lundgreni* Biozone graptolites have been collected from the upper third of the sequence beneath the Mottled Mudstone Member, from localities including Bryndu farm [SO 0881 7705] and the A483 road cutting [SO 0868 7887] 8.5 m below the Mottled Mudstone Member (Figure 40a, locality 9). However, the base of the Nantglyn Mudstones is probably earlier at the eastern margin of the district, for Dimberline and Woodcock (1987) have reported graptolites referable either to the late *riccartonensis* or *rigidus* biozones from near the top of the Penstrowed Grits at Llananno [SO 095 745] (Figure 40a, locality 11) and of the early *riccartonensis* Biozone from Nantglyn Mudstones [at SO 129 713] (Figure 40a, locality 12). Graptolites of the *nassa* Biozone have been

collected from the uppermost part of the lower tongue of the Mottled Mudstone Member just to the north of the district in the A483 road cutting [SO 0861 7893]. The Mottled Mudstone Member also contains a shelly fauna of brachiopods, trilobites and orthocones (see below).

DETAILS

A roadside quarry [SO 0500 7730] (Figure 40a, locality 6; Plate 7a) exposes the top of the predominantly slumped unit that occupies the lower part of the formation south-west of St David's Well. A 9 m-thick unit of slump-folded, thinly interbedded mudstone turbidites (type D) and laminated hemipelagites occurs within undisturbed strata. The basal contact of this slump unit is sharp. The unslumped strata are typical of the basal part of the formation and comprise interbedded turbidite sandstone/mudstone couplets (type C) with subordinate turbidite mudstones (type D). The basal sandstones of the type Ci couplets are up to 25 cm thick.

Typical mudstone turbidites (mainly type Dii with subordinate type Di), thinly interbedded with laminated hemipelagites, are well exposed around Bryndu Farm [SO 0881 7705] (Figure 40a, locality 16). The basal sandstones of the type Di turbidites are up to 2 cm thick but generally are less than 5 mm. A possible *lundgreni* Biozone fauna has been collected from the mudstones and includes *Monograptus* cf. *flemingii compactus* and *M.* cf. *flemingii flemingii*.

The lower tongue of the Mottled Mudstone Member and 75 m of the underlying mudstones are seen in the A483 road cutting [SO 0879 7872 to SO 0860 7894] (Figure 40a, locality 9) at Llanbadarn Fynydd, the uppermost part of this sequence being just beyond the northern limit of the district. The lower half of the section exposes mudstone turbidites (type D) thinly interbedded with laminated hemipelagites. In contrast, in the upper half, below the Mottled Mudstone Member, thin laminated and cross-laminated and locally shelly sandstones, up to 8 cm thick, occur scattered or in bundles within the mudstones, locally comprising 50 per cent of the lithology. Graptolite faunas, including *M.* cf. *flemingii* sensu lato and cf. *Cyrtograptus hamatus* from the base of the road section and *M.* cf. *flemingii flemingii*, *Pristiograptus pseudodubius* and *Monoclimacis flumendosae*?? sensu lato from 8.5 m below the Mottled Mudstone Member, suggest a probable *lundgreni* Biozone age. The Mottled Mudstone Member lower tongue is 17 m thick. The brachiopods *Bracteoleptaena bracteola*, *Eoplectodonta* sp., *Giraldiella* sp., *Skenidioides* sp., *Palaeoneilo* sp., the trilobite *?Leonaspis* sp., and orthocones have been collected from the burrowed mudstones. *Pristiograptus pseudodubius*, *P. dubius dubius*, *Gothograptus nassa* and *G.* aff. *nassa* (broad form) from the uppermost part of the Mottled Mudstone Member lower tongue indicate the *nassa* Biozone.

Depositional model

The latest Llandovery to Wenlock succession contains two types of sedimentary system (Figure 13). The lower part of the Nant-ysgollen Mudstones, composed predominantly of thinly interbedded laminated hemipelagites and turbidite mudstones, is similar to the anoxic facies of earlier slope-apron systems. The base of the formation marks the onset of oxygen-deficient (anaerobic) bottom-water conditions, which were largely sustained throughout the succeeding Wenlock. Brief reversions to oxygenated conditions are recorded by the burrow-mottled mudstone horizons of the Nant-ysgollon Mudstones and the Mottled Mudstone Member. The main anaerobic phase is recognised in other Avalonian and Laurentian basinal successions throughout the Iapetus province (Dimberline et al., 1990), suggesting that it resulted from a eustatic rise in sea level. However, although the early Wenlock is a period of global sea-level highstand, the early part of the period (*spiralis* to *murchisoni* biozones) coincides with a regression in most palaeocontinents (Johnson et al., 1991), thereby showing a eustatic model to be untenable. One possible solution may be that the transgression was local and restricted to the margins of the Iapetus Ocean. Although the evidence for such a transgression on the Midland Platform is conflicting, due to the presence of both regressive and transgressive sequences at this level (Hurst et al., 1978), localised contemporaneous tectonism may have blurred the picture. Other models, apart from a transgression, include major climatic or oceanographic changes, the latter resulting from the contemporary narrowing and restriction of the Iapetus seaway (Harris, 1987).

The incoming of the thicker mudstone turbidites in the upper part of the Nant-ysgollon Mudstones and of the contemporaneous Penstrowed Grits farther east, signalled the abandonment of the slope-apron phase of deposition and the entry of a rapidly expanding, southerly supplied, sandstone-lobe system. The thicker mudstone turbidites of the upper Nant-ysgollon Mudstones represent the muddy fringe facies of this system and the Penstrowed Grits are its principal sandstone-lobe facies.

The Penstrowed Grits are the youngest and most easterly sandstone-lobe facies developed in the Southern Welsh Basin. The packeted sequences of high-matrix sandstone turbidites (type Bii) record the repeated aggradation and abandonment of sandstone lobes supplied by high-concentration turbidity currents and/or dilute debris flows of deltaic derivation (Dimberline and Woodcock, 1987). The presence of abundant bioclastic debris reflects the proximity of this system to the eastern shelf; terrestrial plant remains record increased levels of colonisation in the catchment area. The abundance of coarse mica flakes and phyllite clasts suggests the unroofing of higher-grade metamorphic rocks in the source areas.

Intrabasinal and basin-bounding faults appear to have influenced the growth of the sandstone-lobe system. The westerly regional pattern of onlap seen at the base of the Penstrowed Grits (not easterly onlap as suggested by Dimberline and Woodcock, 1987), demonstrates that the initial (*centrifugus–riccartonensis* biozones) focus for sand deposition lay along a downwarp situated to the west of, and controlled by, a precursor Garth Fault (Figure 40b). However, local eastward onlap in the vicinity of the fault footwall is confirmed by the thicker sequence of Nant-ysgollon Mudstones preserved in the Coedglasson outlier. The westerly younging top and westerly thickening of the Penstrowed Grits suggest that the depositional focus of the system subsequently (*rigidus* to *lundgreni* biozones) shifted westwards in response to reactivation of the Cwmysgawen Fault (Figure 40b). The orientation of slump folds in the Penstrowed Grits confirms the proximity of steep, west-facing gradients to the east of the district during this interval (Dimberline, 1987).

As the sandstone-lobe facies moved westwards, deposition of anoxic slope-apron facies (Nantglyn Mudstones) resumed in the area east of the district as early as the *riccartonensis* Biozone and, after lobe abandonment, extended throughout the district by the *lundgreni* Biozone (Figure 40b). This marked the onset of a period of facies transition in the Southern Welsh Basin, which saw the differentiation of basinal and shelf facies diminish and the evolution of a ramp-like continuum between these shallower and deeper depositional settings along the basin margin. Steep gradients and/or seismic activity, suggested by slumped units in the Nantglyn Mudstones, may record continued intermittent movement on the regional major faults. The shelly, fine-grained sandstone and siltstone 'event beds' in the upper parts of the Nantglyn Mudstones compare closely with lithologies in the early Ludlow Bailey Hill Formation described by Tyler and Woodcock (1987) in east Wales, and may similarly record emplacement below storm wave-base, but from storm-generated flows. The entry of such beds may mark the onset of a eustatic regression in the late Homerian (Johnson et al., 1991), an event which reached its acme during deposition of the Mottled Mudstone Member. This regression saw the return of oxygenated sea-bed conditions and allowed bioturbating infauna and indigenous shelly benthos to colonise the surface of the slope-apron.

CHAPTER EIGHT

Silurian: Llandovery to Wenlock shelf systems

The shelf facies that crops out in the district east of the Garth Fault forms the lowermost part (late Llandovery to mid-Wenlock) of a depositional system initiated in the late Aeronian and ranging through to the late Ludlow.

In the district, the sequence locally rests unconformably on Ordovician rocks (Figures 13, 39 and 42) and mainly comprises distal shelf/ramp mudstones, comparable to the basinal slope-apron mudstones to the west. Three

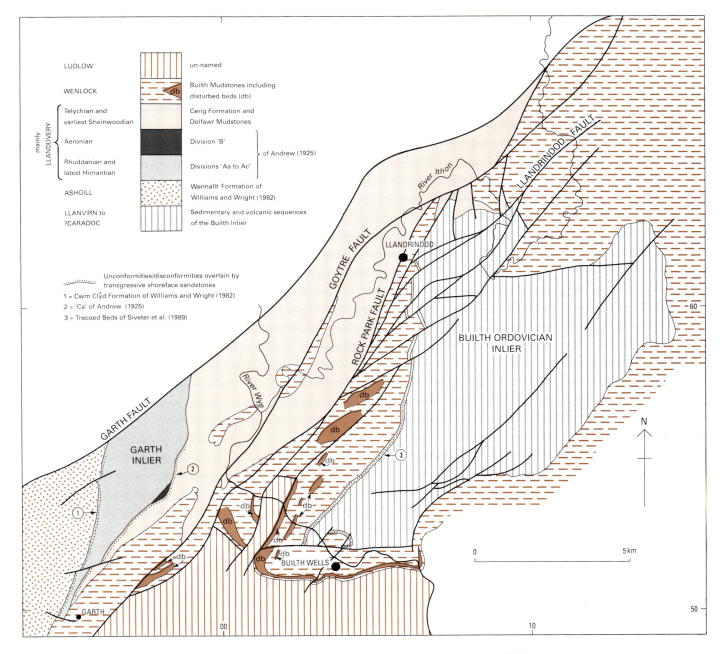

Figure 42 Geological map of Silurian shelf formations in the vicinity of the Builth Inlier; linework south of the district is modified after Jones (1947) and Harris (1987).

formations are recognised, the Cerig Formation, the Dolfawr Mudstones and the Builth Mudstones.

An earlier depositional system (late Hirnantian to Aeronian), cropping out just to the south of the district at Garth, probably underlies the Cerig Formation at depth west of the Builth Ordovician Inlier (Figure 39).

Cerig Formation

The Cerig Formation was defined in the Llandovery district by Cocks et al. (1984). In the district it rests unconformably on the Ordovician rocks of the Builth Inlier and mainly comprises green or grey mudstones, in which colour banding, burrow mottling and diffuse lamination are variably developed. A thin basal unit of shelly sandstones, known to occur to the south of the district, is probably present but not exposed.

The formation equates with Andrew's (1925) Cc and Cd divisions of the Garth area and with the 'Pale Shales' of Jones (1947). In the district there are three principal areas of outcrop, each bounded by faults and each with a different thickness and sequence (Figure 42).

The extensive outcrop of the formation between the Garth and Rock Park faults is largely drift covered, but its thickness, estimated from the Garth area [SN 950 495] (Figure 42) (Ziegler et al., 1968), is in excess of 450 m. In the district the base of the formation is not exposed, but at Garth it overlies a transgressive–regressive, late Aeronian (sedgwickii Biozone) sequence (divisions Ca and Cb of Andrew, 1925) that in turn rests disconformably on earlier Llandovery strata (mainly Rhuddanian) (Andrew, 1925).

In the narrow, drift-covered crop between the Rock Park Fault and the Builth Inlier, the base of the formation is not exposed in the district. However, exposures to the south, in the Builth district (Jones, 1947), demonstrate that it rests with angular unconformity on ?Caradoc, Llandeilo and Llanvirn strata of the Inlier. The conjectural crop of the formation portrayed on the map is probably an oversimplification, for relief on the basal unconformity may give rise to thickness variations and local absence. Less than 15 m are recorded at Trecoed [SO 0525 5520] (Jones, 1947), to the south of the district.

South of Crossgates, between the Rock Park and Llandrindod faults and the Builth Inlier, the formation is up to 100 m thick and rests unconformably on the Ordovician rocks, with marked westward overstep.

The thick sequence of Cerig Formation developed between the Garth and Rock Park faults comprises variably burrow-mottled mudstones and silty mudstones, and rare scattered disturbed beds. A 50 cm-thick unit of laminated and graptolitic hemipelagite has been recorded, immediately to the south of the district, in Estyn Brook [SO 9995 5788].

Three intergradational types of burrow-mottled mudstones are recognised: blocky, pale olive-green mudstones with abundant grey burrow mottles ranging up to 7 mm across, mostly of *Chondrites* but including both narrow and wide diameter forms, as well as the spreiate fills of other feeding burrows; blocky, dark, medium and pale grey to olive-green, colour-banded mudstones, with streaky dark grey laminae, scattered *Chondrites* burrow mottling and phosphate nodules; and fissile, dark grey mudstones which exhibit a diffuse but pervasive lamination, with scattered paler bands and laminae, rare burrow mottling, but common thin, linear, rusty-weathering burrow forms, less than 2 mm across and up to 7 cm long. Scattered siliceous and calcareous nodules, many of cone-in-cone type, are present in all three types of mudstone.

As a result of synsedimentary oxidation and bioturbation in these mudstones, event beds (generated either by storms or by turbidity currents) cannot be distinguished from lithologies of hemipelagic origin. The colour-banded mudstones compare with parts of the Derwenlas Formation and Dolgau Mudstones and may, similarly, contain varying proportions of resedimented material attributable to very low-concentration density flows (cf. type E turbidites), although the bases and tops of such units cannot be recognised in these silt-depleted lithologies. The pale olive-green mudstones and colour-banded mudstones exhibit rapid intergradations, even within individual exposures, appearing to form a continuum of oxic facies, recording random variations in the degree of bioturbation and penecontemporaneous oxidation.

The diffusely laminated mudstones do not resemble the laminated hemipelagic deposits seen in the formation in Estyn Brook and throughout the succeeding Wenlock sequence. However, the lamination suggests reduced levels of bioturbation and resedimented input. The diffusely laminated mudstones are mainly confined to the lower reaches of the River Dulas section [SO 0615 6446 to SO 0638 6406], and may occupy a position low in the formation.

The disturbed beds comprise variably destratified mudstones in which irregular fractures, abundant listric surfaces and randomly orientated siliceous or calcareous nodules are present. Slump folds, exotic clasts and reworked faunas have not been recorded. The disturbed beds locally exhibit sharp bases and lateral contacts, but in other cases their boundaries are very gradational. The stratigraphical distribution of the disturbed beds in the Cerig Formation is not known.

Dark to pale grey and green, burrow-mottled mudstones predominate in the Cerig Formation sequence flanking the northern edge of the Builth Inlier, south of Crossgates. Also present, but only in a tributary of the River Ithon [SO 0843 6335], are mudstones with scattered quartz pebbles up to 1cm in diameter. Burrow-mottled and colour-banded mudstones with abundant brown-weathering moulds of a fragmentary and diverse shelly benthos, occur at several localities in this area. The shells, mainly trilobites and brachiopods, occur in irregular clots and stringers as a result of biogenic disruption of original thin shelly layers (Plate 20). Calcareous nodules, nucleated on the shell-rich clots, are common and weather to fossiliferous 'rottenstones'. The fauna contains components of the deep-water *Clorinda* benthic community (Ziegler, 1965; Ziegler et al., 1968b). Units of these shelly mudstones, over 8 m thick, occur close to the unconformity, suggesting that they may form

Plate 20   Fossils from the *Acidaspis* Limestone facies of the Cerig Formation, Alpine Bridge [SO 0903 6312].

a   Shell-strewn bedding surface (latex cast) (JZ 3294), ×3, (see also e and f).   **b**   *Monograptus* ex. gr. *spiralis* (JZ 3287) ×2.   **c**   *Eoplectodonta* cf. *penkillensis* (latex cast) (JZ 3282), ×5.   **d**   As c, (JZ 3277), ×5.
e   *Laengella* cf. *segmentum* (as a), ×5.   **f**   *Skenidiodes* cf. *lewisii* (as a), ×5.   **g**   as c, (JZ 3294), ×5.
Numbers in round brackets, prefixed by JZ, refer to specimen numbers in BGS collections.

a basal or flanking facies within the Cerig Formation. However, neither the unconformity, nor the passage of the shelly facies into other parts of the formation, is seen. The lithology is identical to that of the *Acidaspis* Limestone, which occurs in the formation east of the Rock Park Fault, though not necessarily at the same horizon.

The attenuated Cerig Formation sequence, east of the Rock Park Fault, is presumed to be similar to that exposed south of the district, notably at Trecoed [SO 0525 5520], where up to 7.5 m of sandstone, locally rich in disarticulated pentamerid and other brachiopod valves, forms a basal unit (Jones, 1947; Ziegler et al., 1968) termed Trecoed Beds by Siveter et al. (1989). It compares lithologically with the Ca and Cb divisions of Andrew (1925) at Garth (Plate 18e). The fauna contains reworked elements of both the *Pentamerus* and *Stricklandia* benthic depth communities of Ziegler et al. (1968b). The sandstone is overlain by unfossiliferous,

variably burrow-mottled, silty mudstones, which are succeeded by the 80 cm-thick *Acidaspis* Limestone of Jones (1947). The latter comprises green mudstones with calcareous nodules containing fragmentary remains of trilobites, brachiopods and rare solitary corals. As in the area south of Crossgates, the lithology includes components of the *Clorinda* depth community of Ziegler et al. (1968b). There is no evidence in the Trecoed section to support the suggestion by Jones (1947), and followed by Ziegler (1968), that there is an unconformity (or non-sequence) at the base of the *Acidaspis* Limestone (Hurst et al., 1978).

BIOSTRATIGRAPHY

The Cerig Formation in its type area at Llandovery ranges from the early *turriculatus* Biozone to the *crenulata* Biozone (sensu Rickards, 1976). The presence of *sedgwickii* Biozone graptolites in the Ca division at Garth, to the south of the district, and of *Monograptus spiralis*, probably indicating the

Table 18   Acritarch taxa in the Cerig Formation (for localities see map on facing page).

ACRITARCH TAXA	① 29742	② 29979	③ 28827	④ 29812	⑤ 29813	⑥ 29978	⑦ 29028	⑧ 29810	⑨ 29811
Sphaeromorphs	•	◆	•	•	•	•	•	•	•
Diexallophasis denticulata	.	◆	◆	•	•	•	•	.	.
Leiosphaeridia spp.	.	•	◆	.	.	.	•	.	.
Micrhystridium spp.	•	.	.	•	•	.	•	.	.
Multiplicisphaeridium spp.	.	•	•	.	.	•	•	.	.
Veryhachium downiei	.	.	.	.	.	.	•	.	.
Micrhystridium inflatum	.	.	.	.	.	.	•	.	.
Estiastra sp.	.	.	.	.	.	.	?	.	.
Domasia elongata	.	.	.	.	•	•	.	.	.
Ammonidium microcladum	•	◆	.	•	•	•	.	.	.
Salopidium granuliferum	◆	.	•	•	•	?	◆	.	.
Salopidium sp.	.	.	.	.	.	.	?	.	.
Oppilatala frondis	.	.	.	.	.	.	cf.	.	.
Oppilatala ramusculosa	.	•	•	.	.	.	•	.	.
Tylotopalla robustispinosa	.	.	•	.	.	.	•	.	.
Diexallophasis sanpetrensis	.	.	•	.	.	.	•	.	.
Lophosphaeridium spp.	.	.	•	.	.	.	•	.	.
Visbysphaera meson	.	.	.	.	.	.	•	.	.
Solisphaeridium nanum	.	.	.	.	.	•	•	.	.
Multiplicisphaeridium fisherii	.	.	.	.	.	.	cf.	.	.
Tunisphaeridium tentaculaferum	.	.	.	.	.	.	cf.	.	.
Cymatiosphaera octoplana	.	.	.	.	.	.	cf.	.	.
Ammonidium listeri	•	◆	•	•	•	◆	?	.	.
Ammonidium sp. A	•	•	•	.	•	.	.	.	.
Oppilatala eoplanktonica	•	•	.	.	•	.	.	.	.
Veryhachium wenlockianum	.	.	.	.	•	.	.	.	.
Diexallophasis granulatispinosa	•	.	.	.	•	.	.	.	.
Domasia symmetrica	.	.	.	.	•	.	.	.	.
Tylotopalla sp.	.	.	.	.	•	.	.	.	.
Veryhachium trispinosum	.	.	.	.	•	.	.	.	.
Salopidium sp. B of Priewalder	.	.	.	?	.	.	.	.	.
Tylotopalla caelamenicutis	.	.	•	.	.	.	.	.	.
Visbysphaera oligofurcata	.	.	?	.	.	.	.	.	.
Domasia trispinosa	•	•	•	.	.	.	.	.	.
Eupoikilofusa striatifera	.	.	•	.	.	.	.	.	.
Tunisphaeridium parvum	.	•	•	.	.	.	.	.	.
Carminella maplewoodensis	.	•	.	.	.	.	.	.	.
Eupoikilofusa aff. rochesterensis	.	•	.	.	.	.	.	.	.
Eupoikilofusa rochesterensis	.	•	.	.	.	.	.	.	.
Micrhystridium inconspicuum	.	•	.	.	.	.	.	.	.
Multiplicisphaeridium paraguaferum	•	•	.	.	.	.	.	.	.
Salopidium sp. C of Priewalder	.	•	.	.	.	.	.	.	.
Dictyotidium sp. A	.	•	.	.	.	.	.	.	.
cf. Ammonidium microcladum	.	•	.	.	.	.	.	.	.
Veryhachium rhomboidium	•	.	.	.	.	.	.	.	.
Multiplicisphaeridium sp. B	?	.	.	.	.	.	.	.	.
RECYCLED ORDOVICIAN ACRITARCHS									
Multiplicisphaeridium cf. continuatum	•	.	.	.	.	.	.	.	.
Multiplicisphaeridium cf. irregulare	•	.	.	.	.	.	.	.	.
Stellechinatum brachyscolum	•	.	.	.	.	.	.	.	.
Vulcanisphaera sp.	•	.	.	.	.	.	.	.	.
Peteinosphaeridium sp.	.	.	.	.	.	.	•	.	.

Legend:
- • less than 10 specimens
- ◆ more than 10 specimens

spiralis Biozone at Pentre Brook [SO 0486 6398], confirms a similar age for the formation in the district (Figure 39). Graptolites indicative of the upper *griestoniensis* Biozone have been obtained from Estyn Brook [SO 9995 5788], immediately south of the district.

The occurrence of the brachiopods *Pentamerus oblongus* and *Stricklandia lens ultima* (= *S. laevis*) at Trecoed (Ziegler et al., 1968), has been taken to indicate a lower *turriculatus* Biozone age for the basal sandstone (cf. Cocks et al., 1984; Siveter et al., 1989). However, new collecting has revealed older elements of the *Stricklandia lens* lineage (*S. lens* cf. *intermedia* and/or *progressa*), suggestive of a pre-*turriculatus* Biozone age. Either the older assemblage is reworked, or the sequence is condensed, correlating wholly or in part with the Ca and Cb divisions of Garth (Figure 39). Reconnaissance work during this survey has established a *sedgwickii* Biozone age for the Ca division at Comin Coch [SN 9594 5177], Garth.

The *Acidaspis* Limestone at Trecoed, where it has traditionally been regarded as Wenlock in age (Jones, 1947; Hurst et al., 1978; Siveter et al., 1989), and comparable facies [SO 0903 6312] in the Crossgates area, both contain the brachiopod *Eoplectodonta* cf. *penkillensis* (Plate 20), suggestive, though not diagnostic, of the late Llandovery (Cocks, 1970). A specimen of *Monograptus* cf. *priodon* from the *Acidaspis* Limestone at Trecoed suggests that it postdates the *turriculatus* Biozone. The likely correlation of the *Acidaspis* Limestone with the Dolfawr Mudstones (see p.173) suggests that it may span the Llandovery–Wenlock Series boundary.

The Cerig Formation has yielded acritarch assemblages of variable diversity, which span the *A. microcladum* to *G. encantador* biozonal interval, indicative of a late Aeronian to Telychian age (Table 18). The trilete spore *Ambitisporites avitus* has been recorded from Clywedog Brook [SO 0849 6576].

DETAILS

Pale green, burrow-mottled mudstones crop out in Pentre Brook [SO 0533 6321 to SO 0564 6313] north of Castell Collen, and in the bed and banks of the River Ithon east of Great Cellws [SO 0821 6448 to SO 0828 6425], south-west of

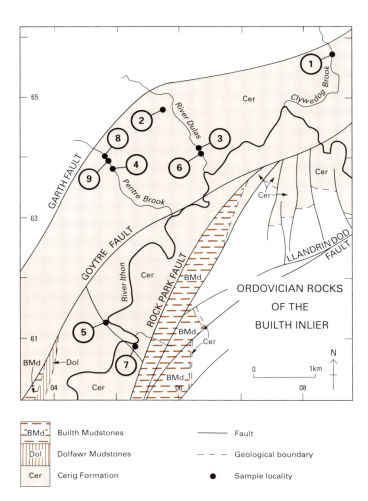

BMd	Builth Mudstones		Fault
Dol	Dolfawr Mudstones	---	Geological boundary
Cer	Cerig Formation	●	Sample locality

Poor exposures in the fossiliferous *Acidaspis* Limestone facies occur along the stream [at SO 0756 6288] north-west of Cefn-coed and in crags [SO 0809 6381]. A track cutting [SO 0903 6312], west of Alpine Bridge, exposes about 8 m of fossiliferous, pale green, pervasively burrow-mottled mudstones. They contain both scattered benthic fossils and shelly nodules, and have yielded the brachiopods *Eoplectodonta* cf. *penkillensis*, *Leangella* cf. *segmentum* and *Skenidiodes* cf. *lewisii*, indeterminate trilobite debris, and the graptolite *Monograptus* ex. gr. *spiralis* (Plate 20).

Dolfawr Mudstones

The Dolfawr Mudstones comprise a transitional sequence, about 130 m thick, between the burrow-mottled mudstones of the Cerig Formation and the laminated mudstones of the Builth Mudstones. The southern bank of the River Ithon [SO 0276 5761 to SO 0264 5768], 500 m east of Dolfawr farm and just south of the district, provides a fully exposed type section. The formation equates with the 'striped shales' of Jones (1947) and the Hafod-yr-ancr Member of Harris (1987). In the district it is restricted to the area west of the Rock Park Fault where it crops out in a predominantly drift-covered, north-north-east-trending tract, west of Lletty [SO 038 600], limited to the north and west by the Goytre Fault. The *Acidaspis* Limestone of Trecoed may be a condensed lateral correlative of the Dolfawr Mudstones (Harris, 1987).

The formation consists predominantly of alternations of burrow-mottled mudstone and laminated mudstone. Units, 0.3 to 2 m thick, comprise thin beds of dark grey, finely laminated, graptolitic silty mudstone (hemipelagite) separated by laminae and beds, less than 1 cm thick, of pale green and grey burrow-mottled mudstone. Also present are beds less than 50 cm thick of pale olive-green to grey-green, pervasively burrow-mottled mudstone, devoid of laminated levels. Such beds are predominant in the lowest part of the formation and decrease in importance upwards. In the upper half of the formation, these burrow-mottled beds locally contain brown-weathering moulds of a scattered and fragmented shelly fauna, comparable with that in the *Acidaspis* Limestone and the shelly mudstone levels in the Cerig Formation. Rare, thin (up to 2 cm), locally cross-laminated, skeletal packstone beds, commonly containing abundant crinoidal debris, occur in the lower part of the formation.

BIOSTRATIGRAPHY

Graptolite assemblages from the lower half of the formation in the River Ithon are referable to the *spiralis–centrifugus* interregnum. Records of *Cyrtograptus centrifugus* and *C. insectus* (Harris, 1987), from the upper part of the formation at the type section, indicate a *centrifugus* Biozone age.

The formation has yielded acritarchs ranging from Telychian to Wenlock in age, including *Ammonidium microcladum*, *Oppilatala ramusculosa*, *Salopidium granuliferum*, *S.* cf. *woolhopense*, and *Tylotopalla wenlockia?*. also recovered were the chitinozoa *Ancyrochitina ancyrea* and *Angochitina longicollis*, the latter of which is elsewhere restricted to strata that span the *griestoniensis–murchisoni* graptolite biozonal interval (Laufeld, 1974; Grahn, 1988).

Cellws [SO 0639 6315], and near Lletty [SO 0437 6008 to SO 0412 5955]. Weakly mottled, pale green mudstones are exposed in the Dyfnant [SO 0309 6067]. Colour-banded mudstones are exposed in the bed and banks of Clywedog Brook [SO 0838 6502 to SO 0815 6490], downstream from Clywedog Bridge. Additional sections in this lithology are seen in quarries [SO 0590 6520, SO 6577 6480] along the Dulas valley, and near Llandrindod sewage works [SO 0500 6045, SO 0498 6043]. A discontinuous section [SO 0615 6446 to SO 0638 6406] along the River Dulas south of Gravel Road consists of dark grey, diffusely laminated mudstones.

Scattered, fine-grained, parallel and cross-laminated micaceous sandstones, up to 10 cm thick, are exposed in a cliff [SO 0496 6121 to SO 0486 6128] along the River Ithon, south of Dol-llwyn-hir, and in Pentre Brook [SO 0482 6404 to SO 0488 6393], where the graptolite *Monograptus spiralis* was collected [SO 0486 6395]. Disturbed beds are exposed in cliff sections [SO 0546 6251 to SO 0567 6254] along the River Ithon, south of Castell Collen.

A tributary [SO 0838 6334 to SO 0855 6338] of the River Ithon, north of Upper Trelowgoed, exposes pale grey, variably burrow-mottled mudstones, with calcareous nodules and abundant pyrite-rich blebs and stringers. The lowest mudstones [SO 0844 6333] contain scattered quartz pebbles up to 1 cm in diameter. Farther west, quarries [SO 0798 6377, SO 0769 6332] expose pervasively burrow-mottled mudstones. The mudstones at the first quarry are darker and display both small and large varieties of *Chondrites* burrows.

Table 19
Distribution of graptolites in the Builth Mudstones in Arlair brook (Rock Park) [SO 0573 6089 to SO 0564 6096], Llandrindod Wells (for localities see map on facing page). For key to range chart symbols see Appendix 3.

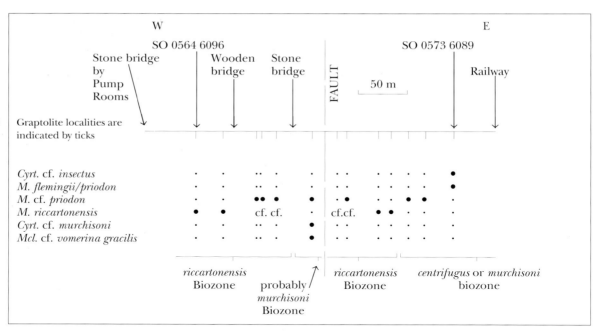

DETAILS

The northern part of the cliff [SO 0369 5956 to SO 0356 5933] along the River Ithon, south-west of Lletty, is the only exposure of the formation in the district. There 33 m of interbedded laminated and burrow-mottled mudstones with rare parallel and cross-laminated crinoidal packstone beds are exposed.

Builth Mudstones

The term Builth Mudstones Formation was first proposed by Harris (1987) to describe the mudstones of Wenlock age described by Jones (1947), west and south of the Builth Inlier. The base of the formation is redefined here to exclude the Hafod yr ancr Member of Harris (1987) and the *Acidaspis* Limestone of Jones (1947). The Builth Mudstones are confined to the south-eastern corner of the district, east of the Goytre Fault. The top of the formation is not observed in the district, where only about the lower 200 m are represented.

The Builth Mudstones mainly comprise variably calcareous, finely laminated graptolitic mudstones but include local disturbed beds. The mudstones are brown-weathering and dark blue-grey in colour with widely spaced paler bands. The lamination comprises an alternation of dark blue-grey, carbonaceous laminae and paler grey silt-rich laminae, all less than 0.5 mm thick. Although the lamination generally persists throughout, in most of the paler bands it is more widely spaced and the silt-rich laminae are more pronounced. Rare, structureless pale bands are also present. Weathered, less calcareous parts of the facies are commonly fissile. Bedding planes are locally strewn with compacted and amorphous carbonaceous debris, abundant graptolites (commonly with stipe lengths of several centimetres), scattered and crushed orthoconic nautiloids, rare articulated bivalves and brachiopods, ostracods, and the spat and disarticulated or comminuted remains of shelly benthos (bivalves, brachiopods, trilobites and crinoids). The fragmentary nature of most of the benthic remains suggests that they are transported, while the graptolites and orthocones, which make up the bulk of the intact components of these assemblages, represent the contemporary planktonic/nektonic fauna. Intact benthic components compare with those reported from the Builth district (Hurst et al., 1978), which belong to the low-diversity, deep-water, *Visbyella trewerna* Community (Calef and Hancock, 1974; Hancock et al., 1974). The more calcareous mudstones are less fissile and more blocky and commonly contain abundant, early diagenetic limestone nodules. The spacing of the lamination preserved in the nodules, suggests a compaction factor of up to 300 per cent for the enveloping mudstones.

The only disturbed bed in the district is exposed in the River Ithon [SO 0835 6411] south of Crossgates, where it comprises a metre-thick zone with complex slump folds and disruptions. However, in the adjacent Builth district, disturbed beds which include exotic shelly and conglomeratic debrites range up to tens of metres in thickness and several kilometres in lateral extent (Jones, 1947; Harris, 1987) (Figure 42).

BIOSTRATIGRAPHY

At Trecoed [SO 0525 5520], in the Builth district, probable *centrifugus* Biozone graptolites have been obtained from the lowest part of the formation, though Harris (1987) assigned these beds to the *murchisoni* Biozone. In the Builth district, Harris (1987) has recorded all the remaining Wenlock biozones from the Builth Mudstones. That part of the formation at outcrop in the present district may, on thickness grounds, range up into the *flexilis* Biozone (Hurst et al., 1978). However, only the *murchisoni* and *riccartonensis* biozones, and possibly the *centrifugus* Biozone, have been proved (Table 19).

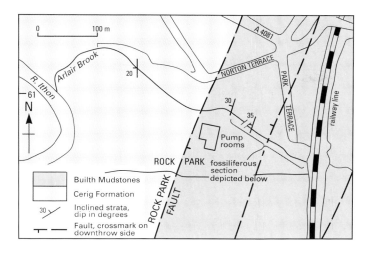

Builth Mudstones

Cerig Formation

Inclined strata,
dip in degrees

Fault, crossmark on
downthrow side

DETAILS

The River Ithon [SO 0830 6420 to SO 0836 6410] south of
Crossgates exposes 30 m of laminated calcareous mudstones.
Limestone nodules, locally coalescing to form beds, are abundant in the lower part of the section, and a 1 m-thick disturbed
bed is present towards the top. There, Roberts (unpublished
maps) recorded graptolites, including 'G. geinitzianus' (Retiolites
geinitzianus), which is not thought to range beyond the
centrifugus Biozone (Rickards, 1976). He also reported this
graptolite from a nearby but now infilled quarry [SO 0815
6375]. Laminated mudstones exposed along a tributary [SO
0866 6368] of the Ithon have yielded Monograptus priodon.

Arlais Brook [SO 0562 6098 to SO 0580 6087] in Rock Park
exposes an almost continuous section through the Builth Mudstones east of the Rock Park Fault. There, laminated mudstones
have yielded graptolites possibly of the centrifugus Biozone, and
definitely those of the murchisoni and riccartonensis biozones
(Table 19). The distribution of these assemblages demonstrates
structural repetition within the section. Farther south, Roberts
(unpublished map) assigned mudstones previously exposed in
a quarry [SO 0576 6003] to the riccartonensis Biozone.

Depositional model

Although the shelf sequence of the district varies in thickness and facies across the Rock Park and Llandrindod
faults, reflecting contemporaneous movements on these
structures, the main controlling influences on sedimentation were contemporary changes in sea level
(Figure 39).

THE MID-AERONIAN REGRESSION

The base of the late Aeronian to Telychian sequence, east
of the Garth Fault, is a disconformity seen in the Garth
area (Andrew and Jones, 1925). Previous authors have
invoked tectonic uplift and erosion (Jones, 1926; George,
1963), loss of strata via slumping (Smith and Long, 1969;
James, 1983) and, as a contributory factor, non-deposition
(Ziegler et al., 1968), to explain this hiatus. However, its
stratigraphical position suggests that it is primarily the
result of a eustatic regression in or around convolutus
Biozone times (Johnson et al., 1981, 1985, 1991),
although local overstep may record local tectonic influences (Andrew, 1925; George, 1963). Erosion following

emergence may also account for the loss of any earlier
Aeronian, Rhuddanian and residual Ashgill strata previously deposited across the Builth Inlier. The regression
is marked in the basin by the prevalence of oxic bottom
conditions during the Derwenlas Formation.

THE LATE AERONIAN TRANSGRESSION

A major transgression, at the onset of sedgwickii Biozone
times, inundated substantial areas of the Silurian shelf
(e.g. Ziegler et al., 1968; Smith and Long, 1969; Bridges,
1975). The apparent absence of strata of this age, at
Garth and Builth, persuaded Ziegler et al. (1968) to
invoke non-deposition over these areas, and invited the
slump interpretation of Smith and Long (1969). Dating
of the Garth succession has demonstrated, however,
(contrary to Ziegler et al., 1968), that beds overlying the
disconformity (Ca of Andrew, 1925) are of sedgwickii Biozone age (p.172; Figure 39). They include well-sorted,
shelly, transgressive shoreface sandstones (Plate 18e). The
succession of benthic communites in the overlying
beds confirms the subsequent deepening, as transgression advanced (Ziegler et al., 1968).

The brachiopods from the flanks of the Builth Inlier at
Trecoed, suggest that initial inundation of this area also
took place during the late Aeronian. The presence of
younger brachiopod faunas (Ziegler et al., 1968) suggests
that the basal Pentamerus-bearing sandstones represent a
condensed or reworked shallow-water sandstone facies
formed across a tectonically sustained topographic high.
This late Aeronian event is marked in the basinal
succession by the widespread M. sedgwickii shales.

Evidence for latest Aeronian regression is provided by
the well-documented sandy Wormwood Formation at
Llandovery (Cocks et al., 1984). The sandstone Cb
division of Garth (Andrew, 1925) probably correlates
with this event.

THE BASAL TELYCHIAN TRANSGRESSION

The onset of Cerig Formation deposition records the
introduction of deeper and more distal shelf or ramp
conditions across the region, following a transgression at
or about the base of the turriculatus Biozone. Shelly
benthic communities were rapidly occluded and replaced
by a soft-bodied ichnofauna adapted to dysaerobic
bottom conditions (cf. Cocks et al., 1984, p.161).
Banding and lamination suggests that sediment input
was both in the form of thin event beds, possibly storm-
generated, low-concentration density flows (cf. type E
turbidites), and suspension fall-out (cf. hemipelagite)
from nepheloid layers or storm-induced suspensate (cf.
Dimberline and Woodcock, 1987).

Pronounced thickness changes between fault segments, and the presence of disturbed beds, suggest penecontemporaneous faulting. Such faulting would have
been coeval with that recognised along the Tywi Lineament and which influenced the sandstone-lobe systems
in the basin to the west. The continuing influence of a
topographically elevated area — the 'Builth High' — is
reflected in the attenuated character of the local Cerig
Formation and the presence of shelly mudstone facies
adjacent to the Builth Inlier. In contrast, west of the

Rock Park Fault, rapid relative subsidence is indicated by the much thicker Cerig Formation sequence.

Tectonic quiescence during late *griestoniensis* and *crenulata* Biozone times, recorded in the progradation of slope-apron facies (Dolgau Mudstones) basinward of the Garth Fault, marked a period when facies differences across the Garth Fault were at a minimum. The Dolgau Mudstones and contemporary parts of the Cerig Formation comprise arbitrary divisions within a distal shelf/ramp to slope-apron facies continuum.

The widely reported mid-Telychian regression (Johnson 1981, 1985) has not been recognised in the Cerig Formation, nor in the basinal sequences; its effects were probably masked by those of intra-Telychian faulting and regional subsidence (Chapters 5 and 6).

THE LATE LLANDOVERY/EARLY WENLOCK TRANSGRESSION

The rapid alternation of burrow-mottled and laminated mudstones in the Dolfawr Mudstones points to deposition in a setting at or close to the dysaerobic–anaerobic boundary (Savrda and Bottjer, 1987), and suggests a greater depositional depth than for the underlying Cerig Formation (but see Berry and Boucot, 1967; Harris, 1987). The same deepening event is marked by the onset of Nant-ysgollen Mudstones deposition in the basin (p.168).

The *Acidaspis* Limestone, at the top of the Cerig Formation, appears to reflect local structural, rather than eustatic influences. The 'Builth High', defined by a syn-sedimentary Rock Park Fault, and periodically sustaining a 'marginal *Clorinda*' or precurser *Visbyella trewerna* Community (Hancock et al., 1974), would here provide the distal shelly benthos, both for local reworking (bioturbated lags of the *Acidaspis* Limestone) and for transportation farther afield (shelly beds of the Dolfawr Mudstones).

Deposition of the Builth Mudstones during the late *centrifugus* Biozone appears to record an abrupt deepening, sufficient to mask the influence of the 'Builth High' and to impose fully anoxic bottom conditions throughout the region. Possible palaeoclimatic and palaeo-oceanographic factors influencing the deposition of the Builth Mudstones have been reviewed by Harris (1987). The laminated mudstones, in common with other anoxic hemipelagites of the district, possibly reflect seasonal variations in sedimentation of suspended material. However, in contrast to the basinal hemipelagites, the high levels of shelly detritus suggest a greater proximity to contemporary shelly communities and increased input of storm-generated skeletal suspensate. The sparse and low-diversity *Visbyella trewerna* Community assemblage reported from the formation (Hurst et al., 1978), first appearing in the Wenlock, records the colonisation of oxygen-deficient environments by specialised benthic (or epibenthic) forms. This assemblage occupied the deepest and most distal portions of a Wenlock shelf/ramp which, despite pene-contemporaneous faulting, probably declined steadily from shallower-water settings to colonising depths thought to be as great as 1000 to 1500 m (Hancock et al., 1974; Hurst et al., 1978).

These distal shelf/ramp facies and the contemporary turbiditic sequences of the basin (Nant-ysgollon Mudstones, Penstrowed Grits, Nantglyn Mudstones) are separated by the Garth Fault. The abrupt facies change may be the result of substantial postdepositional displacement on this fault, but could also be interpreted as the result of syndepositional movements (Dimberline and Woodcock, 1987). Major slump sheets in the Builth Mudstones (Jones, 1947) confirm the presence of steep, contemporary west-facing gradients. It is likely, therefore, that the Builth Mudstones accumulated along the upper reaches of a fault-located slope, which was effective in excluding turbiditic deposits of basinal aspect from coeval distal ramp settings.

NINE

Structure and metamorphism

The Welsh Basin was situated within the Eastern Avalonian microcontinent (Soper, 1986a) and was founded on late Precambrian to early Cambrian sialic 'basement' rocks, which are exposed around the margins of the basin in Anglesey, the Welsh Borderland and Pembrokeshire. The basin was the site of enhanced subsidence from Cambrian to Silurian times; its margins were defined by the north-east-trending Menai Strait and Welsh Borderland fault systems (Gibbons, 1987; Woodcock and Gibbons, 1988). The sedimentary development of the basin and its subsequent uplift were strongly influenced by the reactivation of basement structures (James and James, 1969; Woodcock, 1984a, 1990; Smith, 1987b; Wilson et al., 1992). The oblique collision of Eastern Avalonia with the Laurentian (North American) continent, which took place progressively from late Silurian to mid-Devonian times, resulted in the deformation and inversion of the basin (Soper et al., 1987). The most intense deformation, that of the Acadian Orogeny, occurred during the later part of this interval.

TECTONIC LINEAMENTS

In Central Wales a series of tectonic lineaments trending between north and north-east has been recognised (Figure 1) (Bassett, 1969; James and James, 1969; Woodcock, 1984b, 1990a; Smith, 1987b; Craig, 1987; Woodcock et al., 1988; Woodcock and Gibbons, 1988; Wilson et al, 1992). The lineaments display a variety of features and all were active over an extended time period. They most commonly comprise narrow zones of faults and strong deformation. All had an important influence on sedimentation patterns. Several controlled the form and location of Acadian structures, and most are associated with geophysical anomalies (McDonald et al., 1992). The lineaments are considered to reflect the reactivation and upward propagation of basement faults into the cover sequences. It has been argued that strike-slip deformation was an important factor in their development (Woodcock, 1984b; Smith, 1987b; Craig, 1987). In the district four lineaments have been recognised, which are, from west to east: the Glandyfi, Central Wales, Tywi and Pontesford lineaments (Figure 1).

The location of a tectonic lineament close to the Cardiganshire coast was first recognised in the Aberystwyth district (Glandyfi Tract; Cave and Hains, 1986). There the north-trending Glandyfi Lineament marks a regional divide in fold vergence, with a strong westwards vergence to the west and mainly eastward vergence to the east. The Glandyfi Lineament has been traced southwards onto the Llanilar sheet area, where it is broadly coincident with a zone of upright tight to isoclinal folds,

and a major fracture zone, the Bronnant Fault. The great thickness of the Aberystwyth Grits Group immediately to the west of the Bronnant Fault records contemporaneous movements on the fault (Figure 33; Wilson et al., 1992). Strong Euler aeromagnetic and gravity anomalies (p.204; Figure 55) are associated with the Glandyfi Lineament, and are probably generated from a deep basement feature (McDonald et al., 1992). The course of the lineament south of the district is still uncertain, but there is evidence to suggest that it does not connect with the zone of fold vergence, faulting and transecting cleavage in the Llangranog area, as suggested by Craig (1987).

The Central Wales Lineament comprises a zone of south-east-verging folds, a narrow braided array of steep faults, and transecting cleavage (Smith, 1987b; Woodcock, 1990a). It is situated along the axial zone of the north-north-east- to north-east-trending Central Wales Syncline, a regional Acadian structure. It had an important influence on the early Silurian sedimentation patterns. The marked increase in the thickness of the Rhuddnant Grits immediately west of the lineament is considered to be the result of synsedimentary displacements on the Claerwen Fault (p.148). The weak Euler gravity and stronger aeromagnetic anomalies (Figure 55) situated to the east of the lineament may define the position of the controlling basement structure.

The Tywi Lineament is broadly coincident with the axial zone of the complexly faulted, north-east-trending Tywi Anticline, whose present form is mainly the result of Acadian deformation (Woodcock, 1984a). The lineament was active during the Silurian and possibly earlier. In the Rhayader sheet area the eastern margin of the Tywi Lineament is taken at the Garth Fault (Figure 43a), which separates Ashgill and Silurian shelf successions to the east from basinal successions to the west. The fault also marks the eastern limit of pervasive Acadian deformation. The western margin roughly coincides with the Cwmysgawen Fault. Late Ordovician strata exposed in the core of the Tywi Anticline are disconformably overlain locally by late Llandovery sequences(Chapter 6). This disconformity has been attributed to pre-Acadian folding (Roberts, 1929; George, 1963), and to synsedimentary faulting (James and James, 1969; Ziegler, 1970; Smallwood, 1986). However, a passive, non-depositional slope model has also been proposed to explain the disconformity (James, 1983, 1991). The results of the current survey (Chapter 6) support a submarine, masswasting origin related to an episode of synsedimentary faulting. The Tywi Lineament is associated with a discontinuous Euler gravity anomaly (Figure 55), which may be traced north-eastwards to coincide with the north-western margin of the Cheshire Basin (Carruthers et al., 1992). No Euler aeromagnetic anomaly is coincident

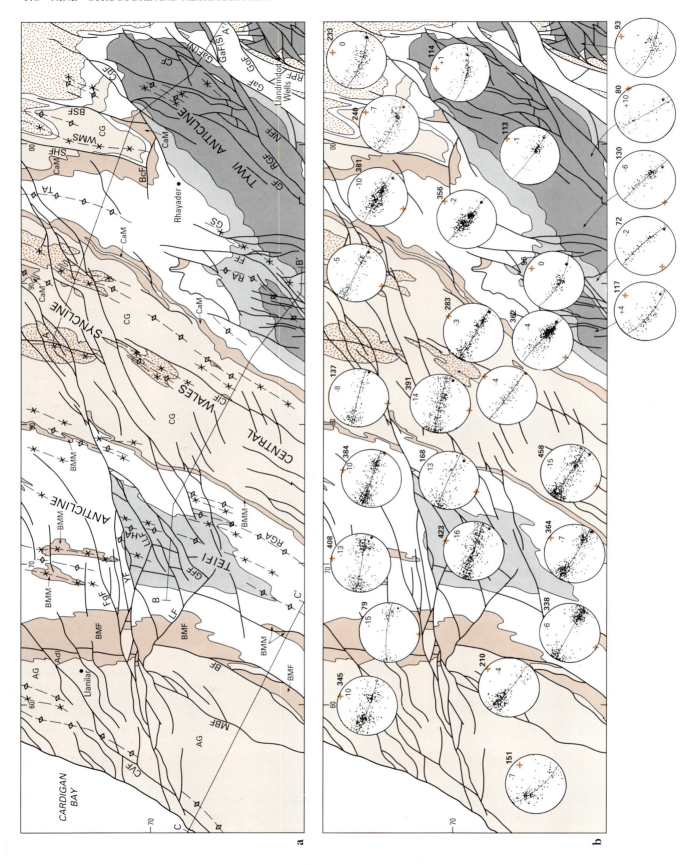

Figure 43   Main structural features of the district. For key see facing page.
a Folds and faults.   **b** Stereograms of bedding and cleavage.

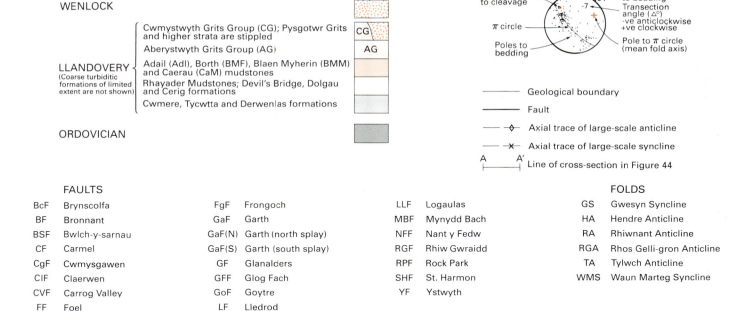

WENLOCK

LLANDOVERY (Coarse turbiditic formations of limited extent are not shown)	Cwmystwyth Grits Group (CG); Pysgotwr Grits and higher strata are stippled	CG	
	Aberystwyth Grits Group (AG)	AG	
	Adail (Adl), Borth (BMF), Blaen Myherin (BMM) and Caerau (CaM) mudstones		
	Rhayader Mudstones; Devil's Bridge, Dolgau and Cerig formations		
	Cwmere, Tycwtta and Derwenlas formations		

ORDOVICIAN

Average pole to cleavage

151 Number of poles to bedding

-7 Transection angle (△°) -ve anticlockwise +ve clockwise

π circle

Poles to bedding

Pole to π circle (mean fold axis)

———— Geological boundary

———— Fault

——◇—— Axial trace of large-scale anticline

——✕—— Axial trace of large-scale syncline

A⊢——⊣A' Line of cross-section in Figure 44

FAULTS

BcF	Brynscolfa		FgF	Frongoch	
BF	Bronnant		GaF	Garth	
BSF	Bwlch-y-sarnau		GaF(N)	Garth (north splay)	
CF	Carmel		GaF(S)	Garth (south splay)	
CgF	Cwmysgawen		GF	Glanalders	
ClF	Claerwen		GFF	Glog Fach	
CVF	Carrog Valley		GoF	Goytre	
FF	Foel		LF	Lledrod	

LLF	Logaulas		GS	Gwesyn Syncline
MBF	Mynydd Bach		HA	Hendre Anticline
NFF	Nant y Fedw		RA	Rhiwnant Anticline
RGF	Rhiw Gwraidd		RGA	Rhos Gelli-gron Anticline
RPF	Rock Park		TA	Tylwch Anticline
SHF	St. Harmon		WMS	Waun Marteg Syncline
YF	Ystwyth			

FOLDS

with the lineament, although it lies along a broad area containing numerous Euler solutions.

The course of the Pontesford Lineament is mainly defined by a series of north-east-trending faults within the inliers of Ordovician rocks at Shelve and Builth, and the folded Silurian rocks of the Clun Forest Disturbance (Woodcock, 1984b and references therein). In the Rhayader sheet area the western margin of the Pontesford Lineament probably abuts against the Tywi Lineament. The lineament was sporadically active from mid-Ordovician to Triassic times. It has been suggested that the lineament was the south-east limit of basinal Ordovician sedimentation (Dewey, 1969; Baker, 1973), and possibly had a controlling influence on the location of the Ordovician volcanism (Rast, 1969). The complex fault systems which cut the Ordovician inliers have been attributed to a period of dextral strike-slip deformation that occurred some time between the late Llandeilo and late Llandovery (Whittard, 1979; Woodcock, 1984b, 1987; Lynas, 1988). The faults were reactivated during the Acadian Orogeny. A discontinuous Euler gravity anomaly coincides with Pontesford Lineament, which also marks the south-eastern limit of a broad area of Euler aeromagnetic solutions.

PRE-ACADIAN STRUCTURES

Evidence for the earliest tectonic activity in the district is found in the Builth Inlier. There, the Llanvirn and Llandeilo strata were faulted and folded prior to the overstep of the late Llandovery sequences. Comparable relationships occur in the Shelve Ordovician Inlier in the adjacent Montgomery district (British Geological Survey, 1994).

In that part of the Builth Inlier which lies within the Rhayader sheet area the bedding generally dips at

moderate angles towards the north-west, but two north-east-trending open folds occur in the extreme south-east. The axial trace of the Cefnllys Anticline crops out between Cwm-brith Bank [SO 087 608] and Castle Bank [SO 090 615], and the north-western limb of the Gilwern Anticline is exposed near Carregwiber Bank [SO 086 596] (Figure 5). The succession is cut by large north-east-trending faults and numerous shorter east-trending and north-trending faults. This fault pattern is consistent with dextral strike-slip movements along the Pontesford Lineament (Woodcock, 1984b). Although this period of deformation and uplift could have occurred at any time between the the Llandeilo and the latest Aeronian, Woodcock and Gibbons (1988) have suggested that it took place during the Pusgillian Stage of the Ashgill (Shelveian event of Toghill, 1992). However, the widespread regional non-sequence that is believed to have developed during that period (Price, 1984) has not been identified in the Tywi Anticline to the west. In contrast, Cave and Hains (in preparation), citing evidence for latest Ashgill/early Llandovery overstep in the Welshpool district (Cave and Dixon, 1993; Bassett et al., 1992), suggest that the comparable unconformity of the Shelve Inlier was the product of a tectonic episode coincident with the late Ashgill glacio-eustatic regression. However, south of the district in the Garth area, there is marked overstep beneath latest Aeronian as well as latest Ashgill transgressive sandstones (Figure 42; Andrew, 1925; Williams and Wright, 1981); both these oversteps appear principally to reflect erosion associated with contemporary eustatic falls in sea level (Chapter 8; Figure 39). In reality, the major unconformities beneath the late Llandovery of the Builth and Shelve inliers may be of composite origin. Localised tilting and folding, associated with movements on the major fractures of the Pontesford Lineament, may have occurred intermittently from Caradoc times onwards. However, the unconformities themselves may record the cumulative effects of several periods of erosion in response to independent eustatic events, principally the late Ashgill and Aeronian regressions.

The presence of conglomerates and slumped beds in the Ashgill sequences along the Tywi Lineament has been cited as evidence for synsedimentary tectonic activity (Smallwood, 1986; James, 1991). It was suggested that coarse detritus accumulated in sediment traps adjacent to dip-slip faults downthrowing to the west (e.g. Rhiw Gwraidd Fault), although the fault tiplines remained buried beneath a thick sedimentary cover. However, the juxtaposition of dissimilar Ashgill sequences across faults, and the lack of correlation between the sandstone and conglomerate units on the opposing limbs of the Tywi Anticline, would indicate that there was significant later, probably Telychian strike-slip faulting along the Tywi Lineament prior to the Telychian overstep. Thus, although Ashgill dip-slip movements may have occurred along the Tywi Lineament, the present distribution of Ashgill stratigraphy is thought to be the result mainly of strike-slip displacements.

During late Llandovery times syndepositional faulting occurred along the Tywi, Central Wales and Glandyfi lineaments. The abrupt changes in the degree of the sub-Telychian overstep across the faults of the Twyi Lineament clearly indicate the presence of contemporaneous faults, downthrowing to the the west, and associated submarine erosion (Chapter 6) (Waters et al., 1992). To the east of the Cwmysgawen/Glanalders fault system over 1000 m of basinal Rhuddanian, Aeronian and early Telychian strata are absent, the whole of the Cwmcringlyn Formation and most of the Yr Allt Formation (some 350 m of strata) are lost to the east of the Carmel Fault, and over 800 m of Tycwtta Mudstones are overstepped across the Nant y Fedw Fault (Figure 43a). The earliest material evidence of fault-induced submarine erosion across the Tywi Lineament is provided by the resedimented deposits of the Cwm Barn Formation, of *utilis* Subzone age of the *turriculatus* Biozone, lying immediately to the west of the Cwmysgawen Fault. Faulting and erosion continued along the lineament into earliest *griestoniensis* Biozone times, when pebbly debrites of the Henfryn Formation were deposited to the east of the Carmel Fault. Slumps in the succeeding Dolgau Mudstones probably reflect continued instability over the lineament during the latest Telychian times.

At approximately the same period as the synerosional faults became active along the Tywi Lineament, a major depositional trough was developed to the west of the Glandyfi Lineament (Wilson et al., 1992). There, the southerly derived Aberystwyth Grits Group, over 2.5 km thick, were largely restricted to an area west of the Bronnant Fault (Chapter 5) (Figure 43a). Enhanced subsidence rates in *turriculatus* Biozone times, due to penecontemporaneous faulting, were initiated during the *gemmatus* Subzone and reached their maximum within the *utilis* Subzone. Activity on the Bronnant Fault declined during the *johnsonae* Subzone and, shortly after, the sandstone turbidite deposystem adjacent to the fault was abandoned.

Throughout much of the late *turriculatus* Biozone the syndepositional fault activity was located farther east, on the Central Wales Lineament. The 1.5 km-thick southerly derived Rhuddnant Grits preferentially accumulated to the west of the Claerwen Fault until late *crispus* times. Subsequent deposition of the thick sequence of Pysgotwr Grits of latest *crispus* to *griestoniensis* biozone age, mainly to the west of the Claerwen Fault, may suggest a reactivation of the structure.

Penecontemporaneous faulting during the deposition of the Telychian shelf successions to the east of the Tywi Lineament resulted in marked thickness changes between fault segments. The Cerig Formation attains its maximum exposed thickness to the east of the Garth Fault but farther east, across the Rock Park Fault, the formation is highly attenuated along the margins of the Builth Inlier.

During the Wenlock, the earlier (*centrifugus–riccartonensis* biozones) sandstone turbidites of the Penstrowed Grits were deposited in a trough to the west of the Garth Fault, but later (*rigidus–lundgreni* biozones) the depositional focus moved westwards and was controlled by the reactivated Cwmysgawen Fault.

ACADIAN STRUCTURES

The folds and cleavage in the Ordovician and Silurian strata of the district were formed during the early to mid-Devonian Acadian Orogeny (Soper et al., 1987). Southeast of the Tywi Lineament the intensity of the Acadian deformation decreases rapidly towards the margin of the stable Midland Platform (Figure 1). In the Rhayader sheet area the eastern limit of pervasive fold and cleavage development broadly coincides with the Garth Fault (Figure 43a). The rocks of the Builth Inlier are not cleaved.

Folds

The folds in the district, with the exception of the pre-Acadian folds in the Builth Inlier, were formed mainly during a single period of deformation. The rare examples of multiple phases of folding are due to localised late Acadian fold modification and fault movement. The relative order of magnitude of the folds in the district varies from broad regional structures to microfolds. The largest (first- and second-order) folds determine the main distribution of the formations on the geological map (Figure 43a and 44). Several lower orders of folds have formed on the larger structures, and it is these that are commonly observed at outcrop. The development of the low-order folds was strongly controlled by the composition of the folded lithologies. In mudstone-dominated formations open folds predominate, whereas tighter folds, of lower order, characterise the thinly interbedded sandstone and mudstone sequences.

FIRST-ORDER FOLDS

The structure of the district is dominated by three first-order folds, which trend north-north-east to north-east:

the Teifi Anticline, the Central Wales Syncline and the Tywi Anticline (Jones, 1912) (Figure 43a). These folds should strictly be referred to as anticlinoria and synclinoria, but the established terms have been retained. Their axial planes lie between 10 and 15 km apart. The Teifi Anticline is an open, upright to steeply inclined, periclinal structure with envelope limb dips of less than 20° (Section B–B′, Figure 44). In the south of the district, near Ystradmeurig [SN 704 677], early to mid-Llandovery strata are exposed in the core of the south-west-plunging anticline. North of the Ystwyth Fault the axial trace is less well defined, but continues in the general direction of the Ordovician inlier of the Plynlimon area (Cave and Hains, 1986). East of the Teifi Anticline the complementary Central Wales Syncline has a similar open structure and preserves a thick sequence of late Llandovery turbidite sandstones in its core (Section B–B′, Figure 44). The Tywi Anticline is a tight, inclined structure which is cored by Ordovician rocks. It has a vertical to overturned eastern limb disrupted by an anastomosing network of strike-faults (Figure 43a and Section A–A′, Figure 44). Towards the north-east the form of the Tywi Anticline becomes less distinct as the structure is traced into the overlying Wenlock strata. Many of the syndepositional faults associated with the Tywi Lineament were reactivated during the development of the Acadian Tywi Anticline.

SECOND-ORDER FOLDS

The second-order folds of the district typically range from open to close upright to steeply inclined structures, with axial plane separations between 1 and 5 km. In the central part of the district they comprise a series of periclinal outliers with opposing shallowly plunging fold axes (Figure 43a). On the western limb of the Teifi Anticline, near Llanafan [SN 688 726], the Blaen Myherin Mudstones Formation are preserved in two periclinal synclines, and along the axial zone of the Central Wales Syncline en échelon outliers of the Pysgotwr Grits occur near Llyn Cerrigllwydion [SN 840 693], Tan-y-berth [SN 914 753] and Craig y Lluest [SN 853 763]. West of Ysbyty Ystwyth [SN 725 715], rocks of the Cwmere Formation are preserved in the core of the second order Hendre Anticline, which can be traced to the north of the Ystwyth Fault at Grogwynion [SN 715 722].

Second-order folds along the north-western limb of the Tywi Anticline have variably spaced axial planes and differing fold styles. In the north, near Pant-y-dwr [SN 984 747], the axial planes of the open north-north-east-trending Tylwch Anticline and Waun Marteg Syncline are separated by approximately 5 km (Section A–A′, Figure 44), whereas farther south, near Llannerch Cawr [SN 902 615], the separation between the close, north-east-trending Rhiwnant Anticline and Gwesyn Syncline is less than 1.5 km (Section B–B′, Figure 44).

LOW-ORDER FOLDS

The form, orientation and facing direction of the low-order folds is variable across the district. Between the Cardiganshire coast and the Bronnant Fault the folds face upwards and to the west, and are generally westward-verging (Section C–C′, Figure 44). The folds are asymmetrical with axial surfaces inclined at up to 45°. The western limbs are steeply inclined to overturned; some can be traced for several kilometres along strike. Cliffs [SN 535 707] at Llanrhystud display a series of open chevron folds with easterly dipping axial planes (Plate 21a), and on the foreshore [SN 576 795] below Alltwen, tight westward-facing folds are exposed (Plate 21b). Many of the lower-order folds are disharmonic flexural-slip structures, locally associated with accommodating faults of small displacement. Fold axes have a variable shallow plunge and range from north-north-east-trending in the north- to north-east-trending in the south (Figure 43).

A progressive regional eastward steepening of the fold axial planes occurs towards the Glandyfi Lineament. Within the lineament, mainly along the crop of the Borth Mudstones, the folds are tight to isoclinal and upward-facing; on the western limb of the Teifi Anticline they are eastward-facing and generally eastward-verging (Section C–C′, Figure 44). This facing divide compares with, and can be traced into the Glandyfi Lineament of the Aberystwyth district (Cave and Hains, 1986).

Across the Teifi Anticline, upright to steeply inclined, open, low-order folds in mudstones of the Derwenlas Formation are exposed in sections near Pont-rhyd-y-groes [SN 741 727 to SN 747 727] and Ysbyty Ystwyth [SN 734 710 to SN 741 708]. In contrast, folds in the nearby thinly interbedded mudstone/sandstone sequences of the Devil's Bridge Formation are generally tighter and asymmetrical. In the Rhuddnant gorge, east of Craig Dolwen [SN 794 782 to SN 805 784], folds in the Devil's Bridge Formation are well displayed on the gently dipping eastern limb of the Teifi Anticline (Figure 45). Axial plane separations of the larger folds are generally less than 50 m, but smaller folds on a scale of metres or centimetres are common in these lithologies (Plate 22c). Folds in the thicker sandstone beds of the Ystrad Meurig Grits Formation, for example in Hendre quarry [SN 720 696], are generally open with rounded hinges; adjacent to strike-faults they have tighter profiles (Plate 22a).

Some of the low-order folds display a marked thinning of the vertical to overturned limbs and a cuspate fold profile (Plates 22b and c). This is the result of flattening during the later stages of fold development. In places, the steep limbs of the larger folds are defined by zones of small-scale tight to isoclinal folds and strike-faults. These zones display a considerable variation in width, and are flanked by more open folds. At Hendre quarry [SN 722 696] one such zone is less than 20 m wide and cut by gently dipping faults and veins (Figure 46 and Plate 22b). To the south of Hafod Mansion [SN 759 733] another zone is approximately 200 m wide and associated with steeply inclined normal and reverse faults (Figure 47).

Across the Central Wales Syncline the predominant folds in the sandstone-rich parts of the Cwmystwyth Grits Group are close, asymmetrical structures with steep north-west-dipping axial planes and vertical to overturned eastern limbs (Plate 23a). Axial plane separations

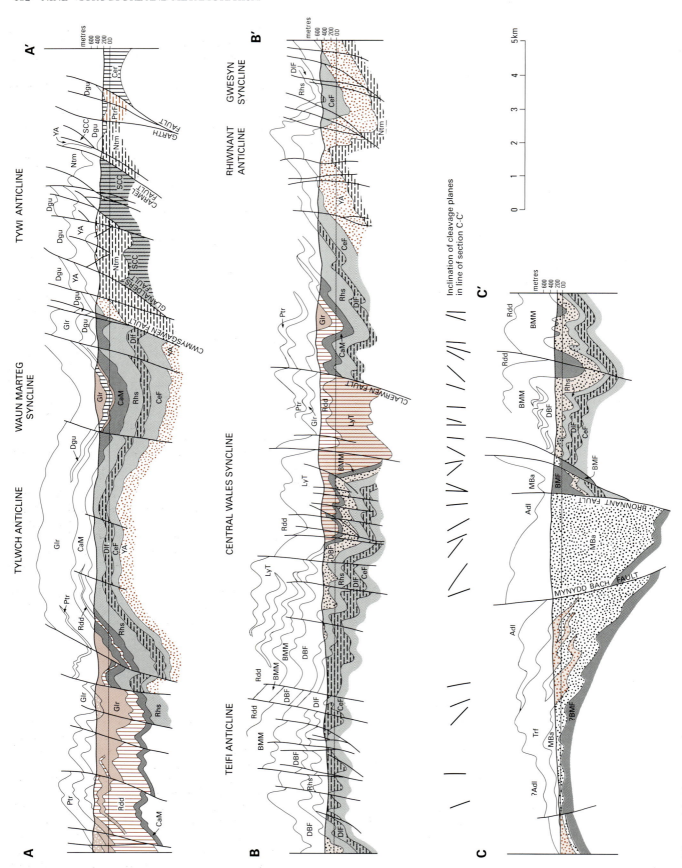

Figure 44   Structural cross-sections across the district. See Figure 43a for location of sections. For key see facing page. Cleavage inclinations are shown above Section C–C'. Horizontal and vertical scales are equal.

Llandovery

	Dolgau Mudstones (Dgu) and Cerig Formation (Cer)
Ptr	Pysgotwr Grits
Glr	Glanyrafon Formation
Rdd	Rhuddnant Grits
LyT	Llyn Teifi Member
	Blaen Myherin (BMM), Adail (Adl) and Caerau (CaM) mudstone formations
Trf	Trefechan Formation
MBa	Mynydd Bach Formation
BMF	Borth Mudstones Formation
DBF	Devil's Bridge Formation
Rhs	Rhayader Mudstones
DlF	Derwenlas Formation
CeF	Cwmere Formation

Ashgill

YA	Yr Allt and Cwmcringlyn formations
PtrF	Pentre Formation
Ntm	Nantmel Mudstones

Caradoc

SCC	St Cynllo's Church Formation

1 km wide, also crop out in the axial zone of the Central Wales Syncline along Nant y Bont [SN 790 601 to SN 797 597]. These structural transitions represent a southwards increase in regional shortening, which is accommodated by tightening of the folds and displacements on steeply dipping strike-faults.

Small-scale folds are present on the second-order folds to the west of the Tywi Anticline. Within the core of the Tylwch Anticline these folds have wavelengths of less than 500 m in the Rhayader Mudstones; they are well exposed along the Wye valley between Dolhelfa [SN 936 733] and Rhayader. The overturned south-eastern limb of the Rhiwnant Anticline, north of Llanerch Cawr [SN 902 618], contains a series of subordinate tight, asymmetrical folds in the sandstones of the Cerig Gwynion Grits. On the moderately dipping north-western limb, several south-east-verging folds with north-east-plunging fold axes crop out along Afon Claerwen [SN 885 627] near Ciloerwynt. Farther east, an anticline and syncline, with axial-plane separations of 500 m, are present near Bryn Moel [SN 930 598], and small-scale, tight eastward-facing folds are exposed in the Nantmel Mudstones [SN 9644 6100], south-east of Nant Cymrun.

In the hinge zone of the Tywi Anticline, near Abbey-cwmhir [SO 052 712], Ashgill and upper Llandovery strata form a series of open to close, eastward-verging folds. The eastern limbs of these folds, in the conglomerate exposed on Camlo Hill [SO 040 695], are overturned (Figure 9). In the overlying Wenlock sequences folds comprise cuspate anticlines, forced above faults, interspersed with open synclines; axial-plane separations range between 0.5 and 1 km (Figure 43a).

In the district, low-order folds are rare on the vertical to overturned south-eastern limb of the Tywi Anticline. Folded dolerite sills define a tight overturned syncline/anticline pair [between SO 0560 6665 and SO 0626 6794] on Baxter's Bank (Figure 11), and a close south-east-facing anticline, in sandstones of the Nantmel Mudstones, is exposed at Carreg-yn-fol Wood quarry [SN 9899 6226].

Fold axes along the Tywi Anticline define a broad arc (Figure 43b). In the hinge zone, near Abbeycwmhir [SO 055 715], and within the overlying Wenlock sequences the folds plunge at about 20° towards the north. South-west of Llanwrthwl [SN 976 637] they plunge very shallowly south-west (Figure 43b).

Cleavage

The majority of the rocks within the district are cleaved. The intensity of the cleavage decreases rapidly south-east of the Garth Fault; the Ordovician rocks of the Builth Inlier are not cleaved. The cleavage in the mudstones varies from a spaced discontinuous fabric in areas of low-grade metamorphism (diagenetic zone and low anchizone) to a closely spaced fabric in areas of higher grade (high anchizone and epizone). The microtextural characteristics of the cleavage are described in the later section on metamorphism. The development of cleavage varies between different lithologies, and at any one locality the mudstones display a more pronounced and

of the larger folds observed at outcrop vary between 100 and 500 m. However, more open, upright folds occur locally (Plate 23b), and along the western limb of the Central Wales Syncline there is a progressive change in fold style. In the north of the district, for example along the Graig Ddu cliff section [SN 810 739 to SN 814 739], belts of steeply inclined folds are separated by more-open folds (Figure 48). Farther south, along the road to Llyn Teifi [SN 786 681], the folding is more uniform with topographical ridges of steeply dipping, overturned strata flanked by low-angle dip slopes (Plate 23c). This fold style is also displayed in a section [SN 773 658 to SN 777 660] on the north side of the Egnant valley, where adjacent hinges of the same order folds define a sheet dip of about 15° to the east. South of the Egnant valley the fold hinges and the shallow limbs are progressively faulted out, leaving wide tracts of steeply dipping, eastward-younging strata. Similar tracts, up to

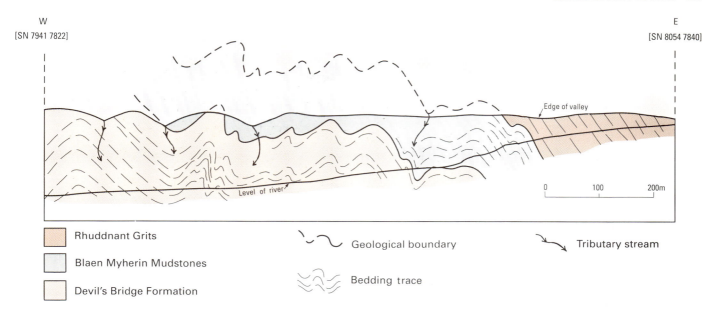

W
[SN 7941 7822]

E
[SN 8054 7840]

Edge of valley

Level of river

0   100   200m

Rhuddnant Grits

Blaen Myherin Mudstones

Devil's Bridge Formation

Geological boundary

Bedding trace

Tributary stream

Figure 45   Sketch of the structures and stratigraphy displayed along the upper reaches of the Rhuddnant valley. The structures have been projected on to an east–west plane.

closer-spaced cleavage than the adjacent sandstones (Plate 24a). Cleavage is poorly developed in rocks where the original bedding fabric has been destroyed by syn-sedimentary slumping. This is most noticeable in the disturbed beds of the Ordovician Yr Allt Formation, which display an irregular anastomosing fabric.

At some localities in rocks of anchizone and epizone grade the main cleavage is cut by a second, more widely spaced, cleavage. The thin silty mudstone units of the Yr Allt Formation in the Rhiwnant Anticline [at SN 8947 6150], and the Rhayader Mudstones along the Claerwen reservoir [at SN 8685 6462] display the second cleavage oriented 20° clockwise to the earlier fabric. The two cleavages are considered not to belong to two separate, regional fold-forming events, but rather to reflect a late stage, local reorientation of the principal stress during the main deformation.

CRENULATION CLEAVAGE

A localised development of a crenulation cleavage, defined by the microfolding of the earlier cleavage, occurs adjacent to some sandstone beds, faults and late quartz veins. The crenulations generally have wavelengths of a few millimetres, but in places are associated with chevron folds with wavelengths of a few centimetres, or in rare cases of up to 3 m (Fitches, 1972). The orientation of the

crenulation cleavage is most commonly subhorizontal and preferentially developed in the steeply inclined limbs of asymmetrical folds. At Y Gorlan [SN 8295 7544] a gently dipping crenulation cleavage is present in a vertical mudstone unit of the Rhuddnant Grits. Subvertical crenulation cleavages occur close to steep quartz veins at Graig Ddu [SN 8109 7381], and adjacent to sandstone beds in Hendre quarry [SN 7202 6992]. The crenulation cleavage is considered to have formed at a late (post-cleavage) stage by localised differential shear in response to fold modification, faulting and veining.

CLEAVAGE ORIENTATION

The strike of the cleavage changes across the district from approximately north-north-east in the north, to north-east in the south (Figure 43b). This variation forms part of the regional arcuation in the strike of the cleavage and trend of the fold axes across the southern part of the Welsh Basin (Woodcock et al., 1988). Local deviation from this pattern is the result of reorientation of the cleavage adjacent to late transverse faults, for example close to the Ystwyth Fault. The direction and amount of dip of the cleavage changes across the fold vergence divide of the Glandyfi Lineament. West of the lineament the cleavage consistently dips at 45° to 60° towards the south-east, but becomes progressively steeper eastwards (Section C–C′, Figure 44). East of the lineament the cleavage generally dips towards the north-west.

On a small scale, cleavage fans occur around many of the low-order folds, being particularly apparent in sequences of thinly interbedded sandstones and mudstones (Plate 24b). Cleavage dips also vary between different lithologies, the angle being least in mudstones and greatest in coarse-grained sandstones (Plate 24a).

Plate 21   West-facing folds in the Trefechan Formation (facing page).

a   Chevron folds. Cliffs [SN 535 707] north of Llanrhystud, viewed from south.   **b**   Tight folds. Foreshore [SN 576 795] below Allt-wen, viewed from north. Faults have developed on the limbs and in the hinge zone of the fold.

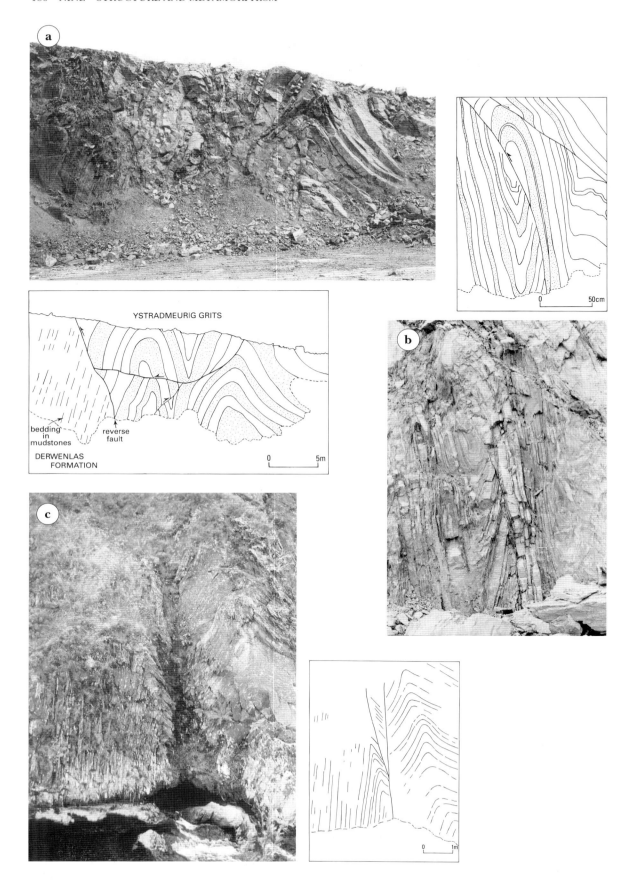

YSTRADMEURIG GRITS

bedding
in
mudstones

reverse
fault

DERWENLAS
FORMATION

0   5m

0   50cm

0   1m

E   W

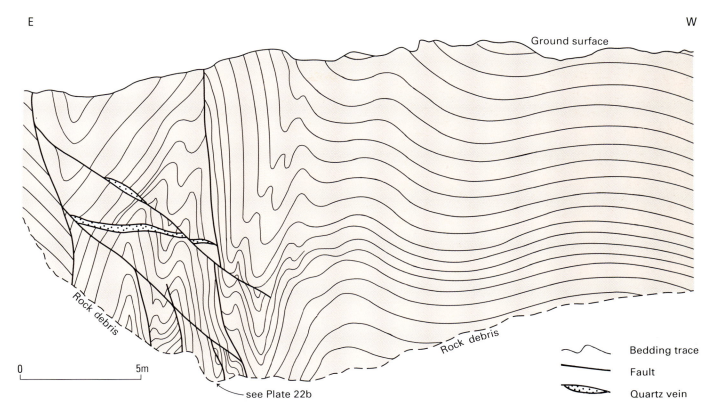

Figure 46   Sketch of the fold structures displayed in Hendre quarry. South face of level 4 [SN 722 695].

CLEAVAGE TRANSECTION

Over most of the district the cleavage is not absolutely parallel to the fold axial planes, but transects them in a clockwise sense (Figure 43b and 49). The minimum angle ($\Delta°$) between the fold axis and the cleavage is a measure of the degree of this transection, and can be derived from stereograms of bedding and cleavage (Borradaile, 1978). Although $\Delta°$ only represents an average value of the transection angles for all orders of folds within a particular area, and is affected by the periclinal form of the folds (Johnson, 1991), the regional trends are considered significant.

Plate 22   Fold styles in the Ystrad Meurig Grits and the Devil's Bridge formations of the Teifi Anticline (facing page).

a   Upright shallowly plunging folds in thickly bedded sandstones of the Ystrad Meurig Grits Formation. Level 3 Hendre quarry [SN 7215 6985], viewed from south. Mudstones of the Derwenlas Formation are exposed to the west of a strike-parallel steep reverse fault. **b**   Tight folds in thinly bedded sandstones and mudstones of the Ystrad Meurig Grits Formation. South face Level 4 Hendre quarry [SN 7205 6965]. The position of these folds on the face is identified in Figure 46. **c**   Asymmetrical east-facing folds in the Devil's Bridge Formation exposed in the gorge of Nant Rhuddnant [SN 7967 7831]. Viewed from the north.

Across the Teifi Anticline and to the west of the Bronnant Fault there is a systematic change in clockwise transection angles from high (10° to 16°) in the north to low (4° to 7°) in the south; this is accompanied by an arcuation in the trend of fold axes from north-north-east to north-east respectively (Figure 43b). These changes are not apparent over the Central Wales Syncline, where clockwise transection is at a maximum (8° to 15°) on the western limb and averages 4° over the axial zone and on the eastern limb. These variations in transection angles across the Central Wales Syncline continue to the south-east of the district along the Afon Berwyn to Camddwr section (Woodcock, 1990). Farther east, on the western limb of the Tywi Anticline, the transection angles decrease markedly, and along the arcuate axial zone of the anticlinorium the cleavage remains near to axial planar. Anticlockwise transection is restricted to the core of the Rhiwnant Anticline (4°) and between the Rhiw Gwraidd and Nant y Fedw faults (10°), close to the eastern limit of cleavage development.

At several localities, cleavage can be observed to transect the axes of the low-order folds. Clockwise transection is well displayed across folds in the Devil's Bridge Formation at Cwm Rheidol [SN 7296 7804] (Plate 24c) and in Rhuddnant Grits at Cwm Ddu [SN 811 738] and Cwm Ystwyth [SN 802 745]. At Cwm Rheidol and Cwm Ddu the transection angles are greatest over tight folds and least over adjacent, more-

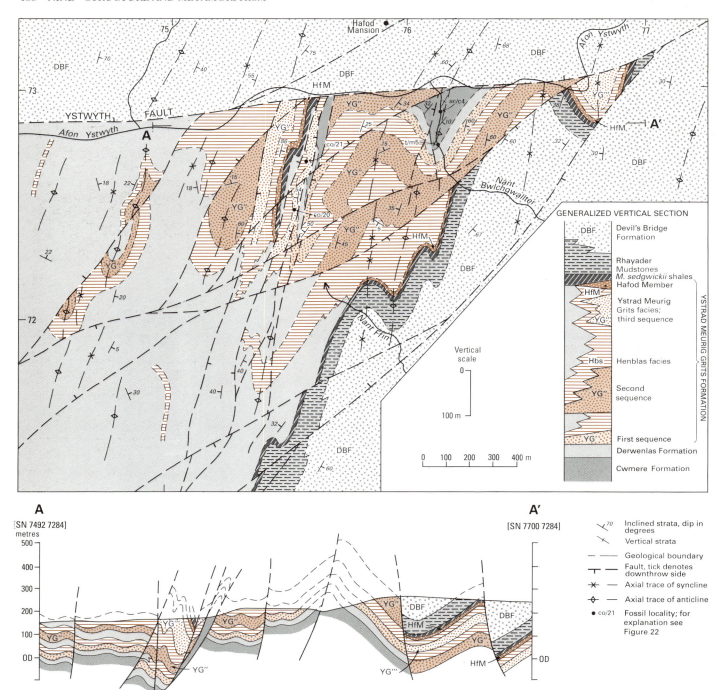

Figure 47 Geological map of the area to the south of the Ystwyth Fault near Hafod Mansion [SN 759 733].

open folds. At Llyn Teifi [SN 7785 6822] an exposed hinge of an asymmetrical fold shows the greatest clockwise transection angle (34°) across the hinge zone and the least (19°) on the limbs; the cleavage is axial planar to secondary minor folds on the subvertical limb of the main fold (Figure 49). These relationships suggest that the secondary folds were associated with the cleavage development, and that there was post-cleavage

tightening of the main fold, which resulted in the change of strike of the cleavage.

It has been widely accepted that transected folds are characteristic of transpressive deformation (Borradaile, 1978; Sanderson and Marchini, 1984; Soper, 1986b), although this has been challenged by other researchers (Treagus and Treagus, 1992). It has been suggested that clockwise transected folds, characteristic of the Welsh

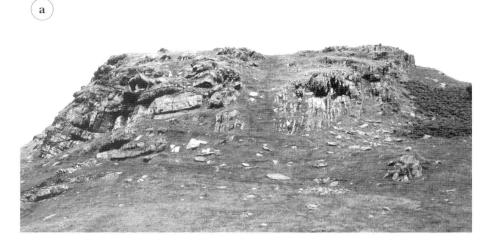

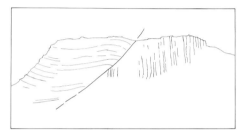

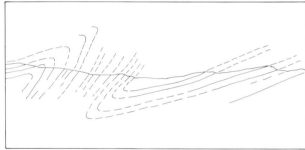

Plate 23   Fold styles in the Cwmystwyth Grits Group of the Central Wales Syncline.

a   Asymmetrical east-facing anticline in the Rhuddnant Grits. [SN 7853 6477], west of Garreglwyd, viewed from south.
b   Open syncline in thick sandstones of the Pysgotwr Grits. [SN 850 730] north-east of Llyn Cerrigllwydion Uchaf, viewed from north.   **c**   Asymmetrical east-facing folds in the Rhuddnant Grits. [SN 783 684] north of Llyn Teifi, viewed from south. The featured topography is defined by gentle dip slopes, and ridges of overturned strata.

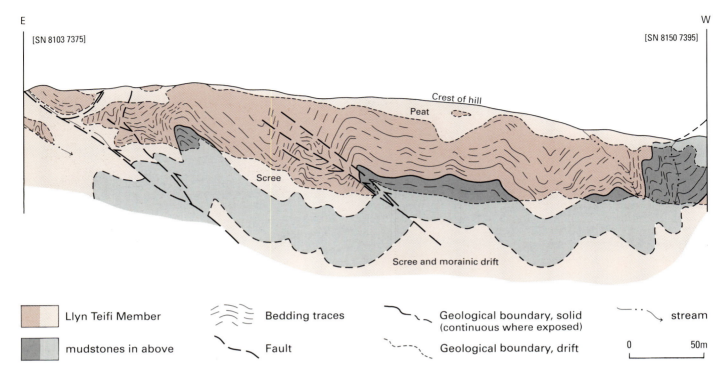

E [SN 8103 7375]   W [SN 8150 7395]

Crest of hill

Peat

Scree

Scree and morainic drift

Llyn Teifi Member

mudstones in above

Bedding traces

Fault

Geological boundary, solid
(continuous where exposed)

Geological boundary, drift

stream

0   50m

Figure 48 Sketch of the structures displayed in the cliffs of Graig Ddu, viewed from the Cwmystwyth road [SN 8125 7484] to the north. The fold profiles on the eastern end of the section are enhanced due to the oblique orientation of the cliff line.

Basin and other Acadian slate belts in the British Isles, were developed during sinistral transpression (Soper et al., 1987; Woodcock et al., 1988). Two transpressive mechanisms have been proposed to account for the clockwise transection: the first suggests that the folds were initiated and rotated prior to the formation of cleavage (Soper, 1986b), and the second that reactivation of basement faults, striking at an oblique angle to the maximum compressive stress, controlled the orientation of the

Cleavage
surface

W

E

Bedding surface

Figure 49 Diagrammatic representation of the fold/cleavage relationships near Llyn Teifi [SN 7785 6822].

folds in the cover (Sanderson and Marchini, 1984; Woodcock, 1990). Variations in degree of transection have been attributed to partitioning of a greater strike-slip component into fault zones, fold cores and steep limbs (Woodcock, 1990), enhanced transpressive strains in the more competent sequences (Stringer and Treagus, 1980; Woodcock, 1990; Pratt and Fitches, 1993), and moulding against curved rigid bodies (Soper et al., 1987).

The pattern of cleavage transection across the district cannot be attributed to one mechanism alone. It is possible that the decrease in transection angles towards the south-east was controlled by the regional north-westerly orientation of the maximum compressive stress relative to the major lineaments: anticlockwise to the

Plate 24 Cleavage relationships in the district (facing page).

a Cleavage refraction in the Rhuddnant Grits. [SN 783 684] north of Llyn Teifi. The mudstones at the base of the picture display a low-dipping slaty cleavage, the ripple-laminated sandstone bed is cut by ill-defined subvertical joints, and a steeply dipping pressure solution cleavage has developed in the overlying, thick, high-matrix sandstone. **b** Fanned cleavage across folds in interbedded mudstones and ripple-laminated sandstones of the Rhuddnant Grits. North shore [SN 8481 6546] of Claerwen reservoir. **c** Clockwise cleavage transection across an anticlinal hinge zone in the Devil's Bridge Formation. Afon Rheidol [SN 7294 7809], south-west of Ystumtuen.

Glandyfi Lineament and nearly orthogonal to the main north-easterly trend of the Tywi Lineament. However, the decrease may also be explained by the relative timing of fold and cleavage development, such that folding preceded the formation of cleavage in the west but became almost synchronous with it along the margins of the Basin in the east. The two mechanisms are not exclusive. An additional controlling factor was the partitioning of the greatest transpression along the steeply inclined to overturned limbs of folds, in particular along the western limb of the Central Wales Syncline.

AGE OF CLEAVAGE FORMATION

Recent studies provide evidence that cleavage was initiated at an early stage in the deformational history of the Welsh Basin. Roberts et al. (in press) have proposed that the mainly depth-related pattern of metamorphism (see below) was imposed on strata which were already undergoing deformation. ^{40}Ar-^{39}Ar plateau ages of 416–420 Ma from cleavage micas, obtained by Dong et al. (in press) from Cambrian bentonites and Ordovician slate in both north and south-west Wales, indicate that a cleavage formed within these rocks as early as mid-Ludlow times. Rb-Sr whole rock isotope analyses from the well-cleaved Rhayader Mudstones (Llandovery) at the Claerwen Reservoir [SN 868 647] have given a broadly comparable regression age of 431 ± 12 Ma, and similar results have been obtained from areas in north Wales (Evans, 1989). It therefore appears that parts of the Welsh Basin were being cleaved at a time when other areas were undergoing active sedimentation. This suggests that cleavage did not instantaneously develop across the basin, but probably spread progressively through the thick sedimentary pile under the influence of compression and a relatively high geothermal gradient (see below). High-strain zones, notably around reactivated faults, may have been particularly susceptible to early cleavage development and probably sustained it longer. The transection relationship of folds and cleavage, and localised development of a later component of the cleavage, both point to cleavage development during a prolonged period of deformation of the basin (cf. Soper, 1986b).

It is generally accepted that the main cleavage-forming event in Wales climaxed in the early to mid-Devonian (Emsian or Eifelian stages), postdating the deposition of the Lower Old Red Sandstone (Woodcock, 1984a; Soper et al., 1987). Cleaved red-beds of Přídolí age (latest Silurian) have been recorded in places along the Pontesford Lineament, south of the district (Cocks et al., 1984), and probably on Anglesey in north Wales (Allen, 1974). The timing of this event is confirmed by Rb-Sr whole rock isotope analyses of volcanic rocks and cleaved mudstones in north Wales, which give mean regression ages from 399 to 409 Ma (Evans, 1989, 1991); a sample from the Nantmel Mudstones (Ashgill) near Llanwrthwl [SN 994 629], with a regression age of 403 ± 15 Ma (Figure 50), supports these data. Closure of the Rb-Sr whole rock system was probably coincident with the eventual uplift of the Welsh Basin, which was broadly synchronous with the Acadian Orogeny of Canada (Soper et al., 1987).

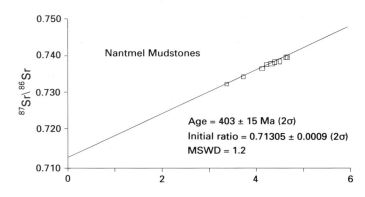

Figure 50 Whole rock Rb/Sr isochron of the Nantmel Mudstones. Samples from a road cutting [SN 976 643] on the A470 near Ashfield.

Faults

STRIKE FAULTS

Strike faults form a north-north-east- to north-east-trending set subparallel to the regional fold trend; some can be traced for over 15 km. The major strike faults have large pre-Acadian displacements (Chapters 5, 6 and 7), but there is little evidence that there were substantial displacements during the Acadian Orogeny. Along the Glandyfi and Central Wales lineaments the amounts of Acadian displacement on the strike faults are poorly constrained. The Glanalders and Carmel faults of the Tywi Lineament, which display reverse throws of up to 1500 m in the Ordovician strata, die out towards the north-east in the overlying Wenlock sequences. Acadian displacements on these faults, including lateral movements, must therefore have been small, although they may increase with depth. On the eastern flank of the Tywi Lineament, the Rhiw Gwraidd, Nant y Fedw and Garth faults were also probably active during the Acadian movements, but the amounts and directions of the displacements are unknown.

Smaller-scale strike faults are numerous throughout the district. They are developed preferentially along the axial planes and steep limbs of the eastward-verging folds. The fault planes generally dip steeply, at over 70°, to the north-west and display both reverse and normal displacements. An anastomosing set of strike faults connects the major faults along the south-eastern limb of the Tywi Anticline to the north of the River Wye (Figure 43), and strike faults with normal displacements cut the steeply dipping, south-eastern limb of the Rhiwnant Anticline (Section B–B', Figure 44). On the eastern limb of the Teifi Anticline, near Hafod [SN 759 733], strike faults with both normal and reverse displacements are associated with a narrow zone of tight folds (Figure 47). Folds seen at outcrop are commonly faulted; for example, near Garreglwyd [SN 7854 6477] the axial plane of an asymmetrical anticline in the Rhuddnant Grits is approximately coincident with a normal fault downthrowing to the north-west (Plate 23a), and at

Hendre quarry a zone of tight folds is flanked by steeply dipping strike faults (Figure 46).

Thrust faults are not common, and no gently dipping decollements of large displacement have been identified in the district. Thrust faults with small displacements have been mapped along the contact between the Caerau Mudstones and Rhuddnant Grits near Craig Goch reservoir [SN 885 690], and at Hendre quarry [SN 722 696] a thrust fault is associated with a hanging wall anticline and sigmoidal extension veins in the footwall (Plate 25). These thrust faults are thought to have formed in response to local bedding-plane slip during folding.

TRANSVERSE FAULTS

The folds, cleavage and faults formed during the main Acadian deformation are cross-cut by four transverse fault sets trending approximately east-north-east, south-east, east, and north (Figure 43a). The first three sets can readily be identified on the satellite image for the district (Plates 26a and b) (Maude, 1987). Additional linear features seen on the image, which are parallel to known faults, probably represent joints or faults with small displacements, not identified during the geological mapping.

The east-north-east-trending faults form the dominant set in the western and central parts of the district. They

Plate 25   Thrust in the Ystrad Meurig Grits Formation.
Gently dipping thrust associated with a hanging wall anticline and sigmoidal tension veins in the footwall. West face of Level 4, Hendre quarry [SN 7205 6975].

dip steeply, usually at over 70°, and display normal down-throws to the north-north-west or south-south-east. Slickensides present on some fault surfaces are subhorizontal or inclined, and indicate at least one phase of oblique-slip movement. For example, in the entrance to an adit [SN 7554 7230] at Ty Coch, slickensides on several closely spaced surfaces within a fault zone plunge at 50° towards the north-east. Adjacent to some east-north-east-trending faults, bedding and cleavage have been reorientated, and locally second generation folds have developed. At Ffrwd yr Ydfran [SN 8226 7636] vertical beds on the eastern

limb of a north-east-trending fold have been refolded into a set of close folds with vertical fold axes.

Few south-east-trending faults have been mapped, although linear features with this orientation are common on the satellite image (Figure 26b). The Lledrod Fault is the most important fault of this set and may be traced from Swyddffynnon [SN 696 658] to north-west of Lledrod [SN 630 709]; it has sinistrally displaced the Bronnant Fault by at least 1.5 km.

The Ystwyth Fault trends 080° and is generally coincident with the steeply incised valleys of Afon Wyre and

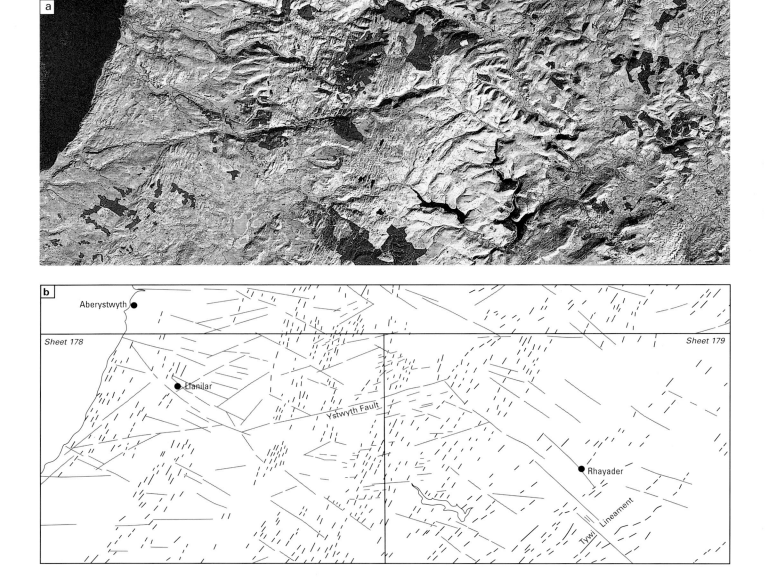

Plate 26 Satellite image and lineament interpretation of the district.

a Satellite image of the district processed by the British Geological survey from Landsat Thematic Mapper scenes for Band 5. If the valleys appear as positive topographic features, it is sometimes possible to obtain a 'correct' view of the landscape by inverting the image. **b** Lineament interpretation of the satellite image (sheet boundaries indicated). Black linears refer to possible bedding and cleavage features,

Afon Ystwyth. It is readily identifiable on the satellite image (Figure 26a). In the west near Lledrod [SN 645 704], the fault is made up of a series of smaller *en-échelon* faults which partition the total displacement across a fault zone. Farther east, in the vicinity of Hafod [SN 759 733] and Cwmystwyth [SN 789 741], the fault converges into a single structure before dividing again into a series of minor fault splays towards the east-north-east. The fault surface is exposed in two sections near Pont-rhyd-y-groes: at the first [SN 7386 7267] it forms a zone of unmineralised crushed and sheared mudstones up to 15 m wide, and at the second [SN 7405 7278 to SN 7419 7282], along the bed of Afon Ystwyth, it is associated with large lenses and stringers of massive calcite. It was also intersected underground in Pugh's Adit at Cwmystwyth Mine [SN 8010 7442], where a section of 'soft ground' 20 m wide was reported containing brecciated blocks of mineralised vein material (Jones, 1922).

An estimate of the displacement on the central part of the Ystwyth Fault has been calculated where it cuts the northern end of the Hendre Anticline [SN 715 721], to the west of Pont-rhyd-y-groes. Fold hinges for particular stratigraphical contacts are displaced 1300 m sinistrally and downthrown 300 m to the north across the fault. The same order and sense of movement would account for the displacement of the Blaen Myherin Mudstones Formation in the Cwmystwyth area [SN 790 741]. To the west, in the Lledrod area [SN 645 705], the combined sinistral displacement of the subvertical Bronnant Fault, across the whole Ystwyth Fault zone, is approximately 1 km. However, this displacement includes both sinistral and dextral movements on east-north-east-trending faults, and sinistral movement on the south-east-trending Lledrod Fault.

A series of north-trending transverse faults cuts the Rhiwnant Anticline and Gwesyn Syncline in the south-east of the Rhayader sheet area (Figure 43a). The faults dip steeply and are normal, with downthrows generally to the west. However, inclined slickensides on a fault plane [at SN 8850 6057] indicate an element of oblique-slip. Several sections of these faults are mineralised but a 3 m-wide fault zone [at SN 8850 6057] consists of angular unorientated blocks of cleaved mudstone set in a matrix of pale grey altered mudstone. A north-trending fault was uncovered during the construction of the foundations of Claerwen dam; the fault zone was up to 2.5 m wide and consisted of a clay gouge (Morgan et al., 1953).

The final displacements on the transverse faults post-date the formation of the main Acadian folds and cleavage. Some of the transverse faults contain mineralised vein systems which have been isotopically dated as approximately early Devonian (see Chapter 11). This indicates that the stresses which formed the transverse faults were the result of a late phase of the Acadian Orogeny. The east-north-east- and south-east-trending faults could belong to a conjugate set, whereas the north-trending faults are more likely to be related to renewed displacements, possibly with a strike-slip component, on the north-east-trending faults along the Tywi Lineament. The last movement on the Ystwyth Fault postdates the emplacement of the mineralised veins (p.214); it may, therefore,

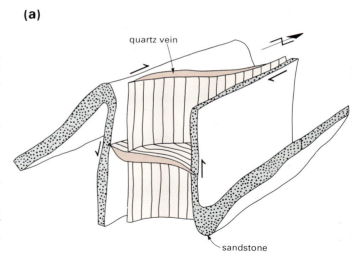

(a)

quartz vein

sandstone

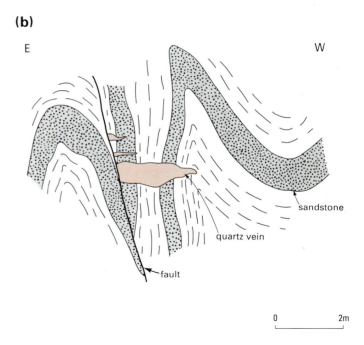

(b)

E   W

sandstone

quartz vein

fault

0   2m

Figure 51 Vein arrays within steeply dipping limbs of asymmetrical folds.

a Schematic diagram of two arrays displayed in the Graig Ddu section [SN 813 739]. **b** Subhorizontal vein array adjacent to a fault. Fault developed along the axial plane of a syncline near Y Gorlan [SN 8295 7543].

have taken place during the early stages of the Hercynian movements.

Veins

The veins in the district are complex and belong to at least three generations. The first formed prior to the main cleavage-forming event, the second was post-cleavage and related to fold modification, and the third was emplaced along the transverse faults.

PRE-CLEAVAGE VEINS

Veins parallel to bedding occur throughout the Silurian basinal succession of the district. They consist of laterally continuous ribbons of calcite and quartz which display variably orientated striations on successive laminations. Tanner (1989, 1990) has suggested that these veins were formed by differential movement across flexural-slip folds during the main tectonic deformation, but prior to the imposition of cleavage. It has also been argued that the veins formed prior to the regional deformation, by a multiple crack-seal process in lithified sediment (Cave and Hains, 1986; Fitches et al., 1986, 1990). Examples are exposed on the foreshore [SN 535 707] north of Llanrhystud.

POST-CLEAVAGE VEINS

Post-cleavage tension veins consist of coarsely crystalline quartz with subordinate carbonate and chlorite. They are commonly confined to the steeply dipping to over-turned limbs of asymmetrical folds. Two vein sets are prominent, gently dipping planar to slightly sigmoidal veins, which range in width from a few millimetres to over 50 cm thick, and subvertical veins oriented at an acute angle to the bedding (Figure 51a). Crenulation of the cleavage along the margins of the veins is common. These veins probably developed as the result of post-cleavage flattening and differential shear across the steep limbs of the folds. Postcleavage tension veins have also developed adjacent to some faults (Figure 51b). At Hendre quarry [SN 7205 6975] sigmoidal quartz veins have developed in the footwall of a thrust (Plate 25).

TRANSVERSE FAULT VEINS

The transverse faults commonly contain quartz veins which display multiple episodes of crystallisation, brecciation and slickenside formation. Many of these veins, in particular the north-north-east-trending set, are mineralised along parts of their lengths. They are described in Chapter 6.

Plate 27 Back-scattered electron image of a cleaved mudstone.

Mudstone (sample BRM 1280) of anchizonal grade from the Rhayader Mudstones near Rhayader [SN 979 688]. Chlorite-mica stacks (cs-chlorite; ms-white mica), grains of quartz (q), albite (ab) and anatase (Ti), and detrital K, Na-mica (mi) are contained in a fine-grained matrix of intergrown white mica and chlorite. Anastomosing domains of oriented white mica and chlorite intergrowths form a discontinuous cleavage, trending approximately vertical on the image. This image was obtained in the Electron Microbeam Laboratory of the Department of Geological Sciences, University of Michigan, USA, courtesy of Professor D R Peacor.

LOW-GRADE REGIONAL METAMORPHISM

The Welsh Basin was affected by low-grade regional metamorphism characterised by grades ranging from the late diagenetic zone, through the anchizone and into the epizone (Roberts, 1981; Bevins and Rowbotham, 1983; Merriman and Roberts, 1985; Robinson and Bevins, 1986; Robinson et al., 1990). The process started with the crystallisation of submicroscopic white micas and chlorite parallel to the bedding (Li et al., 1994), and intensified during the Acadian Orogeny with the growth of coarser-grained mica and chlorite in the cleavage. Throughout the district a broad correlation is found between the grade of metamorphism and the intensity of cleavage development. Cleavage in the mudstones is absent, or only very weakly developed, in the late diagenetic zone; discontinuous spaced cleavage is common in rocks of the low anchizone; and a nearly continuous, more closely spaced, cleavage is present in the high anchizone and epizone. Microtextures characteristic of the discontinuous cleavage show anastomosing domains of oriented metamorphic white mica and chlorite, typically 0.5 to 5.0 μm wide, enclosing chlorite-mica stacks and other detrital grains (Plate 27). In the high anchizone the cleavage is defined by thicker and more numerous microtextural domains of oriented white mica and chlorite. During the formation of cleavage there was an increase in white mica crystallite thickness (Merriman et al., 1990) and modification of the morphology of the detrital grains (Milodowski and Zalasiewicz, 1991).

In samples with two cleavages the metamorphic white mica is concentrated within the earlier cleavage, whereas iron-stained chlorite with subordinate white mica is developed along the younger pressure solution seams.

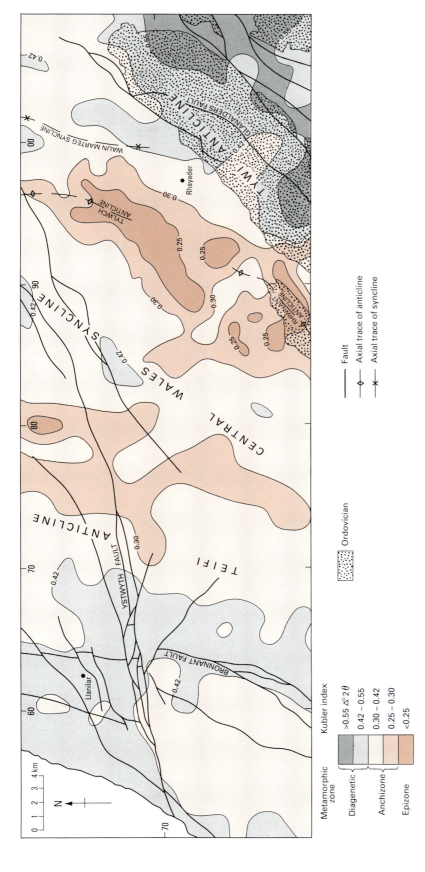

Figure 52 Metamorphic map of the district. The map is based on the Kubler indices of white mica crystallinity of 663 mudstone samples.

Table 20 Kubler indices of white mica crystallinity of mudstones from the Rhayader sheet area. The data are plotted as ranges and means for each formation (n = number of samples; x̄ = one standard deviation).
a. Ashgill, Llandovery and Wenlock formations to the west of the Glanalders Fault.
b. Caradoc to Wenlock strata of the Tywi Anticline, east of the Glanalders Fault.

a.

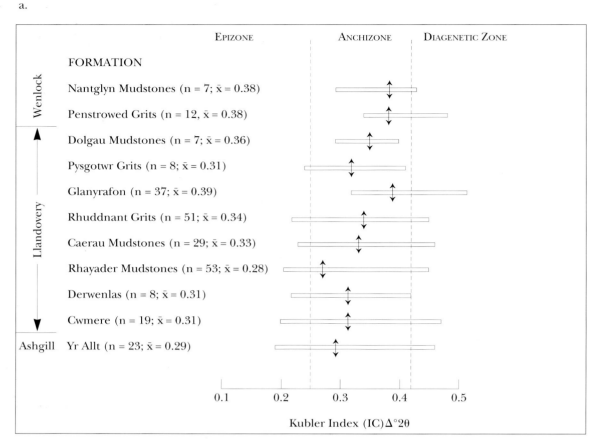

b.

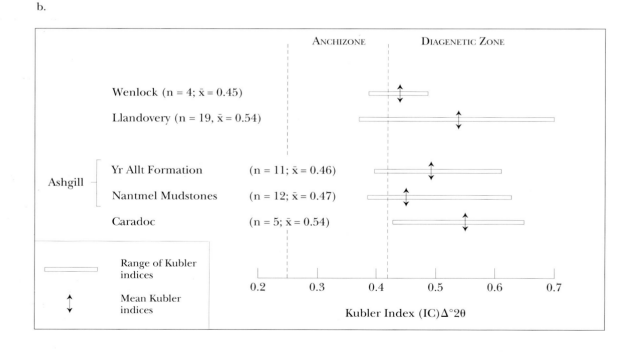

Metamorphic map

Mudstones and hemipelagites were sampled at a density of approximately 1 sample per 1.5 km^2 over the whole district. White mica crystallinity values (Kubler indices) and the mineralogy of the less-than-2 µm fraction of these samples were determined using standard techniques (Kisch, 1990; Roberts et al., 1991; Merriman et al., 1992). The internationally recognised limits of the anchizone (Kubler indices 0.42 and 0.25), together with subdivisions of the late diagenetic zone and anchizone (Kubler indices 0.55 and 0.30) have been used to generate a contoured metamorphic map of the district (Figure 52).

Over most of the district the traces of the isocrysts (lines of equal Kubler index) trend subparallel to the formational boundaries, except close to the Ystwyth Fault in the Llanilar sheet area where they trend east-north-east. The regional pattern shows late diagenetic grades within and to the south-east of the axial zone of the Tywi Anticline and along the western part of the Llanilar sheet area, and higher grades in the central part of the district. In more detail, high anchizonal and epizonal rocks occur in the axial regions of the Teifi Anticline, along the common limb of the Teifi Anticline and Central Wales Syncline, and in the axial regions of the Rhiwnant and Tylwch anticlines. Late diagenetic rocks, to the west of the Tywi Anticline, are mainly restricted to a narrow zone to the east of the Bronnant Fault, along the coastal area to the north of the Ystwyth Fault, and over small areas along the axial region of the Central Wales Syncline.

Conditions of metamorphism

The regional metamorphic pattern across the basinal Silurian sequences shows that the lowest grades are generally associated with the youngest strata. An example of this trend is seen between the north-western limb of the Tywi Anticline and the axial zone of the Central Wales Syncline. There the metamorphic grade, as indicated by the mean Kubler index for each formation, systematically decreases from high anchizone in uppermost Ordovician and lower Llandovery formations, to low anchizone in the upper Llandovery and Wenlock formations (Table 20a). This increase in metamorphic grade at lower stratigraphical levels is broadly similar to that found elsewhere in central Wales (Roberts et al., 1991), and is apparently related to depth of burial.

The metamorphic grade of the Ordovician Yr Allt Formation varies from high anchizone in the Rhiwnant Anticline to late diagenetic zone in the Tywi Anticline (Table 20a and b). This apparent anomaly is explained by lower depths of burial over the the Tywi Lineament, due to the erosion of the mainly early Llandovery succession, in contrast with the preservation of a full basinal Silurian succession over the Rhiwnant Anticline.

Other anomalies in the metamorphic pattern of the district suggest that depth of burial was not the only factor to determine metamorphic grade. The localised high-grade zones situated along the common limb of the Teifi Anticline and the Central Wales Syncline, and along the axial traces of the Rhiwnant and Tylwch anticlines, cannot presently be explained by increased depth of burial alone. It is probable that across these zones there was an additional strain component which increased the background metamorphic grade Roberts et al. (1996).

In the western part of the district, adjacent to the Ystwyth Fault, the isocrysts parallel the fault, with the higher grades lying to the south. This is consistent with the known postmetamorphic (post-cleavage) northward downthrow on the Ystwyth Fault.

SUMMARY OF TECTONIC HISTORY

The following sequence of tectonic events is inferred for the district:

1. Volcanic activity from Llanvirn to Caradoc times, possibly controlled by fault activity along the Pontesford Lineament.
2. Transition from shelf to slope during the Ashgill, related to contemporaneous movements along the Garth Fault. Dip-slip faulting along the Tywi Lineament may have influenced the accumulation of coarse-grained Ashgill sediments.
3. Intermittent faulting, some possibly strike-slip, along the Pontesford Lineament between the Caradoc and the latest Aeronian.
4. Telchyian dip-slip and possibly strike-slip faulting, leading to submarine erosion along the Tywi Lineament and enhanced sedimentation rates, initially to the west of the Glandyfi Lineament and later to the west of the Central Wales and Tywi lineaments. Deposition of Telychian shelf successions, controlled by dip-slip movements on the Llandrindod and Rock Park faults.
5. Wenlock dip-slip movements on the Garth and Cwmysgawen faults controlled deposition of the basinal succession.
6. Low-grade regional metamorphism, initiated by the onset of compression in the mid-Ludlow.
7. Folds and cleavage continued to develop, in response to sinistral transpression and shortening across the basin, during the Acadian Orogeny of early to mid-Devonian age. Areas of high-grade metamorphism developed partly in response to localised increased strain during the deformation.
8. Post-cleavage modification of the folds, associated with the development of accommodation faults, tension veins, and crenulation cleavage.
9. Formation of a transverse fault system, late in the Acadian Orogeny, followed by the emplacement of mineralised veins.
10. Final displacement on the Ystwyth Fault, possibly during the early stages of Hercynian movements.

TEN

Geophysics

Geophysical techniques have been used to facilitate the geological mapping of the district and to gain a fuller understanding of the structure of the Welsh Basin and its basement at depth. Existing gravity and aeromagnetic data have been modelled to provide estimates for the thickness of the sedimentary rocks in the Basin. Distinctive geophysical responses are seen in the surrounding areas and these have been considered in assessing the continuation of major structures across the district (Carruthers et al., 1992). New measurements of physical properties, obtained from exposed Lower Palaeozoic rocks, are summarised in Appendix 4.

SEISMIC DATA

The LISPB refraction line (Bamford et al., 1976) which passed north–south close to the eastern margin of the district, and the marine refraction and reflection data extending into Cardigan Bay, can be extrapolated to provide some indication of the deep crustal structure. Shallow-marine reflection data from Cardigan Bay (Dobson et al., 1973) indicate the extent to which the Lower Palaeozoic platform continues offshore before being overlain by Mesozoic and Tertiary strata. Sonic velocity measurements on a few samples suggest that relatively high values, of about 5 km/s, can be expected within the Lower Palaeozoic sequence (Appendix 4).

The depth to the base of the crust, the Mohorovičić (Moho) discontinuity, interpreted from the LISPB line lies in the range 32 to 35 km across Wales, with a slight reduction southwards. To the west of the district, the crust may thin by as much as 5 km towards the Irish Sea (Meissner et al., 1986). The evidence for this thinning is ambiguous and other interpretations suggest that an anomalous layer of faster velocity may occur at the base of a crust of normal thickness. While both models are consistent with the regional gravity high observed over the Irish Sea (Blundell et al., 1971), uncertainty regarding the location of these changes at depth means that the form of the flanking gravity gradient across the Llanilar sheet area (Figure 53a) cannot be predicted accurately.

The existence of a southerly dipping mid-crustal boundary, possibly representing the Conrad discontinuity, has been inferred from the LISPB line (Manchester, 1983), although the interpreted depth range of 13 to 17 km beneath Wales is only approximate. The increase in velocity across the boundary, from 6 km/s to 6.5 km/s, suggests an associated change in density of about 0.15 Mg/m³. Given that the true strike of the interface may lie closer to the north-east trend of the major faults within the Welsh Basin, it will contribute to the observed increase in gravity anomaly values towards the coast.

The first deep refraction seismic experiment in the district was undertaken by the British Geological Survey in 1992. The seismic line extended in a north-westerly direction from Mynydd Eppynt [SO 000 441] in the adjacent Builth Wells district, through the Tywi Forest to the coast just to the south of Aberystwyth, with over 40 recording stations being deployed along its 60 km length. High explosive emplaced in four 100 m-deep boreholes formed the seismic source on land, and for the extension of the line in Cardigan Bay nearly 900 shots were made from an airgun source. Preliminary interpretation of the data indicates that a north-west-dipping refracting horizon with a velocity of 5.7 km/s is present at a depth of about 2 km beneath the south-eastern margin of the basin, around Llanwrtyd Wells [SN 880 465]. Further refractors with a 'basement' velocity of approximately 6 km/s are present beneath the south-eastern part of the basin, south-east of Drygarn Fawr [SN 862 584], at a depth of about 4 km. These refractors appear to deepen towards the north-west, and probably coincide with a reflecting horizon, at a depth of about 9 km, seen on commercial seismic data collected offshore of Aberystwyth.

GRAVITY DATA

The Bouguer gravity anomaly map (Figure 53a) is based on data collected by the British Geological Survey as part of its national survey programme. This has involved the use of helicopters and hovercraft to access the remoter upland areas and estuaries, so as to achieve a uniform station distibution approaching 1 station per 1.6 km². Marine data for Cardigan Bay (Dobson et al., 1973) are sparser and less uniform than those for the land area, but they indicate the form of the anomaly contours offshore and help to set the regional context.

Density determinations on samples collected within the district are given in Appendix 4. Mudstone grain densities are typically 2.80 to 2.85 Mg/m³, with saturated densities of 2.75 to 2.8 Mg/m³. These are comparable with data from the Lower Palaeozoic rocks in the Lake District (Lee, 1988) and somewhat greater than the values of 2.7 to 2.75 Mg/m³ attributed to rocks of similar age in the Welsh Borderland and Southern Uplands of Scotland. The occurrence of turbidite sandstones, which usually constitute less than 20 per cent of the mudstone-dominated succession, causes a reduction of about 0.05 Mg/m³ in the overall formation densities. Independent estimates of the mean densities of rocks above sea level have been derived from the regional gravity survey data using the correlation between surface elevation and the observed acceleration due to gravity (Turnbull, 1987). These values, up to 2.83 Mg/m³, confirm the

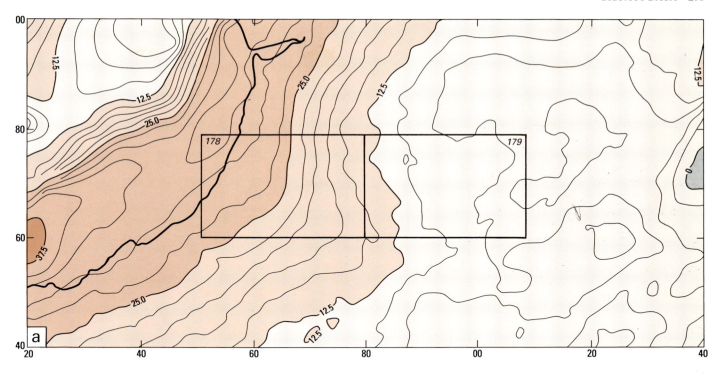

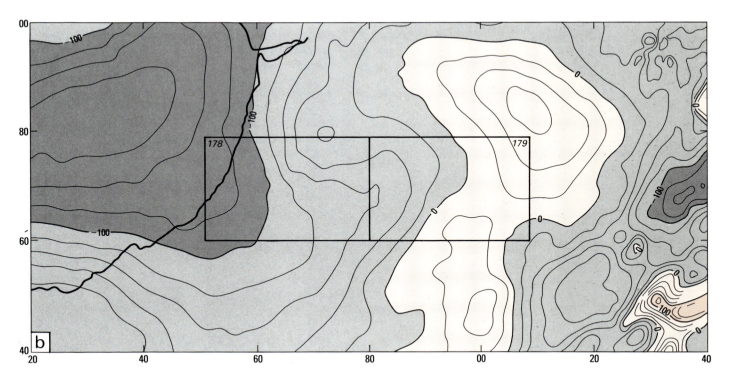

Figure 53 Regional geophysical maps of central Wales.

a Bouguer gravity map. Gravity reduction density = 2.7 Mg/m^3. Contour interval 2.5 mGal.
b Total field aeromagnetic map. Contour interval 20 nT.

evidence from direct measurement that a region of higher density rocks occurs in this part of central Wales.

The gravity anomaly pattern is characterised by a westwards increase in values, with a much steeper gradient across the Llanilar sheet area. The broad scale of this trend is consistent with its origin at either the mid- or base-crustal boundary suggested from the seismic data, and it may well include a contribution from both. The close correspondence between the trend of the gravity anomaly contours and the coastline implies that there is also a direct link with structures seen in the Lower Palaeozoic formations at surface.

Gravity anomaly values reach a maximum offshore, in the north-west corner of the Llanilar sheet area. The reduction in values farther west reflects the presence of younger sedimentary basins in Cardigan Bay, and the higher levels of over 40 mGal reached elsewhere in the Irish Sea are thought to represent the background field. As the anomaly gradient begins to slacken well to the east of the Mesozoic basins, there must be additional effects such as a change in the form of the Lower Palaeozoic basin or a local levelling-out of the deeper crustal boundaries.

AEROMAGNETIC DATA

The total field aeromagnetic data (Figure 53b), which include full coverage of Cardigan Bay, derive from a survey flown in 1960 at a nominal ground clearance of 305 m (1000 ft). North–south flight lines were separated by 2 km, with east–west tie lines every 10 km.

No strongly magnetic rocks have been found within the district, despite the occurrence of intrusions and volcanic rocks associated with the Builth Inlier. The nearest examples of magnetic sedimentary rocks occur to the north, in the Harlech Dome area, where some anomalies relate to pyrrhotite disseminated within specific mudstone units of Cambrian age (I F Smith *in* Allen and Jackson, 1985). Of the different types of Precambrian basement rocks that might be expected at depth here, the Uriconian volcanic rocks and those found near Pembroke have magnetic components, while the Longmyndian rocks, and the outcrops of the Lleyn Peninsula and Anglesey, are essentially non-magnetic.

The subdued appearance of the aeromagnetic anomaly pattern (Figure 53b) is consistent with a cover sequence dominated by sedimentary rocks, and it provides no indication that any magnetic dykes or other intrusions come close to the surface. However, a regional increase in magnetic anomaly values from the Cardigan Bay low towards a maximum developed in the east of the district shows the influence of magnetic rocks at depth. This probably occurs because of a rise in the Precambrian basement towards the margin of the Welsh Basin, though the anomalies may in part represent a lateral change in composition rather than simply a shallower depth to the same magnetic unit.

The broad areal extent and large amplitude of the magnetic low seen over Cardigan Bay indicate that both the basement and the deeper rocks are weakly magnetised overall here, in contrast to the area farther east. The eastern boundary of the nonmagnetic basement is located near the coast, and the anomaly contours run parallel to the northerly trending structures expressed in the Lower Palaeozoic formations. Its north-western margin lies close to the continuation of the Bala fault zone and is probably associated with a major crustal structure, the effects of which will influence the background field to which local anomalies in the district must be referred.

The form of the Cardigan Bay low can be reproduced by assuming a break, 20 to 35 km wide, in the continuity of magnetic basement rocks between south-west Wales and the Lleyn Peninsula to the north. The low occurs as part of the anomaly originating from the edge of the southern magnetic block, while the influence of the Lleyn block accounts for the steepness of the contours on its northern side. However, this type of model still requires a lowering of the magnetic datum by 50 nT or more in order to match the magnitude of the negative values. This shift, which will apply throughout the district, is attributed to a combination of magnetisation contrasts within the lower crust, and a correction to the planar reference field adopted for the UK data compilation (Evans, 1990).

RESIDUAL GRAVITY AND AEROMAGNETIC MAPS

Contour maps of the observed gravity and aeromagnetic data are dominated by the large-scale gradients that cross the district. By removing these long wavelength features it is possible to enhance the smaller anomalies generated by sources nearer the surface. This can be done mathematically to produce the type of residual maps shown in Figure 54. A 'reduction to the pole' transformation was also applied to the aeromagnetic data (in both Figures 53b and 54b) in order to register the anomalies more closely with their sources.

A north–south trend in the residual gravity anomaly contours lies close to the Glandyfi Lineament but the zone of maximum gradient (G1 in Figure 54a) is displaced to the east farther south (G2). The absence of any obvious reduction in gravity anomaly values over the more-arenaceous, less-dense facies seen to the west of the Bronnant Fault, indicates that more-significant density contrasts lie at depth, at or below the base of the Lower Palaeozoic sequence. The magnetic anomaly gradient (M1 in Figure 54b) associated with the Glandyfi Lineament lies on the western flank of a magnetic high (M2) and it dies out rapidly in the southern half of the Llanilar sheet area. The gravity anomaly gradient reduces more slowly to the south, indicating that it is responding to contrasts at a different level within the sequence.

Both gravity and magnetic residual contours show an east-north-east to west-south-west alignment parallel to the valleys of Afon Wyre and Afon Ystwyth (G3 and M3). This is reflected in faulting seen in the north-west corner of the Rhayader sheet area as well as in the Ystwyth Fault itself. The magnetic response is almost certainly associated with a boundary in the underlying basement, but the sharp curvatures seen in the magnetic contours

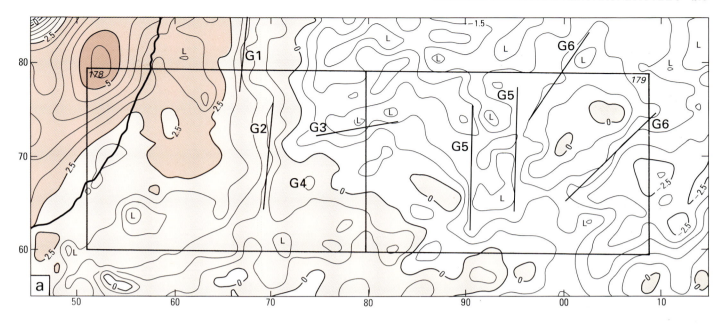

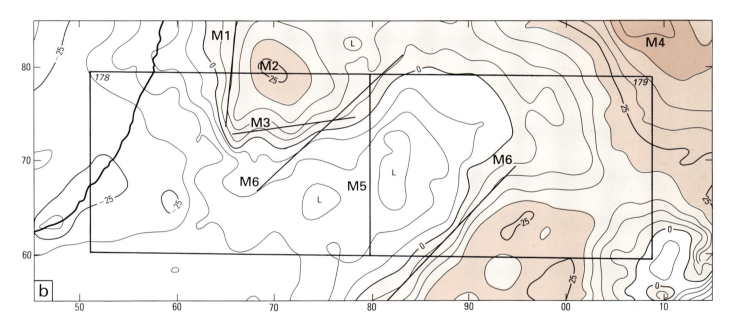

Figure 54 Residual anomaly contour maps of the district, generated by subtracting upwardly continued fields from the observed data (L–low).

a Residual gravity map, from upward continuation to 3 km. Contour interval 0.5 mGal.
b Residual aeromagnetic map, from upward continuation to 4 km. Contour interval 5 nT.

suggest that relatively shallow sources are also present. To the south of the Ystwyth Fault the Teifi Anticline has a weak expression as a gravity low (G4) and units on the margins of this structure may contribute to the local magnetic anomalies.

Undulations in the gravity anomaly in the district are superimposed on a broad area of lower values and they can be linked to folding of the Lower Palaeozoic rocks.

The lack of any strong trend is consistent with the variations in density within this sequence being small, and the number of gravity stations is insufficient to establish any detailed correlation with the geological mapping. However, there is some evidence of north–south segments in the contours (G5), and these are parallel to the trend of faults which host mineralisation to the south-east of Rhayader.

The magnetic high (M2) lies on a salient which can be linked eastwards to a larger feature (M4) near the margin of the basin. As the gravity values remain relatively low, it is probably associated with metamamorphic basement and/or intermediate, rather than basic, volcanic rocks. The embayment of lower magnetic anomaly (M5), which is related to the Central Wales Syncline, probably represents a different rock type within the Precambrian basement. Its margins show a distinctive north-east to south-west alignment (M6) which is also picked up by the gravity anomaly contours (G6), close to the Tywi Anticline.

GEOPHYSICAL INTERPRETATION

Gravity and aeromagnetic Euler solutions

Maps of gravity and magnetic Euler solutions (Figure 55) highlight some of the main trends within the data. The application of Euler deconvolution techniques to potential field data provides estimates of anomaly source locations (Thompson, 1982; Reid et al., 1990). The derived geographical positions are generally more reliable than source depths. Solutions are generated from anomaly gradients and they can usually be associated with the boundaries of well-defined bodies or with the contrasts across fault zones. They are best viewed on a regional scale, when the relation between discontinuous sets of linear (and arcuate) solutions is more apparent. Thus, the pattern of solutions across Wales is attributed to a network of faults within the Precambrian basement, with associated structures having propagated into the cover sequences (McDonald et al., 1992); the boundaries of the major Tertiary/Mesozoic basins are also delineated. Elements of this pattern can be seen in the three Euler solution trends which cross the district.

The most westerly of these trends is a distinctive feature in both the gravity and aeromagnetic data (Figures 55a and 55b). Aeromagnetic anomaly solutions coincide with the Bronnant Fault, which marks the eastern flank of the Glandyfi Lineament; the gravity solutions are offset 3 km to the east. Given that the solutions relate to sources at depth, their displacement with respect to the geological features seen at surface could be due to the migration of faults in the cover sequences. The other two trends are less clear: a weak set of gravity and aeromagnetic solutions occurs east of the Central Wales Lineament, and some gravity solutions follow the Tywi Lineament. The latter set may be traced north-eastwards into the strong feature which flanks the Cheshire Basin. A lack of solutions only implies that the contrast in physical properties between the juxtaposed rocks is small; it does not mean that the structures themselves are discontinuous.

Model studies show that solution depths tend to cluster near the centre of a step representing a fault, but the depths show a scatter of about 50 per cent around the mid-point value. In the case of the Glandyfi Lineament, where the thickness of the Llandovery sequence is known to increase by approximately 2.5 km

to the west, the solutions suggest a maximum depth of about 8 km to magnetic basement. However, this might be an underestimate as it does not allow for the possibility of interference from shallower sources. The fact that gravity solutions are typically shallower by 1 km may be explained by the presence of clastic, nonmagnetic Cambrian rocks which are less dense than the overlying mudstones.

Gravity and aeromagnetic profile models

While there is no simple relation between the densities of the Precambrian and Lower Palaeozoic formations throughout Wales as a whole (Powell, 1955), a more consistent contrast can occur on a smaller scale. The densities within the Precambrian range from as low as 2.65 Mg/m^3 where acid Arvonian/Monian volcanic rocks predominate, to as high as 2.85 Mg/m^3 in the Longmyndian rocks, and more in some components of the Mona Complex. Values of 2.75 tp 2.8 Mg/m^3 measured on the Lower Palaeozoic mudstones within the district (Appendix 4) are near the top of this range and they are likely to exceed the mean density of the underlying basement rocks.

East–west profiles along grid lines 75N and 65N indicate the main characteristics of the anomaly patterns across the district (Figure 56). Modelling the full gravity anomaly in terms of the density contrast between Lower Palaeozoic rocks and a granitic continental crust indicates that the thickness of the basin may exceed 10 km, depending on the values adopted. The models shown assume that there is a contribution from an interface in the mid to lower crust which would influence the regional field over the whole of Wales.

The Bouguer gravity anomaly profile reproduces the measured values directly, but a constant regional field was taken from the magnetic anomaly data to increase values by 100 nT. This adjustment is to allow both for the influence of the negative magnetic anomaly in Cardigan Bay and for an adjustment to the background field used in the original data reduction. The regional field almost certainly rises towards the east but as its form is unknown, any such effect has been incorporated within the model.

Both the increase in magnetic values and the decrease in gravity anomaly to the east of the profiles are associated with a rise in (magnetic) basement, against which the Lower Palaeozic basin thins. However, the two sets of anomaly gradients cannot be matched using a single interface at this level and it is necessary to distinguish the base of the denser rocks from the top of the magnetic layer. This can be attributed to the presence of nonmagnetic Cambrian sediments with near-background density, as noted above, or to a different style of Precambrian basement. A south-west Wales (or Uriconian) affinity is assumed for the magnetic basement, using an analogy from the Hayscastle and St David's anticlines, where magnetic anomalies are associated with less-dense Precambrian rocks (Cornwell and Cave, 1986). The basement is only resolved where it has distinctive magnetic properties and so the termination of the magnetic

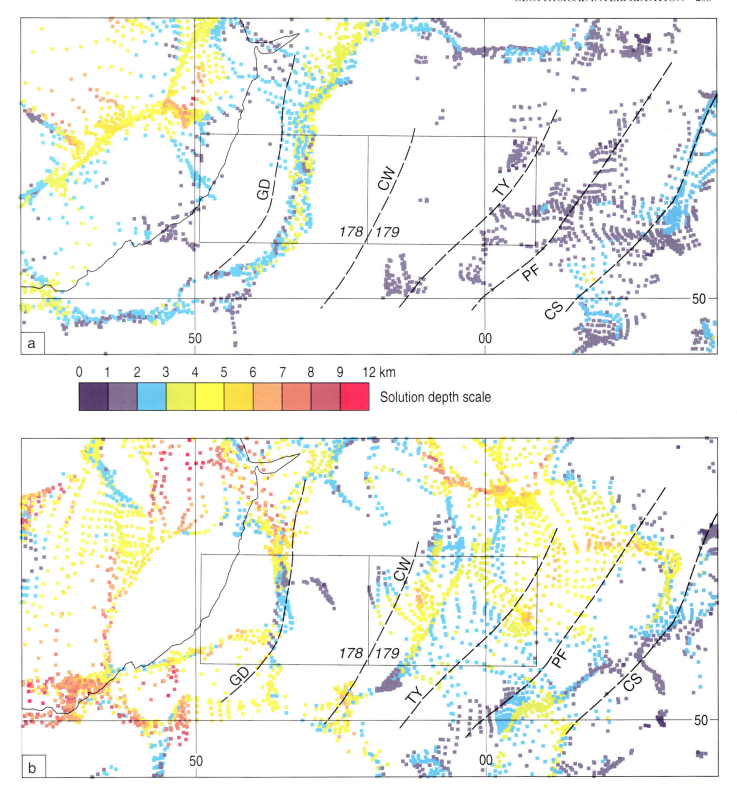

Figure 55   Euler solution maps of the district.

Structural index = 0.5; window size = 15 km; solution depth uncertainty limit = 50%. Tectonic lineaments identified by dashed lines and letters; GD — Glandyfi, CW — Central Wales, TY — Tywi, PF — Pontesford, CH — Church Stretton.
a   Derived from gravity data.   **b**   Derived from aeromagnetic data.

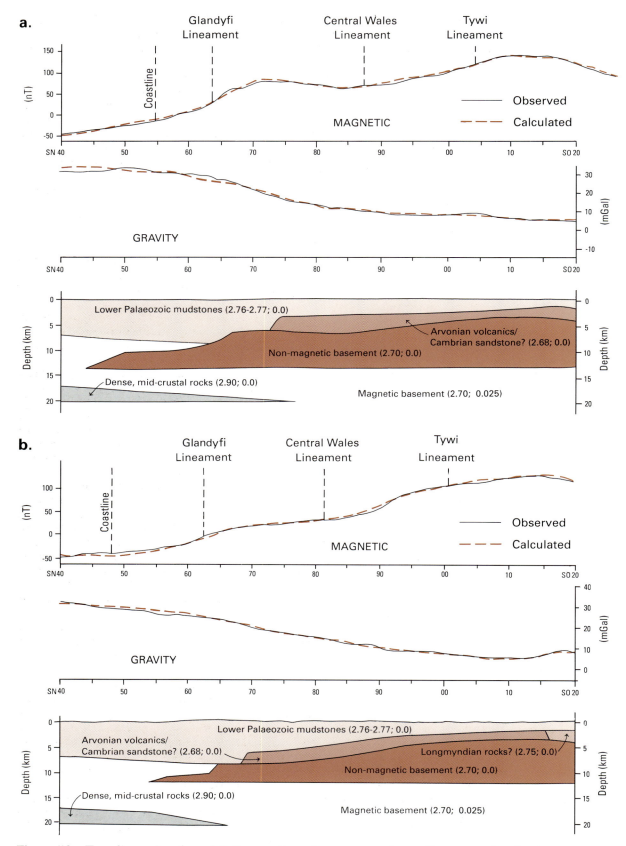

Figure 56   Two-dimensional models for gravity and aeromagnetic profiles across the district.
a   East–west grid line 75N.   **b**   East–west grid line 65N.
Physical property values: density in Mg/m³; magnetic susceptibility in SI units.

block towards the west implies no more than a lateral change in the character of the Precambrian rocks: whether this is due to faulting or some other form of contact is not defined.

The increase in gravity anomaly at the eastern end of profile 65N (Figure 56b) is attributed to a buried extension of the dense Longmyndian rocks seen at outcrop on the western side of the Church Stretton fault zone. Localised gravity anomalies on both profiles must have sources close to the surface and they can be matched in more detail than has been shown. An example of such features is seen between the Central Wales and Tywi lineaments on profile 65N (Figure 56b); these may relate to the north–south trends associated with the mineralised faults in the Rhiwnant area (G5 in Figure 54a). Gravity anomalies of small amplitude are also seen to be associated with the Glandyfi Lineament on profile 75N.

Profile 75N shows a steep gravity gradient east of the Glandyfi Lineament (Figure 56a). This implies a rapid westward thickening of the basin at the expense of Cambrian or non-magnetic Precambrian rocks. The depth to the magnetic basement increases west of the Glandyfi Lineament. A reduction in gravity values seen at the western end of profile 75N (Figure 56a) is too close to the coast to be attributable to the Mesozoic basin in Cardigan Bay. The marine data are less reliable in defining the anomaly gradient here but again there appears to be a contribution from within the Lower Palaeozoic sequence. This may be linked to the margin of the magnetic basement.

GROUND GEOPHYSICAL SURVEYS

Geophysical measurements taken along a 32 km reconnaissance traverse line crossing the regional geological strike were used to assist the geological mapping programme (Carruthers et al., 1989). The line trended in a north-west to south-east direction from near Ysbyty Ystwyth [SN 751 715] to Drygarn Fawr [SN 863 584] and thence eastwards towards Llandrindod Wells, stopping 1 km before the River Wye [SN 999 599]. Magnetic, self potential (SP), very low frequency electromagnetic (VLF-EM) and radiometric data were collected over the full length of the traverse, with induced polarisation (IP) and resistivity data only from the Rhiwnant Anticline. Additional data, primarily IP/resistivity and SP, are also available from two local grids (Carruthers et al., 1991).

Systematic variations in the geophysical measurements can be related to the degree of weathering (i.e. to porosity and clay content) or to the presence of specific accessory minerals. These may be associated with specific rock types or with structural effects such as faulting. In general, the observed IP and SP anomalies reflect stratigraphic control, although their continuity is intermittent. The fact that survey line separations of less than 100 m are needed in order to be confident of any correlations probably results from a combination of causes, including changes in mineralogy along strike, the effects of cross-cutting faults, and variable drift cover.

Differences in geophysical characteristics between formations are demonstrated in the results across the Rhiwnant Anticline near the contact between the siltstones of the Yr Allt Formation and the anoxic facies mudstones of the Cwmere Formation (Figure 57). Over the north-western limb there is an increase in IP response (apparent chargeability) and a significant decrease, of 200 mV, in the background SP level. Farther west, the chargeability decreases even more across the overlying oxic facies mudstones of the Derwenlas Formation. The largest chargeability and SP anomalies are found both near the axis of the anticline and close to the contact of the Yr Allt and Cwmere formations on the south-eastern limb. The former is thought to respresent a pod-like body, rich in either graphite or pyrite. The VLF response (not shown) was more variable over the Yr Allt Formation but it did not highlight any particular features and probably reflects a variable thickness of overburden.

Cross-sections of the IP survey data (e.g. Figure 58) are related qualitatively to the distribution of different rock types. The vertical scale is proportional to the separation between transmitter and receiver dipoles used in taking the measurements and is simply related to depth. However, model studies can be used to relate the anomaly pattern to the shape and location of the sources. High resistivity values are given by the sandstones and conglomerates, while the mudstones are more conductive. Zones of lower resistivity and relatively high chargeability, a combination which is emphasised in high specific capacity values, suggest the presence of conductive accessory minerals such as pyrite or graphite. Pyrrhotite would not be expected in view of the absence of magnetic anomalies over these zones although a weak (5 nT) high was detected within the Rhiwnant Anticline between Drygarn Fawr and the station 1000 m north-west on the reconnaissance line. Specific capacity values clearly distinguish the anoxic-facies mudstones both from the conglomerates and sandstones of the Caban Conglomerate Formation, and the oxic-facies mudstones of the Derwenlas Formation.

The largest VLF anomaly observed on the main traverse was centred [SN 815 649] between Llyn Gynon and the western end of Claerwen reservoir. There is no obvious feature at the surface which would account for this response, and the Rhuddnant Grits within which it occurs, did not produce anomalies of this magnitude elsewhere. It represents an extensive but weakly conductive zone, which can be related to the point of convergence of the faults trending south-south-west from the outlier of Pysgotwr Grits. Radiometric data showed very low ratemeter counts throughout the length of the traverse, although significant changes in level were detected. The anomalies could be explained mainly in terms of variations in the drift cover, with exposed rock usually giving the highest readings; vegetation cover also produced some smaller effects.

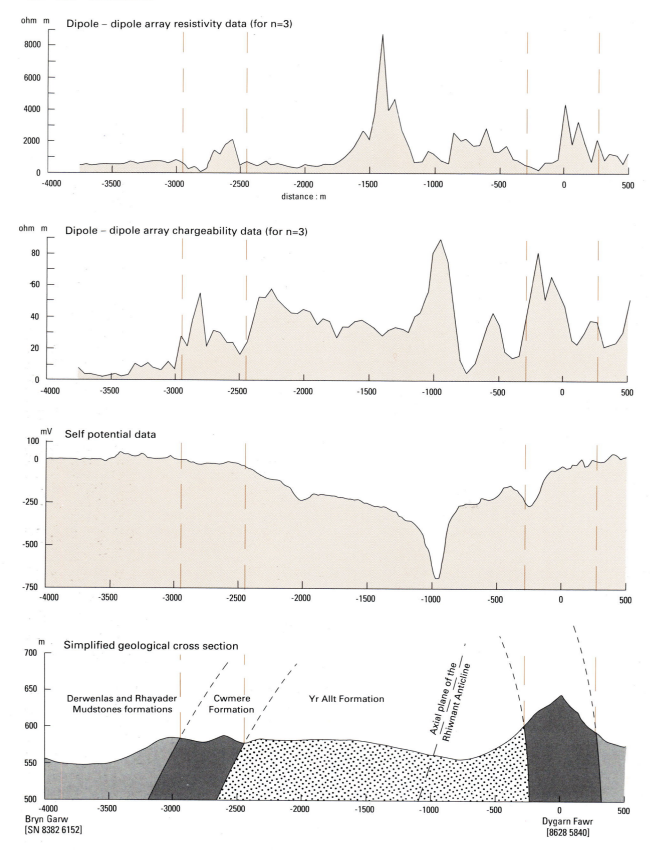

Figure 57 Geophysical profiles across the Rhiwnant Anticline. Distances measured from the base station at Drygarn Fawr [SN 8628 5840].

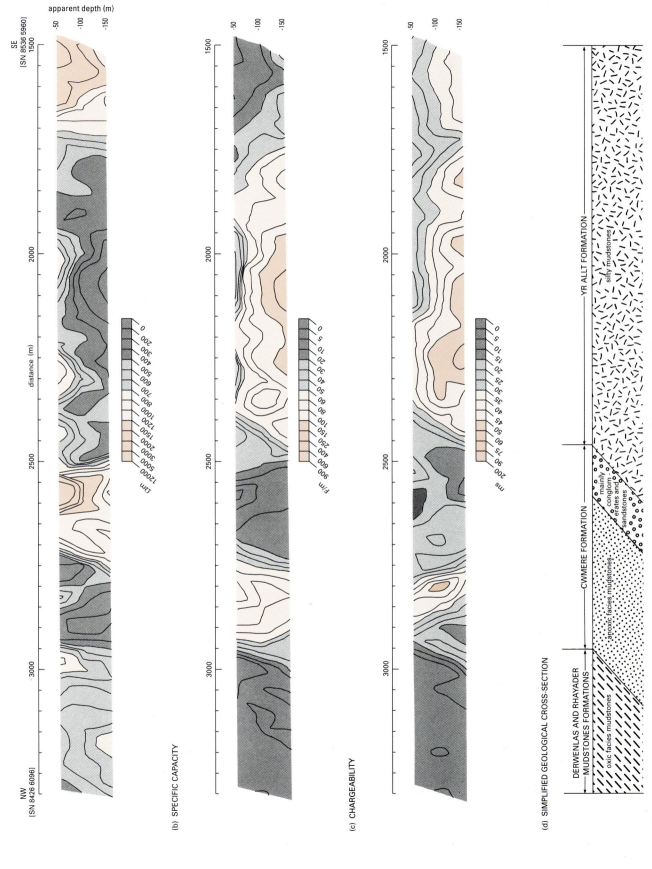

Figure 58   Apparent IP cross-sections across the northern limb of the Rhiwnant Anticline.
Dipole length — 50 m. Distances measured from the base station at Drygarn Fawr [SN 8628 5840]
a Apparent resistivity.   **b** Specific capacity.   **c** Chargeability.   **d** Simplified geological cross-section.

ELEVEN

Economic geology

VEIN MINERALISATION

The district lies within the Central Wales Mining Field (Jones, 1922; Institute of Geological Sciences, 1974). The earliest known mining activity in the area was at Bryn Copa [SN 811 752], near Cwmystwyth, where radiocarbon analyses of charcoal collected from tip material have yielded dates ranging from 1800 to 1000 BC (Timberlake, 1989). Although the Romans are thought to have worked the Dolaucothi Gold Mine,

50 km to the south, there is only circumstantial evidence that they mined lead in the district (Lewis, 1967; Hughes, 1981). During the 12th century, monks from Strata Florida Abbey carried out limited lead mining, but it is not clear which mines they worked. The first specific reference to a mine at Cwmystwyth [SN 746 804] is in a mining lease granted in 1535 (Hughes, 1981). It is possible that the Welsh silver bullion produced during the early part of the 17th century was mined in the district. A number of mines (Figure 59) were opened

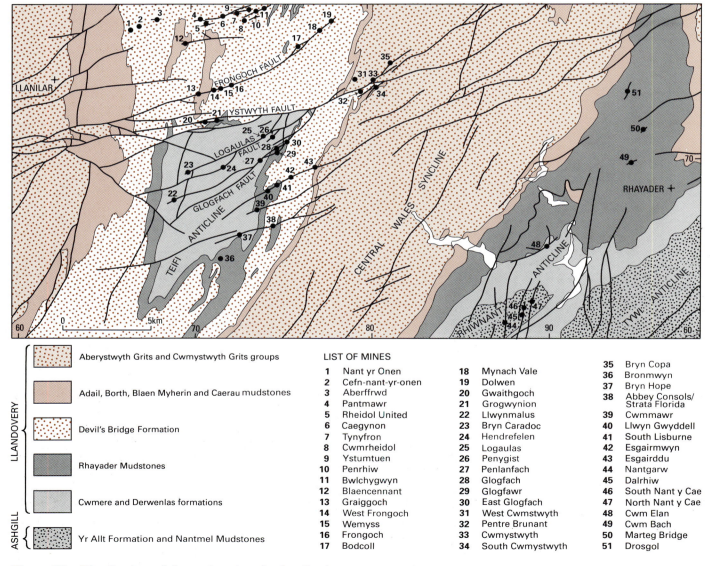

		LIST OF MINES			
Aberystwyth Grits and Cwmystwyth Grits groups	1	Nant yr Onen	18	Mynach Vale	35 Bryn Copa
	2	Cefn-nant-yr-onen	19	Dolwen	36 Bronmwyn
Adail, Borth, Blaen Myherin and Caerau mudstones	3	Aberffrwd	20	Gwaithgoch	37 Bryn Hope
	4	Pantmawr	21	Grogwynion	38 Abbey Consols/ Strata Florida
Devil's Bridge Formation	5	Rheidol United	22	Llwynmalus	39 Cwmmawr
	6	Caegynon	23	Bryn Caradoc	40 Llwyn Gwyddell
	7	Tynyfron	24	Hendrefelen	41 South Lisburne
Rhayader Mudstones	8	Cwmrheidol	25	Logaulas	42 Esgairmwyn
	9	Ystumtuen	26	Penygist	43 Esgairddu
	10	Penrhiw	27	Penlanfach	44 Nantgarw
Cwmere and Derwenlas formations	11	Bwlchygwyn	28	Glogfach	45 Dalrhiw
	12	Blaencennant	29	Glogfawr	46 South Nant y Cae
	13	Graiggoch	30	East Glogfach	47 North Nant y Cae
Yr Allt Formation and Nantmel Mudstones	14	West Frongoch	31	West Cwmstwyth	48 Cwm Elan
	15	Wemyss	32	Pentre Brunant	49 Cwm Bach
	16	Frongoch	33	Cwmystwyth	50 Marteg Bridge
	17	Bodcoll	34	South Cwmystwyth	51 Drosgol

LLANDOVERY / ASHGILL

Figure 59 Distribution of the main mines in the district.

over the next 200 years, the most important period of production being from 1850 to 1870 (Jones, 1922). Subsequent decline in metal prices led to mine closures, and in 1923 Cwmystwyth, the largest mine in the district (Plate 28), finally ceased operation. Esgair Mwyn Mine continued to extract ore until 1927, but more recent ventures to reopen the mine and to treat the waste dumps have proved uneconomic (Bick, 1974). Mineral exploration and drilling have been carried out in the district during the last twenty years, but no new ore bodies, or extensions to known ore bodies, have been discovered.

The total recorded production from the Central Wales Mining Field (Dunham, 1943–44) was 486 296 tonnes of lead concentrates averaging about 75 per cent lead (1845 to 1938), 153 193 tonnes of zinc concentrates (1854 to 1938), and 15 153 tonnes of copper (Foster-Smith, 1978; 1979). The lead ores from several of the mines contained silver in amounts that ranged from 80 to 1250 g/tonne.

Vein distribution and form

The mineralised veins are largely confined to the central part of the district, between the Glandyfi and the Tywi tectonic lineaments (Figure 59). The majority are situated along the axial zone and eastern flank of the Teifi Anticline, and are hosted by the Derwenlas, Rhayader Mudstones and Devil's Bridge formations. There was a marked decrease in the amount of ore in these veins with depth, and mining rarely continued into the underlying Cwmere Formation. Although only a few mineralised veins cut rocks of the Cwmystwyth Grits Group, the most productive mine of the district, Cwmystwyth Mine (Plate 28), is hosted by the basal sandstone-rich Llyn Teifi Member of that group. In the eastern part of the district a few mineralised veins cut the Ashgill silt-stones in the core of the Rhiwnant Anticline, but they do not extend into the overlying Cwmere Formation.

Most of the mineralised veins are developed in steeply dipping, east-north-east-trending, late, normal faults and joints. The margins of some veins are defined by slickensided surfaces which indicate both dip-slip and inclined displacements. However, in general, the magnitude of the displacements across of veins is small. Many of the faults can be traced along strike for several kilometres, but only relatively short sections contain economic mineralisation. The veins of the Teifi Anticline are orientated at an acute angle to the prominent Ystwyth Fault (Figure 59), which contains little mineralisation and has a final period of movement that postdates the emplacement of the main mineralised veins.

In addition to the east-north-east-striking veins, a north-striking vein and a subhorizontal vein are present at the Glogfawr [SN 749 707] and Cwmystwyth mines respectively. In the Rhiwnant valley [SN 890 616] a set of three north-trending mineralised faults cuts the Ashgill sedimentary rocks in the core of the Rhiwnant Anticline (Figure 59).

The average width of the high-grade sections of the mineralised veins was generally less than one metre, but short sections of almost pure sulphide, up to 2 m wide, were encountered at Cwmystwyth Mine and a few other mines. Commonly, the zone of mineralisation was much wider, up to 30 m, but only the high-grade veins within this were worked. The lodes usually form a parallel array of breccia zones and mineral veins, flanked by a narrow selvedge of stockwork veining in the country rock. The majority of the breccias consist of angular clasts of country rock and mineralised vein material cemented by later gangue and sulphide minerals (Plate 29). The brecciation is thought to have resulted from several phases of in-situ fragmentation by fluids under extremely high

Plate 28 Old mine buildings and dressing sheds at Cwmystwyth Mine.

The opencast on the Kingside and Comet veins is at the base of the sandstone cliffs, half-way up the prominent spur beyond the dressing sheds. (A 14724).

Plate 29 Polished sample of ore from Frongoch Mine.

Brecciated fragments of mudstone (light grey) and sphalerite (black) are cemented by later sphalerite and white quartz. The width of the specimen is 15 cm.

pressures (Phillips, 1972; 1986). The central parts of some veins consist of an intrusive breccia made up of widely separated rounded rock fragments, abraded during high rates of fluid flow, and cemented by quartz and sulphides.

Vein mineralogy

The most important minerals worked in the mining field were galena (PbS), sphalerite (ZnS) and, to a lesser degree, chalcopyrite ($CuFeS_2$). Lead was the principal metal produced, with silver as a by-product. The records for zinc production fluctuate widely from mine to mine and bear little relation to the amount of lead produced; where zinc was extracted in quantity the Pb:Zn (metal) ratio varied from 0.1 to 5.3. However, the price of zinc was low during the main production period, and much of the sphalerite therefore may have been left unworked. Other sulphide minerals include crystalline and framboidal pyrite (FeS_2), and trace amounts of arsenopyrite (FeAsS), cobaltite (CoAsS) (Raybould, 1974) and ullmanite (NiSbS) (Rust, 1990).

Quartz is the ubiquitous gangue mineral throughout the mining field. It generally forms massive layers on the vein walls and around rock fragments in the vein breccias, or euhedral crystals in vughs. The quartz is commonly intergrown with disseminated sulphides, but in some veins layers of sulphide-rich quartz alternate with layers containing little or no quartz (Raybould, 1974). Carbonate, generally ankerite ($Ca(Mg,Fe)(CO_3)_2$), is found in minor quantities in most of the veins. However, in the Rhiwnant valley area coarse-grained calcite and ankerite are the dominant gangue minerals.

The general paragenetic sequence of sulphide minerals in the mining field is pyrite-chalcopyrite-sphalerite-arsenopyrite-galena (Raybould, 1974). The paragenetic position of cobaltite has not been established, but it is clearly earlier than the arsenopyrite that encloses it. Mineral zonation along the strike of the individual veins occurs at some mines. On the Comet Vein at Cwmystwyth

Mine and on the Logaulas Vein the proportion of galena increases eastwards (Jones, 1922). No consistent regional zonation pattern can be distinguished across the district.

Secondary carbonate, sulphate and phosphate minerals, formed by oxidation of the primary sulphides, occur in most of the mine tips. For example, cerussite ($PbCO_3$) occurs at Logaulas Mine [SN 740 715], smithsonite ($ZnCO_3$) and pyromorphite ($Pb_5(PO_4)_3Cl$) on Bryn Copa [SN 811 752] at Cwmystwyth Mine and malachite ($Cu_2(CO_3)(OH)_2$) at the Rhiwnant valley mines [SN 890 615] (Jones and Moreton, 1977).

Studies on the geochemistry of the country rocks adjacent to mineralised veins has shown little evidence for widespread wallrock alteration (Kakar, 1971; Ashton, 1978). At Cwmystwyth Mine rock samples were collected at 20 m intervals from the walls of Bonsall Adit which accesses the Kingside Lode. Lead and zinc show considerable enhancement, to 1500 ppm and 900 ppm respectively, adjacent to the veins, but elsewhere remain at background levels (10 ppm Pb and 110 ppm Zn). The samples were also analysed for ammonium which has been used as a pathfinder for some types of mineralisation (Ridgway and Appleton, 1990). Although all the samples gave high levels, between 800 to 1200 ppm, there was no enhancement close to the veins, and variation is considered to be due to the chemistry of the host rock.

Mine descriptions

The mines described below are representative of the structure, style, mineralogy and grades of mineralisation to be found in that part of the Central Wales Mining Field which lies within the district. A comprehensive description of the mines in the district is given by Jones (1922), who visited many of them when they were in production. The history of the development of the mines and additional mining information are given by Bick (1974), Foster-Smith (1978, 1979), Hall (1971, 1989), and Jones and Moreton (1977).

FRONGOCH MINE

Frongoch Mine [SN 721 744] is situated about 2 km north-west of Pont-rhyd-y-groes and is one of several mines developed along the east-north-east-trending Frongoch Fault. The vein is hosted by thinly interbedded mudstone and sandstone turbidites of the Devil's Bridge Formation. The most productive section of the fault was about 450 m long at Frongoch Mine where it was mined to a depth of 260 m. The overall width of the vein was about 11 m and this remained fairly constant with depth. The vein consisted of two galena-rich lodes separated by barren rock, together with a sphalerite-rich lode along the footwall.

Frongoch Mine was worked almost continuously from 1798 to 1902. The main product was galena up to 1878, with 23 027 tonnes of concentrate being mined from 1859 to 1876 (Jones, 1922); in contrast from 1879 to 1903 sphalerite production totalled over 50 000 tonnes, with only 7191 tonnes of galena. The galena:sphalerite ratio, calculated from production records, was about 1:2. The average silver content was 100 g per tonne of refined lead metal.

LOGAULAS MINE

At Logaulas Mine [SN 740 715], near Ysbyty Ystwyth, the zone of mineralisation follows a prominent north-east-trending normal fault, which cuts the Derwenlas Formation and has a downthrow of some tens of metres to the south-east. The ore was mainly won from two subparallel veins about 30 m apart, which were followed for a strike length of about 1 km, and to a maximum depth of 240 m (Figure 60). In places the intervening mudstones were mineralised and were also worked, but grades were much lower. The veins consist of mudstone fragments cemented by quartz, calcite, galena and sphalerite; the latter became dominant at the western end of the vein. The marked decrease in mineralisation in the lowest two levels has been attributed to the possible proximity of the underlying Cwmere Formation, which is generally barren (Jones, 1922). The total output of the mine between 1834 and 1891 was about 39 000 tonnes of lead with a silver content of about 140 g per tonne of metal.

GLOGFAWR AND GLOGFACH MINES

At Glogfawr [SN 7486 7062] and Glogfach [SN 7490 7097] mines the ore was won mainly from two subparallel north-west-dipping veins separated by approximately 300 m. The veins follow north-east-trending faults which cut mudstone and sandstone turbidites of the Rhayader Mudstones and Devil's Bridge Formation. The Glogfawr Vein was worked to a depth of about 220 m over a strike length of just over a kilometre; the Glogfach Vein was worked over a strike length of 500 m to a depth of 260 m. Both veins are intersected by a north-north-east-striking cross-fault dipping at 60° to the west, which is defined by a band of sheared mudstone. The Glogfawr Vein and the cross-fault are displaced across the zone of intersection (inset in Figure 61), and the cross-fault is strongly mineralised to the north and south of the Glogfawr Vein. Min-

PLAN OF MINE

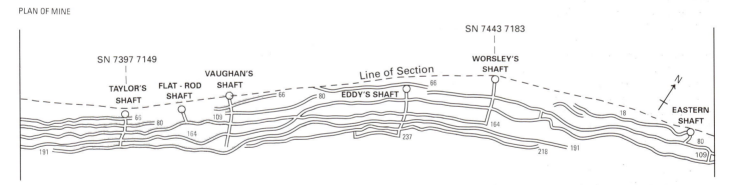

PROJECTED VERTICAL SECTION

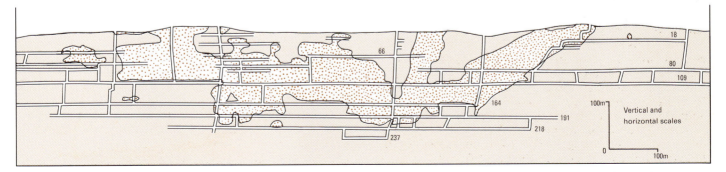

Figure 60 Simplified plan and projected vertical section of workings in the main lode of the Logaulas Mine. On the vertical section the worked out mineralisation is indicated by stippling. The figures are depths in metres of the main levels.

eralisation on the Glogfach Vein is restricted to the west of the cross-fault which, at Glogfach, is not mineralised. The north-north-east-trending cross-fault belongs to series of faults initiated during the regional fold episode (Acadian). It is probable, therefore, that it was reactivated during the later north-east-trending fault activity and emplacement of the mineralised veins.

The total production of lead between 1838 and 1917 from the Glogfawr and Glogfach mines was about 14 000 and 7000 tonnes respectively. Both mines were rich in silver with an average content of about 260 g per tonne of metal. There is no recorded zinc production from either mine.

CWMYSTWYTH MINE

Cwmystwyth Mine [SN 746 804] is situated on the steep northern valley side of the upper reaches of Afon Ystwyth, just to the north of the east-north-east-trending Ystwyth Fault (Plate 28). Three main veins were worked within the sandstone-rich Llyn Teifi Member of the Rhuddnant Grits: Kingside, Comet and Mitchell's (Figure 62). The Kingside and Comet veins dip south at 50° to 60° and come together in the main section of the mine over a strike length of 300 m, to form a single lode up to 20 m wide. The Comet Vein varies in width from 4 to 13 m and consists mainly of fragments of country rock cemented by quartz, sulphides and minor calcite. Lodes of galena, sphalerite and pyrite within the vein reach 3 m in width. The Kingside Vein has an average width of about 1 m and typically consists of rounded clasts of sphalerite, up to 7 cm in diameter, with chalcopyrite clots rimmed by crystalline quartz, set in a matrix of coarse-grained quartz with interstitial galena and subordinate sphalerite. In the western part of the workings the Kingside Vein was associated with a subhorizontal mass of pure galena 2 m thick.

In the central part of the mine the Comet and Kingside veins are displaced across Mitchell's Vein (Plate 30), which is developed along a normal fault dipping north at 45° to 60° (Figure 62). Mitchell's Vein has an average width of 40 cm and varies in composition from pure sphalerite to interlayered galena and sphalerite (Jones, 1922); sphalerite is the more abundant at depth.

South of the mine workings the veins are truncated by the Ystwyth Fault, which consists of a 30 m-wide zone of clay gouge, containing rare blocks of brecciated vein material. Although minor vein mineralisation occurs to the south of the Ystwyth Fault, there is no evidence to suggest that it represents the continuation of the main mineralisation.

The veins at Cwmystwyth have been mined from prehistoric times to the present century. The production of lead and zinc concentrates from 1848 to 1916 was 33 000 and 19 000 tonnes respectively, and the total amount of silver extracted has been estimated at approximately 3000 kg (Jones, 1922; Hughes, 1981).

RHIWNANT VALLEY MINES

In the Rhiwnant valley a set of three mineralised veins has been worked at North Nant y Cae [SN 898 616],

South Nant y Cae [SN 885 609], Dalrhiw [SN 885 607] and Nantygarw mines [SN 875 606] (Hall, 1971). Although the production from these mines was small, they are scientifically important because they show a different orientation and style of mineralisation to the rest of the mining field. The veins are contained in north-trending faults which cut Ashgill siltstones and sandstones in the core of the Rhiwnant Anticline. These faults are unmineralised where they pass into the overlying Cwmere Formation. The veins are generally less than 2 m wide and consist of brecciated fragments of country rock cemented by abundant coarse-grained calcite and iron carbonate, amorphous white quartz, and sulphides. The dominant ore mineral is chalcopyrite, with subordinate quantities of galena. There is no record of sphalerite production, although rare samples occur in mine tips. Traces of secondary malachite and native copper are present.

The vein at North Nant y Cae Mine dips west at 70° and has been worked from a series of shafts and shallow open-cuts over a strike length of about 500 m. The maximum depth of the workings is reported to have been about 100 m (Ball and Nutt, 1976). The main vein varies between 2 and 3 m in width with a network of mineralised veins extending up to 1 m into the footwall. Well-developed slickensides plunge south at 45° on the footwall contact. The vein consists of amorphous quartz, banded parallel to the vein margin along the footwall, and a combination of stockwork quartz, carbonate and sulphide veins, together with mineralised breccia in the main part of the vein.

Isotopic- and fluid-inclusion studies

The composition, temperature of crystallisation, age, and source of the hydrothermal fluids that formed the epithermal veins of the Central Wales Mining Field, have been the subject of considerable debate for over 80 years (Finlayson, 1910; Jones, 1922; Phillips, 1972; Raybould, 1974). Lead and sulphur isotopic analyses of galena, and analyses of fluid inclusions in quartz crystals from the veins, have been used to elucidate the problem.

LEAD ISOTOPES

Lead-isotopic analyses of galenas from a representative selection of mineralised veins across the Central Wales Mining Field fall into two clusters on the Pb^{206}/Pb^{204} versus Pb^{207}/Pb^{204} plot (Figure 63). Galena collected from a tension vein associated with a thrust fault in Hendre quarry [SN 721 697] plots in the least radiogenic cluster (Figure 63). These clusters can be explained in two ways. First, they might represent two periods of mineralisation sourced from a common homogeneous parent; using the lead isotope growth curve (Cumming and Richards, 1975), the model ages for these periods of mineralisation are approximately 390 Ma (early Devonian) and between 330 and 360 Ma (early Carboniferous). Second, there might have been only one period of mineralisation, with hydrothermal fluids derived from two sources; the fluids with the

Figure 61
Geological map of Glogfawr and Glogfach mines. The inset mine plan shows the outcrop and worked levels on the Glogfawr Vein and the cross-fault. The figures are depths in metres of the main levels below adit 'A' shown on the map.

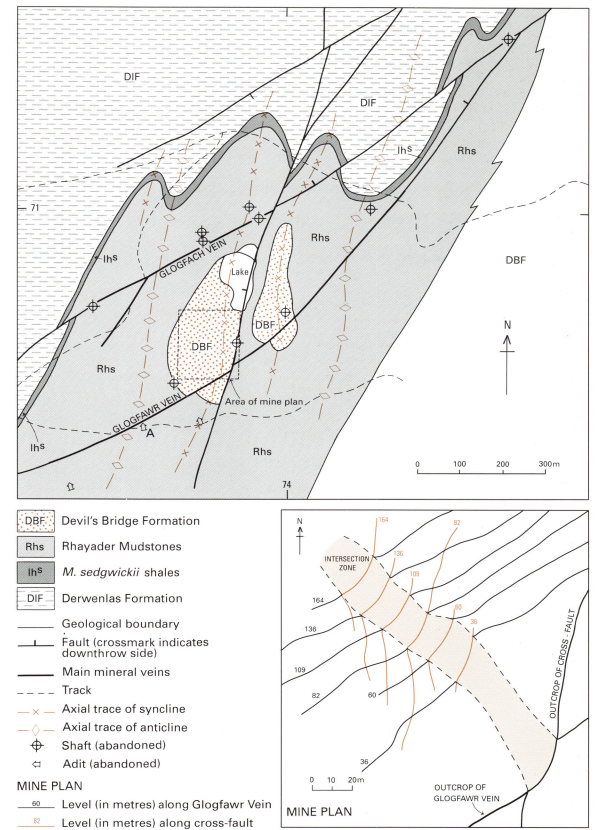

	DBF	Devil's Bridge Formation
	Rhs	Rhayader Mudstones
	Ihˢ	*M. sedgwickii* shales
	DIF	Derwenlas Formation

Geological boundary

Fault (crossmark indicates downthrow side)

Main mineral veins

Track

Axial trace of syncline

Axial trace of anticline

Shaft (abandoned)

Adit (abandoned)

MINE PLAN

60 — Level (in metres) along Glogfawr Vein

82 — Level (in metres) along cross-fault

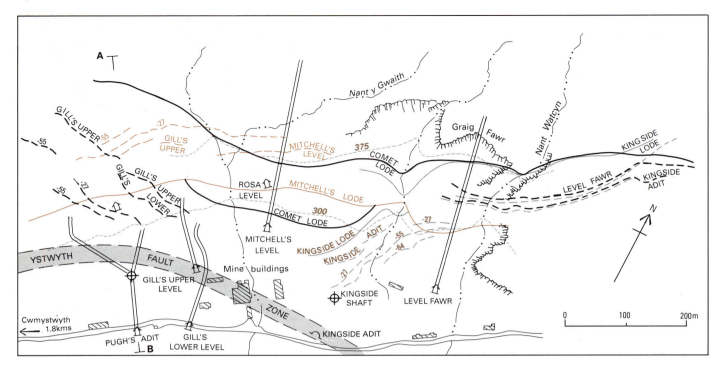

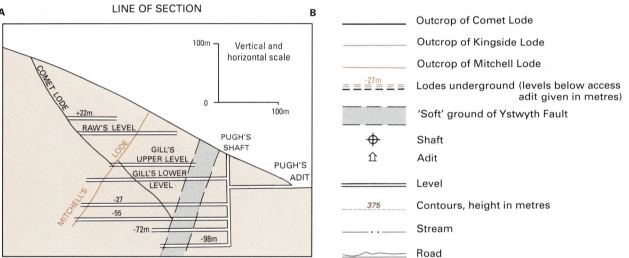

Figure 62 Plan and section of Cwmystwyth Mine [SN 746 804]. The eastward extensions of the Comet and Mitchell's veins are conjectural and projected from underground workings. (After Jones, 1922, p.114.)

younger model ages would in this case have been contaminated with a radiogenic component. The first explanation is considered to be more plausible because the galenas from the two clusters are not characterised either by geographical position or by age of host rock (Fletcher et al., 1993). The generation and separation of two isotopically distinct hydrothermal cells within such a small area is not considered likely.

All the Central Wales lead isotope data lie on the model growth curve (Figure 63) which would imply that the source of the galena was well mixed and multicycled

sediments, and that no ancient isotopically depleted basement, or very little sediment from such a basement, was available to the mineralising processes. The hydrothermal fluids were, therefore, confined to, and probably sourced from, the Lower Palaeozoic sediments, and did not mix with any component from the underlying basement.

SULPHUR ISOTOPES

The most common primary stable sulphide phase in unaltered sediments is pyrite, produced by bacterial

Plate 30 Underground workings on Kingside Adit Level, Cwmystwyth Mine.

Two lodes on the Comet Vein are truncated by a steeply dipping fault, on the left, associated with Mitchell's Vein (photograph courtesy of R Jones, National Museum of Wales).

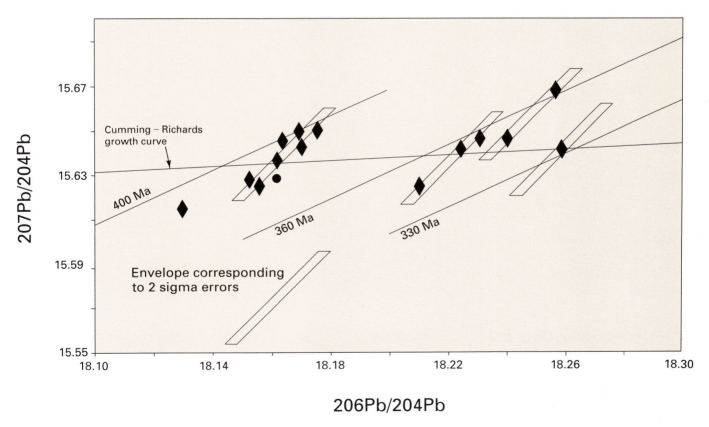

Figure 63 Plot of lead isotopic ratios of galenas from veins in Lower Palaeozoic rocks of central Wales. Galena from Hendre quarry is plotted as a solid circle. The Cumming-Richards growth curve and representative isochrons are also shown.

sulphate reduction of coeval sea-water. The $\delta^{34}S$ values of five diagenetic pyrites from the Ordovician and Silurian sedimentary rocks of the district range from -8.7 to $+41.1$. This very large range could be the result of closed-system reduction of Ordovician and Silurian sea-water in the sediments during diagenesis.

The sulphur isotopes of galenas from six veins gave a range of $\delta^{34}S$ values from 0 to -14. There is no systematic difference between the sulphur isotope compositions from the two lead-isotope clusters (see previous section); this suggests that the sulphides were derived from a single source.

The limited sulphur-isotope data available are consistent with the sulphur of the galena having been leached from the Ordovician and Silurian mudstones and sandstones during epithermal mineralisation.

K-Ar isotopes

K-Ar dating of the clay minerals from vein gouges and the altered wallrocks has given ages of 345 ± 5 Ma from Cwmystwyth Mine and 358 ± 6 Ma from Glogfach Mine (Ineson and Mitchell, 1975). Dates from the whole mining field have a mean of 354 ± 7 Ma (early Carboniferous). These dates are in agreement with the field evidence that the mineralised veins cross-cut the folds and cleavage formed during the Acadian Orogeny. However, the K-Ar dates can only be regarded as minimum ages, as argon loss can occur even at low temperatures.

Fluid-inclusion studies

Fluid inclusions from a total of 35 vein quartz samples from the Central Wales Mining Field have been analysed by thermometric and gas mass spectrometric techniques. The quartz veins fall into two catagories: largely unmineralised late-metamorphic veins associated with fold development, and cross-cutting epithermal mineralised veins. Homogenisation temperatures range from $125°$ to $205°C$, while salinity varies from 3 to 25 wt% NaCl equivalents for both categories. The salinity values are bimodal, with groupings 3 to 10 and 18 to 25 wt%. The lower-temperature fluids carry a higher component of $CaCl_2$ and below $160°C$ constitute most of the higher salinity measurements. The inclusion volatiles (H_2O, CO_2, CH_4, N_2) for both quartz vein categories have high total contents of dissolved gases, containing up to 5.5 mol% ($CO_2 + CH_4 + N_2$). Such concentrations are typical of quartz veins in low-grade metamorphic environments, but unusual for epithermal quartz veins. In most samples CO_2 is the dominant nonaqueous volatile; N_2/CH_4 ratios are close to unity and CO_2/CH_4 greater than 10 (Figure 64). Where the late-metamorphic veins cut Caradoc graptolitic mudstones, the fluid compositions are significantly methanoic (up to 11 mol% CH_4) with CO_2/CH_4 ratios less than 0.2; this is consistent with graphite buffering by the enclosing wallrocks. The chemical similarity of the low-grade metamorphic fluids and those responsible for the epithermal mineralisation implies a common source. By comparision with other low-grade metamorphic terrains (Mullis, 1987) the

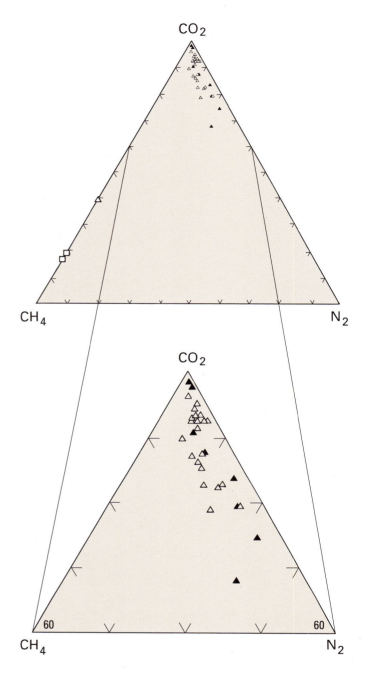

▲ Mineralised epithermal quartz veins

△ Unmineralised metamorphic quartz veins

☐ Metamorphic quartz veins hosted in Caradoc graptolitic mudstones

Figure 64 Plot of the volatile composition of the fluid inclusions from epithermal and metamorphic quartz veins in the district.

fluids were most probably expelled from the early Palaeozoic sediments during the final stages of deformation of the basin.

Origin of the vein mineralisation

Various hypotheses have been proposed for the source of the mineralising fluids in mid-Wales. Early ideas favoured a magmatic hydrothermal source (Finlayson, 1910; Jones, 1922), but since there are no igneous intrusions in the vicinity of the mines, and geophysical evidence suggests that none exists at depth, such a source is unlikely. A preferred model, supported by new sulphur and lead isotope data and fluid inclusion evidence, invokes leaching of metals from the Lower Palaeozoic sediments during dewatering and low-grade metamorphism. Raybould (1974) has suggested that the hydrothermal solutions carried the metals mainly as bisulphide complexes. The brecciation along faults was probably caused by hydraulic fracture during repeated episodes of pressure release (Phillips, 1972).

The mineralised veins of the district are largely confined between the Glandyfi and Tywi lineaments (Figure 59), which were formed by the propagation of faults into the Lower Palaeozoic sequences above major basement structures. The upward and outward migration of the pore and metamorphic fluids from the deep axial parts of the basin during the final stages of basin inversion may, therefore, have been channelled by the structures and sedimentary facies variations associated with these lineaments.

STRATIFORM MINERALISATION

At Pen-rhiw Frank [SO 0843 6005], stratiform lead mineralisation occurs along the interface of a basalt lava and reworked tuffs in the Builth Volcanic Formation (Marshall et al., 1987). The gently dipping zone of mineralisation, proved in boreholes (Figure 5) consists of secondary lead minerals disseminated in a clay matrix. The residual soils overlying the zone contain anomalous concentrations of lead (2000 ppm) and zinc (340 ppm). The maximum thickness of the zone, proved in boreholes, is 5.8 m, with an average grade of 0.45 per cent Pb. Although the mineralisation appears to be of limited aerial extent, there is a possibility that base metal mineralisation of greater significance is associated with the volcanic rocks.

It has been suggested that the Ordovician and Silurian mudstones of central Wales could host stratiform base metal mineralisation (Badham, 1981). Although nodular masses and cubes of diagenetic pyrite occur throughout the mudstones, none has been found to be associated with base metals. At a few localities the coarse-grained sandstones and conglomerates contain sulphides. For example, at Hendre Quarry [SN 721 697] the top of a sandstone bed contains abundant pyrite with minor sphalerite. This mineralisation could be the result of increased fluid flow in the more permeable sandstone bed during basin dewatering.

INDUSTRIAL MINERALS

Aggregate

The lowest sandstone sequence (Ystrad Meurig Grits facies) of the Ystrad Meurig Grits Formation and a variable proportion of the interbedded mudstones of the Derwenlas Formation are extracted from Hendre quarry [SN 722 694], the only working quarry in the district. The crushed material is used for hardcore, building aggregate and cement building blocks. Until recently good quality roadstone and aggregate were extracted from the thick sequence of massive sandstones (Cerig Gwynion Grits facies) within the Caban Conglomerate Formation at Cerrig Gwynion quarry [SN 971 656], near Rhayader. Many of the old mine dumps in the district have also been used as an aggregate resource, for example at Frongoch Mine [SN 721 744], but the material contains much sulphide and is therefore of little use in the building industry.

Building stone

There are numerous small abandoned quarries throughout the district, which supplied building stone for local construction use. The sandstones from the Aberystwyth Grits Group and the Caban Conglomerates Formation were particularly favoured. The conglomerates (second sequence of Caban Conglomerates facies) from the Caban Conglomerate Formation, adjacent to the Caban-coch reservoir [SN 924 647], were extensively quarried for the construction of dams in the Elan valley. In addition, the small dolerite bodies within the Llanfawr Mudstones were quarried near Llandrindod Wells for building stone. Some of the well-cleaved Llandovery mudstones were used for flooring slabs, but were unsuitable for roofing slates.

Sand and gravel

Apart from the river deposits within the main drainage systems, there are no extensive sand and gravel resources within the district. Limited extraction is presently undertaken from the river gravels [SN 667 787] of Afon Rheidol near Capel Bangor. Several small glacial sand and gravel deposits are located along the Ystwyth, Teifi, Elan and Dulas valleys, and some have been intermittently worked, for example at Caerfagu quarry [SO 0445 6540] and near Tregaron [SN 6674 5804].

Peat

Until as recently as the 1930s peat was cut in quantity from the large peat bog of Cors Caron, north-east of Tregaron, where many of the old cuts are over 2 m deep. Many of the thinner peat deposits found throughout the district, especially in the upland areas, were probably also used as a local source of fuel.

HYDROGEOLOGY

The high rainfall over the Cambrian Mountains together with the steeply incised valleys have made the district an ideal catchment area. The five Elan valley reservoirs provide most of the water supply for Birmingham. The Caban-coch, Carreg-ddu, Penygarreg and Craig Goch reservoirs were commissioned in 1906, and the Claerwen reservoir in 1952. There have been plans to extend the reservoir system, but old mine workings may cause metal pollution problems, for example in the Rheidol and Ystwyth valleys.

The mineral springs of Llandrindod Wells have been used for medicinal purposes since the 18th century. There were formerly ten springs in the vicinity of the Rock Park Fault [SO 056 609], one of a series of faults along the western margin of the Builth Inlier. The springs were classified on their presumed mineral content into chalybeate (iron-rich), saline, magnesium-rich, and sulphurous. Recent analyses of the spring waters have shown that the two most saline sources (chalybeate and saline) contain 2530 mg/l chlorides, but it is suspected that the waters were more saline in the past (Edmunds and Robins, 1991). No evidence has been found for any of the springs being richer in magnesium than the others. The content of sulphur in the so-called sulphurous springs has varied over the years, and some of the springs classified as saline also have a strong sulphurous smell. Analytical evidence suggests that the saline waters are the primary source and that the other types are formed by the mixing of the saline source with shallow groundwaters. The saline water is characterised by elevated concentrations of lithium, iron, manganese, barium and nickel, and the near absence of sulphate. The estimated flow of the saline water at the Park Rock site is about 1 or 2 litres per second with discharge temperatures ranging between 11°C and 13°C.

The saline water has arisen by deep circulation of meteoric water from the recharge area east of the town, via the relatively permeable Ordovician volcanic and sedimentary rocks of the Builth Inlier (Edmunds and Robins, 1991). The meteoric water has probably mixed with ancient saline waters, which may have had a residence time of at least several thousand years, before being channelled along the Park Rock Fault. The origin of the salinity is likely to have been the organic-rich shales within the Lower Palaeozoic sequences of the district.

TWELVE

Quaternary deposits

The Quaternary period was characterised by rapid climatic fluctuations, when intervals of extreme cold alternated with temperate conditions similar to those of the present day. During cold periods, icecaps were established over much of northern Europe, including Britain, giving rise to a distinctive series of landforms and deposits. Periglacial conditions preceded the advance of the ice and followed its retreat, with deposits mainly resulting from intense freeze–thaw and solifluction. In general, the early periglacial landforms and deposits have been extensively modified or destroyed by ice advance. The present landscape is largely the result of the latest (Late Devensian) glaciation and subsequent periglaciation, together with fluvial processes that were active during the period of climatic oscillation known as the Late Devensian Late-glacial (Lowe and Walker, 1984; Campbell and Bowen, 1989). Postglacial (Holocene) deposits have accumulated on this remnant glacial landscape.

At least two glacial advances, separated by an interglacial period, have been recognised from deposits in south Wales (Campbell and Bowen, 1989; Bowen, 1991), but evidence of only the final (Late Devensian) glaciation is preserved in mid-Wales. It has previously been suggested that mid-Wales was mostly free of ice during the Late Devensian, and that the deposits are mainly the result of periglacial action (Watson, 1970, 1977; Mitchell, 1960, 1972; Synge, 1964). However, this idea has largely been refuted (Bowen, 1974; Potts, 1971; Macklin and Lewin, 1986) and it is now accepted that Late Devensian glacial landforms and deposits occur throughout mid-Wales, but have been subject to extensive periglacial modification (Cave and Hains, 1986; Campbell and Bowen 1989, and references therein).

The Late Devensian deposits of the district are mostly locally sourced tills and morainic deposits, with minor amounts of glaciofluvial sands and gravels. South-west of Llanrhystud [SN 538 697] these are overlain, in places, by till derived from coeval ice of Irish Sea origin, which progressively impinged on the coastlands of Cardigan Bay as far as Pembrokeshire (Bowen, 1973a, 1974) (Figure 65). These tills are not separately distinguished on Sheet 178.

The latest Devensian and Holocene deposits in the district are mainly alluvial sediments, peat and marine deposits, but also include landslips, localised on steep valley slopes, and recent man-made features.

Till

The deposits shown as till on the geological maps of the district are mostly the ground moraine of Welsh and Irish Sea ice sheets, and include the mixture of deposits previously classified as 'morainic drift', as well as true till ('boulder clay') (Cave and Hains, 1986). They also include periglacial deposits ('head') and scree, which could not be readily distinguished from till during the rapid mapping of the district. Till deposits (in this broad sense) occur throughout the district, being generally preserved in topographic depressions and within the major valley systems; continuous spreads of till cover the coastal tract (Figure 66 and 67; Plate 31) west of Mynydd Bach [SN 610 670], the area around Ffair Rhos [SN 740 680] and parts of the east of the Rhayader sheet area.

Till deposits (sensu stricto) from the Welsh ice sheet are generally structureless, bluish grey, stony, silty and sandy clays which weather ochreous brown and yellow. The clasts are dominantly of Lower Palaeozoic turbidite sandstones, subangular to subrounded and poorly sorted, ranging in size from pebbles to large boulders (Plate 32a) and, locally, forming up to 40 per cent by volume of the deposit. Till of Irish sea derivation, exposed on the coast between Llanrhystud [SN 530 699] and Llanon [SN 509 674], is generally greenish grey, locally shelly, and contains a more varied clast assemblage including igneous rocks, quartzites, flint and jasper (Watson, 1970).

The major river valleys contain a range of heterogeneous deposits, previously classified as 'morainic drift' (Cave and Hains, 1986), including till, head, clayey structureless gravels and subordinate bedded sand and gravel. They are locally deeply dissected, but extend up to 20 m above the valley floor in places, assuming the general form of a high terrace. River incision has, locally, produced a series of erosional benches. From their general disposition, and the waterlaid nature of some of the sands and gravels, it has been suggested that they represent kame deposits of valley glaciers, formed during the waning stage of glaciation (Cave and Hains, 1986).

Periglacial processes have considerably modified glacial landforms in the district. Sections through till and morainic drift deposits reveal crude grading, imbrication and alignment of clasts in their upper part, indicative of varying degrees of solifluction. Terraces of soliflucted till occur in the upper reaches of the Elan valley [SN 871 733], around St Harmon [SN 9890 7280] and to the east of Castle Hill [SO 013 676]. They have also been observed around Rhos Gelli-gron [SN 724 634], at Treflyn [SO 015 644] and in Nant Cymrun [SO 965 613], where the terrace fronts are marked by trains of large boulders, a feature consistent with emplacement by mass movement processes (Embleton and King, 1975). Linear ridges immediately north-east of Cilcennin [SN 5220 6070 to SN 5235 6040] and in the vicinity of Cors Caron (Tregaron Bog) [SN 686 635] are interpreted as moraines. A conspicuous arcuate morainic ridge bor-

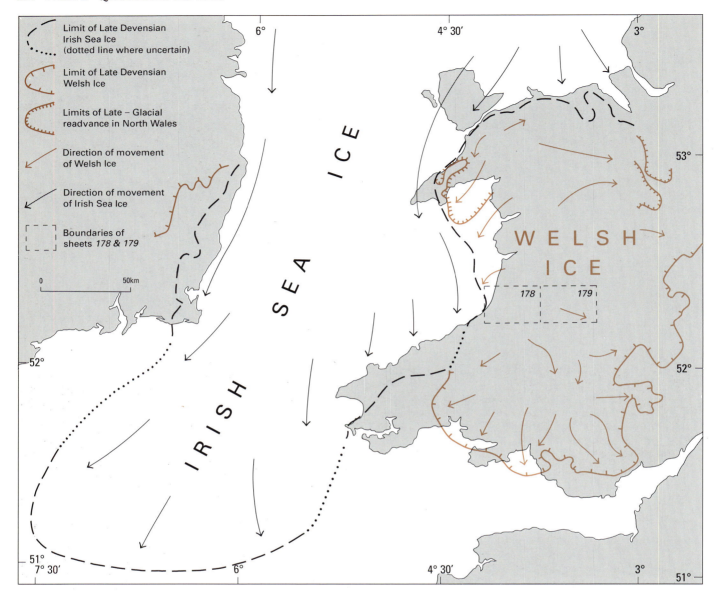

Figure 65 The direction of movement and maximum extent of Welsh and Irish Sea ice during the last (Late Devensian) glaciation. The limits of local readvance during the Late Devensian Late-glacial are also shown.

dering the south-western end of Cors Caron (Figure 70), comprising tills and intercalated gravels with glaciofluvial gravels to the south (Plates 32d and e), was previously regarded as the maximum extent of Devensian ice in the Teifi valley (Charlesworth, 1929). However, it has subsequently been shown that the distribution of Devensian glacial drift is more extensive and that the limits of glaciation lay farther south (Woodland and Evans, 1964; Bowen, 1973b, 1981); thus, the Tregaron moraine probably represents a stillstand, or local readvance, that took place during the retreat phase of the last glaciation.

Throughout the district, sections of deposits mapped as till commonly reveal alternations of glacial and peri-

glacially worked material, including irregular accumulations of a basal head which formed in advance of the main glaciation (Bowen, 1973a, 1973b). The most notable example of interbedded till and head deposits occurs on the coast in the vicinity of Morfa Bychan [SN 566 771] (Campbell and Bowen, 1989, and references therein). The cliffs there (Figure 67) expose up to 40 m of drift deposits overlying, and banked against, a steep preglacial bedrock topography of the Trefechan Formation (the 'fossil cliff' of Wood, 1959). The drift succession comprises a basal head overlain, in turn, by soliflucted till, an upper head deposit and loess. The lowermost unit (the 'Yellow Head' of Watson, 1977) con-

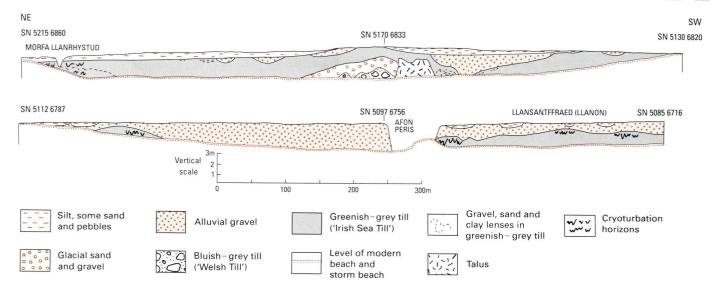

Figure 66 Drift deposits in the sea cliffs between Morfa Llanrhystud and Llansantffraid. The lower section starts approximately 200 m to the south-west of the upper section.

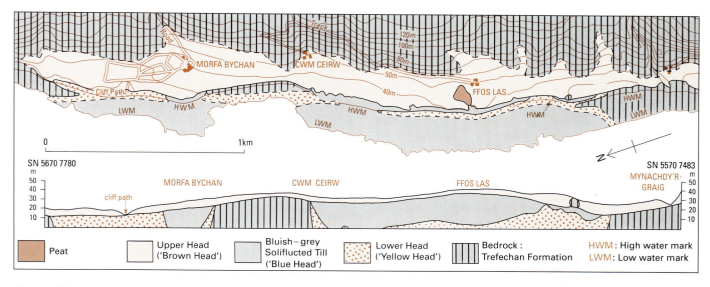

Figure 67 Drift deposits in the sea cliffs between Morfa Bychan and Mynachdy'r-graig. The distribution of the drift deposits below high water mark is based partly on Watson and Watson (1967). Descriptions in brackets are those of Watson (1977).

sists of aligned angular clasts of sandstone and mudstone, locally up to 2 m across, in a matrix of yellowish brown sandy clay. The clasts are derived from the adjacent Trefechan Formation, the larger boulders being generally confined to a zone in close proximity to bedrock. The proportion of matrix is variable, and the finer head grades locally to scree. Between Cwm-ceirw [SN 5624 7657] and Ffos-lâs [SN 5604 7575], a stratification of tabular pebbles is inclined westwards at up to 30° and there is evidence, locally, of cryoturbation. The overlying

soliflucted till ('Blue Head' of Watson, 1977), up to 30 m thick and resting with sharp contact on the head deposit, is distinguished from the latter by its larger, more rounded clasts, supported in a bluish grey clay matrix. Lenses of imbricated gravel occur in places, particularly towards the top of the unit between Cwm ceirw and Ffos lâs. Aligned pebbles and a seaward-dipping stratification of up to 20° within this unit, defined by gravel lenses and clast concentrations, led to the suggestion that the entire sequence was of periglacial origin (Watson and Watson,

Plate 31 Drift sequence in cliffs near Morfa Bychan.

The cliffs in the foreground show bluish grey till with seaward-dipping lenses of gravel in its upper part, overlain by head deposits and buff silts. View northwards from near Ffos Las [SN 5599 7605].

1967; Watson, 1970, 1982). However, striated clasts and glacially abraded quartz grains confirm the glacial origin of the unit (Bowen, 1977; Vincent, 1976), and it is now generally accepted as till which soliflucted soon after its deposition; the included gravels are thought to represent contemporaneous slopewash deposits. The soliflueted till is sharply overlain by a distinctive, coarse-grained head deposit (the 'Brown Head' of Watson, 1977) comprising angular, unweathered sandstone clasts in a sandy clay matrix, and revealing evidence of cryoturbation, including involutions and vertical stone structures (Watson and Watson, 1967). It oversteps the till on to bedrock between Morfa Bychan and Cwm-ceirw (Figure 67) and, between Cwm-ceirw and Ffos-lâs, is succeeded by a structureless brown silt with localised gravelly pockets (the 'Loess' of Watson, 1977).

Scree deposits have not been separately distinguished on Sheets 178 and 179, but locally form aprons to the steeper slopes, notably within the main river valleys. They comprise highly angular, platy rock fragments, aligned parallel to the slope, with a variable but usually small matrix content. They result from periglacial processes postdating the Late Devensian glaciation and locally overlie till deposits, as at Bwlchyddwyallt [SN 7086 6280]. Sections through scree, in places, reveal a crude threefold stratification, with unsorted debris overlain by a layered unit ('stratified scree'; Potts, 1971) of coarse and fine material overlain, in turn, by more unsorted scree. It has been suggested that these divisions record climatic fluctuations during Late Devensian Late-glacial times (Older Dryas–Allerød–Younger Dryas; 13 500– 10 000 BP) (Watson, 1965; Potts, 1971; Lowe and Walker, 1984).

Nivation cirques on north-facing slopes have been recorded in the Rhayader district at Cwmddu [SN 739 811], Cwm Tinwen [SN 831 748], Moelfryn [SN 898 665] and near Cerrig Gwinau [SN 852 647] (Watson,

1966; Potts, 1971). They are considered to have formed from the accumulation of perennial snowfields during late-glacial times, rather than as true glacial features (Watson, 1966), and are generally floored by a range of periglacially reworked deposits. Cwmddu (Figure 68, Plate 33) contains a solifluction fan composed of a series of head terraces, each capped by slopewash deposits recording periodic degeneration of the snowfield (Watson, 1966). A feature of the Cwm Tinwen cirque is a moraine-like structure (Figure 68), considered to be a protalus rampart (Watson, 1966; Campbell and Bowen, 1989), formed by the glissading of frost-shattered debris down the snow slope.

Glacial sand and gravel

The deposits shown as glacial sand and gravel on Sheets 178 and 179 include sands and gravels of indeterminate origin within morainic drift, as well as those of glaciofluvial outwash. The deposits are not extensive, but have been recorded at widely scattered localities throughout the district. The larger occurrences are those around Tregaron [SN 680 600] (Figure 70; Plate 32d and e) and Ystradmeurig [SN 710 673], in the Elan valley near Dolafallen [SN 954 666], and in the Dulas valley

Plate 32 Glacial deposits of the district (facing page).

a Bluish grey till derived from 'Welsh Ice', containing abundant, locally derived sandstone clasts. Cliff section, south-west of Llanrhystud [SN 5170 6833]. **b** Bluish grey till overlain by alluvial gravels. Cliff section [SN 5177 6836] between Llanrhystud and Llanon. **c** Detail of b showing lens of cross-laminated sand in glacial gravels. **d** Steeply inclined foreset bedding in glacial outwash gravels in a pit [SN 6674 5804] near Tregaron. **e** Glaciotectonic structures in glacial outwash gravels. Location as d.

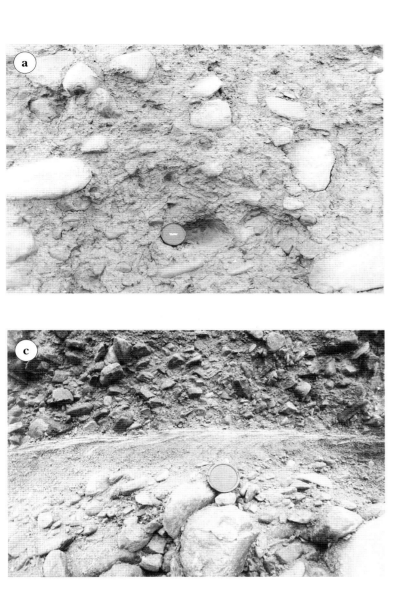

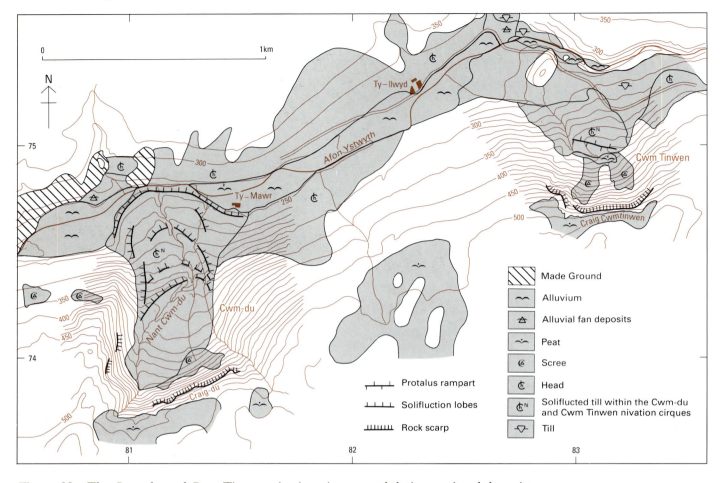

Figure 68   The Cwm-du and Cwm Tinwen nivation cirques and their associated deposits.

between Daverneithen [SO 016 660] and Caerfagu abattoir [SO 045 653]. Smaller deposits occur near Rhyd-rosser [SN 565 677], in the Ystwyth valley between Pont-rhyd-y-groes [SN 740 725] and Cwmystwyth [SN 789 740], and around Cerrigcroes [SO 037 615]; a notable sequence of late-glacial fluvial gravels is exposed on the coast near Llanon [SN 5095 6745].

The glacial sand and gravel deposits in the district range from poorly sorted clayey gravels to moderately well-sorted pebble-cobble gravels, with beds and lenses of laminated and cross-laminated silt and sand (Plates 32c and d). Pebbles are mostly of sandstone with subordinate mudstone and vein quartz. Pebble imbrication is a feature of some of the glaciofluvial gravels, and cryoturbation structures have been recorded locally. Large-scale foreset bedding, inclined to the south-east, is exposed in a pit at Caerfagu abattoir, in gravels representing the deposits of a small outwash fan.

The coastal drift sections at Llanon reveal a succession of late-glacial gravels overlying tills and intercalated gravels of Irish Sea and Welsh provenance (Plate 32b). The principal section [SN 5096 6748 to SN 5069 6683] lies to the south-west of the River Peris; to the north, a continuous section [SN 5097 6756 to SN 5115 6805] in

the gravels is exposed for 550 m, beyond which gravel pockets rest on an uneven surface of greenish grey till. The deposits are mainly imbricated pebble-cobble gravels with a few sand lenses. Fine examples of involutions and vertical stone structures occur within the gravels and adjacent tills. Truncation of earlier-formed involutions and cryoturbation effects within the overlying gravels provide evidence for several phases of periglaciation. The gravels appear to represent the deposits of an alluvial fan system from precursors of the rivers Peris and Clydan (Watson, 1977) and, thus, record a fluvial system that has existed from late-glacial times until the present day.

Alluvium

River and stream deposits are grouped together in a single drift category, Alluvium, on Sheets 178 and 179. This includes floodplain alluvium, river terrace deposits and alluvial fans, mostly of Holocene age. The floodplain alluvium of the district generally comprises brown and grey silts and clays, with interbedded sands and gravels and, locally, peat lenses. All the major river valleys contain broad alluvial tracts with, in places, associated dis-

Plate 33 View of the Cwm Ddu nivation cirque. (A14719).

In the foreground, a series of ridges of soliflucted morainic deposits extend across the slope. See also Figure 68.

continuous river terrace deposits of coarse pebble to cobble gravel; narrow, irregular deposits of alluvium also occur in their higher reaches, and in many of the smaller river valleys. Alluvial fans, comprising interbedded coarse gravels, pebbly sands and silt, are generally found at the confluence of tributaries with the main river valleys, often grading imperceptibly into the alluvium; they probably accumulated under conditions of high stream flow, during Late Devensian to early Holocene times (Cave and Hains, 1986), and most are not being actively deposited. A combination of fluvial and solifluction processes has formed narrow alluvial cones on some steep valley sides; these have mostly been included with head and soliflucted till.

Peat

Extensive peat deposits are developed in parts of the district, mainly in areas of high precipitation and restricted or enclosed drainage. They are widespread in the upland area of Mynydd Bach and the Cambrian Mountains west of Rhayader and, west of Mynydd Bach, they blanket interfluves and fill hollows in till deposits. Extensive peat bogs, overlying lacustrine clays in the Teifi valley north of Tregaron and at the head of the Elan valley, are important in revealing a history of climatic change since the Late Devensian (Godwin and Mitchell, 1938; Hibbert and Switsur, 1976; Moore and Chater, 1969; Moore, 1970).

The characteristic linear, ridged topography of the Llandovery sandstone turbidite sequences has commonly restricted the drainage and allowed peat to accumulate in elongate hollows. Over upland areas and interfluves where there is little topographical relief, blanket peat has generally formed on an impermeable substrate of till, but is mostly less than 1 m thick. Radiocarbon dating of blanket peat from the upland areas of mid-Wales indicates that it initially formed during the early Holocene (Welin et al., 1974; Cave and Hains, 1986). It has been suggested that blanket bogs started to develop in response to the increased precipitation during regional pollen zone VIIa, (c. 7500–5000 BP; Bowen, 1974); however, pollen studies of a site on the north side of Llyn Gynon [SN 8010 6480] (Moore and Chater, 1969) suggest that, in places, blanket peat accumulation began as early as zone VI (c. 9000–7500 BP).

A series of shallow boreholes through the Elan valley bog (Gors Lwyd) [SN 857 754] (Moore and Chater, 1969; Moore, 1970) (Figure 69) revealed a succession of peat deposits, up to 5 m thick, overlying organic-rich muds and blue lacustrine clay. Pollen biozonation of the bog demonstrates that deposition of the lake clay and organic muds mainly occurred during late-glacial times (regional pollen zones I–III), with a marked increase in birch, juniper and scrubby vegetation in zones I–II corresponding to the Allerød climatic amelioration (Moore, 1970; Campbell and Bowen, 1989). Accumulation of peat in the bog began during the early Holocene (c. 10 000 BP; pollen zone IV) and has continued to the present day.

Cors Caron (Tregaron bog) [SN 685 635] (Figure 70) is one of the largest areas of raised peat bog in Britain and its deposits have been studied in detail (Campbell and Bowen, 1989 and references therein). It comprises three separate areas of raised bog, convex in morphology but modified by peat cutting, rising up to 8 m above the level of the River Teifi (Godwin and Conway, 1939). Boreholes have proved at least 8 m of peat overlying organic-rich muds with, at the base, a stiff blue-grey lacustrine clay which was not penetrated (Godwin and Mitchell, 1938); the lacustrine clay probably accumulated during late-glacial times, in water impounded by the moraine at Tregaron. The marginal areas of the bog

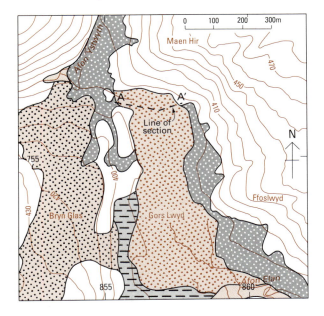

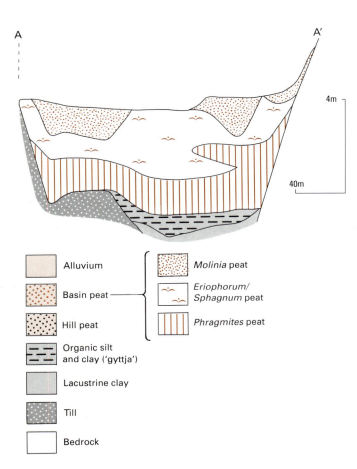

Alluvium

Basin peat

Hill peat

Organic silt and clay ('gyttja')

Lacustrine clay

Till

Bedrock

Molinia peat

Eriophorum/ Sphagmum peat

Phragmites peat

reveal intercalations of slopewash deposits, but peat overlies till along its western side near Llwynbeudy [SN 6737 6491] and Maesglas [SN 6712 6229]. Radiocarbon dating of the organic clay underlying the peat gave a latest Devensian age of 10 200 ± 220 BP (Hibbert and Switsur, 1976); local pollen assemblages from the peat record a series of vegetational and climatic changes to the present day.

Marine deposits

The marine deposits shown on Sheet 178 include beach, storm beach and intertidal deposits, mostly between Llanon [SN 5085 6720] and Carreg Ti-pw [SN 5350 7080]. The beach and intertidal deposits generally comprise sands and pebble-cobble gravels, and are backed by a storm beach of cobble gravel which forms a prominent feature for about 1 km south-west of Morfa [SN 5282 6960].

Landslip

Small-scale landslips have been recorded at several localities in the Rhayader sheet area. They generally occur on steep valley slopes in both solid formations and drift, ranging from rotational slips, (e.g. near Garth [SN 984 608]), to unconsolidated slides (e.g. in Nant yr Iau [SN 850 630]). Landslips along the banks of the River Ithon near Crossgates [SO 082 642 and SO 086 638] are due to undercutting by the river. It is possible that landslipping occurred as early as the Late Devensian but there is no direct evidence for this; one landslip, in the Wye valley near Banc Dolhelfa [SN 922 749] is known to have moved around 1850.

Made ground

The deposits shown as made ground on the maps are man-made deposits, comprising tipped materials of variable thickness and composition. Throughout the district there are many small tips, mainly from lead workings and quarries. The larger spoil heaps include those of the Frongoch [SN 7220 7446], Glogfawr [SN 7474 7048], Esgairmwyn [SN 7550 6912] and Cwmystwyth mines [SN 8038 7460]. Substantial amounts of waste from quarrying operations occur at Hendre quarry [SN 7226 6960] and in the Elan valley below the Caban-coch reservoir [SN 9286 6468].

Figure 69   The drift geology of the Elan Valley Bog (Gors Lwyd) [SN 858 753]. Section simplified from Moore and Chater (1969).

Figure 70 The drift geology of Tregaron Bog (Cors Caron) and surrounding area. The contours at 1 m intervals on the western bog are based on a survey by Godwin and Conway (1939).

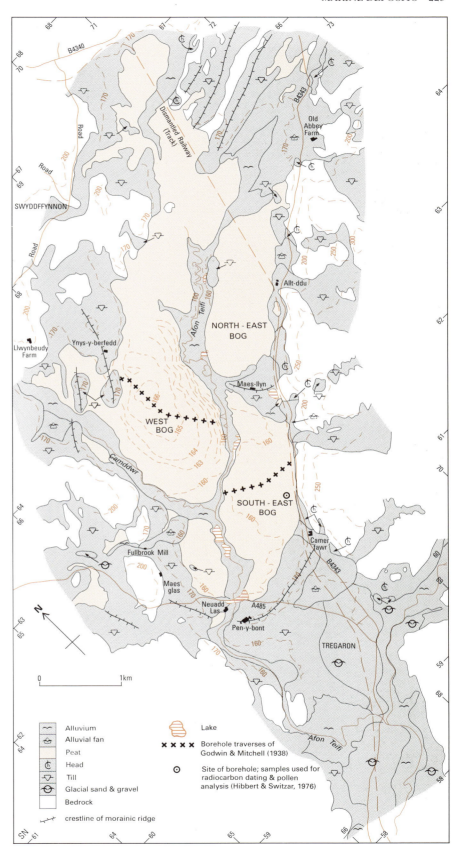

REFERENCES

Most of the references listed below are held in the Library of the British Geological Survey at Keyworth, Nottingham. Copies of the references can be purchased subject to the current copyright legislation.

ALDRIDGE, R J, DORNING, K J, HILL, P J, RICHARDSON, J B, and SIVETER, D J. 1979. Microfossil distribution in the Silurian of Britain and Ireland. 433–438 in The Caledonides of the British Isles — reviewed. HARRIS, A L, HOLLAND, C H, and LEAKE, B E (editors). *Special Publication of the Geological Society of London*, No. 8.

ALLEN, J R L. 1974. The Devonian rocks of Wales and the Welsh Borderland. 47–84 in *The Upper Palaeozoic and post-Palaeozoic rocks of Wales*. OWEN, T R (editor). (Cardiff: University of Wales Press.)

ALLEN, P M, and JACKSON, A A. 1985. Geology of the country around Harlech. *Memoir of the British Geological Survey*, Sheet 135 (England and Wales).

ANDREW, G. 1925. The Llandovery rocks of Garth (Breconshire). *Quarterly Journal of the Geological Society of London*, Vol. 81, 389–406.

ANDREW, G, and JONES, O T. 1925. The relations between the Llandovery rocks of Llandovery and those of Garth. *Quarterly Journal of the Geological Society of London*, Vol. 81, 389–406.

ANKETELL, J M, and LOVELL, J P B. 1976. Upper Llandoverian Grogal Sandstones and Aberystwyth Grits in the New Quay area, Central Wales: a possible upwards transition from contourites to turbidites. *Geological Journal*, Vol. 11, 101–108.

BADHAM, J P N. 1981. Shale-hosted Pb-Zn deposits: products of exhalation of formation waters? *Transactions of the Institution of Mining and Metallurgy*, Vol. 90, Section B, B70–B76.

BAILEY, R J. 1969. Ludlovian sedimentation in south central Wales. 283–304 in *The Precambrian and Lower Palaeozoic rocks of Wales*. WOOD, A (editor). (Cardiff: University of Wales Press.)

BAKER, J W. 1973. A marginal late Proterozoic ocean basin in the Welsh region. *Geological Magazine*, Vol. 110, 447–455.

BALL, T K, DAVIES, J R, WATERS, R A, and ZALASIEWICZ, J A. 1992. Geochemical discrimination of Silurian mudstones according to depositional process and provenance within the Southern Welsh Basin. *Geological Magazine*, Vol. 129, 567–572.

BALL, T K, and NUTT, M J C. 1976. Preliminary mineral reconnaissance of Central Wales. *Report of the Institute of Geological Sciences*, No. 75/14.

BAMFORD, D, FABER, S, JACOB, B, KAMINSKI, W, NUNN, K, PRODEHL, C, FUCHS, K, KING, R, and WILLMORE, P. 1976. A lithospheric profile in Britain, i. Preliminary results. *Geophysical Journal of the Royal Astronomical Society*, Vol. 44, 145–160.

BASSETT, D A. 1955. The Silurian rocks of the Talerddig district, Montgomeryshire. *Quarterly Journal of the Geological Society of London*, Vol. 111, 239–264.

BASSETT, D A. 1963. The Welsh Palaeozoic Geosyncline: A review of recent work on stratigraphy and sedimentation. 35–69 in *The British Caledonides*. JOHNSON, M R W, and STEWART, F H (editors). (Edinburgh: Oliver and Boyd.)

BASSETT, D A. 1969. Some of the major structures of early Palaeozoic age in Wales and the Welsh Borderland: an historical essay. 67–116 in *The Precambrian and Lower Palaeozoic rocks of Wales*. WOOD, A (editor). (Cardiff: University of Wales Press.)

BASSETT, M G. 1974. Review of the stratigraphy of the Wenlock Series in the Welsh Borderland and South Wales. *Palaeontology*, Vol. 17, 745–77.

BASSETT, M G. 1989. Brachiopods. 232–242 in A global standard for the Silurian System. HOLLAND, C H, and BASSETT, M G (editors). *National Museum of Wales, Cardiff, Geological Series*, No. 9.

BASSETT, M G, BLUCK, B J, CAVE, R, HOLLAND, C H, and LAWSON, J D. 1992. Silurian. 37–56 in Atlas of palaeogeography and lithofacies. COPE, J C W, INGHAM, J K, and RAWSON, P F (editors). *Memoir of the Geological Society of London*, No. 13.

BATES, D E B. 1982. The Aberystwyth Grits. 81–90 in *Geological excursion in Dyfed, south-west Wales*. BASSETT, M G (editor). (Cardiff: National Museum of Wales.)

BERRY, W B N, and BOUCOT, A J. 1967. Pelecypod–graptolite associations in the old world Silurian. *Geological Society of America Bulletin*, Vol. 78, 1515–1522.

BERRY, W B, and WILDE, P. 1978. Progressive ventilation of the oceans — an explanation for the distribution of the Lower Palaeozoic black shales. *American Journal of Science*, Vol. 278, 257–275.

BEVINS, R E, and ROWBOTHAM, G. 1983. Low-grade metamorphism within the Welsh sector of the paratectonic Caledonides. *Geological Journal*, Vol. 18, 141–168.

BICK, D E. 1974. *The old metal mines of Mid-Wales. Part 1. Cardiganshire south of Devil's Bridge*. (Newent: The Pound House.)

BJERRESKOV, M. 1975. Llandoverian and Wenlockian graptolites on Bornholm. *Fossils and Strata*, Vol. 8, 1–94.

BLUNDELL, D J, DAVEY, F J, and GRAVES, L J. 1971. Geophysical surveys over the South Irish Sea and Nymphe Bank. *Journal of the Geological Society of London*, Vol. 127, 339–375.

BORRADAILE, G J. 1978. Transected folds: a study illustrated with examples from Canada and Scotland. *Geological Society of America Bulletin*, Vol. 89, 481–493.

BOSWELL, P G H. 1926. A contribution to the geology of the eastern part of the Denbighshire moors. *Quarterly Journal of the Geological Society of London*, Vol. 82, 556–585.

BOUCEK, B. 1953. Biostratigrafie, vyvoj a korrelace zelkovickych a motolskych vrstev ceskeho siluru. *Sbornik Ustredniho ustavu geologickeho*, Vol. 20, 421–484.

BOUMA, A H. 1962. *Sedimentology of some flysch deposits — a graphic approach to facies interpretation*. (Amsterdam/New York: Elsevier.)

BOWEN, D Q. 1973a. The Pleistocene succession of the Irish Sea. *Proceedings of the Geologists' Association*, Vol. 84, 249–272.

BOWEN, D Q. 1973b. The Pleistocene history of Wales and the Borderland. *Geological Journal*, Vol. 8, 207–224.

BOWEN, D Q. 1974. The Quaternary of Wales. 372–426 in *The Upper Palaeozoic and post-Palaeozoic rocks of Wales.* OWEN, T R (editor). (Cardiff: University of Wales Press.)

BOWEN, D Q. 1977. *Wales and the Cheshire–Shropshire Lowland.* INQUA X Congress. Guidebook for excursions A8 & C8. (Norwich: Geo Abstracts.)

BOWEN, D Q. 1981. The 'South Wales end-moraine': fifty years after. 60–67 in *The Quaternary of Britain.* NEALE, J, and FLENLEY, J (editors). (Oxford: Pergamon Press.)

BOWEN, D Q. 1991. Time and space in the glacial sediment systems of the British Isles. 3–12 in *Glacial deposits in Great Britain and Ireland.* EHLERS, J, GIBBARD, P L, and ROSE, J (editors). (Rotterdam: A A Balkema.)

BRENCHLEY, P J. 1988. Environmental changes close to the Ordovician–Silurian boundary. 377–385 in A global analysis of the Ordovician–Silurian boundary. COCKS, L R M, and RICKARDS, R B (editors). *Bulletin of the British Museum (Natural History), Geology*, Vol. 43.

BRENCHLEY, P J, and CULLEN, B. 1984. The environmental distribution of associations belonging to the *Hirnantia* fauna — evidence from North Wales and Norway. 113–125 in Aspects of the Ordovician System. BRUTON, D L (editor). *Palaeontological Contributions from the University of Oslo*, No. 295.

BRENCHLEY, P J, and NEWALL, G. 1984. Late Ordovician environmental changes and their effect on faunas. 65–79 in Aspects of the Ordovician System. BRUTON, D L (editor). *Palaeontological Contributions from the University of Oslo*, No. 295.

BRIDGES, P H. 1975. The transgression of a hard substrate shelf: the Llandovery (Lower Silurian) of the Welsh Borderland. *Journal of Sedimentary Petrology*, Vol. 45, 79–94.

BRITISH GEOLOGICAL SURVEY. 1993. Corwen. England and Wales Sheet 120. Solid and Drift Geology, 1:50 000. (Keyworth, Nottingham: British Geological Survey.)

BRITISH GEOLOGICAL SURVEY. 1994. Montgomery. England and Wales Sheet 165. Solid Geology, 1:50 000. (Keyworth, Nottingham: British Geological Survey.)

BULMAN, O M B. 1958. The sequence of graptolite faunas. *Palaeontology*, Vol. 1, 159–173.

CALEF, C E, and HANCOCK, N J. 1974. Wenlock and Ludlow marine communities in Wales and the Welsh Borderland. *Palaeontology*, Vol. 17, 779–810.

CAMPBELL, S, and BOWEN, D Q. 1989. *Quaternary of Wales.* Geological Conservation Review Series A4.1. (Peterborough: Nature Conservancy Council.)

CAMPBELL, S D G, HOWELLS, M F, SMITH, M, and REEDMAN, A J. 1988. A Caradoc failed-rift within the Ordovician Marginal Basin of Wales. *Geological Magazine*, Vol. 125, 257–266.

CARRUTHERS, R M, EVANS, R B, GIBBERD, A J, and ROYLES, C P. 1989. A preliminary account of geophysical fieldwork undertaken in the Rhayader/Elan Valley district of central Wales. *Project Note of the Regional Geophysics Group of the British Geological Survey*, No. 89/9.

CARRUTHERS, R M, FLETCHER, C J N, McDONALD, A J W, and EVANS, R B. 1992. Some constraints on the form of the Welsh Basin from regional gravity and magnetic data, with particular reference to Central Wales. *Geological Magazine*, Vol. 129, 515–522.

CARRUTHERS, R M, ROLLIN, K E, ROYLES, C P, and EVANS, R B. 1991. Geophysical investigations for minerals over the Rhiwnant Inlier, near Rhayader. *Project Note of the Regional Geophysics Group of the British Geological Survey*, No. 91/2.

CARTER, R M. 1975. A discussion and classification of subaqueous mass-transport with particular application to grain-flow, slurry-flow and fluxoturbidites. *Earth Science Reviews*, Vol. 11, 145–177.

CASE, J E. 1974. Major basins along the contintental margin of Nothern South America. 733–741 in *The geology of continental margins.* BURK, C A, and DRAKE, C L (editors). (New York: Springer-Verlag.)

CAVE, R. 1965. The Nod Glas sediments of Caradoc age in North Wales. *Geological Journal*, Vol. 4, 279–298.

CAVE, R. 1975. Glan-fred Borehole, Dol-y-bont. P.7 in Institute of Geological Sciences. IGS Boreholes 1974. *Report of the Institute of Geological Sciences*, No.75/7.

CAVE, R. 1976. P.24 in *Institute of Geological Sciences, annual report for 1975.* (London: Institute of Geological Sciences.)

CAVE, R. 1979. Sedimentary environments of the basinal Llandovery of mid-Wales. 517–526 in The Caledonides of the British Isles — reviewed. HARRIS, A L, HOLLAND, C H, and LEAKE, B E (editors). *Special Publication of the Geological Society of London*, No. 8.

CAVE, R. 1991. How it all began. *Mine and Quarry*, Vol.20, 21–23.

CAVE, R. 1992. The Ystrad Meurig Grits, an early-Silurian turbiditic sandstone lobe in mid-Wales. *Journal of the Earth Science Teachers' Association*, Vol.17, 87–95.

CAVE, R, and DIXON, R J. 1993. The Ordovician and Silurian of the Welshpool area. 52–84 in *Geological excursions in Powys, Wales.* WOODCOCK, N H, and BASSETT, M G (editors). (Cardiff: University of Wales Press.)

CAVE, R, and HAINS, B A. 1986. Geology of the country between Aberystwyth and Machynlleth. *Memoir of the British Geological Survey*, Sheet 163 (England and Wales).

CAVE, R, and HAINS, B A. In preparation. Geology of the country around Montgomery. *Memoir of the British Geological Survey*, Sheet 165 (England and Wales).

CAVE, R, and PRICE, D. 1978. The Ashgill series near Welshpool, North Wales. *Geological Magazine*, Vol. 115, 183–194.

CHARLESWORTH, J K. 1929. The South Wales end-moraine. *Quarterly Journal of the Geological Society of London*, Vol. 85, 335–358.

CLAYTON, C. 1992. The sedimentology of a confined turbidite system in the early Silurian Welsh Basin. Unpublished PhD thesis, University of Cambridge.

COCKS, L R M. 1970. The Silurian brachiopods of the superfamily *Plectambonitacea*. *Bulletin of the British Museum, Natural History (Geology)*, Vol. 19, 139–203.

COCKS, L R M, and FORTEY, R A. 1982. Faunal evidence for oceanic separations in the Palaeozoic of Britian. *Journal of the Geological Society of London*, Vol. 139, 467–480.

COCKS, L R M, WOODCOCK, N H, RICKARDS, R B, TEMPLE, J T, and LANE, P D. 1984. The Llandovery Series of the type area. *Bulletin of the British Museum (Natural History) Geology*, Vol. 38, 131–182.

CORNWELL, J D, and CAVE, R. 1986. An airborne geophysical survey of part of west Dyfed, South Wales, and some related ground surveys. *Mineral Reconnaissance Programme Report of the British Geological Survey*, No. 63.

CRAIG, J. 1987. The structure of the Llangrannog Lineament, West Wales: a Caledonian transpression zone. *Geological Journal*, Vol. 22 (Thematic Issue), 167–181.

CRIMES, T P, and CROSSLEY, J D. 1980. Inter-turbidite bottom current orientation from trace fossils, with an example from the Silurian flysch of Wales. *Journal of Sedimentary Petrology,* Vol. 50, 821–830.

CRIMES, T P, and CROSSLEY, J D. 1991. A diverse ichnofauna from Silurian flysch of the Aberystwyth Grits Formation, Wales. *Geological Journal,* Vol. 26, 27–64.

CUMMING, G L, and RICHARDS, J R. 1975. Ore lead isotope ratios in a continuously changing earth. *Earth and Planetary Science Letters,* Vol. 28, 155–171.

CURTIS, C D. 1980. Diagenetic alteration of black shales. *Journal of the Geological Society of London,* Vol. 137, 189–194.

CURTIS, C D. 1987. Mineralogical consequences of organic matter degradation in sediments; inorganic/organic diagenesis. 108–123 in *Marine clastic sedimentology.* LEGETT, J K, and ZUFFA, G G (editors). (London: Graham and Trotman.)

DAVIES, J H. 1980. A suggested re-interpretation of the Lower Palaeozoic stratigraphy around Llanfan Fawr, Central Wales. *Geological Journal,* Vol.15, 131–133.

DAVIES, K A. 1926. The geology of the country between Drygarn and Abergwesyn (Breconshire). *Quarterly Journal of the Geological Society of London,* Vol. 82, 436–64.

DAVIES, K A. 1928. Contributions to the geology of central Wales. I. —Notes on the geology of the southern portion of central Wales. II. — The geology of the country between Rhayader (Radnorshire) and Abergwesyn (Breconshire). *Proceedings of the Geologists' Association,* Vol. 39, 157–168.

DAVIES, K A. 1929. Notes on the graptolite faunas of the Upper Ordovician and Lower Silurian. *Geological Magazine,* Vol. 66, 1–27.

DAVIES, K A. 1933. The geology of the country between Abergwesyn (Breconshire) and Pumpsaint (Carmarthenshire). *Quarterly Journal of the Geological Society of London,* Vol. 89, 172–201.

DAVIES, K A, and PLATT, J I. 1933. The conglomerates and grits of the Bala and Valentian rocks of the district between Rhayader (Radnorshire) and Llansawel (Carmarthenshire). *Quarterly Journal of the Geological Society of London,* Vol. 89, 202–220.

DAVIES, W, and CAVE, R. 1976. Folding and cleavage determined during sedimentation. *Sedimentary Geology,* Vol. 15, 89–133.

DEWEY, J F. 1969. Structure and sequence in paratectonic British Caledonides. *Memoir of the American Association of Petroleum Geologists,* Vol. 12, 309–335.

DIMBERLINE, A J. 1987. The sedimentology and diagenisis of the Wenlock turbidite system, Wales, U K. Unpublished PhD thesis, University of Cambridge.

DIMBERLINE, A J, BELL, A, and WOODCOCK, N H. 1990. A laminated hemipelagic facies from the Wenlock and Ludlow of the Welsh Basin. *Journal of the Geological Society of London,* Vol. 147, 693–701.

DIMBERLINE, A J, and WOODCOCK, N H. 1987. The southeast margin of the Wenlock turbidite system, Mid-Wales. *Geological Journal,* Vol. 22, 61–71.

DOBSON, M R, EVANS, W E, and WHITTINGTON, R. 1973. The geology of the south Irish Sea. *Report of the Institute of Geological Sciences,* No. 73/11.

DONG, H, HALL, C M, HALLIDAY, A N, PEACOR, D R, MERRIMAN, R J, and ROBERTS, B. In preparation. ^{40}Ar-^{39}Ar dating of Acadian metamorphism using re-equilibrated epizonal illite from the Lower Palaeozoic Welsh Basin. Submitted to *Geology.*

DORNING, K J. 1981a. Silurian acritarchs from the type Wenlock and Ludlow of Shropshire, England. *Review of Palaeobotany and Palynology,* Vol. 34, 175–203.

DORNING, K J. 1981b. Silurian acritarch distribution in the Ludlovian shelf sea of South Wales and the Welsh Borderland. 31–36 in *Microfossils from recent and fossil shelf seas.* NEALE, J W, and BRASIER, M D (editors). (Chichester: Ellis Horwood.)

DORNING, K J, and BELL, D G. 1987. The Silurian carbonate shelf microflora: acritarch distribution in the Much Wenlock Limestone Formation. 266–287 in *Micropalaeontology of carbonate environments.* HART, M B (editor). (Chichester: Ellis Horwood.)

DUNHAM, K C. 1943–44. The production of galena and associated minerals in the northern Pennines; with comparative statistics for Great Britain. *Transactions of the Institution of Mining and Metallurgy, London,* Vol. 53, 181–214.

DUNKLEY, P N. 1979. Ordovician volcanicity of the SE Harlech Dome. 597–601 *in* The Caledonides of the British Isles —reviewed. HARRIS, A L, HOLLAND, C H, and LEAKE, B E, (editors). *Special Publication of the Geological Society of London,* No. 8.

EDMUNDS, W M, and ROBINS, N S. 1991. The mineral waters at Llandrindod Wells and Central Wales. *British Geological Survey Technical Report,* WD/91/60C.

ELLES, G L. 1940. The stratigraphy and faunal succession in the Ordovician rocks of the Builth–Llandrindod inlier, Radnorshire. *Quarterly Journal of the Geological Society of London,* Vol. 95, 385–445.

ELLES, G L, and WOOD, E M R. 1908. A monograph of British graptolites. *Monographs of the Palaeontological Society.*

EMBLETON, C, and KING, C A M. 1975. *Periglacial geomorphology.* (London: Edward Arnold.)

ENOS, P. 1977. Flow regimes in debris flow. *Sedimentology,* Vol. 24, 133–142.

EVANS, A D. 1990. Magnetic anomaly values in UK — a cautionary note. *Project Note of the Regional Geophysics Group of the British Geological Survey,* No. 90/6.

EVANS, J A. 1989. Short paper: a note on Rb-Sr whole rock ages from cleaved mudrocks in the Welsh Basin. *Journal of the Geological Society of London,* Vol. 146, 901–904.

EVANS, J A. 1992. Geochemical and isotope composition of pebbles from the Caban Conglomerate Formation and their bearing on the source of Welsh Palaeozoic sedimentary rocks. *Geological Magazine,* Vol. 129, 581–587.

FENSOME, R A, WILLIAMS, G L, SEDLEY BARSS, M, FREEMAN, J M, and HILL, J M. 1990. Acritarchs and fossil Prasinophytes: an index to genera, species and infraspecific taxa. *American Association of Stratigraphic Palynologists,* Vol. 25, 1–771.

FINLAYSON, A M. 1910. The metallogeny of the British Isles. *Quarterly Journal of the Geological Society of London,* Vol. 66, 281–298.

FITCHES, W R. 1972. Polyphase deformation structures in the Welsh Caledonides near Aberystwyth. *Geological Magazine,* Vol. 109, 149–155.

FITCHES, W R, CAVE, R, CRAIG, J, and MALTMAN, A J. 1986. Early veins as evidence of detachment in the Lower Palaeozoic rocks of the Welsh Basin. *Journal of Structural Geology,* Vol. 8, 607–620.

FITCHES, W R, CAVE, R, CRAIG, J, and MALTMAN, A J. 1990. The flexural-slip mechanism: discussion. *Journal of Structural Geology,* Vol. 12, 1081–1087.

FLETCHER, C J N, SWAINBANK, I G, and COLMAN, T B. 1993. Evolution of lead isotopes in Wales. *Journal of the Geological Society of London*, Vol. 150, 77–82.

FORTEY, R A, and COCKS, L R M. 1986. Marginal faunal belts and their structural implications, with examples from the Lower Palaeozoic. *Journal of the Geological Society of London*, Vol.143, 151–160.

FORTEY, R A, and OWENS, R M. 1987. The Arenig Series in South Wales. *Bulletin of the British Museum (Natural History), Geology*, Vol. 41, 69–307.

FOSTER-SMITH, J R. 1978. The mines of Montgomery and Radnorshire. *British Mining*, No. 10.

FOSTER-SMITH, J R. 1979. The mines of Cardiganshire. *British Mining*, No. 12.

FURNESS, H. 1978. A comparative study of Caledonian volcanics in Wales and west Norway. Unpublished PhD thesis, University of Oxford.

GALLOWAY, W E. 1989. Genetic stratigraphic sequences in basin analysis 1: Architecture and genesis of flooding-surface bounded depositional units. *Bulletin of the American Association of Petroleum Geologists*, Vol.73, 125–142.

GEOLOGICAL SURVEY OF GREAT BRITAIN. 1850. Old Series one-inch Sheet 56 NE. (London: Royal Engineers.)

GEORGE, T N. 1963. Palaeozoic growth of the British Caledonides. 1–33 in *The British Caledonides*. JOHNSON, M R W, and STEWART, F H (editors). (Edinburgh: Oliver & Boyd.)

GEORGE, T N. 1970. *British regional geology: South Wales* (3rd edition). (London: HMSO for Institute of Geological Sciences.)

GIBBONS, W. 1987. Menai Strait fault system: An early Caledonian terrane boundary in North Wales. *Geology*, Vol. 15, 744–747.

GODWIN, H, and CONWAY, V M. 1939. The ecology of a raised bog near Tregaron, Cardiganshire. *Journal of Ecology*, Vol. 27, 313–359.

GODWIN, H, and MITCHELL, G F. 1938. Stratigraphy and development of two raised bogs near Tregaron, Cardiganshire. *The New Phytologist*, Vol. 37, 425–454.

GRAHN, Y. 1988. Chitinozoan stratigraphy in the Ashgill and Llandovery. *Bulletin of the British Museum (Natural History)*, Geology Series, Vol. 43, 317–323.

HALL, G W. 1971. *Metal mines of southern Wales*. (Westbury-on-Severn: G W Hall.)

HALL, G W. 1989. The last 100 years of mining at Pontrhydygroes, Dyfed. *Mining Magazine*, February, 116–121.

HANCOCK, N J, HURST, J M, and FURSICH, F T. 1974. The depth inhabited by Silurian brachiopod communities. *Journal of the Geological Society of London*, Vol. 130, 151–156.

HARRIS, J H. 1987. The geology of the Wenlock Shales around Builth Wells. Unpublished PhD thesis, University of Cambridge.

HEDE, J E. 1915. Skånes Colonusskiffer. *Lunds Universitets Årsskrift*, N.F.Afd, Vol. 11 (No. 6), 66pp and 4 plates.

HEIN, F J, and WALKER, R G. 1982. The Cambro-Ordovician Cap Enrage Formation, Quebec, Canada: conglomerate deposits of a braided submarine channel with terraces. *Sedimentology*, Vol.29, 309–329.

HIBBERT, F A, and SWITSUR, V R. 1976. Radiocarbon dating of Flandrian pollen zones in Wales and Northern England. *New Phytologist*, Vol. 77, 793–807.

HILL, P J. 1974. Stratigraphic palynology of acritarchs from the type area of the Llandovery and the Welsh Borderland. *Review of Palaeobotany and Palynology*, Vol. 18, 11–23.

HILL, P J, and DORNING, K J. 1984. Acritarchs. 174–176 *in* The Llandovery Series of the type area. COCKS, L R M, WOODCOCK, N H, RICKARDS, R B, TEMPLE, J T, and LANE, P D (editors). *Bulletin of the British Museum (Natural History) Geology*, Vol. 38.

HOLROYD, J. 1978. The sedimentologic and geotectonic significance of Lower Palaeozoic flysch-rudites. Unpublished PhD thesis, University of Wales, Swansea.

HOWELLS, M F, REEDMAN, A J, and CAMPBELL, S D G. 1991. *Ordovician (Caradoc) marginal basin volcanism in Snowdonia (north-west Wales)*. (London: HMSO for the British Geological Survey).

HUGHES, C P. 1969. The Ordovician trilobite faunas of the Builth–Llandrindod inlier, central Wales. *Bulletin of the British Museum (Natural History), Geology*, Vol. 18, 39–103.

HUGHES, C P. 1971. The Ordovician trilobite faunas of the Builth–Llandrindod Inlier, central Wales. Part II. *Bulletin of the British Museum (Natural History), Geology*, Vol. 20, 115–182.

HUGHES, C P. 1979. The Ordovician trilobite faunas of the Builth–Llandrindod Inlier, central Wales. Part III. *Bulletin of the British Museum (Natural History), Geology*, Vol. 32, 109–181.

HUGHES, R A. 1989. Llandeilo and Caradoc graptolites of the Builth and Shelve Inliers. *Monograph of the Palaeontographical Society*, No. 577.

HUGHES, S J S. 1981. The Cwmystwyth mines. *British Mining*, No. 17.

HUGHES, T McKENNY. 1879. On the Silurian rocks of the valley of Clwyd. *Quarterly Journal of the Geological Society of London*, Vol. 35, 694–698.

HURST, J M, HANCOCK, N J, and McKERROW, W S. 1978. Wenlock stratigraphy and palaeogeography of Wales and the Welsh Borderland. *Proceedings of the Geologists' Association*, Vol. 89, 197–226.

HUTT, J E. 1974–75. The Llandovery graptolites of the English Lake District. *Palaeontographical Society (Monograph)*, Vol. 128. 129 and 137pp.

INESON, P R, and MITCHELL, J G. 1975. K-Ar isotopic age determinations from some Welsh mineral localities. *Transactions of the Institution of Mining and Metallurgy*, (Section B), Vol. 84, B7–B16.

INSTITUTE OF GEOLOGICAL SCIENCES. 1972. Newtown Special Sheet. 1:10 560. (Southampton: Ordnance Survey for Institute of Geological Sciences.)

INSTITUTE OF GEOLOGICAL SCIENCES. 1974. Central Wales Mining Field. 1:100 000. (Southampton: Ordnance Survey for Institute of Geological Sciences.)

INSTITUTE OF GEOLOGICAL SCIENCES. 1977. Llandrindod Wells Ordovician Inlier. Solid. Parts of 1:25 000 Sheets SO 05, 06, 15 and 16. Classical areas of British Geology. (Southampton: Ordnance Survey for Institute of Geological Sciences.)

JACOBSON, S R, and ACHAB, A. 1985. Acritarch biostratigraphy of the *Dicellograptus complanatus* graptolite zone from the Vaureal Formation (Ashgillian), Anticosti Island, Quebec, Canada. *Palynology*, Vol. 9, 165–198.

JAMES, D M D. 1983. Observations and speculations on the northeast Towy 'axis', mid-Wales. *Geological Journal*, Vol. 18, 283–296.

JAMES, D M D. 1986. The Rhiwnant inlier, Powys, Mid-Wales. *Geological Magazine*, Vol. 123, 585–587.

JAMES, D M D. 1991. A late Ordovician/early Silurian non-depositional slope and perched roll-over basin along the Tywi Anticline, Mid Wales. *Geological Journal*, Vol. 26, 7–26.

JAMES, D M D, and JAMES, J. 1969. The influence of deep fractures on some areas of Ashgillian–Llandoverian sedimentation in Wales. *Geological Magazine*, Vol. 106, 562–582.

JENKYNS, H C. 1980. Cretaceous anoxic events: from continents to oceans. *Journal of the Geological Society of London*, Vol. 137, 171–188.

JENKYNS, H C. 1986. Pelagic environments. 343–397 in *Sedimentary environments and facies* (2nd edition). READING, H G (editor). (Oxford: Blackwell Scientific.)

JONES, J A, and MORETON, N J M. 1977. *The mines and minerals of mid-Wales (North Cardigan and West Montgomery).* (Leicester: Bailey, Biddles and Halford.)

JONES, O T. 1909. The Hartfell-Valentian succession in the district around Plynlimon and Pont Erwyd (North Cardiganshire). *Quarterly Journal of the Geological Society of London*, Vol. 65, 463–537.

JONES, O T. 1912. The geological structure of Central Wales and the adjoining regions. *Quarterly Journal of the Geological Society of London*, Vol. 68, 328–344.

JONES, O T. 1922. Lead and zinc. The mining district of north Cardiganshire and west Montgomeryshire. *Memoirs of the Geological Survey, Special Reports on the Mineral Resources of Great Britain*, Vol. 20.

JONES, O T. 1938. Anniversary Address on the evolution of a geosyncline. *Quarterly Journal of the Geological Society of London*, Vol. 94, 1x–cx.

JONES, O T. 1947. The geology of the Silurian rocks west and south of the Carneddau range, Radnorshire. *Quarterly Journal of the Geological Society of London*, Vol. 103, 1–36.

JONES, O T. 1949. The geology of the Llandovery district. Part II: the northern area. *Quarterly Journal of the Geological Society of London*, Vol.105, 43–63.

JONES, O T. 1956. The geological evolution of Wales and the adjacent regions. *Quarterly Journal of the Geological Society of London*, Vol. 111, 323–351.

JONES, O T, and PUGH, W J. 1916. The geology of the district around Machynlleth and the Llyfnant Valley. *Quarterly Journal of the Geological Society of London*, Vol. 71, 343–385.

JONES, O T, and PUGH, W J. 1941. The Ordovician rocks of the Builth district. A preliminary account. *Geological Magazine*, Vol. 78, 185–191.

JONES, O T, and PUGH, W J. 1948a. A multilayered dolerite complex of laccolithic form near Llandrindod Wells, Radnorshire. *Quarterly Journal of the Geological Society of London*, Vol. 104, 43–70.

JONES, O T, and PUGH, W J. 1948b. The form and distribution of dolerite masses in the Builth–Llandrindod Inlier, Radnorshire. *Quarterly Journal of the Geological Society of London*, Vol. 104, 71–98.

JONES, O T, and PUGH, W J. 1949. An early Ordovician shore-line in Radnorshire, near Builth Wells. *Quarterly Journal of the Geological Society of London*, Vol. 105, 65–99.

JONES, W D V. 1945. The Valentian succession around Llanidloes, Montgomeryshire. *Quarterly Journal of the Geological Society of London*, Vol. 100, 309–332.

JOHNSON, M E, COCKS, L R M, and COPPER, P. 1981. Late Ordovician–Early Silurian fluctuations in sea level from eastern Anticosti Island, Quebec. *Lethaia*, Vol. 14, 73–82.

JOHNSON, M E, RONG, J-Y, and YANG, X-C. 1985. Intercontinental correlation by sea-level events in the Early Silurian of North America and China (Yangtze Platform). *Bulletin of the Geological Society of America*, Vol.96, 1384–1397.

JOHNSON, M E, KALJO, D, and RONG, J-Y. 1991. Silurian eustasy. 145–163 in The Murchison Symposium; proceedings of an international conference on the Silurian System. BASSETT, M G, LANE, P D, and EDWARDS, D (editors). *Special Papers in Palaeontology*, Vol. 44.

JOHNSON, T E. 1991. Nomenclature and geometric classification of cleavage-transected folds. *Journal of Structural Geology*, Vol. 13, 261–274.

KAKAR, S D. 1971. A study of the trace elements associated with the lead-zinc ores of Mid-Wales. Unpublished PhD thesis, University of Wales (Aberystwyth).

KEEPING, W. 1881. The geology of central Wales. With an appendix on some new species of Cladophora by C Lapworth. *Quarterly Journal of the Geological Society of London*, Vol. 37, 141–177.

KELLING, G. 1964. The turbidite concept in Britain. 75–92 in *Turbidites*. BOUMA, A H, and BROUWER, A (editors). (Amsterdam: Elsevier.)

KELLING, G, and HOLROYD, J. 1978. Clastsize, shape and composition in some ancient and modern fan gravels. 138–159 in *Sedimentation in submarine canyons, fans and trenches*. STANLEY, D J, and KELLING, G (editors). (Stroudsburg, Pennsylvania: Dowden, Hutchinson & Ross.)

KELLING, G, and WOOLLANDS, M A. 1969. The stratigraphy and sedimentation of the Llandoverian rocks of the Rhayader district. 255–282 in *The Pre-Cambrian and Lower Palaeozoic rocks of Wales*. WOOD. A (editor). (Cardiff: University of Wales Press.)

KENNEDY, R J. 1989. Ordovician (Llanvirn) trilobites from SW Wales. *Monograph of the Palaeontographical Society, London*, (Publication no. 576, part of vol. 141 for 1987).

KISCH, H J. 1990. Calibration of the anchizone: a critical comparison of illite 'crystallinity' scales used for definition. *Journal of Metamorphic Geology*, Vol. 8, 31–46.

KOKELAAR, B P. 1988. Tectonic controls of Ordovician arc and marginal basin volcanism in Wales. *Journal of the Geological Society of London*, Vol. 145, 759–775.

KOKELAAR, B P, HOWELLS, M F, BEVINS, R E, ROACH, R A, and DUNKLEY, P N. 1984. The Ordovician marginal basin of Wales. 245–269 in Marginal basin geology, volcanic and associated sedimentary and tectonic processes in modern and ancient marginal basins. KOKELAAR, B P, and HOWELLS, M F (editors). *Special Publication of the Geological Society of London*, No. 16.

KUENEN, Ph H. 1953. Graded bedding, with observations on Lower Palaeozoic rocks of Britain. *Verhandelingen Koninklijke Nederlandse Akademie van Wetenschappen Afedeling Natuurkunde*, Section 1, Vol. 20, 2–47.

LAPWORTH, C. 1879. On the tripartite classification of the Lower Palaeozoic rocks. *Geological Magazine*, Vol. 6, 1–15.

LAPWORTH, H. 1900. The Silurian sequence of Rhayader. *Quarterly Journal of the Geological Society of London*, Vol. 56, 67–137.

LAUFELD, S. 1974. Silurian Chitinozoa from Gotland. *Fossils and Strata*, No. 5.

LEE, M K. 1988. Density variations within Lake District granites and Lower Palaeozoic rocks. *Technical Report of the Regional Geophysics Group of the British Geological Survey*, No. WK/98/1.

LEGGETT, J K. 1978. Eustasy and pelagic regimes in the Iapetus Ocean during the Ordovician and Silurian. *Earth and Planetary Science Letters*, Vol. 41, 163–169.

LEGGETT, J K. 1980. British Lower Palaeozoic black shales and their palaeo-oceanographic significance. *Journal of the Geological Society of London*, Vol. 137, 139–156.

LEGGETT, J K, MCKERROW, W S, COCKS, L R M, and RICKARDS, R B. 1981. Periodicity in the early Palaeozoic marine realm. *Journal of the Geological Society of London*, Vol. 138, 167–176.

LEGGETT, J K, MCKERROW, W S, and SOPER, N J. 1983. A model for the crustal evolution of southern Scotland. *Tectonics*, Vol. 2, 187–210.

LEWIS, W J. 1967. *Lead mining in Wales*. (Cardiff: University of Wales Press.)

LI, G, PEACOR, D R, MERRIMAN, R J, and ROBERTS, B. 1994. The diagenetic to low-grade metamorphic evolution of matrix white micas in the system muscovite-paragonite in a mudrock from Central Wales, UK. *Clays and Clay Minerals*, Vol. 42, 369–381.

LOCKLEY, M G. 1978. New evidence of the age of some Lower Palaeozoic rocks near Llanafan-Fawr, Central Wales. *Geological Journal*, Vol. 13, 15–24.

LOCKLEY, M G. 1980. The geology of the Llanuwchllyn to Llanymawddwy area, South Gwynedd, North Wales. *Geological Journal*, Vol. 15, 21–41.

LOCKLEY, M G. 1983. Brachiopods from a Lower Palaeozoic mass flow deposit near Llanafan Fawr, central Wales. *Geological Journal*, Vol. 18, 93–99.

LOCKLEY, M G, and WILLIAMS, A. 1981. Lower Ordovician Brachiopoda from mid and southwest Wales. *Bulletin of the British Museum (Natural History), Geology*, Vol. 35, 1–78.

LONG, G H. 1966. Investigations into the sedimentation and sedimentary history of the Talerddig Grits (Upper Llandoverian) and their lateral equivalents in Central Wales. Unpublished PhD thesis, University of London.

LOVELL, J P B. 1970. The palaeogeographical significance of lateral variations in the ratio of sandstone to shale and other features of the Aberystwyth Grits. *Geological Magazine*, Vol. 107, 147–158.

LOWE, D R. 1976. Subaqueous liquefied and fluidized sediment flows and their deposits. *Sedimentology*, Vol. 23, 285–308.

LOWE, D R. 1982. Sediment gravity flows: II. Depositional models with special reference to the deposits of high-density turbidity currents. *Journal of Sedimentary Petrology*, Vol. 52, 279–297.

LOWE, J J, and WALKER, M J C. 1984. *Reconstructing Quaternary environments*. (London: Longman.)

LOYDELL, D K. 1989. Middle and Upper Llandovery graptolites from western mid-Wales and southern Sweden. Unpublished PhD thesis, University College of Wales, Aberystwyth.

LOYDELL, D K. 1991a. The biostratigraphy and formational relationships of the upper Aeronian and lower Telychian (Llandovery, Silurian) formations of western mid-Wales. *Geological Journal*, Vol. 26, 209–244.

LOYDELL, D K. 1991b. Dob's Linn — the type locality of the Telychian (Upper Llandovery) *Rastrites maximus* Biozone? *Newsletters in Stratigraphy*, Vol. 25, 155–161.

LOYDELL, D K. 1992. Upper Aeronian and Lower Telychian (Llandovery) graptolites from western mid-Wales. Part 1. *Monograph of the Palaeontographical Society, London*, Vol. 146, 1–55.

LOYDELL, D K, and CAVE, R. 1993. The Telychian (Upper Llandovery) stratigraphy of Buttington Brick Pit, Wales. *Newsletters in Stratigraphy*, Vol. 29, 91–103.

LYNAS, B D T. 1988. Evidence for strike-slip faulting in the Shelve Ordovician inlier, Welsh Borderland: implications for the south British Caledonides. *Geological Journal*, Vol. 23, 39–57.

MABILLARD, J E, and ALDRIDGE, R J. 1985. Microfossil distribution across the base of the Wenlock Series in the type area. *Palaeontology*, Vol. 28, 89–100.

MACKIE, A H, and SMALLWOOD, S D. 1987. A revised stratigraphy of the Abergwesyn–Pumpsaint area, mid-Wales. *Geological Journal*, Vol. 22, 45–60.

MACKLIN, M G, and LEWIN, J. 1986. Terraced fills of Pleistocene and Holocene age in the Rheidol Valley, Wales. *Journal of Quaternary Science*, Vol. 1, 21–34.

MANCHESTER, R J. 1983. The crustal structure of Wales from interpretation of gravity and magnetic measurements. Unpublished MSc thesis, University of Birmingham.

MARSHALL, T R, LEAKE, R C, and ROLLIN, K E. 1987. A mineral reconnaissance survey of the Llandrindod Wells/Builth Wells Ordovician Inlier, Powys. *Mineral Reconnaissance Programme British Geological Survey*, No. 92.

MAUDE, R. 1987. Lineaments in enhanced Landsat images from a portion of West Wales. *Geological Journal*, Vol. 22, Thematic Issue, 107–118.

MCCANN, T, and PICKERING, K T. 1989. Palaeocurrent evidence of a northern structural high to the Welsh Basin during the Late Llandovery. *Journal of the Geological Society of London*, Vol. 147, 887–891.

MCDONALD, A J W, FLETCHER, C J N, CARRUTHERS, R M, and EVANS, R B. 1992. Interpretation of the regional gravity and magnetic surveys of Wales, using shaded relief and Euler deconvolution techniques. *Geological Magazine*, Vol. 129, 523–531.

MCDONALD, D I M. 1986. Proximal to distal sedimentological variation in a linear turbidite trough: implications for the fan model. *Sedimentology*, Vol. 33, 243–259.

MCKERROW, W S. 1979. Ordovician and Silurian changes in sea level. *Journal of the Geological Society of London*, Vol. 136, 137–145.

MCKERROW, W S, and SOPER, N J. 1989. The Iapetus Suture in the British Isles. *Geological Magazine*, Vol.126, 1–8.

MEISSNER, R, MATTHEWS, D, and WEVER, Th. 1986. The 'Moho' in and around Great Britain. *Annales Geophysicae*, Vol. 4, B, 659–664.

MERRIMAN, R J, and ROBERTS, B. 1985. A survey of white mica crystallinity and polytypes in pelitic rocks of Snowdonia and Llyn, North Wales. *Mineralogical Magazine*, Vol. 49, 305–319.

MERRIMAN, R J, ROBERTS, B, and PEACOR, D R. 1990. A transmission electron microscope study of white mica crystallite size distribution in a mudstone to slate transitional sequence, North Wales, UK. *Contributions to Mineralogy and Petrology*, Vol. 106, 27–40.

MILODOWSKI, A E, and ZALASIEWICZ, J A. 1991a. The origin and sedimentary, diagenetic and metamorphic evolution of chlorite-mica stacks in Llandovery sediments of Central Wales, UK. *Geological Magazine*, Vol. 128, 263–278.

MILODOWSKI, A E, and ZALASIEWICZ, J A. 1991b. Redistribution of rare earth elements during diagenesis of turbidite/hemipelagite mudrock sequences of Llandovery age from Central Wales. 101-124 in Developments in sedimentary provenance studies. MORTON, A C, TODD, S P, and HANGTON, P D W (editors). *Special Publication of the Geological Society of London*, No. 57.

MITCHELL, G F. 1960. The Pleistocene history of the Irish Sea. *British Association for the Advancement of Science*, Vol. 17, 313–325.

MITCHELL, G F. 1972. The Pleistocene history of the Irish Sea: second approximation. *Scientific Proceedings of the Royal Dublin Society*, Series A, Vol. 4, 181–199.

MORGAN, H D, SCOTT, P A, WALTON, R J C, and FALKINER, R H. 1953. The Claerwen Dam. *Proceedings of the Institution of Civil Engineers*, Paper No. 5912, 254–262.

MOORE, P D. 1970. Studies in the vegetational history in mid-Wales: II. The late-glacial period in Cardiganshire. *New Phytologist*, Vol. 69, 363–375.

MOORE, P D, and CHATER, E H. 1969. Studies in the vegetational history of mid-Wales: I. The post-glacial period in Cardiganshire. *New Phytologist*, Vol. 68, 183–196.

MORTON, A C, DAVIES, J R, and WATERS, R A. 1992. Heavy minerals as a guide to turbidite provenance in the Lower Palaeozoic Southern Welsh Basin: a pilot study. *Geological Magazine*, Vol. 129, 573–580.

MULLIS, J. 1987. Fluid inclusion studies during very low grade metamorphism. In *Low temperature metamorphism*. FREY, M (editor). (Glasgow and London; Blackie.)

MURCHISON, R I. 1839. *The Silurian System*. (London: John Murray.)

MURPHY, F C, and HUTTON, D H W. 1986. Is the Southern Uplands of Scotland really an accretionary prism? *Geology*, Vol. 14, 354–357.

MUTTI, E. 1979. Turbidites et cones sous-marins profonds. 353–419 in *Sedimentation detritique (fluviatile, littorale et marine)*. HOMWEOOD, P (editor). (Fribourg: Institut de Geologie, Universite de Fribourg, Suisse.)

MUTTI, E. 1985. Turbidite systems and their relations to depositional sequences. 65–93 in *Provenance of arenites*. ZUFFA, E G (editor). NATO-ASI Series. (Reidel Publishing Company.)

MUTTI, E, and GHIBAUDO, G. 1972. Un esempio di torbiditi di conoide sottomarina esterna — Le Arenarie di San Salvatore (Formazione di Bobio, Miocene) nell' Appenino di Piacenza. *Memorie dell' Accademia della Scienze di Torino, Classe di Scienze Fisiche, Matematiche e Naturali*, Serie 4, No.16, 40.

MUTTI, E, and NORMARK, W R. 1987. Comparing modern and ancient turbidite systems. 1–38 in *Marine clastic sedimentology: concepts and case studies*. LEGGETT, J K, and ZUFFA, G G (editors). (London: Graham and Trotman.)

NARDIN, T R, HEIN, F J, GORSLINE, D S, and EDWARDS, B D. 1979. A review of mass movement processes, sediment and acoustic characteristics and contrasts in slope and base-of-slope systems versus canyon-fan-basin floor systems. 61–73 in Geology of continental slopes. DOYLE, L J, and PILKEY, O H (editors). *Society of Economic Palaeontology, Special Publication*, No. 27.

NORMARK, W R, PIPER, D J W, and HESS, G R. 1979. Distributary channels, sand lobes and mesotopography of Navy Submarine Fan, California Borderland, with applications to ancient fan sediments. *Sedimentology*, Vol. 26, 249–774.

OKADA, H, and SMITH, A J. 1980. The Welsh 'geosyncline' of the Silurian was a fore-arc basin. *Nature, London*, Vol. 290, 492–495.

OZELCI, H F. 1960. Magnetic anomalies in North Wales. Unpublished MSc thesis, University of Birmingham.

PHILLIPS, W E A, STILLMAN, C J, and MURPHY, T. 1976. A Caledonian plate tectonic model. *Journal of the Geological Society of London*, Vol. 132, 579–609.

PHILLIPS, W J. 1972. Hydraulic fracturing and mineralisation. *Journal of the Geological Society, London*, Vol. 128, 337–359.

PHILLIPS, W J. 1986. Hydraulic fracturing effects in the formation of mineral deposits. *Transactions of the Institution of Mining and Metallurgy, (Section B)*, Vol. 95, B17–B24.

PICKERING, K T. 1987. Wet-sediment deformation in the Upper Ordovician Point Leamington Formation: an active thrust-imbricate system during sedimentation, Notre Dame Bay, north-central Newfoundland. 213–239 in Deformation of sediments and sedimentary rocks. JONES, M E, and PRESTON, R M F (editors). *Special Publication of the Geological Society of London*, No. 29.

PICKERING, K T, HISCOTT, R N, and HEIN, F J. 1989. *Deep marine environments*. (London: Unwin Hymen.)

PICKERING, K T, STOW, D A V, WATSON, M P, and HISCOTT, R N. 1986. Deep water facies, processes and models: a review and classification scheme for modern and ancient sediments. *Earth Science Reviews*, Vol. 23, 75–174.

POTTS, A S. 1971. Fossil cryonival features in central Wales. *Geografiska Annaler*, Vol. 53A, 39–51.

POWELL, D W. 1955. Gravity and magnetic anomalies in North Wales. *Quarterly Journal of the Geological Society of London*, Vol. 111, 375–397.

PRATT, W T. 1990. Tectonics and Caradocian sedimentation in the Tywyn-Corris area of mid-Wales. Unpublished PhD thesis, University College of Wales, Aberystwyth.

PRATT, W T, FITCHES, W R. 1993. The significance of transected folds along the Bala lineament, Wales. *Journal of Structural Geology*, Vol.15, 55–68.

PRATT, W T, WOODHALL, D G, and HOWELLS, M F. 1995. The geology of the country around Cadair Idris. *Memoir of the British Geological Survey*, Sheet 149. (England and Wales).

PRICE, D. 1984. The Pusgillian Stage in Wales. *Geological Magazine*, Vol. 121, 99–105.

PRICE, D, and MAGOR, P M. 1984. The ecological significance of variation in the generic composition of Rawtheyan (late Ordovician) trilobite faunas from North Wales, UK. *Geological Journal*, Vol. 19, 187–200.

PRIEWALDER, H. 1987. Acritarchen aus dem Silur des Cellon-Profils (Karnische Alpen, Osterreich). *Abhandlungen der Geologischen Bundesanstalt*, Band 40, 1–121.

PUGH, W J. 1923. The geology of the district around Corris and Aberllefenni (Merionethshire). *Quarterly Journal of the Geological Society of London*, Vol. 85, 242–306.

RAAF, J F M De, BOERSMA, J R, and GELDER, A Van. 1977. Wave generated structures and sequences from a shallow marine succession, Lower Carboniferous, County Cork, Ireland. *Sedimentology*, Vol. 24, 451–483.

RAST, N. 1969. The relationship between Ordovician structure and volcanicity in Wales. 305–335 in *The Precambrian and Lower Palaeozoic rocks of Wales*. WOOD, A (editor). (Cardiff: University of Wales Press.)

RAYBOULD, J G. 1974. Ore textures, paragenesis and zoning in the lead-zinc veins of mid-Wales. *Transactions of the Institution of Mining and Metallurgy, (Section B)*, Vol. 83, B112–B119.

READE, T M. 1897. Notes on the drift of the mid-Wales coast. *Proceedings of the Liverpool Geological Society*, Vol. 7, 410–419.

REID, A B, ALLSOP, J M, GRANSER, H, MILLETT, A J, and SOMERTON, I W. 1990. Magnetic interpretation in three dimensions using Euler deconvolution. *Geophysics*, Vol. 55, 80–91.

RICKARDS, R B. 1976. The sequence of Silurian graptolite zones in the British Isles. *Geological Journal*, Vol. 11, 153–188.

RIDGWAY, J, and APPLETON, J D. 1990. Ammonium geochemistry in mineral exploration — a comparison of results from the American Cordillera and the southwest Pacific. *Applied Geochemistry*, Vol. 5, 475–489.

RIGBY, J E. 1980. Taxonomy and key identification systems for graptolites. Unpublished PhD thesis, University of Cambridge.

ROBERTS, B. 1981. Low grade and very low grade regional metabasic Ordovician rocks of Llyn and Snowdonia, Gwynedd, North Wales. *Geological Magazine*, Vol. 118, 189–200.

ROBERTS, B, and MERRIMAN, R J. 1985. The distiction between Caledonian burial and regional metamorphism in metapelites from North Wales: an analysis of isocryst patterns. *Journal of the Geological Society of London*, Vol. 142, 615–624.

ROBERTS, B, MERRIMAN, R J, HIRON, S R, FLETCHER, C J N, and WILSON, D. 1996. Synchronous very low grade metamorphism, compression and inversion in the central part of the Welsh Lower Palaeozoic Basin. *Journal of the Geological Society of London*, Vol. 153, 277–285.

ROBERTS, B, MERRIMAN, R J, and PRATT, W T. 1991. The influence of strain, lithology and stratigraphical depth on white mica (illite) crystallinity in mudrocks from the vicinity of the Corris Slate Belt, Wales: implications for the timing of metamorphism in the Welsh Basin. *Geological Magazine*, Vol. 128, 633–645.

ROBERTS, R O. 1927. The igneous and associated Ordovician rocks of Baxters Bank, Radnorshire. *Geological Magazine*, Vol. 64, 289–298.

ROBERTS, R O. 1929. The geology of the district around Abbey-cwmhir (Radnorshire). *Quarterly Journal of the Geological Society of London*, Vol. 85, 651–676.

ROBINSON, D, and BEVINS, R E. 1986. Incipient metamorphism in the Lower Palaeozoic marginal basin of Wales. *Journal of Metamorphic Geology*, Vol. 4, 101–113.

ROBINSON, D, WARR, L, and BEVINS, R E. 1990. The illite 'crystallinity' technique: a critical appraisal of its precision. *Journal of Metamorphic Geology*, Vol. 8, 333–344.

RUSHTON, A W A. 1990. Ordovician graptolite biostratigraphy in the Welsh Basin: a review. *Journal of the Geological Society, London*, Vol. 147, 611–614.

RUST, S A. 1990. Ullmanite from Hendrefelen Mine, Ysbyty Ystwyth, Dyfed, Wales, UK. *Journal of Mines and Minerals*, No. 8, 47.

SANDERSON, D J, and MARCHINI, D. 1984. Transpression. *Journal of Structural Geology*, Vol. 6, 449–458.

SAVRDA, C E, and BOTTJER, D J. 1987. The exaerobic zone, a new oxygen-deficient marine biofacies. *Nature, London*, Vol. 327, 54–56.

SHELDON, P R. 1987a. Trilobite evolution and faunal distribution in some Ordovician rocks of the Builth inlier, central Wales. Unpublished PhD thesis, Cambridge University.

SHELDON, P R. 1987b. Parallel gradualistic evolution of Ordovician trilobites. *Nature, London*, Vol. 330, 561–563.

SHELDON, P R. 1988. Trilobite size-frequency distributions, recognition of instars, and phyletic size changes. *Lethaia*, Vol. 21, 293–306.

SIVETER, D J, OWENS, R M, and THOMAS, A T. 1989. Silurian field excursions: a geotraverse across Wales and the Welsh Borderland. *National Museum of Wales, Cardiff, Geological Series*, No. 10.

SKEVINGTON, D. 1973. Ordovician graptolites. 27–35 in *Atlas of palaeobiogeography*. HALLAM, A (editor). (Amsterdam: Elsevier.)

SMALLWOOD, S. 1986. Sedimentation across the Tywi Lineament, mid Wales. *Philosophical Transactions of the Royal Society, London*, Vol. A317, 279–288.

SMITH, A J, and LONG, G H. 1969. The Upper Llandoverian sediments of Wales and the Welsh Borderland. 239–253 in *The Precambrian and Lower Palaeozoic rocks of Wales*. WOOD, A (editor). (Cardiff: University of Wales Press.)

SMITH, R D A. 1987a. The *Griestoniensis* Zone turbidite system, Welsh Basin. 89–107 in *Marine clastic sedimentation*. LEGGETT, J K, and ZUFFA, G G (editors). (London: Graham and Trotman.)

SMITH, R D A. 1987b. Structure and deformation history of the Central Wales synclinorium, northeast Dyfed: implications for long-lived basement structure. *Geological Journal*, Vol. 22, 183–198.

SMITH, R D A. 1987c. Early diagenetic phosphate cements in a turbidite basin. 141–156 in Diagenesis of sedimentary sequences. MARHSALL, J D (editor). *Special Publication of the Geological Society of London*, No. 36.

SMITH, R D A. 1988. A sedimentological analysis of the Late Llandovery Welsh Basin. Unpublished PhD thesis, University of Cambridge.

SMITH, R D A. 1990. Discussion on the palaeocurrent evidence of a northern structural high to the Welsh Basin during the Llandovery. *Journal of the Geological Society of London*, Vol. 147, 886–887.

SMITH, R D A, and ANKETELL, J M. 1992. Welsh Basin 'contourites' reinterpreted as fine-grained turbidites; the Grogal Sandstones. *Geological Magazine*, Vol. 129, 609–614.

SMITH, R D A, WATERS, R A, and DAVIES, J R. 1991. *Late Ordovician and early Silurian turbidite systems in the Welsh Basin*. 13th International Sedimentological Congress, Nottingham, Field Guide No. 20. (Cambridge: British Sedimentological Research Group.)

SOPER, N J. 1986a. The Newer Granite problem: a geotectonic view. *Geological Magazine*, Vol. 123, 227–236.

SOPER, N J. 1986b. Geometry of transecting, anastomosing solution cleavage in transpression zones. *Journal of Structural Geology*, Vol. 8, 937–940.

SOPER, N J, and HUTTON, D H W. 1984. Late Caledonian sinistral displacements in Britain: implications for three-plate collision model. *Tectonics*, Vol. 3, 781–794.

SOPER, N J, WEBB, B C, and WOODCOCK, N H. 1987. Late Caledonian (Acadian) transpression in north-west England: timing, geometry and geotectonic significance. *Proceedings of the Yorkshire Geological Society*, Vol. 46, 175–192.

SOPER, N J, and WOODCOCK, N H. 1990. Silurian collision and sediment dispersal patterns in southern Britain. *Geological Magazine*, Vol. 127, 527–542.

STAMP, L D A, and WOOLDRIDGE, S W. 1923. The igneous and associated rocks of Llanwrtyd (Brecon). *Quarterly Journal of the Geological Society of London*, Vol. 79, 16–46.

STOW, D A V. 1977. Late Quaternary stratigraphy and sedimentation on the Nova Scotia outer continental margin. Unpublished PhD thesis, Dalhousie University.

STOW, D A V. 1985. Fine-grained sediments in deep water: An overview of processes and facies models. *Geo-Marine Letters*, Vol. 5, 17–23.

STOW, D A V. 1986. Deep clastic seas. 399–444 in *Sedimentary environments and facies*. READING, H (editor). (Oxford: Blackwell.)

STOW, D A V, ALAM, M, and PIPER, D J W. 1984. Sedimentology of the Halifax Formation, Nova Scotia: Lower Palaeozoic fine-grained turbidites. 127–144 in Fine-grained sediments: deep-water processes and facies. STOW, D A V, and PIPER, D J W (editors). *Special Publication of the Geological Society of London*, No. 15.

STOW, D A V, and BOWEN, A J. 1980. A physical model for the transport and sorting of fine grained sediment by turbidity currents. *Sedimentology*, Vol. 2, 31–46.

STOW, D A V, HOWELL, D G, and NELSON, C H. 1985. Sedimentary, tectonic and sea level controls. 15–22 in *Submarine fans and related turbidite systems*. BOUMA, A H, NORMARK, W R, and BARNES, N E (editors). (New York: Springer.)

STOW, D A V, and PIPER, D J W. 1984. Deep-water fine-grained sediments: facies models. 611–646 in Fine-grained sediments: deep-water processes and facies. STOW, D A V, and PIPER, D J W (editors). *Special Publication of the Geological Society of London*, No. 15.

STRACHAN, I. 1986. The Ordovician graptolites of the Shelve district, Shropshire. *Bulletin of the British Museum (Natural History) Geology Series*, Vol. 40, 1–58.

STRAHAN, A, CANTRILL, T C, DIXON, E E L, and THOMAS, H H. 1909. The geology of the South Wales Coalfield. Part X. The country around Carmarthen. *Memoir of the Geological Survey of Great Britain*, Sheet 229.

STRAHAN, A, and JONES, O T. 1914. The geology of the South Wales Coalfield. Part XI. The country around Haverfordwest. *Memoir of the Geological Survey of Great Britain*, Sheet 228.

STRINGER, P, and TREAGUS, J E. 1980. Non-axial planar S₁ in the Hawick Rocks of the Galloway area, Southern Uplands, Scotland. *Journal of Structural Geology*, Vol. 2, 317–331.

STRONG, G E. 1979. An Upper Llandoverian turbidite sequence of the Bwlch-glas area, near Aberystwyth, Dyfed. *Geological Journal*, Vol. 14, 99–106.

SYNGE, F M. 1963. A correlation between the drifts of south-east Ireland with those of west Wales. *Irish Geographer*, Vol. 4, 360–366.

SYNGE, F M. 1964. The glacial succession in west Caernarfonshire. *Proceedings of the Geologists' Association*, Vol. 75, 431–444.

TANNER, P W G. 1989. The flexural-slip mechanism. *Journal of Structural Geology*, Vol. 11, 635–655.

TANNER, P W G. 1990. The flexural-slip mechanism: reply. *Journal of Structural Geology*, Vol. 12, 1085–1087.

TEMPLE, J T. 1987. Early Llandovery brachiopods of Wales. *Monograph of the Palaeontographical Society, London*, Publication No. 572, part of Vol. 139 for 1985.

TEMPLE, J T. 1988. Ordovician–Silurian boundary strata in Wales. 65–71 in A global analysis of the Ordovician–Silurian boundary. COCKS, L R M, and RICKARDS, R B (editors). *Bulletin of the British Museum (Natural History), Geology Series*, Vol. 43.

TEMPLE, J T, and CAVE, R. 1992. Preliminary report on the geochemistry and mineralogy of the Nod Glas and related sediments (Ordovician) of Wales. *Geological Magazine*, Vol. 129, 589–594.

THOMAS, A T. 1979. Trilobite associations in the British Wenlock. 447–451 in The Caledonides of the British Isles — reviewed. HARRIS, A L, HOLLAND, C H, and LEAKE, B E (editors). *Special Publication of the Geological Society of London*, No. 8.

THOMPSON, D T. 1982. EULDPH: A new technique for making computer-assisted depth estimates from magnetic data. *Geophysics*, Vol. 47, 31–37.

THORNTON, S E. 1984. Basin model for hemipelagic sedimentation in a tectonically active continental margin: Santa Barbara Basin, California Continental Borderland. 377–394 in Fine-grained sediments: deep-water processes and facies. STOW, D A V, and PIPER, D J W (editors). *Special Publication of the Geological Society of London*, No. 15.

THORPE, R S. 1979. Late Precambrian igneous activity in southern Britian. 579–584 in The Caledonides of the British Isles — reviewed. HARRIS, A L, HOLLAND, C H, and LEAKE, B E (editors). *Special Publication of the Geological Society of London*, No. 8.

TIMBERLAKE, S. 1989. Excavations and fieldwork on Copa Hill, Cwmystwyth. *In* Early mining in the British Isles. CREW, P, and CREW, S (editors). *Plas Tan y Bwlch Occasional, Paper*, No. 1. (Maentwrog.)

TOGHILL, P. 1992. The Shelveian event, a late Ordovician tectonic episode in southern Britain (Eastern Avalonia). *Proceedings of the Geologists' Association*, Vol. 103, 31–35.

TREAGUS, J E, and TREAGUS, S E. 1992. Folds and the strain ellipsiod: a general model. *Journal of Structural Geology*, Vol. 14, 361–367.

TUNNICLIFF, S P. 1989. An early record of probable nowakiid tentaculitoids from Wales. *Palaeontology*, Vol. 32, 685–688.

TURNBULL, G. 1987. Bouguer densities from mainland Britain determined from linear regression of gravity anomalies with topographic relief. *Project Note of the Regional Geophysics Group of the British Geological Survey*, No. 87/3.

TURNER, R E. 1984. Acritarchs from the type area of the Ordovician Caradoc Series, Shropshire. *Palaeontographica, Abteilung B*, Vol. 190, 85–157.

TYLER, J E, and WOODCOCK, N H. 1987. The Bailey Hill Formation: Ludlow Series turbidites in the Welsh Borderland re-interpreted as distal storm deposits. *Geological Journal*, Vol. 22, 73–86.

VINCENT, P J. 1976. Some periglacial deposits near Aberystwyth, Wales, as seen with a scanning electron microscope. *Biuletyn Peryglacjalny*, Vol. 25, 59–64.

VON RAD, U, HINZ, K, SARNTHEIN, M, and SEIBOLD, E (editors). 1982. *Geology of the northwest African continental margin*. (New York: Springer-Verlag, Berlin Heidelberg.)

WALKER, R G. 1965. The origin and significance of the internal sedimentary structures of turbidites. *Proceedings of the Yorkshire Geological Society*, Vol. 35, 1–32.

WALKER, R G. 1967. Turbidite sedimentary structures and their relationship to proximal and distal depositional environments. *Journal of Sedimentary Petrology*, Vol. 37, 25–43.

WARREN, P T, PRICE, D, NUTT, M J C, and SMITH, E G. 1984. Geology of the country around Rhyl and Denbigh. *Memoir of the British Geological Survey,* Sheets 95 and 107. (England and Wales).

WATERS, R A, DAVIES, J R, FLETCHER, C J N, and WILSON, D. 1992. Discussion of 'A late Ordovician/early Silurian non-depositional slope and perched basin along the Tywi Anticline, Mid Wales' by D M D James. *Geological Journal,* Vol. 27, 285–294.

WATERS, R A, DAVIES, J R, FLETCHER, C J N, WILSON, D, ZALASIEWICZ, J A, and CAVE, R.. 1993. Llandovery basinal and slope sequences of the Rhayader district. 155–182 in *Geological excursions in Powys, Wales.* WOODCOCK, N H, and BASSETT, M G (editors). (Cardiff: University of Wales Press.)

WATSON, E. 1965. Grèzes litées ou eboulis ordonnés tardiglaciaires dans la région d'Aberystwyth, au centre du Pays de Galles. *Bulletin of the Association of French Geographers,* Vol. 338, 16–25.

WATSON, E. 1966. Two nivation cirques near Aberystwyth, Wales. *Biuletyn Peryglacjalny,* Vol. 15, 79–101.

WATSON, E. 1970. The Cardigan Bay area. 125–145 in *The glaciations of Wales and adjoining regions.* LEWIS, C A (editor). (London: Longman.)

WATSON, E. 1977. *Mid and North Wales.* INQUA X Congress. Guidebook for excursion C9. (Norwich: Geo Abstracts.)

WATSON, E. 1982. Periglacial slope deposits at Morfa-bychan, near Aberystwyth. 313–325 in *Geological excursions in Dyfed, south-west Wales.* BASSETT, M G (editor). (Cardiff: National Museum of Wales.)

WATSON, E, and WATSON, S. 1967. The periglacial origin of the drifts at Morfa-bychan, near Aberystwyth. *Geological Journal,* Vol. 5, 419–440.

WELIN, E, ENGSTRAND, L, and VACZY, S. 1974. Institute of Geological Sciences, Radiocarbon dates V. *Radiocarbon,* Vol. 16, 95–104.

WHITTARD, W F. 1979. An account of the Ordovician rocks of the Shelve Inlier in west Salop and part of north Powys. *Bulletin of the British Museum (Natural History), Geology Series,* Vol. 33, 1–69.

WILLIAMS, A, LOCKLEY, M G, and HURST, J M. 1981. Benthic palaeocommunities represented in the Ffairfach Group and coeval Ordovician successions of Wales. *Palaeontology,* Vol. 24, 661–694.

WILLIAMS, A, STRACHAN, I, BASSETT, D A, DEAN, W T, INGHAM, J K, WRIGHT, A D, and WHITTINGTON, H B. 1972. A correlation of Ordovician rocks of the British Isles. *Special Report of the Geological Society of London,* No. 3.

WILLIAMS, A, and WRIGHT, A D. 1981. The Ordovician–Silurian boundary in the Garth area of southwest Powys, Wales. *Geological Journal,* Vol. 16, 1–39.

WILLIAMS, K E. 1927. 1. The glacial drifts of western Cardiganshire. 2. The geology of the Pont-rhyd-y-groes country, Cardiganshire. Unpublished MSc thesis, University of Wales.

WILSON, D, DAVIES, J R, FLETCHER, C J N, and WATERS, R A. 1993. The Ordovician rocks of the Rhayader district. 183–208 in *Geological excursions in Powys, Wales.* WOODCOCK, N H, and BASSETT, M G (editors). (Cardiff: University of Wales Press.)

WILSON, D, DAVIES, J R, WATERS, R A, and ZALASIEWICZ, J A. 1992. A fault-controlled depositional model for the Aberystwyth Grits turbidite system. *Geological Magazine,* Vol. 129, 595–607.

WOOD, A, and SMITH, A J. 1959. The sedimentation and sedimentary history of the Aberystwyth Grits (Upper Llandoverian). *Quarterly Journal of the Geological Society of London,* Vol. 114, 163–195.

WOOD, E M R. 1906. On the Tarannon Series of Tarannon. *Quarterly Journal of the Geological Society of London,* Vol. 62, 644–701.

WOODCOCK, N H. 1984a. Early Palaeozoic sedimentation and tectonics in Wales. *Proceedings of the Geologists' Association,* Vol. 95, 323–335.

WOODCOCK, N H. 1984b. The Pontesford Lineament, Welsh Borderland. *Journal of the Geological Society of London,* Vol. 141, 1001–1014.

WOODCOCK, N H. 1987. Kinematics of strike-slip faulting, Builth Inlier, Mid-Wales. *Journal of Structural Geology,* Vol. 9, 353–363.

WOODCOCK, N H. 1990a. Transpressive Acadian deformation across the Central Wales Lineament. *Journal of Structural Geology,* Vol. 12, 329–337.

WOODCOCK, N H. 1990b. Sequence stratigraphy of the Palaeozoic Welsh Basin. *Journal of the Geological Society of London,* Vol. 147, 537–547.

WOODCOCK, N H, AWAN, M A, JOHNSON, T E, MACKIE, A H, and SMITH, R D A. 1988. Acadian tectonics in Wales during Avalonian/Laurentia convergence. *Tectonics,* Vol. 7, 483–495.

WOODCOCK, N H, and GIBBONS, W. 1988. Is the Welsh Borderland Fault System a terrane boundary? *Journal of the Geological Society of London,* Vol. 145, 915–923.

WOODCOCK, N H, and SMALLWOOD, S D. 1987. Late Ordovician shallow marine environments due to glacio-eustatic regression: Scrach Formation, Mid-Wales. *Journal of the Geological Society of London,* Vol. 144, 393–400.

WOODLAND, A W, and EVANS, W B. 1964. The geology of the South Wales Coalfield, Part IV, the country around Pontypridd and Maesteg. *Memoir of the Geological Survey of Great Britain,* Sheet 248.

WOOLLANDS, M A. 1970. The stratigraphy and sedimentary history of the Llandovery rocks between Llandovery and Rhayader. Unpublished PhD thesis, University of London.

ZALASIEWICZ, J A. 1990. Silurian graptolite biostratigraphy in the Welsh Basin. *Journal of the Geological Society of London,* Vol. 147, 619–622.

ZIEGLER, A M. 1965. Silurian marine communities and their environmental significance. *Nature, London,* Vol. 207, 270–272.

ZIEGLER, A M. 1970. Geosynclinal development of the British Isles during the Silurian period. *Journal of Geology,* Vol. 78, 445–479.

ZIEGLER, A M, COCKS, L R M, and BAMBACH, R K. 1968. The composition and structure of Lower Silurian marine communities. *Lethaia, Oslo,* Vol. 1, 1–27.

ZIEGLER, A M, COCKS, L R M, and McKERROW, W S. 1968. The Llandovery transgression of the Welsh Borderland. *Palaeontology,* Vol. 11, 736–782.

APPENDIX 1

Selected borehole data

a   Pen-rhiw Frank boreholes

Four boreholes were drilled in the Pen-rhiw Frank area, to the east of Llandrindod Wells, as part of a mineral reconnaissance programme. Marshall et al. (1987) give additional geochemical data on the whole rock samples collected from the core.

Pen-rhiw Frank Borehole No. 1 [SO 0870 6011] (vertical)
Surface level c. 350 m above OD

	Thickness m	Depth m
BUILTH VOLCANIC FORMATION		
Coarse-grained tuff with vesicular clasts 0.5 to 1.5 cm	5.42	5.42
Bedded, medium- to coarse-grained tuff	4.90	10.32
Coarse-grained tuff, clasts up to 7 cm, faintly bedded in part	85.98	96.30

Pen-rhiw Frank Borehole No. 2 [SO 0856 6002] (vertical)
Surface level c. 362 m above OD

	Thickness m	Depth m
BUILTH VOLCANIC FORMATION		
Medium- and coarse-grained tuff	8.86	8.86
Basalt, pale grey to green, vesicular in part	9.69	18.55
Coarse-grained tuff	2.83	21.38
Basalt	2.02	23.43
Coarse-grained tuff	28.27	51.70

Pen-rhiw Frank Borehole No. 3 [SO 0843 6006] (azimuth 183° inclination 60°)
Surface level c. 351 m above OD

	Thickness (apparent) m	Depth m
Drift	2.72	2.72
BUILTH VOLCANIC FORMATION		
Basalt, vesicular	17.52	14.80
Coarse-grained tuff, clasts up to 3 cm	21.24	36.04
Bedded coarse-grained tuff	20.28	56.32
Coarse-grained tuff, clasts up to 6 cm	71.98	128.30

Pen-rhiw Frank Borehole No. 4 [SO 0882 6001] (vertical)
Surface level c. 360 m above OD

	Thickness m	Depth m
BUILTH VOLCANIC FORMATION		
Coarse-grained tuff, clasts up to 3 cm	39.46	39.46

b   Cwm Rheidol No. 1 Borehole [SN 7302 7833] (azimuth 350° inclination 45°)
Surface level 173.7 m above OD

	Thickness (apparent) m	Depth m
DEVIL'S BRIDGE FORMATION		
Thinly interbedded turbidite mudstones and fine-grained sandstones. Fragment of *Streptograptus* cf. *filiformis* at 156.73 m	406.41	406.41

RHAYADER MUDSTONES

Very thinly bedded turbidite mudstones with very thin, fine-grained sandstone beds and laminae; gradational base	53.86	460.27

YSTRAD MEURIG GRITS FORMATION

Henblas facies, very thinly interbedded turbidite sandstones and mudstones; gradational base	14.89	475.16
Ystrad Meurig Grits facies (fourth sequence), thin- to thick-bedded, fine- to medium-grained turbidite sandstones with thin beds and laminae of turbidite mudstone. Fragment of *M.* cf. *sedgwickii* at 498.01 m displays a stipe width consistent with a *sedgwickii/halli* biozonal age (locality 35 on Table 13 and Figure 23)	31.32	506.47

DERWENLAS FORMATION

Very thinly bedded turbidite mudstones with fine-grained turbidite sandstone laminae. Between 513.79 and 526.00 m interbedded dark grey laminated hemipelagites contained *Monograptus convolutus, M. lobiferus, M.* aff. *denticulatus?, Pribylograptus leptotheca?, Rastrites* cf. *longispinus, Ra.* aff. *longispinus,* cf. *Cephalograptus cometa extrema, Petalograptus* cf. *minor, Normalograptus scalaris?* and *?? Metaclimacograptus* sp., an assemblage indicative of the upper part of the *convolutus* Biozone (Locality 28 on Table 13 and Figure 23); seen to	27.71	534.18

c   Cwm Rheidol No. 2 Borehole (borehole drilled on same site and in same direction as Cwm Rheidol No. 1 but at an inclination of 30°)

	Thickness (apparent) m	Depth m
DEVIL'S BRIDGE FORMATION		
Thinly interbedded turbidite mudstones and fine-grained sandstones. Between 124.20 and 199.43 m interbedded dark grey laminated hemipelagites contained *Monograptus halli,* cf. *M. turriculatus?, M. contortus?, Pristiograptus* cf. *renaudi, Pristiograptus* sp. (slender form), *Glyptograptus* cf. *fastigatus* and *Petalograptus* sp., an assemblage that suggests the *renaudi* Subzone; and between 326.27 and 347.15 m *Monograptus turriculatus* s.l., *M. gemmatus, Streptograptus* aff. *pseudoruncinatus?* and a hooked monograptid, which suggests the *gemmatus* Subzone. Gradational base	351.11	351.11
RHAYADER MUDSTONES		
Very thinly bedded turbidite mudstones with sandstone laminae; seen to	16.33	367.44

d   Frongoch Borehole [SN 7231 7434] (vertical)
Surface level 243.8 m above OD

	Thickness m	Depth m
DEVIL'S BRIDGE FORMATION	251.29	251.29

APPENDIX 2

Distribution of acritarchs

2a   In Caradoc and Ashgill rocks of the Rhayader sheet area

Sample locations are given in **2aiv**

Abbreviations:
SCC — St Cynllo's Church Formation
PtrF — Pentre Formation
CgF — Cwmcringlyn Formation

Cnt — Cefnnantmel Member
cg — conglomerate and sandstone facies
lh — hemipelagite facies
DCM — Dol-y-fan Conglomerate.

2ai   Indigenous acritarch taxa.

SCC	Nantmel Mudstones				Yr Allt		PtrF	CgF	Formation
	Cnt		cg	lh		DCM			Member/facies
2	1	13	3	2	15	8	5	2	Number of samples
									Acritarch taxa
cf.									Actinotodissus longitaleosus
cf.									•Coryphidium elegans
x		x							Orthosphaeridium quadrinatum
x		x							Ordovicidium heteromorphicum
x		x							••Micrhystridium aremoricanum
x		x							•Striatotheca principalis var. parva
x			x						Villosacapsula setosapellicula
cf.		aff.			cf.				Diexallophasis sanpetrensis
aff.					aff.				?Striatotheca scabrata
cf.	x	x			x				Baltisphaerosum bystrentos
x		x			cf.	cf.			Multiplicisphaeridium irregulare
x						?			Navifusa sp.
cf.		x				cf.			•Vogtlandia multiradialis
x		x			x	x			Striatotheca spp.
x			x		x	x			Orthosphaeridium spp.
x		x			x	x			Peteinosphaeridium trifurcatum ssp. intermedium
cf.						x			•Arkonia virgata
x							x		Veryhachium wenlockianum
x		x			x	x	x		Leiosphaeridia spp.
x		x			x	x		cf.	Stellechinatum celestum
x		x			x			x	Goniosphaeridium polygonale
x	x	x	x	x	x	x	x	x	Goniosphaeridium spp.
x		x			x	x	cf.	x	Veryhachium valiente
x		x			x	x		x	Diexallophasis denticulata
x		x	x		x	x	x	x	Baltisphaeridium spp.
x		x			x	x	x	x	Villosacapsula irrorata
x		x				x	x	x	Peteinosphaeridium trifurcatum ssp. breviradiatum
x					x		x	x	Arkonia spp.
cf.		cf.			cf.	cf.		cf.	•Arkonia tenuata
x	x	x		x	x	x	cf.	x	Stellechinatum brachyscolum
	x	x			x	x	x	x	Veryhachium trispinosum
	cf.				cf.				Cheleutochroa diaphorosa
	x	x			x				Diexallophasis spp.
	x	x			x	x		x	Stellechinatum helosum
	x	x	x		x	x	x	x	Micrhystridium spp.
	x	x	x	x	x	x	x	x	Multiplicisphaeridium spp.
		x	cf.		x	x	x	x	Peteinosphaeridium trifurcatum ssp. trifurcatum
		?							Estiastra magna
		cf.							Multiplicisphaeridium continuatum
		cf.							Cheleutochroa homoi
		x							•Dicrodiacrodium normale
		x							Baltisphaeridium annelieae

• these taxa were recorded as recycled by Turner (1982), but are now considered indigenous ;
•• these taxa may be recycled.

2ai   *continued.*

SCC	Nantmel Mudstones				Yr Allt		PtrF	CgF	Formation
	Cnt		cg	lh		DCM			Member/facies
2	1	13	3	2	15	8	5	2	Number of samples
									Acritarch taxa
		x							*Baltisphaeridium longispinosum* ssp. *delicatum*
		x							*Dictyotidium* sp.
		x							*Elektoriskos* sp.
		x							*Fractoricoronula trirhetica*
		x							*Goniosphaeridium* sp. A Turner 1984
		x							*Multiplicisphaeridium paraquaferum*
		x							*Multiplicisphaeridium wrightii*
		x							*Navifusa similis*
		x							*Ordovicidium* sp.
		x							*Veryhachium downiei*
		x							*Micrhystridium equispinosum*
		aff.							*Stellechinatum llandeilum*
		cf.			cf.				*Veryhachium subglobosum*
		cf.			cf.				•*Striatotheca principalis*
		x			cf.				*Ordovicidium elegantulum*
		x			x				*Peteinosphaeridium nanofurcatum*
		x			x				*Peteinosphaeridium nudum*
		x			x				•*Striatotheca frequens*
		x			x	cf.			*Baltisphaeridium accinctum*
		x				cf.			•*Frankea breviuscula*
		aff.				cf.			*Tylotopalla robustispinosa*
		x				x			*Baltisphaerosum dispar*
		x				x			*Elektoriskos williereae*
		x			x	x			*Lophosphaeridium* spp.
		x				x			*Moyeria cabottii*
		x					x		•*Coryphidium bohemicum*
		x			x		x		*Micrhystridium inflatum*
		x			x		x		*Polygonium* spp.
		cf.	cf.					aff.	*Tylotopalla caelamenicutis*
		cf.						cf.	*Domasia* sp. A
		x			x	cf.		cf.	*Solisphaeridium nanum*
		x			cf.	cf.		cf.	*Actinotodissus crassus*
		x						x	*Cheleutochroa meionia*
		x			x	x		x	••*Coryphidium* spp.
		x		x	x	x	x	x	*Peteinosphaeridium* spp.
		x						x	•*Striatotheca quieta*
		x						x	*Veryhachium hamii*
		cf.			x			x	*Multiplicisphaeridium raspum*
			x	x	x	x		x	*Veryhachium* spp.
					cf.				*Baltisphaeridium oligopsakium*
					x				*Marrocanium* sp.
					x				*Excultibrachium* sp.
					cf.				*Priscotheca diadela*
					x				*Veryhachium rhomboidium*
					cf.				*Cheleutochroa gymnobrachiata*
					cf.				*Multiplicisphaeridium imitatum*
					cf.				*Tylotopalla digitifera*
					?				*Rhiptosocherma* sp.
					?				*Diexallophasis granulatispinosa*
					?				*Estiastra* sp.
					x	x			*Veryhachium reductum*
					x	x			*Eupoikilofusa striatifera*
					cf.	?			*Tunisphaeridium eisenackii*
						x			*Uncinisphaera* sp.
						x			*Gorgonisphaeridium* sp.
						x			*Orthosphaeridium chondrododora*
						?			*Orthosphaeridium insculptum*
						cf.			*Peteinosphaeridium velatum*
						cf.		cf.	*Multiplicisphaeridium bifurcatum*

• these taxa were recorded as recycled by Turner (1982), but are now considered indigenous ;
•• these taxa may be recycled.

2ai   *continued.*

SCC	Nantmel Mudstones				Yr Allt	PtrF	CgF	Formation
	Cnt		cg	lh	DCM			Member/facies
2	1	13	3	2	15	8	5	2	Number of samples
									Acritarch taxa
					cf.		cf.	*Goniosphaeridium elongatum*
						?		*Helosphaeridium citrinipeltatum*
							x	*Frankea* sp.
							x	*Orthosphaeridium ternatum*
							x	*Stellechinatum* sp.
							x	*Veryhachium oklahomense*
							cf.	*Tunisphaeridium parvum*

2aii   Recycled mid-Cambrian to early Ordovician acritarch taxa.

SCC	Nantmel Mudstones				Yr Allt	PtrF	CgF	Formation
	Cnt		cg	lh	DCM			Member/facies
2	1	13	3	2	15	8	5	2	Number of samples
									Recycled mid-Cambrian–lower Ordovician taxa
cf.								*Stelliferidium philippotii*
x		cf.						*Stelliferidium cortinulum*
cf.					x			*Stelliferidium stelligerum*
x	x	x	x		x	x	x	x	*Acanthodiacrodium* spp.
		?						*Lophodiacrodium* sp.
		?						*Cymatiogalea bellicosa*
		x				x	x		*Vulcanisphaera* spp.
		x			x	x		x	*Cymatiogalea* spp.
		x	x		x	x	x	x	*Stelliferidium* spp.
					cf.				*Stelliferidium fimbrium*
					x	x			*Vulcanisphaera africana*
						x			*Vavrdovella* sp.
						x			*Timofeevia phosphoritica*
						x			*Vulcanisphaera cirrita*
						?			*Cymatiogalea cristata*
						?			*Cymatiogalea membranispina*
						x		x	*Vulcanisphaera britannica*

2aiii   Recycled probable Arenig to Llanvirn acritarch taxa.

SCC	Nantmel Mudstones				Yr Allt	PtrF	CgF	Formation
	Cnt		cg	lh	DCM			Member/facies
2	1	13	3	2	15	8	5	2	Number of samples
									Recycled Arenig–Llanvirn taxa
x								"*Culcitispina brevis*" of Booth 1979
x		x						*Striatotheca rarirrugulata*
x		cf.			x	cf.		*Arbusculidium filamentosum*
cf.						cf.		*Vogtlandia ramifurcata*
	x	x						*Adorfia* spp.
		cf.						*Adorfia prolongata*
		x						*Dicrodiacrodium* sp.
		x						*Multiplicisphaeridium* sp. D Booth 1979
		x						*Vogtlandia* sp.
		?						*Pirea* sp.
		x				cf.		cf.	*Adorfia firma*
		x						x	*Arbusculidium* sp.

2aiv   Location of Caradoc and Ashgill acritarch samples.

Sample number (MPA)	Grid reference	Sample number (MPA)	Grid reference
St Cynllo's Church Formation (SCC)			
29814	SO 0542 6659		
29970	SO 0644 6898		
Nantmel Mudstones (Ntm)			
Bioturbated mudstone facies			
29477	SO 0555 6653	29816	SO 0412 6535
29478	SO 0590 6638	29971	SO 0647 6898
29479	SO 0597 6616	29973	SO 0634 6905
29480	SO 0609 6604	29976	SO 0367 6484
29495	SO 0393 6483	29977	SO 0375 6482
29496	SO 0375 6482	30712	SO 0578 6857
Conglomerate and sandstone facies (Cg)			
29818	SN 9973 6260		
29974	SO 0608 6908		
29975	SO 0338 6462		
Hemipelagite facies (lh)			
31265	SO 0557 6966		
31269	SO 0318 6963		
Cefnnantmel Member (Cnt)			
31268	SO 0360 6703		
Yr Allt Formation (YA)			
Silty mudstone facies			
26183	SN 8983 6155	29019	SO 0090 6083
26184	SN 8983 6155	29020	SO 0094 6073
26185	SN 8983 6155	29815	SO 0760 6957
27626	SN 8919 6205	29817	SO 0023 6218
27627	SN 8855 6238	33504	SO 0115 6089
28110	SO 0584 7156	33505	SO 0115 6089
29018	SO 0089 6086		
Dol-y-fan Conglomerate (DCM)			
29518	SO 0123 6098	33506	SO 0125 6099
29519	SO 0123 6098	33507	SO 0124 6098
29520	SO 0272 6212	33508	SO 0124 6100
29521	SO 0294 6253	33509	SO 0120 6096
Pentre Formation (PtrF)			
28639	SO 0885 6808		
28640	SO 0878 6786		
29739	SO 0840 6663		
32347	SO 0886 6808		
32348	SO 0886 6808		
Cwmcringlyn Formation (CgF)			
28111	SO 0532 7188		
28112	SO 0732 7290		

2b   In Silurian rocks

Sample locations are given in **2bv**.

Abbreviations:

BMM — Blaen Myherin Mudstones
CaM — Caerau Mudstones (′ lower tongue; ″ upper tongue)
Cbn″ — Caban Conglomerates facies (second sequence)
Cef — Cwmere Formation
Cwn — Cwm Barn Formation
CyG — Cerig Gwynion Grits facies
DBF — Devil's Bridge Formation
Dgu — Dolgau Mudstones (′ lower tongue; ″ upper tongue)
Dfn — Dyffryn Flags facies
DLF — Derwenlas Formation
Glr — Glanyrafon Formation (′ lower tongue; ″ upper tongue)
Hen — Henfryn Formation
lh — units with laminated hemipelagite (lh^s — *M. sedgwickii* Shales)
LyT — Llyn Teifi Member
MMb — Mottled Mudstone Member
Ptr — Pysgotwr Grits (including pebbly sandstones sa)
PdG — Penstrowed Grits
Rdd — Rhuddnant Grits
Rhs — Rhayader Mudstones
YA/db — Yr Allt Formation/disturbed beds
YG′ — Ystrad Meurig Grits facies (first sequence).

Key to sample localities:
• isolated samples
•
• samples taken along a continuous traverse
•

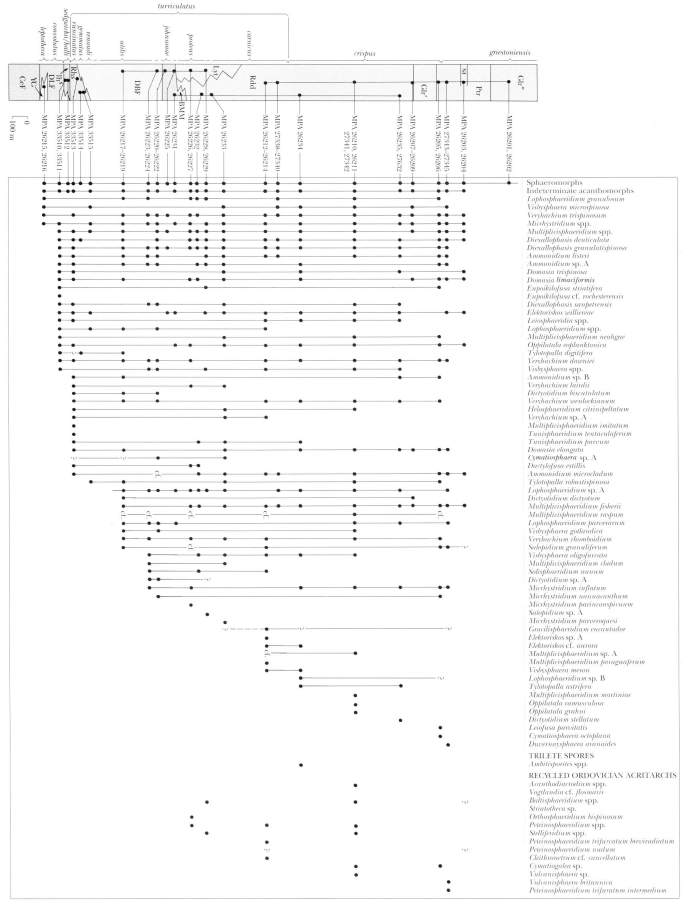

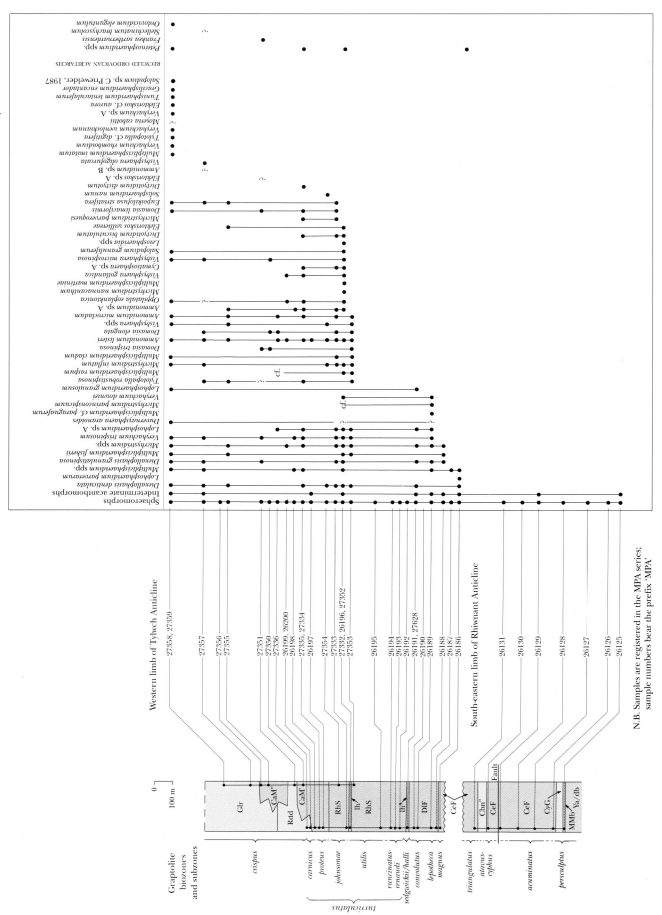

2bii   Acritarch occurrences in the latest Ordovician and Llandovery succession of the south-eastern limb of the Rhiwnant Anticline and western limb of the Tylwch Anticline.

2biii Acriarch occurrences in the latest Ordovician to Llandovery succession on the eastern limb of the Tylwch Anticline.

Graptolite biozones

Graptolite biozone labels: *persculptus*, *acuminatus*, *atavus*, *cyphus/acinaces*, *triangulatus*, *magnus*, *leptotheca*, *convolutus*, *sedgwickii/halli*, *turriculatus*, *crispus*, *griestonensis*, *crenulata/spiralis*, upper *ricartonensis*

Formation units: CxG, MMb, Dfn, CeF, Dfr, Rhs, Cwmysgawen Fault, Cwn, CaM, Gfr, Dgu', Dgu'', Nyg, PdG

Sample numbers: 27360, 27361, 27362, 27363, 27364, 33500, 27365, 27366, 27367, 33501, 33502, 33503, 27630, 27629, 30490, 30491, 29287, 27633, 29287, 29288, 28068, 28071, 29747, 27861, 27866, 28067, 28074, 27863, 33493, 28066, 33494–5, 27867, 27868

Scale: 0 — 100 m

N.B. Samples are registered in the MPA series; sample numbers bear the prefix 'MPA'

Sphaeromorphs
Indeterminate acanthomorphs
Visbysphaera microspinosa
Tylotopalla caelamenicutis
Lophosphaeridium parverarum
Tylotopalla robustispinosa
Diexallophasis granulatispinosa
Micrhystridium spp.
Dictyotidium dictyotum
Solisphaeridium nanum
Multiplicisphaeridium imitatum
Multiplicisphaeridium raspum
Diexallophasis denticulata
Lophosphaeridium spp.
Micrhystridium acerbum
Lophosphaeridium granulosum
Multiplicisphaeridium fisherii
Micrhystridium inflatum
Multiplicisphaeridium spp.
Veryhachium trispinosum
Veryhachium rhomboidium
Veryhachium wenlockianum
Veryhachium downiei
Veryhachium lairdii
Multiplicisphaeridium neahgae
Multiplicisphaeridium paraguaferum
Helosphaeridium citrinipeltatum
Tunisphaeridium parvum
Diexallophasis sanpetrensis
Oppilatala eoplanktonica
Oppilatala singularis
Eupoikilofusa striatifera
Tylotopalla digitifera
Visbysphaera gotlandica
Visbysphaera meson
Leiosphaeridia spp.
Leiofusa cf. *tumida*
Micrhystridium nannacanthum
Cymatiosphaera densisepta
Leiofusa parvitatis
Oppilatala sp. A
Moyeria cabottii
Multiplicisphaeridium cladum
Elektoriskos williereae
Domasia limaciformis
Domasia trispinosa
Domasia symmetrica
Duvernaysphaera aranaides
Lophosphaeridium sp. A
Veryhachium sp. A
Ammonidium sp. B
Micrhystridium parinconspicuum
Tylotopalla spp.
Cymatiosphaera octoplana
Salopidium granuliferum
Ammonidium listeri
Salopidium echinodermum
Ammonidium microcladum
Ammonidium sp. A
Oppilatala frondis
Domasia elongata
Helosphaeridium echinoforme
cf. *Ammonidium microcladum*
Multiplicisphaeridium martiniae
Tylotopalla astrifera
Visbysphaera oligofurcata
Veryhachium reductum
Micrhystridium parveroquesi
Gracilisphaeridium encantador
Oppilatala ramusculosa
Multiplicisphaeridium sp. A
Multiplicisphaeridium aff. sp. A
Dictyotidium sp. A
Visbysphaera brevifurcata
Lophosphaeridium cf. *hauskae*
Dictyotidium stellatum
Domasia sp. A
Quadraditum fantasticum
Oppilatala grahni
Multiplicisphaeridium cf. *rochesterense*
Cymatiosphaera cuba
Salopidium sp. A
Visbysphaera erratica
Carminella maplewoodensis
Multiplicisphaeridium sp. B
Deunffia monospinosa
Oppilatala sp. B

RECYCLED ORDOVICIAN ACRITARCHS
Stelliferidium spp.
Orthosphaeridium sp.
Goniosphaeridium spp.
Baltisphaeridium spp.
Stellechinatum cf. *brachyscolum*
Acanthodiacrodium spp.
Peteinosphaeridium cf. *nanofurcatum*
Peteinosphaeridium spp.
Vulcanisphaera cf. *pila*
Vulcanisphaera sp.
Cymatiogalea sp.
Cymatiogalea cristata
Stelliferidium simplex
Arkonia tenuata
Arkonia virgata
Stellechinatum helosum
Multiplicisphaeridium continuatum
Orthosphaeridium bispinosum

2biv Acritarch and spore occurrences in the Llandovery succession between the Garth and Cwmysgawen faults.

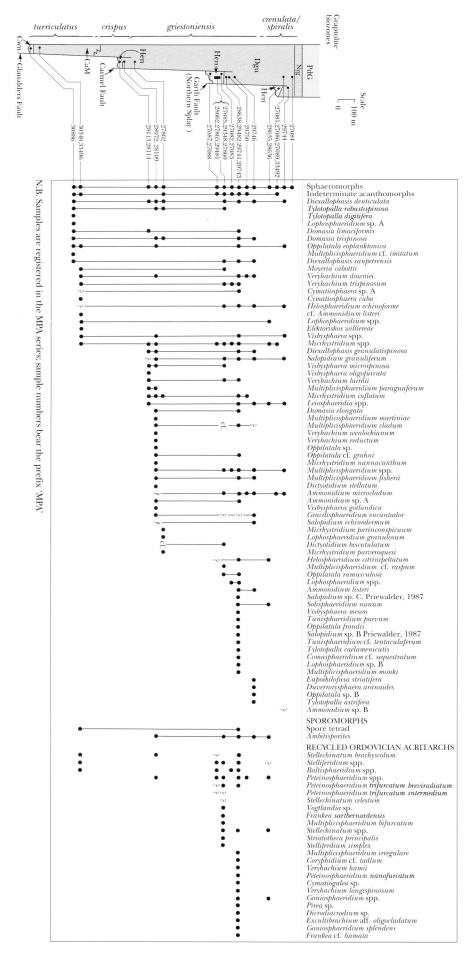

N.B. Samples are registered in the MPA series; sample numbers bear the prefix 'MPA'

2bv   Location of Silurian acritarch samples.

Sample No. (MPA)	National Grid reference	Sample No. (MPA)	National Grid reference	Sample No. (MPA)	National Grid reference
26125	SN 8916 6096	26288	SN 779 684	27868	SO 0165 7740
26126	SN 8917 6083	27081	SO 0832 7227	28062	SO 0779 6903
26127	SN 8919 6076	27082	SO 0767 7092	28066	SO 0092 7707
26128	SN 8917 6062	27083	SO 0767 7092	28067	SN 9975 7477
26129	SN 8919 6049	27084	SO 0847 7208	28068	SN 9798 7079
26130	SN 8923 6017	27085	SO 0777 6907	28071	SN 9875 7575
26131	SN 8925 6010	27086	SO 0833 7262	28072	SO 0568 7235
26186	SN 8720 6341	27087	SO 0771 7011	28074	SO 0097 7152
26187	SN 8720 6341	27088	SO 0774 7019	28109	SO 0478 7199
26188	SN 8712 6351	27089	SO 0836 7214	28113	SO 0666 7302
26189	SN 8706 6393	27332	SN 8629 6520	28114	SO 0657 7306
26190	SN 8707 6398	27333	SN 8610 6504	28635	SO 0809 6738
26191	SN 8707 6404	27334	SN 8562 6535	28636	SO 0823 6709
26192	SN 8723 6422	27335	SN 8562 6535	28638	SO 0782 6903
26193	SN 8706 6444	27336	SN 8486 6550	28827	SO 0634 6418
26194	SN 8687 6468	27338	SN 8310 7536	29028	SO 0530 6087
26195	SN 8658 6491	27339	SN 8310 7536	29196	SN 8629 6520
26197	SN 8596 6516	27340	SN 8310 7536	29287	SN 9994 6989
26198	SN 8525 6549	27341	SN 8380 7549	29288	SO 0006 7001
26199	SN 8521 6562	27342	SN 8380 7549	29348	SO 0777 6907
26200	SN 8521 6562	27343	SN 8488 7563	29481	SO 0610 6602
26201	SN 8541 7585	27344	SN 8488 7563	29482	SO 0613 6601
26202	SN 8541 7585	27345	SN 8488 7563	29741	SO 0830 6638
26203	SN 8505 7571	27350	SN 9344 7419	29742	SO 0849 6576
26204	SN 8505 7571	27351	SN 9339 7428	29743	SO 0755 6788
26205	SN 846 757	27352	SN 9428 7422	29744	SO 0805 6760
26206	SN 846 757	27353	SN 9432 7413	29746	SO 0845 6847
26207	SN 842 756	27354	SN 9422 7446	29747	SO 0238 7077
26208	SN 842 756	27355	SN 9295 7435	29750	SO 0616 6602
26209	SN 842 756	27356	SN 9278 7424	29810	SO 0483 6404
26210	SN 838 755	27357	SN 9255 7430	29811	SO 0487 6395
26211	SN 838 755	27358	SN 9228 7443	29812	SO 0491 6390
26212	SN 829 754	27359	SN 9228 7443	29813	SO 0488 6124
26213	SN 829 754	27360	SN 9680 6575	29978	SO 0637 6405
26214	SN 829 754	27361	SN 9809 6716	29979	SO 0576 6480
26215	SN 702 683	27362	SN 9773 6739	30340	SO 0234 7067
26216	SN 702 683	27363	SN 9773 6739	30490	SN 9987 6977
26217	SN 757 681	27364	SN 9771 6739	30491	SN 9978 6964
26218	SN 757 681	27365	SN 9761 6747	30886	SO 0216 7045
26219	SN 757 681	27366	SN 9761 6747	33492	SO 0817 7410
26220	SN 768 681	27367	SN 9761 6747	33493	SO 0296 7456
26221	SN 768 681	27628	SN 8667 6354	33494	SO 0090 7709
26222	SN 768 681	27629	SN 9625 6655	33495	SO 0093 7706
26223	SN 707 681	27630	SN 9745 6781	33496	SO 0220 7063
26224	SN 707 681	27632	SN 841 778	33500	SN 9768 6740
26225	SN 770 681	27633	SO 0019 7002	33501	SN 9760 6751
26226	SN 776 682	27860	SO 0779 6903	33502	SN 9759 6757
26229	SN 779 684	27861	SO 0353 7166	33503	SN 9744 6770
26231	SN 796 774	27862	SO 0523 7217	33510	SN 6859 6608
26232	SN 808 785	27863	SO 0213 7348	33511	SN 6861 6607
26233	SN 816 787	27864	SO 0275 7140	33512	SN 6815 6570
26234	SN 827 783	27865	SO 0779 6903	33513	SN 6744 6705
26235	SN 841 779	27866	SO 0354 7167	33514	SN 7109 6309
26277	SN 776 682	27867	SO 0165 7740	33515	SN 7109 6311

APPENDIX 3

Ranges of latest Ordovician and Silurian graptolite taxa in the district

	Biozones									
	persculptus	*acuminatus*	*atavus*	*acinaces*	*cyphus*	*triangulatus*	*magnus*	*leptotheca*	*convolutus*	*sedgwickii/halli*
Normalograptus? persculptus	X	L	.	.	.	.	.	.	.	.
N? parvulus	A	L	.	.	.	.	.	.	.	.
Normalograptus normalis	A	A	X	X	.	.	.	.	.	.
Normalograptus angustus	X	X	X	X	.	.	.	.	.	.
Glyptograptus? avitus	X	.	.	.	.	.	.	.	.	.
Normalograptus medius	X	X	X	cf.	?	.	.	.	.	.
'*Diplograptus*' aff. *modestus primus*	.	X	.	.	.	.	.	.	.	.
Akidograptus ascensus	.	L	.	.	.	.	.	.	.	.
Parakidograptus acuminatus	.	U	.	.	.	.	.	.	.	.
'*Orthograptus*' *cabanensis*	.	U	.	.	.	.	.	.	.	.
Atavograptus atavus	.	.	X	X	X	.	.	.	.	.
Atavograptus gracilis	.	.	X	.	.	.	.	.	.	.
Normalograptus rectangularis	.	.	X	X	X	.	.	.	.	.
Lagarograptus acinaces	.	.	.	A	X	.	.	.	.	.
Glyptograptus aff. *incertus* sensu Hutt, 1974	.	.	.	*	*	.	.	.	.	.
Dimorphograptus confertus confertus	.	.	.	U	.	.	.	.	.	.
Cystograptus vesiculosus	.	.	.	U	.	.	cf.	.	.	.
Metaclimacograptus hughesi sensu Bulman & Rickards	.	.	.	U	*	.	X	?	.	.
Pribylograptus sandersoni	.	.	.	U	.	.	.	.	.	.
Pristiograptus fragilis pristinus	.	.	.	U	.	.	.	.	.	.
Glyptograptus tamariscus tamariscus	.	.	.	U	.	.	cf.	cf.	X	.
'*Orthograptus*' cf. *mutabilis*	.	.	.	U	X	.	.	.	.	.
Rhaphidograptus toernquisti	.	.	.	U(A)	A	A	X	.	.	.
Coronograptus cyphus cyphus	.	.	.	U	X	.	.	.	.	.
Pribylograptus incommodus	.	.	.	U	.	.	.	.	.	.
Atavograptus cf. *strachani*	.	.	.	.	X	?	.	.	.	.
Monograptus ex. gr. *revolutus/austerus*	.	.	.	.	X	X	.	.	.	.
Monograptus austerus cf. *austerus*	.	.	.	.	X	.	.	.	.	.
Monograptus cf. *sudburiae*	.	.	.	.	X	.	.	.	.	.
?Monoclimacis sp.	.	.	.	.	R	.	.	.	.	.
Coronograptus gregarius aff. *minisculus*	.	.	.	.	X	.	.	.	.	.
Normalograptus wyensis	.	.	.	.	X	.	.	.	.	.
Glyptograptus tamariscus cf. *varians*	.	.	.	.	X	.	.	.	.	.
cf. *Pribylograptus argutus argutus*	.	.	.	.	X	X	?	?	.	.
Metaclimacograptus undulatus	.	.	.	.	X	X	X	X	X	A
Coronograptus gregarius gregarius	.	.	.	.	?U	X	X	X	L	.
Monograptus triangulatus triangulatus	.	.	.	.	.	X	.	.	.	.
Monograptus triangulatus extremus	.	.	.	.	.	R	.	.	.	.
Monograptus aff. *walkeri?*	.	.	.	.	.	X	.	.	.	.
Monograptus sp. 1 of Zalasiewicz & Tunnicliff	.	.	.	.	.	X	.	.	.	.
Pristiograptus concinnus	.	.	.	.	.	X	.	.	cf.	.
Monograptus ex. gr. *revolutus*	.	.	.	.	.	X	.	.	.	.
Monograptus communis communis	.	.	.	.	.	X	.	.	.	.
Glyptograptus? alternis	.	.	.	.	.	X	.	.	.	.
aff. *Pseudoglyptograptus vas?*	.	.	.	.	.	X	.	.	.	.
Rastrites sp.	.	.	.	.	.	X	.	.	.	.
Monograptus communis rostratus	.	.	.	.	.	U	.	.	.	.
Glyptograptus cf. *enodis enodis*	.	.	.	.	.	?	X	.	.	.
Monograptus triangulatus fimbriatus	.	.	.	.	.	U	A	.	.	.
Normalograptus? magnus	.	.	.	.	.	.	A	.	.	.
Rastrites longispinus	.	.	.	.	.	.	X	?	.	.
aff. *Atavograptus strachani*	.	.	.	.	.	.	R	.	.	.
Monograptus chrysalis	.	.	.	.	.	.	X	.	.	.
Pseudoglyptograptus barriei	.	.	.	.	.	.	A	?	.	.

Biozones and subzones

| | magnus | leptotheca | convolutus | sedgwickii/halli | turriculatus | | | | |
					runcinatus	gemmatus	renaudi	utilis	johnsonae
Pribylograptus leptotheca	X	X	X	.	.	.	.	.	.
Petalolithus ovatoelongatus	X	.	X	.	.	.	.	.	.
Monograptus pseudoplanus	U	.	.	.	.	.	.	.	.
Monograptus argenteus	.	A	.	.	.	.	.	.	.
Rastrites peregrinus	.	X	aff.	.	.	.	.	.	.
'Glyptograptus' sinuatus sinuatus	.	X	.	.	.	.	.	.	.
Orthograptus insectiformis	.	X	.	.	.	.	.	.	.
Monograptus millepeda	.	X	.	.	.	.	.	.	.
Metaclimacograptus hughesi	.	X	L	.	.	.	.	.	.
Monograptus imago	.	X	L	.	.	.	.	.	.
Normalograptus? scalaris	.	cf.	X	.	.	.	.	X	.
Monograptus denticulatus sensu Sudbury, 1958	.	X	X	.	.	.	.	.	.
Monograptus lobiferus lobiferus	.	?	A	.	.	.	.	.	.
Monograptus involutus	.	cf.	X	X	.	.	.	X	.
Monograptus aff. convolutus of Storch, 1980	.	.	L	.	.	.	.	.	.
Monograptus denticulatus	.	.	L	.	.	.	.	.	.
Monograptus lobiferus aff. harpago	.	.	L	.	.	.	.	.	.
Monograptus ?aff. gemmatus	.	.	L(R)	.	.	.	.	.	.
Monograptus aff. argenteus	.	.	L	.	.	.	.	.	.
aff. Pseudoglyptograptus sp. 1 Rickards	.	.	L	.	.	.	.	.	.
Glyptograptus incertus	.	.	L	.	.	.	.	.	.
Glyptograptus tamariscus aff. angulatus	.	.	L	.	.	.	.	.	.
Monograptus urceolinus	.	.	L(R)	.	.	.	.	.	.
Rastrites spina sensu Rickards, 1970	.	.	L(R)	.	.	.	.	.	.
Orthograptus bellulus	.	.	L	.	.	.	.	.	.
Monograptus convolutus	.	.	X	.	.	.	.	.	.
Monograptus decipiens	.	.	X	.	.	.	.	.	.
Monograptus limatulus	.	.	X	.	.	.	.	.	.
Rastrites hybridus hybridus	.	.	X	.	.	.	.	.	.
Monograptus capillaris	.	.	X	.	.	.	.	.	.
Monograptus lobiferus cf. harpago	.	.	X	.	.	.	.	.	.
Monograptus cf. undulatus	.	.	X	.	.	.	.	.	.
Pristiograptus aff. jaculum	.	.	X	.	.	.	.	.	.
Pristiograptus regularus regularis	.	.	X	.	.	.	.	.	.
Clinoclimacograptus retroversus	.	.	X	.	.	.	.	.	.
Pseudoretiolites perlatus	.	.	X	.	.	.	.	.	.
Monograptus sp. 2	.	.	X	.	.	.	.	.	.
Petalolithus cf. minor	.	.	R	.	.	.	.	.	.
Lagarograptus sp.	.	.	R	.	.	.	.	.	.
Cephalograptus cometa extrema	.	.	U	.	.	.	.	.	.
Streptograptus sp.	.	.	U	.	.	.	.	.	.
Monograptus sedgwickii	.	.	.	A	?	?	?	?	.
Lagarograptus tenuis	.	.	.	L	.	.	.	.	.
Pristiograptus regularis latus	.	.	.	L	.	.	.	.	.
Monograptus contortus	.	.	.	X	.	.	.	.	.
Pristiograptus variabilis	.	.	.	X	.	.	.	.	.
Streptograptus ansulosus	.	.	.	X	.	.	.	.	.
Monograptus urceolus	.	.	.	X	.	.	.	.	.
Monograptus capillaris sensu Loydell	.	.	.	X	.	.	.	.	.
Petalolithus cf. elongatus	.	.	.	X	.	.	.	.	.
Monograptus halli	.	.	.	U	*	X	X	X	.
Pristiograptus aff. variabilis	.	.	.	U(R)	.	.	.	.	.
Pristiograptus sp. 2 of Loydell	.	.	.	U(R)	.	.	.	.	.

Biozones and subzone

	sedgwickii/halli	runcinatus	gemmatus	renaudi	utilis	johnsonae	proteus	carnicus	galaensis	crispus	sartorius	lower	upper	crenulata
		turriculatus							*crispus*			*griestoniensis*		
Pristiograptus nudus s.l.	?	.	.	.	X	X	X	X	X	X	X	X	X	X
Monograptus turriculatus s.l.	.	*	X	X	X	X	X	X	X	?	.	.	.	.
Monograptus runcinatus	.	*	X	X	.	.	.	.	.	.	.	.	.	.
Pristiograptus renaudi	.	*	X	A	.	.	.	.	.	.	.	.	.	.
Petalograptus kurcki	.	*	X	.	.	.	.	.	.	.	.	.	.	.
Monograptus sp. 3	.	*R	*R	*R	.	.	.	.	.	.	.	.	.	.
Streptograptus plumosus	.	*cf	*cf	X	X	.	.	.	.	.	.	.	.	.
Monograptus bjerroskovae	.	*	*	X	X	X	X	?	.	.	.	.	.	.
Streptograptus pseudoruncinatus	.	*cf	X	.	?	.	.	.	.	.	.	.	.	.
Monograptus gemmatus	.	.	X	.	.	.	.	.	.	.	.	.	.	.
Rastrites fugax	.	.	X	.	.	.	.	.	.	.	.	.	.	.
Glyptograptus fastigatus	.	.	A	.	.	.	.	.	.	.	.	.	.	.
Glyptograptus aff. *fastigatus* (broad form)	.	.	R	.	.	.	.	.	.	.	.	.	.	.
Glyptograptus sp.	.	.	R	.	.	.	.	.	.	.	.	.	.	.
Streptograptus sp. of Loydell 1993	.	.	X	.	.	.	.	.	.	.	.	.	.	.
Rastrites cf. *maximus*	.	.	R	.	.	.	.	.	.	.	.	.	.	.
Petalolithus altissimus s.l.	.	.	X	.	X	.	.	.	.	.	.	.	.	.
Glyptograptus cf. *elegans*	.	.	X	.	.	.	.	.	.	.	.	.	.	.
Streptograptus aff. *filiformis*	.	.	X	.	.	.	.	.	.	.	.	.	.	.
Normalograptus? nebula	.	.	X	.	X	?	.	X	.	X	.	X	X	.
Monograptus cavei	.	.	.	X	X	.	.	.	.	.	.	.	.	.
Monograptus marri	.	.	.	X	A	A	X	cf	cf	cf	.	?.	.	.
Monograptus utilis	.	.	.	.	A	.	.	.	.	.	.	.	.	.
Monograptus planus	.	.	.	.	X	.	.	.	.	.	.	.	.	.
Rastrites linnaei	.	.	.	.	R	.	.	.	.	.	.	.	.	.
Streptograptus storchi	.	.	.	.	X	X	X	A	.	.	.	.	.	.
Pristiograptus bjerringus	.	.	.	.	X	.	.	.	.	.	.	.	.	.
Petalolithus conicus	.	.	.	.	R	.	.	.	.	.	.	.	.	.
Streptograptus barrandei	.	.	.	.	X	.	.	.	.	.	.	.	.	.
Streptograptus johnsonae	.	.	.	.	R	A	.	.	.	.	.	.	.	.
Petalolithus aff. *tenuis*	.	.	.	.	R	.	.	.	.	.	.	.	.	.
Monograptus rickardsi	.	.	.	.	.	?	X	X	X	X	.	.	.	.
Petalolithus tenuis s.l.	.	.	.	.	.	.	X	X	X	X	.	.	.	.
Monograptus proteus	.	.	.	.	.	.	X	X	.	.	.	.	.	.
Streptograptus pseudobecki	.	.	.	.	.	.	R	.	.	.	.	.	.	.
Streptograptus tenuis	.	.	.	.	.	.	R	.	.	.	.	.	.	.
Monoclimacis? galaensis	.	.	.	.	.	.	?	X	X	.	.	.	.	.
Monograptus cf. *acus*	.	.	.	.	.	.	R	.	.	.	.	.	.	.
Monograptus becki?	.	.	.	.	.	.	R	.	.	.	.	.	.	.
Monograptus aff. Rickardsi (slender form)	.	.	.	.	.	.	.	X	.	.	.	.	.	.
Monograptus carnicus	.	.	.	.	.	.	.	X	X	.	.	.	.	.
Streptograptus whitei	.	.	.	.	.	.	.	X	.	.	.	.	.	.
Monograptus aff. *crispus*	.	.	.	.	.	.	.	R	X	.	.	.	.	.
Streptograptus exiguus	.	.	.	.	.	.	.	X	A	X	X	X	.	.
Monograptus clintonensis	.	.	.	.	.	.	.	?	X	X	X	X	.	.
Streptograptus mustadi	.	.	.	.	.	.	.	.	R	.	.	.	.	.
Streptograptus crispus	.	.	.	.	.	.	.	.	X	A	.	.	.	.
Monograptus discus	.	.	.	.	.	.	.	.	X	X	X	X	X	X
Petalolithus sp. aff. *kurcki?*	.	.	.	.	.	.	.	.	X	.	.	.	.	.

	crispus	sartorius	griestoniensis lower	griestoniensis upper	crenulata	spiralis	spiralis-centifugus interregnum	centifugus	murchisoni	riccartonensis	rigidus	linnarssoni (= flexilis)	ellesae	lundgreni	nassa
Streptograptus loydelli	A	?													
Retiolites geinitzianus subspp.	X	X	X	X	X	X	X	X							
Monograptus pergracilis?	aff		X												
Monograptus aff. *pragensis*		R													
Monograptus pragensis		X	X												
Monograptus aff. *sartorius*		X													
Monograptus sartorius		X	X												
Monograptus tullbergi spiraloides		?	X	X											
Normalograptus? aff. *scalaris*		R													
Monoclimacis cf. *griestoniensis* sensu Elles & Wood			A	R											
Monoclimacis griestoniensis			R	A											
Monograptus tullbergi tullbergi			X	?	X										
Streptograptus aff. *loydelli*			X												
Monograptus priodon			X	X	X	X	X	X							
Monograptus pseudocommunis			X												
Pristiograptus initialis			?	?	X										
Monograptus aff. *pragensis?* (slender form)				X											
Monoclimacis vomerina vomerina					A	X	X	X							
Monoclimacis crenulata sensu Elles & Wood					X										
Monoclimacis linnarssoni sensu Bjerreskov						X									
Monograptus aff. *falx*						X									
Monograptus parapriodon						X		?							
Pristiograptus aff. *initialis* (broad form)						X									
Monograptus aff. *minimus cautleyensis*						X									
Monoclimacis vomerina gracilis s.l.							X	X	cf						
Monograptus aff. *priodon* (narrow form)							X								
Monograptus aff. *cultellus?*							X								
?Pseudoplegmatograptus sp.							X								
Cyrtograptus cf. *centrifugus*								X							
Cyrtograptus cf. *insectus*								X							
Monograptus flemingii-priodon transients								X	X	X					
Monoclimacis vomerina cf. *basilica*								X							
Cyrtograptus cf. *murchisoni*									X						
Monograptus riccartonensis										X					
Pristiograptus cf. *dubius*										X	*	*	*	X	X
Monograptus radotinensis inclinatus										X					
Monoclimacis flumendosae flumendosae										X		X			
Monograptus flexilis belophorus?										X					
Monograptus firmus sedburghensis?										X					
Pristiograptus aff. *meneghini meneghini*										*	*	*	*		
Plectograptus? bouceki										*	*	*			
Monoclimacis flexilis flexilis											X				
Cyrtograptus rigidus cautleyensis											X				
Cyrtograptus linnarssoni											X				
Monograptus flemingii cf. *flemingii*														X	
Monograptus flemingii cf. *compactus*														X	
cf. *Cyrtograptus hamatus*														X	
Pristiograptus pseudodubius														X	X
Gothograptus nassa															X
Gothograptus aff. *nassa* (broad form)															X

Key to range chart symbols:

X or • — present, A — abundant, R — rare, U — present in upper part of biozone/subzone, L — present in lower part of biozone/subzone, cf. — qualified identification, ? — doubtful identification, ?? — very doubtful identification, aff. — identity not exact (observed morphological differences), * — range of species not known exactly (present within interval indicated).

APPENDIX 4

Physical property determinations on rock samples

a   Density and porosity

Those samples collected on the adjacent Aberystwyth Sheet (163) are indicated with an asterisk.

Group or Formation	Lithology	Grid reference	Saturated density (t/m^3)	Grain density (t/m^3)	Effective porosity (%)
Borth Mudstones	Mudstone	SN 605 888*	2.75	2.82	3.4
Aberystwyth Grits	Mudstone	SN 583 830*	2.77	2.81	2.1
			2.76	2.82	3.2
			2.76	2.83	3.7
	(arenaceous)		2.64	2.76	6.7
Aberystwyth Grits	Mudstone	SN 576 797*	2.77	2.80	1.8
			2.78	2.82	1.9
	(arenaceous)		2.71	2.73	0.9
	(arenaceous)		2.71	2.72	1.1
Aberystwyth Grits	Mudstone	SN 612 796	2.6	2.82	7.6
			2.66	2.82	9.0
	(arenaceous)		2.74	2.78	2.0
Aberystwyth Grits	Mudstone	SN 543 697	2.66	2.76	5.3
			2.66	2.74	5.1
			2.70	2.72	1.2
	(arenaceous)		2.67	2.72	2.8
Aberystwyth Grits	Mudstone	SN 588 621	2.70	2.80	5.1
			2.73	2.80	4.0
	(arenaceous)		2.64	2.73	4.8
Derwenlas Formation	Mudstone	SN 701 679	2.76	2.78	1.2
			2.76	2.78	1.1
	(arenaceous)		2.77	2.79	0.9
	(arenaceous)		2.74	2.75	0.8
Derwenlas Formation	Mudstone	SN 742 728	2.77	2.82	2.7
Devil's Bridge Formation	Mudstone	SN 746 729	2.78	2.79	0.8
	(arenaceous)		2.70	2.75	3.0
Devil's Bridge Formation	Mudstone	SN 764 760	2.75	2.80	2.6
			2.72	2.78	3.2
			2.77	2.81	2.0
	(arenaceous)		2.71	2.76	2.4
Bryn Glas Formation	Mudstone	SN 754 815*	2.78	2.81	1.8
Cwmere Formation	Mudstone	SN 793 828*	2.72	2.77	3.0
			2.73	2.77	2.2
			2.74	2.78	2.6
Blaen Myherin Mudstones Formation	Mudstone	SN 768 681	2.69	2.77	4.8
			2.69	2.76	3.8
			2.72	2.77	2.9
Dolgau Mudstones	Mudstone	SN 030 745	2.69	2.85	8.9
Rhayader Mudstones	Mudstone	SN 952 715	2.78	2.79	1.0
	Sandstone		2.78	2.80	1.2

b   Sonic velocity and density

Group, Formation or facies	Sample type	Grid reference	Saturated density (t/m^3)	Grain density (t/m^3)	Effective porosity (%)	Sonic velocity (km/s)
Cerig Gwynion Grits facies	Mudstone Sandstone Sandstone	SN 968 657	2.71 2.67 2.77	2.73 2.69 2.80	1.1 0.7 1.3	5.76 5.78 4.57
Caban Conglomerate Formation	Mudstone	SN 924 646	2.68	2.69	0.5	4.80
Penstrowed Grits	Sandstone	SO 037 755	2.70	2.71	0.8	5.51

Magnetic susceptibility measurements on these samples were within the range 0.0001–0.0004 SI units.

FOSSIL INDEX

GENERAL INDEX

BRITISH GEOLOGICAL SURVEY

Keyworth, Nottingham NG12 5GG
0115 936 3100

Murchison House, West Mains Road, Edinburgh
EH9 3LA   0131-667 1000

London Information Office, Natural History Museum
Earth Galleries, Exhibition Road, London SW7 2DE
0171-589 4090

The full range of Survey publications is available through the Sales Desks at Keyworth and at Murchison House, Edinburgh, and in the BGS London Information Office in the Natural History Museum (Earth Galleries). The adjacent bookshop stocks the more popular books for sale over the counter. Most BGS books and reports can be bought from The Stationery Office and through The Stationery Office agents and retailers. Maps are listed in the BGS Map Catalogue, and can be bought together with books and reports through BGS-approved stockists and agents as well as direct from BGS.

The British Geological Survey carries out the geological survey of Great Britain and Northern Ireland (the latter as an agency service for the government of Northern Ireland), and of the surrounding continental shelf, as well as its basic research projects. It also undertakes programmes of British technical aid in geology in developing countries as arranged by the Overseas Development Administration.

The British Geological Survey is a component body of the Natural Environment Research Council.

The Stationery Office publications are available from:

The Stationery Office Publications Centre
(Mail, fax and telephone orders only)
PO Box 276, London SW8 5DT
Telephone orders 0171-873 9090
General enquiries 0171-873 0011
Queuing system in operation for both numbers
Fax orders 0171-873 8200

The Stationery Office Bookshops
49 High Holborn, London WC1V 6HB
(counter service only)
0171-873 0011   Fax 0171-831 1326
68–69 Bull Street, Birmingham B4 6AD
0121-236 9696   Fax 0121-236 9699
33 Wine Street, Bristol BS1 2BQ
0117-9264306   Fax 0117-9294515
9 Princess Street, Manchester M60 8AS
0161-834 7201   Fax 0161-833 0634
16 Arthur Street, Belfast BT1 4GD
01232-238451   Fax 01232-235401
71 Lothian Road, Edinburgh EH3 9AZ
0131-228 4181   Fax 0131-229 2734
The Stationery Office Oriel Bookshop,
The Friary, Cardiff CF1 4AA
01222-395548   Fax 01222-384347

The Stationery Office's Accredited Agents
(see Yellow Pages)

And through good booksellers